T0073139

THE ALLADI *Diary*

Memoirs of Alladi Ramakrishnan

Table of Contents

Published by

World Scientific Publishing Co. Pte. Ltd.
5 Toh Tuck Link, Singapore 596224
USA office: 27 Warren Street, Suite 401-402, Hackensack, NJ 07601
UK office: 57 Shelton Street, Covent Garden, London WC2H 9HE

Library of Congress Cataloging-in-Publication Data
Names: Ramakrishnan, Alladi, author. | Alladi, Krishnaswami, editor.
Title: The Alladi diary : memoirs of Alladi Ramakrishnan / edited by
    Krishnaswami Alladi (University of Florida, USA).
Description: New Jersey : World Scientific, 2019. |
    Includes bibliographical references and index.
Identifiers: LCCN 2019007553 | ISBN 9789811202872 (hc : alk. paper)
Subjects: LCSH: Ramakrishnan, Alladi. | Physicists--India--Biography.
Classification: LCC QC16.R255 R36 2019 | DDC 530.092 [B] --dc23
LC record available at https://lccn.loc.gov/2019007553

British Library Cataloguing-in-Publication Data
A catalogue record for this book is available from the British Library.

For any available supplementary material, please visit
https://www.worldscientific.com/worldscibooks/10.1142/11346#t=suppl

Desk Editors: Anthony Alexander/Tan Rok Ting

Typeset by Stallion Press
Email: enquiries@stallionpress.com

Printed in Singapore

THE ALLADI *Diary*

Memoirs of Alladi Ramakrishnan

Edited by

Krishnaswami Alladi
University of Florida, USA

World Scientific

NEW JERSEY · LONDON · SINGAPORE · BEIJING · SHANGHAI · HONG KONG · TAIPEI · CHENNAI · TOKYO

PART I
Ekamra Nivas to Matscience

The author with Nobel Laureate Niels Bohr at Ekamra Nivas, Jan 1960.

"Ekamra Nivas," the mansion built by Sir Alladi Krishnaswami Iyer, was where several sections of the Constitution of India were written, and is also the womb of MATSCIENCE, The Institute of Mathematical Sciences.

To my mother
VENKALAKSHMI

The smile that illuminated Ekamra Nivas was no ordinary smile; its mere memory warms our hearts, cleanses our thoughts, cheers our spirits, enlightens our minds, and dispels our gloom to reveal the presence of God to our mortal eyes.

With my mother Lady Alladi.

Sir Alladi Krishnaswami Iyer.

Editor's Preface

My father, the late Professor Alladi Ramakrishnan, was an internationally known scientist who had a life remarkable in many ways. This autobiography describes chronologically the many extraordinary events that influenced his life and shaped his career. It emphasises four themes:

(i) the phenomenal legal career of his father Sir Alladi Krishnaswami Iyer and his role in drafting the Constitution of India,

(ii) the tremendous efforts of my father to introduce modern physics in Madras by conducting a Theoretical Physics Seminar at our family home Ekamra Nivas with dozens of eminent scientists from around the world as speakers, that ultimately led to the creation of MATSCIENCE, The Institute of Mathematical Sciences, in Madras in 1962, with him as the Director,

(iii) his visits to about 200 academic centres of research worldwide to lecture on his research in theoretical physics, and

(iv) a general discussion of his fundamental work in stochastic processes, elementary particle physics, matrix theory, and Einstein's special theory of relativity, the last of which kept him intellectually active till the end.

Part I

As the son of Sir Alladi Krishnaswami Iyer, one of India's most eminent lawyers who served on the Drafting Committee of the Constitution of India, my father was fortunate to be a direct witness to many events relating to my grandfather's legal career, and in particular to the writing of the Indian Constitution; many of these events took place in *Ekamra Nivas*, a palatial home in Madras that my grandfather built, and where my father lived all his life. This first of two parts contains a description of several aspects of Sir Alladi's glorious career, the persons of eminence with whom he interacted, and the key role he played in drafting the Constitution. Under the advice and influence of his father, my father initially obtained a law degree, but then he eschewed a lucrative legal profession to pursue his passion — scientific research. My grandmother Lady Alladi played a crucial role in convincing her husband to permit my father to pursue a scientific career, and in gratitude, my father has dedicated this book to his mother.

Although my father did very fundamental work right from the beginning as a PhD student, his scientific career was strewn with obstacles that could have easily deterred him. But then with his firm resolve, his initiative, and by a series of incredible events each as improbable as the other, he overcame each of these obstacles. The crowning success was his creation of MATSCIENCE, the Institute of Mathematical Sciences, in Madras, on January 3, 1962. The creation of MATSCIENCE is not just an incredible saga in Indian science but in the world of science. This entire story, starting with his life as a student under Homi Bhabha in the early days of the Tata Institute, then going to Manchester, England (for his PhD) where his fundamental work in probability was recognised by the eminent statisticians M. S. Bartlett and D. G. Kendall, his return to India to take up a faculty position at Madras University, his visit to the Institute for Advanced Study in Princeton at the invitation of its Director Robert Oppenheimer and the influence it had on him, his launch of the Theoretical Physics Seminar at his family home Ekamra Nivas — a seminar that attracted scientific luminaries from around the world, the positive impression of the seminar that Nobel Laureate Niels Bohr conveyed to India's Prime Minister Jawaharlal Nehru, the meeting of my father and his students with the Prime Minister in October 1961, and the creation of MATSCIENCE on January 3, 1962 with the crucial help of Education Minister C. Subramaniam — all this is described in glorious detail in Part I. In this effort to run the seminar at home, my mother (Lalitha Ramakrishnan) played the crucial role of a gracious host to the eminent visitors and the students. The speech that my father gave at the inauguration of MATSCIENCE entitled "The miracle has happened", which is included as an Appendix, is a model of English diction, and describes in a few pages the incredible events leading to the creation of MATSCIENCE.

In discussing the influences on his scientific career in Part I, my father also describes his work and of his students in the areas of probability and stochastic processes, cosmic rays, and elementary particle physics, in the context of the great advances made by leaders in these fields worldwide. Such discussion is non-technical and would appeal to non-experts.

Part I actually begins with a description of life in Madras and at Ekamra Nivas during the period 1930–45, providing a contrast of the leisured pace of life in South India and the turmoil elsewhere in the world caused by World War II. My father has discussed various aspects of the Hindu way of life in the first half of the twentieth century that would be of interest not just to non-Indians but also to people in India and persons of Indian origin living overseas, since that charming way of life has since disappeared due to various influences.

Part I also has several appendices of important letters and documents. In the original publication of The Alladi Diary by East-West Books, Madras, India, these documents and letters were retyped for publication in the volume. In editing the Alladi Diary, I have replaced most of the retyped versions of some of these letters and documents with their scanned originals. A fuller set of letters, documents, articles and photographs related to the Alladi Diary may be found at the website [1].

The original Volume I published by East-West Books contained four chapters of his international travels in 1962, 63, 64, and 65, after the inauguration of MATSCIENCE. In editing the Alladi Diary, I have moved these four chapters to Part II and concluded Part I with the inauguration of MATSCIENCE, to be consistent with the title "Ekamra Nivas to MATSCIENCE" that my father had for Volume I.

Part II

Part II focuses on my father's worldwide academic trips and the eminent visiting scientists to MATSCIENCE, especially to the Anniversary Symposia held each year in January. My father's global travels and the international visitors to MATSCIENCE are inter-related. Many of those who visited were impressed with MATSCIENCE and invited him to their institutions. Similarly on his academic trips abroad, he made new contacts and invited active researchers to MATSCIENCE. The visiting scientists program was most active during the first decade of his directorship when government authorities both in Madras and in the Centre, enthusiastically gave the requested support.

With regard to his academic trips, he was like Paul Erdös (but on a smaller scale), visiting a dozen or more universities in a span of two to three months each year. Just as for Erdös, among the multitude of universities he visited, there were a few where he lectured most frequently, my father visited the Naval Research Laboratory, the Rockwell Science Center, and the University of Texas at Dallas, almost annually. Hence in his dedication to Part II, he has included his hosts there — Maurice Shapiro, John Richardson, and Erwin Fenyves. As Director, he decided to combine all his annual international lecturing assignments (apart from conferences) in a single trip of about three month duration in the summer so that he could maximise his stay at MATSCIENCE for the rest of the year. He always took my mother on these international trips, except when he travelled just to attend a conference. Since these trips were in the summers, I had the privilege of accompanying him until I left for the United States for my PhD in 1975. Thus I had the opportunity to observe his interaction with leading scientists all over the world.

My father believed that it was crucial for scientists to reach out to the public at large and disseminate knowledge about the great advances in their fields of activity through lectures. He was an electrifying speaker, and with his highly visible position as Director of MATSCIENCE, he was in great demand in South India for general lectures on science and society — for the All India Radio, at Rotary clubs, at schools and colleges, and at various public functions. He accepted these invitations happily, and these engagements did not affect the time he devoted to research or to its quality. He was one of the most effective communicators of science and therefore a great ambassador for science in India. In Part II he refers to many of his public lectures on science.

My father had wide ranging interests well beyond science. He was passionately interested in Carnatic music — interest that he developed after marriage to my mother, and owing to his knowledge of Telugu, he was an authority on the music of Saint Thyagaraga, the greatesr of the Carnatic music composers. Thus he gave lectures on the "Spiritual Diction of Thyagaraja" at the Alladi Foundation, at the Annual Conference of the Music Academy, and at other forums. Great practitioners of music attended his lectures with a keen interest. He often talks about events relating to Carnatic music in his narrative.

Most Indians of his generation believed in God and in their religion. My father's faith in God was so strong that he attributes all his successes, and all the good events in his life to divine Grace. People may ask how a scientist who is so rational in his thinking and in the study of Nature, can believe in the existence of a Supreme Being? Most people from India, including scientists believe in God, but they may not openly admit that in the way he frankly does. Before each of his trips overseas, and before and after every major event in his life, my father went to Tirumala-Tirupathi to offer prayers to Lord Venkateswara, our family deity. His was not shy to emphasise this in his narrative. My father had a special devotion to Lord Rama, which was why he launched his Theoretical Physics Seminar on Rama's birthday (Rama Navami) in 1959 even though, Navami in the lunar calendar is considered inauspicious. He was influenced strongly by the ideals emphasised in The Ramayana. His devotion to Rama was yet another reason that he enjoyed the compositions of Saint Thyagaraja. He performed Sandhyavandanam (daily prayers to the Sun and other Gods) every morning, and never missed his Amavasya Tharpanam (prayers each new moon day to departed ancestors) even during his travels abroad.

My father was an aviation enthusiast. He loved travelling by aeroplanes and he expresses his joy of flying throughout the narrative, providing details of airlines, aircraft, and even flight numbers. His love of aviation infected me, and I am passionately interested in aviation. On a scientific level, he mentions his visits to the Boeing and Douglas Aircraft Companies and their advanced research labs.

My father took pleasure in many things he did with his family and friends — going on holiday trips, his visits to the wildlife sanctuaries, watching the sunsets at Waikiki Beach in Honolulu, to name a few. He shares these pleasures with us with childlike simplicity.

It is impressive that even though he retired ar the age of 60, his research productivity continued strongly up to his 80th year. His last research paper on a new "Rod Approach to Relativity" appeared in 2000 in the Bellman Memorial Issue of the Journal of Mathematical Analysis and Applications, fifty years after his very first paper; he felt that his last paper was as significant as his first, and he was justly proud of both. I have therefore included copies of the first page of his first and last papers (and a few others) in Appendices to Parts I and II respectively.

Even in his retirement, he was invited regularly to give talks on his research in Europe and the USA, although these were not as numerous as when he was Director of MATSCIENCE. So on his trips abroad after 1983, he spent the bulk of his time with my family, making visits in between to various academic institutions. He enjoyed the company of all my friends in Florida, and they in turn looked forward to conversations with him due to his wide ranging knowledge and radiant optimism.

My father's scientific activities and achievements inspired me to take to an academic career in science, and he encouraged me every step of the way. Thus in his narrative, he refers to many of my activities as a mathematician. He also encouraged my mother in her pursuit of Carnatic music, and so he refers to her concert performances, which in later years was jointly with my wife Mathura.

My father has two passions in sports — tennis and cricket, and so periodically he discusses exciting events in the world's of cricket and tennis. He played tennis into his late seventies, and enjoyed both the game and the company of his tennis mates.

Finally, I want to stress that this is a personal diary as the title of the book indicates; he describes his life story chronologically — not just year by year, but almost month by month within each year. This is more apparent in Part II than in Part I. He was able to describe events chronologically because he meticulously maintained annual diaries and preserved his academic correspondence over a period of five decades! The reader has to understand, that since events are described chronologically and not thematically, there would be some repetition in discussing certain themes.

In editing Part II, I have added some pictures to what he already had in Volume II, included a sample of the first pages of some important papers of his, his list of publications, and the list of his PhD students. Captions for the pictures selected by my father and by me are indicated suitably in Part II as by the Author and by the Editor. Most importantly, I have added Notes at the end of each chapter, to provide a context for many things in the narrative. The Notes are intended to be helpful to readers both within and outside India. Also, I have included an Index at the end of Part II but for both parts of the book.

―――――――――

The Alladi Diary is indeed a kaleidoscopic and chronological description of the fascinating and multi-faceted life of an eminent and influential scientist. When my father first had it printed in the early eighties, it was "for private circulation only". Many friends and scientists who read that, urged him to get it published. So he published it in India with East-West Books. After he passed away in 2008, I made the decision to get it published through an international publisher of world repute and I am thankful to World Scientific for their interest in doing so. I hope this edited version will appeal to a wide spectrum of readers.

Krishnaswami Alladi
Gainesville, Florida

About the Editor

KRISHNASWAMI ALLADI is the son of Alladi Ramakrishnan and Lalitha Ramakrishnan. He is Professor of Mathematics at the University of Florida, where he was department chairman during 1998–2008. He received his PhD from the University of California, Los Angeles, in 1978. His research is in number theory where he has made notable contributions. He is the Founder and Editor-in-Chief of *The Ramanujan Journal* (Springer), devoted to all areas of mathematics influenced by the Indian mathematical genius Srinivasa Ramanujan. He helped launch the SASTRA Ramanujan Prize given to very young mathematicians for outstanding contributions to areas influenced by Ramanujan and has chaired the Prize Committee since the inception of the prize in 2005. He is married to Mathura and has two daughters Lalitha and Amritha.

Editor's Acknowledgements

Just as my father was a witness to many significant events relating to his father's career, I was a witness as a young boy to the creation of MATSCIENCE at our family home Ekamra Nivas, where I saw the enormous effort my father made to bring scientists of world repute to his Theoretical Physics Seminar, and my mother Lalitha Ramakrishnan graciously hosting these eminent scholars. Thus it has been an immense pleasure to edit his memoirs (The Alladi Diary) for international circulation. In doing so, I have had the benefit of comments from, and support of, my mother, my wife Mathura and my daughters Lalitha and Amritha. I am most appreciative of my colleague Frank Garvan at the University of Florida who gave me technical help that enabled me to work with WSPC tex-style files, and my son-in-law Jis Joseph, who compiled the pictures in a suitable form for World Scientific. I must pay special thanks to my dear friend G. P. Krishnamurthy not only for helping my father when he was writing the Alladi Diary, but also assisting me by providing important information about many topics relating to my grandfather in Part I.

I wish to thank the following agencies for permission to scan the first page of various papers of significance to my father's scientific career:

The Royal Society for Homi Bhabha's 1950 paper in the Proceedings of the Royal Society (Appendix 4B, Part I).

The Cambridge Philosophical Society for Alladi Ramakrishnan's 1950 paper in the *Proceedings of the Cambridge Philosophical Society* (Appendix 4C, Part I).

Elsevier for permission regarding Alladi Ramakrishnan's papers in the *Journal of Mathematical Analysis and Applications* — two papers in 1967, and one each in 1973 and 2000 (Appendices 7, 8, 9 and 13, Part II).

Elsevier for permission regarding Norman Levinson's 1974 paper in the Journal of Mathematical Analysis and Applications (Appendix 10, Part II).

Springer for permission regarding Costa de Beauregard's 1986 paper in Foundations of Physics (Appendix 11 of Part II).

We also thank:

IOP Science for permission regarding Alladi Ramakrishnan's 1995 paper in Physics Education (Appendix 15, Part II).

East-West Books for permission to publish this Edited Version of The Alladi Diary Vols. I and II as a single volume in two parts by World Scientific Publishing Company.

All photographs for which the source and/or permission are not explicitly acknowledged, are courtesy of Krishnaswami Alladi.

Finally I must express my appreciation to several members of World Scientific Publishing Company (WSPC) — Professor K. K. Phua (Chairman and Editor-in-Chief), Mr. Max Phua (Managing Director), Ms. Ranjana Rajan, Ms. Nisha Rahul, Ms. Dipasri Sardar, Mr. Anthony Alexander and their colleagues at Aces (WSPC) Chennai, India, and to Ms. Lai Fun, Ms. Rok Ting, Mr. Rajesh Babu and Ms. Peng Ah Huay of WSPC Singapore — for their interest in, and their help with, publishing the Edited Version of The Alladi Diary, for their patience in allowing me time to make the edits, and for the superb production.

Krishnaswami Alladi
Gainesville, Florida

Author's Preface to Part I

This is a kaleidoscopic story of our times, running through seven decades of events as seen and felt by me through personal experience, but focussing on two incredible stories — (i) the brilliant legal career of my father which led him to be on the Drafting Committee of the Indian Constitution, and (ii) the creation of an advanced institute for mathematical research in Madras due to my efforts of running a theoretical physics seminar at my family home. All this is described against the background of the placid period of peace and complacency in India under the British rule, the holocaust of the Second World War which altered the course of human civilisation as no convulsion did before, the dawn of Indian freedom after a bleeding night of civil strife, and the post-war years of the fantastic triumphs of science and technology altering the pace of human lives and leading Man into inter-planetary worlds. There are several reasons for me to write these memoirs: My boyhood and youth were spent in a traditional atmosphere of a Hindu way of life which knew not the excitements of the Western world, the pressures of ambition and the competitive spirit of scientific endeavour. That world has almost disappeared, and there may be some historical inquisitiveness on the part of the post-war generation to have a glimpse of such a world. Then came the war and we in India for the most part watched as at an arena without involvement, taking a distant view of the convulsion that altered the course of human history and in turn the destiny of our own country. As a direct witness to the deliberations of the Constituent Assembly and the Drafting Committee of the Constitution of India of which my father was a member, I feel an inner urge to recount the events of those stirring years. In the post-war era, opportunities came in a flood when my travels took me through thirty countries and two hundred centres of advanced learning in the world. I want to share with you my observations on the scientific work done at these great centers in a most stimulating atmosphere. And I want to describe my efforts in conducting a Theoretical Physics Seminar in my family home *Ekamra Nivas* to create an atmosphere in Madras similar to that at the Institute for Advanced Study in Princeton, that finally bore fruit and led to creation of MATSCIENCE, The Institute of Mathematical Sciences. This miracle was the direct consequence of the positive impressions of Nobel Laureate Niels Bohr after his visit to my home and the support of the far sighted Prime Minister of India, Jawaharlal Nehru, and the enthusiastic Minister for Education C. Subramaniam. My wife Lalitha played a solicitous host to all the visiting scientists many of whom were our house guests and lectured in the main hall of Ekamra Nivas. In the course of transmitting my scientific experiences, I shall also discuss my own research contributions — first in the theory of probability and stochastic processes, next in elementary particle physics, and finally in the special theory of relativity — by means of which I have been able to introduce new ideas in domains that have been dominated by leading scientists like Einstein, Dirac, Chandrasekhar, Gell-Mann and Feynman.

I want the readers to share with me the thrill of my experience. The story starts with life with my father, and leads to my life as a father.

The idea of publishing the Diary was suggested by Professor Richard Eberhart, Poet Laureate of America, during my stay with my son Krishna in Florida in 1987.

We start with my father's message, the spirit of which animates the whole Diary.

I am thankful to my personal assistant R. Ganapathi for patiently typing the manuscript in spite of frequent changes I made before publication.

Alladi Ramakrishnan
1.1.2000

Acknowledgements

"The Hindu" has played a very significant role in the creation and growth of MATSCIENCE, by the publication of Professor Niels Bohr's statement about our work in Ekamra Nivas which attracted the attention of the Prime Minister of India Pandit Nehru, and by fully reporting the proceedings of the inaugural function of MATSCIENCE and the activities of the visiting scientists. More recently, its distinguished Editor N. Ravi was keen on even including my new work on Einstein and Lorentz in the Science and Technology Supplement of The Hindu (Jan 21, 1999) so as to reach the public at large; this is included [1] and a discussion of my work in Special Relativity is contained in Part II. Throughout my career, my wife Lalitha gave unstinted support of all my efforts and even attended my lectures at various centres around the world with diligent interest!

The revelation of the 'key to space-time unity' in the form of a "Rod Approach" (reported in Part II) occurred at the residence of my son Krishnaswami Alladi, during my stay with him, his wife Mathura and his daughters Lalitha and Amritha in Gainesville, Florida. With loving pre-science, my grand daughters had presented me with a book on Einstein on Father's Day a year before!

Alladi Ramakrishnan
1.1.2000

Notations and Conventions

1. English spellings of Indian names

Most Indian and Hindu names in this book are either names in Sanskrit or in the vernacular languages like Tamil, Telugu, etc. When these names are written in English, they sometimes take different spellings although the pronunciation would be the same. For example, *Iyer*, a sub-caste among the Brahmins, can be spelt as *Iyer, Aiyer, Aiyar, or Ayyar*. Similarly, the sub-caste *Iyengar*, can be spelt as *Iyengar, Aiyangar, or Ayyangar*.

Also, *swami* which means God, and which appears often in many South Indian names, is also commonly spelt as *swamy*.

When quoting others, the spellings that they have used is followed. In the text, other possible spellings are sometimes indicated in parenthesis.

2. English spellings of words and usage

Alladi Ramakrishnan has used English spellings such as *specialised* or *programme*. Thus English spellings are used throughout the book instead of American spellings like *specialized* or *program*. Also English usage of words is adopted such as *spelt* instead of *spelled*.

3. Notation for Notes and References

For each Chapter, the Notes are at the end and are numbered. In the text for each Chapter, the Notes are indicated by a superscript. For example *Drafting Committee*[5] in Chapter 10 means that Item 5 in the Notes of that Chapter is on the Drafting Committee.

All books and articles referred to are listed at the end. In the text, a reference is indicated by numbering within square brackets such as [4].

A Father's Message*

21st May 1953

On this my seventy-first birthday,[1] which according to the Bible is the full age, many thoughts cross my mind. Unfortunately in this country of ours, the talents of many people run to waste without being noticed or having had a full opportunity to be utilised. On the whole, I should think Providence has been exceptionally kind to me in spite of the ailments and physical sufferings I have been subject to. I have had more than my share of success in life. I cannot lay to my credit any tangible public service or notable sacrifice though I have kept in view certain ideals in the conduct of my life. I was the eighth among my parents' children. It is said that having regard to the penury in which my parents were placed, my mother cursed herself when she was carrying me nearly seven and a half years after the birth of my elder brother. It is said however that my father was proud of my birth and entertained great hopes of me[3] and named me Krishna,[2] being the eighth of the children born to my parents. I have had more than my legitimate share of success in life but alas in our country, how much talent runs to waste! It is this thought that has been the main spring of socialistic thought in the West though the reading of the Puranas[4] convinces me it is possible to conceive of another state of society in which talent meets the necessary reward and recognition. What pains me most is that many of those who come up from penury seldom think of the lot of their poor brethren. I am reminded of the illustration given of a person crossing the channel or river thinking of his achievement in crossing the channel and not of many others who may die in the process of crossing the channel. Prosperity following adversity may bring about two kinds of feeling in a man. It may make him vain and self conscious of his great achievements in life; it may make him humble and better realise the lot of those who are not able to fight their way in life. I have striven after the second ideal, though how far I have succeeded in achieving that ideal is for others to judge and not for me. I shall close these disjointed remarks with this thought that may God in the few years vouchsafed for me,[5] strengthen the ideas rooted in me.

FATHER

---

* This message was dictated by Sir Alladi Krishnaswami Iyer to his second son Alladi Ramakrishnan on his seventy first birthday when he was lying down like Sage Bhishma[6] physically infirm but definitely alert as ever.

Notes

1) Sir Alladi was born on 14 May, 1883. So he dictated this in 1953 one week after his birthday. In referring to it as his seventy first birthday, Sir Alladi was counting the day he was born — 14 May, 1883 — as his first birthday, a practice not uncommon in India.

2) It is common for Hindus to be named after Gods. The idea is that if people are named after Gods, then by calling their names one would be saying the names of Gods, and this would generate good *karma*. Lord Krishna was the eighth of the children to his parents. Since Sir Alladi was the eighth born to his parents, he was named Krishnaswami. In Sanskrit, swami means God.

3) Sir Alladi's father Ekambara Sastri (sometimes spelt at Ekamra Sastri) was a priest with a very modest income in the village of Pudur sixty miles north of Madras. He felt that his young son Krishnaswami was brilliant and would have a bright career. Ekambara Sastri knew that his son deserved an English medium education in a city and so he moved to Madras for the sake of Krishnaswami's education (see Seshachalapathi's article in [1]). Sir Alladi was very appreciative of this.

4) Puranas are Hindu holy scriptures which describe stories of Gods emphasising the deeds they performed. Puranas also can be stories associated with places or how certain temples came into existence. These are called *sthala puranas*, but even in these, it is the acts of Gods which are emphasised.

5) Sir Alladi died a few months later the same year. Before he died, in 1953 he started dictating the story of his legal career to Ramakrishnan. But he was only able to cover the initial part of his career up to around 1910. See "My Story" by Sir Alladi in [1].

6) In the great battle described in the Hindu epic "The Mahabharatha", the famous warrior Bhishma gave his advice to friends and foes as he lay wounded and dying.

Chapter 1

Early Memories

I was born on August 9, 1923, as the eighth child of my parents Sir Alladi Krishnaswami Iyer and Lady Alladi Venkalakshmamma and I am grateful for naming me Ramakrishnan according to the Hindu custom after the Avatars of Ramayana and Mahabharatha,[1] the twin epics coextensive with our ancient culture and tradition.

The earliest recollections of my boyhood relate to my primary school at the Mylapore Convent near the southern end of the Madras Marina Beach. In those days it was customary for the members of the middle class to send their children to convent school for so much prestige was attached to speaking good English and to discipline in childhood. I remember very clearly getting up a long and wide flight of steps to the prayer class when all the children knelt down as various extracts from The Bible were read by the teacher. Most of the teachers were European nuns who were extremely friendly to the children, at the same time insisted on strict conformity to school rules and regulations. We used to play under the shade of sprawling trees within the convent compound. There were only a few children who came by car and I was one of them, but we looked with envy at the magnificent horse-drawn coach in which one of my classmates came to school.

A lesson which still stays fresh in my mind is the exercise in the geography class when we were asked to draw diagrams of the streets of Mylapore[2] and of Madras and mark the location of important buildings. We were provided with excellent notebooks with imported ivory smooth paper the like of which I have not seen today in our schools or colleges. The most exciting experience was buying new books at the opening of the school; each one of us got a thrill in buying the school bag and clean brown wrapping paper. It was important to have a well sharpened pencil and I preferred the bright red Elephant brand made in Germany, and available for about five annas[3] a dozen in a lovely case marked Johann Faber. One day I found the pencil missing in my bag and I was so afraid to meet my teacher that I ran all the way from the school to a small shop on Kutchery road and returned to the class almost breathless with the pencil in my hand.

The convent school was a co-ed institution only for the primary classes and boys were not allowed to continue there after the mature age of eight or nine! At that time the P. S. High School[4] was considered the best boys school in the locality and I was greatly excited to shift there in 1931 to the fifth class after three years in the convent. This amounted to a double promotion since the standard in the convent was quite high.

The P. S. High School had a very imposing building with a very spacious sports ground attached to it. But the fifth class was conducted in an all too modest zinc shed. Our class teacher Sarangapani Iyengar wielded the cane, convinced of the doctrine that 'to spare the rod was to spoil the child'. The class room instruction is not worth mentioning except for the classes of Satya Godavari Sharma, a unique personality among the teaching community. He was extraordinarily handsome and unusually proud of his profession. His classes commenced with the simple Sanskrit sloka (= verse):

Hare Rama, Hare Rama, Rama Rama Hare Hare
Hare Krishna, Hare Krishna, Krishna Krishna Hare Hare

Little did I imagine that forty years later, in Los Angeles, this verse would caste a hypnotic spell over the minds and hearts of new converts to the Krishna cult among the younger generation of Americans!

At home I was taught the first few hundred verses of the Valmiki Ramayana[5] and my father took pleasure in asking me to recite them before his friends. My lifelong devotion the Hindu tradition and faith which was undiminished through all my global travels, stemmed essentially from the music and magic of these verses from the Ramayana.

My family belonged to the Tamil Brahmin Brihacharanam community[6] whose ancestors moved a few centuries ago to a region sixty miles north of Madras. It was a sort of custom for members of our community[7] to read and write Telugu in school but speak Tamil at home.

My father was the leading figure in the legal world at Madras and had built a spacious house of exquisite architectural beauty on Luz Church road in Mylapore, the strong hold of the educated middle class. "Ekamra Nivas",[8] splendid and elegant for comfort and convenience, was not only home of a legal luminary of legendary fame, but was to become later the womb of an institute of advanced research of international reputation.

As I look back, I realise that South India had a social structure undisturbed by external influences. Everyone was conscious and complacent about his place in that structure. Elaborate religious rituals characterised Hindu social life, while in the realm of economics and politics, Britain dominated the life and spirit of the literate classes. There was little or no interest in America or in other countries of Europe, except perhaps France. It was assumed that world consisted of India and the omniscient and omnipotent British Empire. Britain was the source and seat of power and it was accepted that the sun could never set on the British Empire. India was its proudest possession, its teeming millions content in poverty, and its upper class zealous and secure about their preferred and well guarded position in the social hierarchy, closest to the patronage of the benign British Government. The pinnacle of prestige was the British Civil Service and talented Indians vied with one another to secure a place in that steel frame. The most significant feature of British rule was the completeness of its sovereignty with almost no interference in the social structure of India. This was of course the secret of colonial stability.

To the educated Indian, Heaven's gift to civilisation was the English gentleman. Eton and Harrow, Oxford and Cambridge, inspired our middle class, an adulation matched only by the veneration for the Privy Council and the British judicial system. The Governor symbolised British power while the Viceroy represented the Emperor himself. The aim of our high officials was to qualify for a seat at the governor's table, our public men to aspire for coveted titles. Even the exalted Maharajah who hunted with African cheetahs, would wait on the British resident's approving smile. 'The whispers of the residency were the thunders of the palace.' The Viceroy's splendour at the Imperial capital had remained undiminished since the spacious days of Lord Curzon.

It was during this period that Mahatma Gandhi's[9] influence suddenly engulfed the entire Indian scene. The frail half-naked saint came with the reputation of fighting single handed the titanic and vicious power of racial prejudice in South Africa. He had emerged as the moral victor of a heroic struggle against a government which knew no mercy and yielded no quarter to the teeming black population. Under the magic wand of his influence, the complacency and torpor of centuries were shaken and ancient India was quickened to a new life. The name of the Mahatma became a house-hold word, and in the true Hindu tradition of saint and disciple like Ramakrishna and Vivekananda, people spoke of a young and handsome Nehru as the favourite follower of Mahatma Gandhi. Slowly it became a pride to speak of non-violent resistance to the armed might of the British Empire and people dreamt of a free India.

Sathyagraha or moral indignation became a religion and a creed. The Indian Civil Servant, hitherto an awesome symbol of British power, was looked upon as an agent of foreign oppression. Such was the atmosphere in the country during the years of my high school education.

Among the teachers I remember very well in the First Form[10] in the P. S. High School was a tall disciplinarian who was as effective with his knuckles as with his cane in inflicting pain and punishment on the erring student. He was a contrast to Lakshmana Iyer who smiled when he did not talk, and when he talked, he flattered a young boy's aspirations. The only classes I remember very well in my First Form are the Ramayana Lectures in the evening by the venerable Sri Ramaswamy Sastrigal whose life after retirement as a Judge was almost as long and eventful as before it.

In the Second Form I was interested in the arithmetic classes on 'time and work' solved by the arrow method dinned into the unwilling ears of students with pertinacity.

My interest in sport at that time was cricket. It dominated our recreational life since in the southern part of Madras there were many vacant lots used by boys who formed small clubs, the monthly subscription ranging from four annas to a couple of rupees. A good English cricket ball cost barely two rupees and we frequently spent excited hours in anticipation of playing with such new balls. A handsome cricket bat made of English willow was available for twenty rupees. I remember with pleasure buying the Gradidges Driver[11] and preparing it for play by soaking the bat in linseed oil, a clumsy ritual we enjoyed immensely.

It was then that the tales of the incredible exploits of Don Bradman[12] of Australia reached us through the sports columns of the daily newspapers. The names of Australian cricketers Orielly and Grimmett, Woodful and Ponsford, Oldfield and McCabe, and of Englishmen Larwood[12] and Bowes, Hobbs and Sutcliffe, Hammond and Verity, were daily currency during our own school matches. There were endless discussions even at school whether Bradman really dreaded Larwood or whether 'bodyline' was really cricket.[12] C. K. Nayudu the tall dusky Indian had just then made his legendary reputation for hitting sixers during his captaincy against England in the famed Lords Cricket Grounds in London.

Though forty years have passed, I remember in remarkable detail every ball and every stroke, the style and personality of various players of two matches at the M.C.C.[13] grounds in Madras in February 1934 — MCC vs India (one day First Class Match) and the England vs India Test Match. Even recent matches I do not remember half so well. Clark the fast left arm bowler from Nottingham skittled the wicket as one of our local heros nervously stepped back a few paces to avoid the whistling missile. The masculine Amar Singh, India's effective medium pace bowler, delighted the crowds with sparkling sixers. The elegant Bakewell partnered Walters as the opening batsman. The wily wizard and spinner Verity wiped out the Indian batsmen off the field. The British team was captained by the imperturbable Douglas Jardine,[12] a symbol of British prestige and tenacity, confidence and power — qualities for which that sceptered isle inspired the envy of less happier lands! And all this on one of the most beautiful cricket grounds in the world — Chepauk in Madras — fringed by lush tropical trees. The spacious elegance of the M.C.C. Pavilion symbolised the leisured comfort of the English middle class for whom 'afternoon tea' was the hallmark of a preserved way of life. The lowest ticket cost four annas which allowed one to watch either sitting or standing a few yards from the boundary line. The chairs near the Pavilion cost Rs. 5 or Rs. 20 for the Season. One could order lunch for Rs. 3 and have it served by bearers in colonial style. Only a privileged few could sit inside the Pavilion.

The most interesting classes in the Third Form were the English classes taught by G. Srinivasachariar. He instilled in me a love for English literature and for delicious poetry and sparkling prose. I remember vividly how effectively he brought out the liveliness and loveliness in the poem *Brook* by Alfred Lord Tennyson. I became convinced of Alexander Pope's dictum that words must seem an echo to the sense.

In one of the English classes, I was asked to describe my concept of a happy moment. I immediately wrote a essay describing the preparations for a journey to Bangalore! A trip to Bangalore and Mysore by car had always been an attraction for me all my life. At that time we had a magnificent Buick[14] which would have been the credit to an oil baron in Kuwait or of a Maharaja of an Indian state. Looking at the sleek automobiles of the United States today, I still can say that General Motors has not produced a similar vehicle to this day. It was a seven seater luxury limousine and my desire was to travel to Bangalore and Mysore in that car.

That summer we went to Ootacamund and during hours of pleasant leisure, my father would read out to me many passages from famous books and also current literature. Once he read from a book on 'Science and Society' emphasising the internationalism of science contrasted with the nationalism of politics. It had such a stupendous effect on my imagination that I nursed a rising ambition to become a scientist, for I felt that the profession of law or politics was confined within the national boundaries.

I dreamt of an ideal life where one could work fundamental problems in physics and enjoy enlightened leisure listening to music. He also read to me Haldane's 'Dedicated Life' which has made an enduring impression upon my views on education (see Appendix 1). To this day I read those passages over and over with never failing interest.

In Ootacamund we stayed in a lovely house *Silverton* in Fernhill and I remember vividly how my mother used to assist my father in dressing him in a black suit for the banquets at the Government House.[15]

It was in the Fourth Form that my first interest in science was aroused. It took two years before it became a passion for mathematics and the mathematical sciences. The most significant event of the year was the visit of Jawaharlal Nehru to Madras and his famous address at Thilak Ghat on the Madras beach. Over one hundred thousand people gathered on the beach to hear their dear Jawaharlal when the population of Madras was only about four hundred thousand. The phrase 'our nation, crushed under the iron heel of the British Empire' rings in my ears to this day. For the first time in our school education, we had a sense of participation in the great freedom struggle.

It was at that time that the fame of Sir Sarvepalli Radhakrishnan[16] as a great orator had spread throughout our land. I attended some lectures of Sir Radhakrishnan and I particularly remember the address at the University Examination Hall when my father presided. In 1936 Radhakrishnan was invited to take up the Spalding Professorship of Eastern Ethics and Oriental Philosophy at Oxford University. My father was a frequent visitor to Radhakrishnan's house, and Sir S. himself came to Ekamra Nivas many times. The most characteristic feature of his conversation was radiant optimism. His very presence instilled a desire for achievement in the younger generation. His son S. Gopal[16] who was my classmate in the Fourth form, was to accompany his father to England and attend one of the public schools[17] there. Sir S. suggested that I should accompany them to England and have a public school education along with Gopal. It was estimated that schools in like Eton and Harrow, the cost of education and residence would be about Rs. 400/month. It was then considered very expensive even for the privileged upper class. I remember my father telling me that Nehru had his early education in Harrow. But my father ruled out this suggestion from Sir S. on domestic grounds and I felt very disappointed.

In the Fifth Form, the emphasis was essentially on English. In the English class we had a teacher who emphasised fluency in speech. He practised his fluency with obvious effort, and made heroic attempts to avoid pauses! Sir S. Radhakrishnan and Right Hon'ble Srinivasa Sastri were the idols at that time (for public speaking). Sir S. resembled the magnificent torrent of the Niagara whereas Srinivasa Sastri was like a cool limpid stream with crystal clear water flowing — slow, steady, and transparent. It was during that period that I read the works of the famous English novelists Scott, Stevenson, and Dickens.

It was in the Sixth Form that my first interest in mathematics was keenly aroused. The best part of the year's education was in geometry, and I attribute my interest in the subject to our teacher Narasimhachari (Mathematics classes were of two types — optional (Algebra and Geometry) and elementary (Arithmetic). The pedal line theorem and its applications were the piece-de-resistance of my efforts. I enjoyed doing problems outside the syllabus and I got acquainted with concurrency and collinearity — which was supposed to be dealt with only at the college level. I developed an avidity for doing geometry riders[18] by shorter methods.

I also remember very well how I used to do problems on mixtures without the use of the x-method. The teacher was so angry at this profanity that he used to send me out of the class at the slightest pretext! I did not have any feeling of guilt for I knew he was annoyed at the quickness with which I was doing the problems. However, in the English class the situation was different. Under our teacher Shamanna, we developed a taste for delectable phrases. His hero was Stevenson and his ideal book was 'Kidnapped'.

The period 1936–39 was an eventful time in the political history of India. The new Government of India Act of 1935 had received the Royal assent on 4 August, 1935. It represented the first major step towards self government in India by gradual and constitutional means for which the British genius was so famous. It had all the features of the beginning of an Indian Constitution — popular representation which went back to 1892, dyarchy and ministerial responsibility which can be traced to 1921, provincial autonomy and a new federal structure with safeguards for minorities with the addition of a popular government for the provinces. It was the most suitable time for action by enlightened representatives of the educated classes who wished to stand between the stark asceticism of Mahatma Gandhi and the vaulting idealism of Jawaharlal Nehru. In Madras at that time there was a thinker, savant, and administrator, who was to play this role — C. Rajagopalachari[19] or C.R. as millions knew him later. He was a great friend and associate of my father to whom constitutional law was a natural instinct. C.R. was able to convince the forward looking Congress that no compromise will be made in the objectives by accepting to work the Indian constitution. It was not even a case of stooping to conquer but working and waiting till the great moment for action arose. Thus in 1937 the first popular ministries in India took office and C.R. assumed the Chief Ministership in Madras. In his team was V. V. Giri who was later to become the President of India by popular acclamation. Among C.R.'s followers was C. Subramaniam who was later to become Cabinet Minister. During this period, my father was Advocate General of Madras and there were many evenings when C.R. came to Ekamra Nivas for consultations on new legislation. It was a pleasant encounter of wits between a genius par excellence (my father) and an administrator (C.R.) whose insight into human affairs, sharpness of intellect and flame-like purity became part of the South Indian tradition.

Even as a schoolboy, I could not just be a spectator to this drama on the Indian scene. Thus we started the Mylapore Students Union, in which political and even intellectual problems were discussed with seriousness but without getting involved in the controversies. To us the Madras Ministry was a replica of the British Cabinet Government which held the admiration of young and old, school boy and civilian, worker and politician alike. Our education had trained us to think like Disraeli and Gladstone, Lloyd George and Baldwin, as luminaries in an illuminated stage of history. Indians may fight the British, or resent their rule, but veneration for the British institutions and the admiration for the English language was part of Indian heritage and a way of life. So for about two years we had all the trappings of a cabinet system of government and a Federal constitution. Reserved power of the Governor in the provinces and the Governor General ensured the all embracing sovereignty of the Government.

The 1935 Act was a massive constitutional document, and one of my father's assistants, N. Rajagopala Iyengar,[20] had mastered the details to such an extent that he wrote a commentary which was used as a text by constitutional lawyers of that day. To my father it was the first occasion and opportunity to display and utilise his immense knowledge of federal constitutions. The problems that arose between the provinces and the centre gave some amplitude for the use of his legal prowess. I remember watching with pride my father's frequent journeys to Delhi in connection with important work involving constitutional questions.

So the actors moved and played under the towering shadow of Lord Linlithgow, the Viceroy who held the country in a loose rein, fully conscious that he could tighten it harder than a hangman's noose at the slightest sign from Whitehall and Downing Street. Meanwhile the Mahatma watched with a saint's disdain, the faint and feeble show of illusory power by the popular government.

Notes

1) Coincidentally Ramakrishnan was the eighth (born) child to his parents just as his father was. But among the children who survived and lived through adulthood, he was the fifth. The name *Ramakrishnan* combines the names of the two most prominent avatars (= incarnations) of Lord Vishnu, namely Rama and Krishna.

2) Mylapore was the subdivision of the city of Madras where Ramakrishnan lived and where this convent school was located. "Mayil" in Tamil means peacock, and the name Mylapore which literally means the "place of peacocks" derives from the fact that long time ago, peacocks abounded in this locality.

3) Sixteen annas made a Rupee. Annas do not exist anymore.

4) The Pennathur Subramaniam (PS) High School was the most famous boys high school in Mylapore. The school has on its alumni several luminaries from Madras, because Mylapore was the intellectual center of Madras and most Mylaporeans sent their boys to P. S. High School in those days.

5) The two great epics of the Hindu religion are the Ramayana and the Mahabharatha, with the Ramayana being the older of the two. There are several versions of the Ramayana, but the original is due to Valmiki, considered to be the very first poet scholar (*Adi Kavi*) in Sanskrit.

6) The Brahmins in the Hindu religion represented the priest caste. In South India, the Brahmins could be classified as either Vaishnavites who are followers primarily of Lord Vishnu, and Smarthas who worship all Hindu Gods, but with Lord Shiva as the principal deity. Ramakrishnan's family members were Smarthas. One of the subgroups among the Smarthas is Brihacharanam.

7) The specific community of Smarthas to which Ramakrishnan's family belonged were called Pudur Dravidas. The Pudur Dravidas are Tamils (hence the term Dravidas) who settled in Pudur (currently in the state of Andhra Pradesh, Telugu speaking area) nearly 500 years ago. They could speak and write in Telugu. The Pudur Dravida community has produced several outstanding scholars.

8) *Ekamra Nivas* is the name of the family home built by Sir Alladi. Ramakrishnan grew up there. Sir Alladi named the house after his father Ekambara Sastri out of respect and love. "Nivas" means abode, and so Ekamra Nivas is the abode of Ekambara Sastri (also spelt as Ekamra Sastri). The house was constructed with brick and (lime) mortar, and the ceilings are supported by massive (Burma) teak beams; such a construction of the ceiling is called a Madras terrace. Burma teak is considered to be the highest quality teak, totally resistant to termites.

9) Mahatma Gandhi (1869–1948) led the Indian independence movement against the British Rule. After spending some years in South Africa where he protested against racial prejudice, he arrived in India in 1915 and became the leader of the Indian National Congress in 1921. Thus as a young boy Ramakrishnan heard of various actions of Mahatma Gandhi for India's independence including Satyagraha.

10) Elementary or Primary School was up to Class 5. From Class 6 to 11 it was High School. The classes in High School were termed as First Form through Sixth Form.

11) Gradidges was one of the most famous makers in England of cricket bats. These cricket bats made of willow had to be brushed with linseed oil so that they would have the proper resilience. English willow was always considered superior to Kashmir willow.

12) Among all test matches between various test playing countries, the England-Australia rivalry has the most hallowed tradition. Sir Donald Bradman of Australia

is considered the greatest batsman on all time and known as the "run machine". In order to curb Bradman, the English captain Douglas Jardine encouraged the fiery fast bowler Harold Larwood on his team to bowl the ball in such a way as to strike the batsman's body. This "bodyline" style of bowling curbed not only Bradman but the entire Australian cricket team during the test series of 1932–33 in which England crushed Australia. Both bodyline and its proponent Jardine, faced severe criticism. Bodyline was banned and Jardine relieved of his captaincy a year later. The test match played in Madras between England (under Jardine's captaincy) and India, February 10–13, 1934, which Ramakrishnan saw was the last test match played by Jardine. Prior to this, on February 7, 1934, the English team under Jardine played a First Class Match in Madras, also referred to by Ramakrishnan in his narrative. In the First Class matches played in that 1933–34 tour of India, the English team were known as the MCC.

13) With regard to the venue of the matches in Madras, M.C.C. refers to the Madras Cricket Club located in the Chepauk subdivision of Madras. The famous M.C.C. in England is the Marylebone Cricket Club located in the Lords cricket grounds in London, and which until 1993 was the governing body of cricket. The MCC Team refers to the team of the Marylebone Cricket Club.

14) The Buick that Sir Alladi had was a special limited edition 1933 Model 90L, L for Limousine. Only a few were produced — 338 to be exact — and just a handful send to India. This was the most expensive of the Buicks of that era.

15) The British developed Hill Stations in India where they would retreat during the summer. They had their official residences there, including the Government House for the Governor. Ootacamund was the hill station in the Madras Presidency.

16) Sir S. Radhakrishnan (1888–1975) was a statesman and scholar of world repute. He was a dear friend of Sir Alladi. Sir S. was Vice-President of India from 1952 to 62, and then President of India from 1962 to 67. Radhakrishnan's only son S. Gopal (1923–2002) was of the same age as Ramakrishnan and the two were classmates in the P. S. High School. Radhakrishnan's son-in-law, M. Seshachalaphi (known for his mastery of the English language) was especially close to Sir Alladi who treated him almost like his own son. He has written a charming account of Sir Alladi (see [1]).

17) Public schools in UK are the same as private schools in the USA.

18) The term *rider*, typically used in high school geometry, means a problem based on a major theorem. The British text books on geometry such as by Hall and Stevens, Hall and Knight, and Pierpont, had an excellent set of riders after each theorem.

19) Chakravarthi Rajagopalachari (1878–1972), popularly known as Rajaji or C.R., was a great statesman, scholar, and writer of twentieth century India. He and Sir Alladi were very close as contemporaries. Rajaji served as the Chief Minister of the Madras Presidency during 1937–39, which is when he would call on Sir Alladi regularly at Ekamra Nivas in the late evenings to discuss legal issues pertaining to the Presidency. See the article "C. R. and Alladi — unison in contrast" by Alladi Ramakrishnan [1].

Here is what Rajagopalachari said (see [7]) on the occasion of Sir Alladi's 60th birthday: "More than all the achievements of intellect, Sir Alladi's simplicity of heart is even more bewitching than the superb qualities of head for which he is admired. I have known him for thirty years. There are not many lovable people in the world like him."

20) N. Rajagopala Iyengar began his legal training in 1926 under T. R. Venkatarama Sastriar and Sir Alladi. He was a lawyer in the Madras High Court where he was appointed as judge in 1953. From 1960 to 64 he was a justice of the Supreme Court. In 1962, for the Madras High Court Centenary, he wrote an article on Sir Alladi (see [1]).

Chapter 2

Ekamra Nivas (1930–38)

Ekamra Nivas to me is not just a family home, spacious, elegant, old-fashioned, and comfortable — it is a way of life, a haunting memory, and a living inspiration. It represents the spacious days of my past, my childhood and boyhood, a tranquil life with my parents, peace and orderliness under the British government, a Hindu way of life, placid, calm, complacent, and serene. About thirty of us lived under one single roof with no financial problems or worries, since my father was an affluent lawyer when the good things in life were inexpensive, discontent among the poor was absent due to the wide prevalence of personal charity, and when the sweet enjoyment of sharing was considered one of the principal pleasures of life. Ekamra Nivas was an island in the sun and each of us had our place in it.

Our life moved around our two great parents, each great in his or her own way inspired by three noble passions — thirst for knowledge, longing for friendship, and a concern for the suffering of others. My father was a lawyer par-excellence, with a legendary reputation, the *facile princeps* of the learned profession, respected by all — rich and poor clientele, princes and zamindars, landlords and wealthy traders, poor vedic scholars, and poorer servants and peasants. In private life he was simple, dressed in a plain cotton shirt, unbuttoned most of the time, his *dhothi*[1] wound around carelessly, almost showing his knees. He never touched a coin or a currency note, he never handled money or spoke about the cost of things, or settled accounts. He earned in just the same manner as Don Bradman scored runs!

It was a fortunate circumstance that in the first stage of his earning career, his elder brother Seetharamayya took care of our household. He was manager, caretaker, trustee, designer, planner, policy maker, and 'chief' of Ekamra Nivas — Peddanayana (= elder uncle) as he was called. He had an unerring pragmatic sense towards worldly affairs — a man of sterling honesty, born to command and get things done. He had a high sense of ethics and morality and was almost intolerant of the absence of these qualities in others. To him religion was part of a way of life, not an obsession. He did not attach much importance to ritual, yet he was a conformist, for he believed in discipline, order, and rectitude of conduct. Under his aegis, my father acquired considerable landed property sixty miles north of Madras in Pudur, the village of his birth. There were also investments in established companies. At that time there were not many industries and Indians had not learnt the art of making money by floating companies and using public funds for enterprises under private management. The few that existed like the Tata Steel, (B & C) Mills, Indian Iron, etc., were managed well by British expertise or Western trained businessmen.

When I joined school, my uncle had shifted his headquarters to our village to supervise agricultural operations there. My mother was in charge of the household. She was small in stature, fair of colour, and had a beautiful face of such softness, calmness and grace, that it haunts me like a passion to this day. She was impeccably clean, always looking like a person who had just taken a bath. She dressed always in traditional style in an Indian silk saree which accentuated the lily tincture of her face and the dignity of her bearing. She was soft spoken and spoke very clearly and wisely. She had only a primary school education, but her understanding of human nature, her perception of the role of science, her concept of an ordered government, and her view of family life, were of such high order, that she would have been at home in any society in Cambridge or Harvard, Princeton or Paris. Endowed beyond the reach of art, she was at ease with human nature and tolerant of the foibles of others. I never could remember her get angry or speak a harsh word to anyone. Her displeasure at any wrong done in her presence was shown by a tinge of sadness in her face which to a sensitive mind was more expressive than words. She was a living symbol of chastity, devotion and affection towards a husband who denied her nothing. If my father was the sun, she was the moon of the little universe of Ekamra Nivas.

In the thirties, Madras had a population of three and a half lakhs,[2] and we lived in Mylapore — the area surrounding the Kapaleeswara Temple. Except for the spacious homes of the lawyers and judges, and the huts of the poorer classes, the rest of Mylapore was all coconut groves or empty spaces. Luz Church Road on which Ekamra Nivas stood, was a broad avenue lined with beautiful trees. Madras had some spacious and elegant buildings with only two stories amidst sprawling gardens. The most impressive was the Government House in the centre of the city and the summer residence in Guindy on a two thousand acre deer park!

Except for buses, trains, automobiles and electricity, technology had not made any impact on our country, particularly South India. There were only six classes among the Indian intellectuals — lawyers and judges, civil servants and clerks, school and college teachers. The main features of life were domestic happiness and cultural activities. There was no native industry worth the name and agriculture was the main source of wealth. Money lending was by far the most important way of multiplying wealth and treated as a dignified method of investment. Litigation was part of a way of life and mostly it related to debts, land, and inheritance.

We were seven children of our parents. In addition, we had our cousins at home who were less fortunately placed — who came to stay with their aunt and uncle (my parents) for their school education. Besides them, there was the band of servants, assistants, and clerks — two cooks, two gardeners, two odd job male servants, two maids, two cleaners, two drivers (chauffeurs), a resident milkman with an assistant, two nurses (Aayyas) for young children, a personal attendant and two clerks for my father, and two private tutors. We also had a family barber whose duty was to attend on anyone who showed his head or chin to him for the cost of eight annas (= half a rupee). So also a *dhobi* (washerman) who collected all the dirty linen and brought them back white as snow but in various stages of damage after boiling them in cauldrons near Red Hills, where water was plentiful. The laundering cost was incredibly cheap — three rupees for one hundred pieces, independent of the size and texture — the effective cost being much greater since the dhothis and shirts and their colours could not survive his special treatment for more than five visits!

My father's office was full of 'juniors', lawyers, and clientele — a colourful medley of persons dressed mostly in a white shirt and dhothi. Among the clients was a good sprinkling of Zamindars and Rajas whose extensive properties and extended families created enough problems for litigation. Whatever the status of the clients, they came to call on my father personally — so great was his reputation,[3] so deep was the faith he inspired.

Among his standing juniors were two remarkably competent men, Ramamurthy Iyer and Rajagopala Iyengar. The former became an affluent lawyer and the other a Supreme Court Judge in later life. The spacious verandah — one hundred feet long — was thronged by men attached to the principal clients. My father was a poor sleeper and an early riser. He was up by five o'clock and would go to his office room, curl up in his easy chair, and start reading the 'causes' or law reports in the silence before dawn. Our cook Krishna Iyer would bring him coffee — all milk plus the first 'decoction', which means the first drops of viscous coffee essence from the percolator. By seven in the morning the clients and their lawyers would be streaming in by car, rickshaw, or walk.

Ekamra Nivas is a palatial building with almost a dozen living rooms, each the size of a small apartment in a modern building complex. But privacy in the Western sense was in those days almost non-existent. With the exception of my father's office and his bed room, the rest was 'open ground'. Besides the members of our family, twice the number of close relatives were moving around. We rambled over the entire house, studied and played wherever we liked. The kitchen area, the size of a big bungalow, was separated from the main building by a covered verandah which is still the best part of the house to relax while reading a newspaper.

Among the staff of of Ekamra Nivas, Krishna Iyer the cook, led the rest. He was as versatile as the legendary Jeeves, for he was cook, valet, contractor, odd errands man, companion and caretaker, all rolled into one.[4] He must have weighted less than a hundred pounds, just as small as my father, for he would wear my father's old clothes, or try out his new suits with equal facility! He produced the most delicious food that I have ever tasted. The ingredients he needed were few but well chosen. He acquired the reputation of being able cook in a *jutka* — a moving pony carriage! Actually he was an expert in cooking in a train compartment when my father used to reserve a four berth compartment to ensure privacy. The fame and flavor of his *Sambhar* and *Rasam* had spread all over South India where Alladi was a household name. My father ate sparingly, but he would appreciate Krishna Iyer's culinary genius.

After cooking in the morning, Krishna Iyer would get ready the *tiffin* for the court, which meant snacks like Dosa and Vada for at least ten persons who would surround my father in the High Court chambers. On holidays when the office room in the house was full of clients and lawyers, he would prepare over one hundred plates of snacks for their refreshment.

Second in importance was Subbanna, the personal attendant of my father. He was a distant relative of such varied gifts which could not be contained or constrained by the discipline of school or college, that he remained a 'versatile intellect at large'! He was a man of all seasons and for all odd jobs. He was valet, body guard, personal assistant, chaperon, and bell-boy, all combined in one. He was on easy friendly terms with everyone who met my father, from the Chief Justice to the poor student, from the high official to the client's clerk. There was no subject which did not engage his attention. He picked up knowledge by conversation and surmise. He was one of the fastest walkers I have ever met and it was a matter of pride that he weighed less than a hundred pounds! He felt himself an authority on dress, travel, and society, and claimed friendship with everyone whom he had the opportunity to meet. He had definite opinions on every issue and would supply information on the most obscure matters. He could even 'create' news by suitable surmise which made the conversation interesting but sometimes misled those who relied on him. He was never bothered about politics and the weighty issues of the hour; he was always concerned with people, their habits and manners, and the art of getting things done. He was never concerned with the state of the Indian Railway system, he just had to get the reservation on an appointed date, get the extra weight accepted with or without the extra charge. He was lovable and loving with such talents and trifles, all the more so because he was deeply attached to my father who commanded the veneration and respect of those around him. My father's memory was remarkable, but he would forget names of places and sometimes of persons and Subbanna was always there to prompt. Many times this saved delicate situations, but now and then created some! He was both ubiquitous and evanescent. He could slither into high places with incredible ease by sleight of motion or trick of tongue, his watchful eye and wakeful ear receptive to the faintest suggestions and rumours. He would vanish without warning when his presence was most needed to the helpless annoyance of my lenient father.

It is impossible to think of my father without Krishna Iyer or Subbanna. He was aware of their faults and foibles and loved them all the more for such failings. These assistants attended to his every need, even helping him put on his shoes, coat or trousers. He was no snob or aristocrat, but he was oblivious of his surroundings; he was just a brain and a heart in a frail physical frame, invested with qualities like friendship, genius, memory, and affection. His hands knew only to make notes on the legal papers or mark his favourite phrases in books. In power of thought, in force of intellect, in depth of memory, in keenness of perception, and in his judgement of human minds, he was unsurpassed.

Till I was ten, I never travelled outside of Madras. My father of course used to go out to the *mofussil*[5] like Madura in the South, Nellore in the north. Our only knowledge of the outside world was through the newspapers — *The Hindu* and *The Mail*. No one in our circles dreamt of travelling outside India; even to go to England was given to a chosen few. We heard of a few who had gone to England to pass the Indian Civil Service, and legends, true or bizarre, had grown around them. The I.C.S. represented the high watermark of authority; the few Indians who qualified for it were treated as a super-privileged class. With the growth of the national movement, these products of the western education were reckoned as bulwarks of imperial domination. Some were hated, some admired, but still their influence over society was great, and their power, immense.

It was in my early school days, that we heard of the 'talkie' or motion pictures with sound. The American films were our windows to the new world, a land of dreams, of wealth and prosperity, initiative and enterprise.

To my father America represented a democratic federal constitution, different in style but similar in spirit to the British. He was well acquainted with American politics and economics, society and culture, for he read books voraciously. What fascinated him was American democracy and its legal system, always in comparison with that of the British Commonwealth. He was least concerned with the high standards of life, the glamour and glitter of the urban life in America. I remember his reading to me some passages from a biography of Roosevelt by Emil Ludwig who had also written a book on Abraham Lincoln. He explained to me how Roosevelt tided over his physical infirmity — paralysis. Though weak in health, my father would never accept ill health as an excuse for laziness. He was one of the most hard working intellectuals I had seen in my life. He admired the triumph of intellect over brute force, of will over physical infirmity. To him Roosevelt represented the irrepressible energy, the creative exuberance, and the vaulting ambition of the New World.

My first recollections of my early boyhood are that we had two cars — a Studebaker and a Dodge, both tourers with soft collapsible tops. I remember how proud I was to get down at school during the first or second form in a new green Ford with yellow spoked wheels which replaced the Dodge. I can still hear the peculiar whine of the early Ford engine. Anyone could use the second car, but the first one was to be used only by my father. The senior driver was Sundararaj, a short, stocky and dignified person who allowed none to take liberties with him. He was always clean shaven, with a spotlessly clean shirt, and a while 'mull' dhothi. He was an expert mechanic who could take to pieces and put it together again with absolute confidence. In America he would have risen to the position of works manager in a General Motors factory. He had great faith in the GM cars Chevrolet and Buick — a preference which soon prevailed in our house. He conceded that Dodge, Plymouth and Chrysler, were close seconds in reliability. In contrast, the second driver (Chauffeur) was a plebian, a friendly ruffian, sudden and quick in quarrel, with an unhappy knack of getting into trouble with policemen whom he regarded as wasteful inconvenience.

Complete democracy prevailed over the use of the second car. Every morning, with five gallons of petrol costing just five rupees, the car was ready for a hundred miles of short trips in the city anywhere for anyone, its free use controlled only by the uncertain temper of our driver.

In the mornings we worked under the guidance of our tutors. By temperament I was a loner who liked to work alone. Our English teacher Subhabhat was a most lovable disciplinarian. He did not teach me in a routine fashion, but supervised my reading. He insisted on high standards but felt proud that I was at the top of my class. He wanted me to be way ahead — a difficult thing to do for my school drew the finest of talent from the most intellectual part of Madras.

British history was part of the curriculum in the fourth form. I preferred it to Indian history since it was a chronological record of kings and wars. Our private teacher would say that for an ordinary student, it was enough to know the names of two wives of Henry VIII — Catherine of Aragon and Anne Boleyn, but the advanced student with History as an optional subject, had to know the names of all the wives of that resplendent tyrant sovereign. My own irrepressible admiration for Britain and its traditions can be traced to my reading of British history during my boyhood.

My father was too engrossed in his work to worry about the details of our school education. But he frequently called me and read a few passages from the books he read. Though surrounded by friends, clients, and lawyers, he was easily accessible — the quality of truly great men.

During the summer our teacher would take us to the famous book shops of MacMillan and Oxford University Press on Mount Road. We felt we were on the soil of England inside those bookshops where the finest imported books from England would be unpacked. With a gleam in his eye, he would tell us that we should soon qualify to read the originals. He took a special pride in claiming the ten percent discount for teachers.

In 1934 when I was in the Second Form, a salesman from Byramshaw, a local motor company dealing in Buicks, came to our house and informed us that a very special limousine Buick seven seater with two 'Dickies',[6] deep blue in colour, with nickel plated wheels and prominent stepneys on the side, was available for sale. It was expected that only one of two persons could, and would, buy it — my father or Sir C. P. Ramaswami Iyer. As a boy of eleven, enthusiastic about cars, I was consulted by my father and we bought it for Rs. 12,500, an immense sum in those days when a Chevrolet or Ford was being sold for Rs. 3000. It was a magnificent Buick which would have been coveted by a Saudi Arabian oil king! And Sounderraj our chauffeur, was its sole guardian and manager. For twenty years he preserved the shine of its paint, the gleam of its nickel, and the smoothness of its performance. No wonder he considered himself the King of Chauffeurs and a chauffeur of Kings! The car had a unique horn — a short metallic sound which could be heard half a mile away, forewarning us that father was returning home from the High Court or the beach. I almost hear it now.

Every year from 1935 we used to go to the 'Hills' or to Bangalore for the summer, a relief from the heat of Madras. Krishna Iyer would first go with the entire luggage to arrange accommodation. It was delight watching him load the *jutkas* (= pony carriages) and bargain with the coachmen. He had a way of dealing with human beings — ticket collectors, weighing inspectors, taxi drivers — a frail man cheerfully doing the job of a dozen able bodied workers, often times with a rhyme on his lips.

I remember every hour of our expedition to Ooty in 1935. The journey by the Buick through Bangalore and Mysore to Ooty, and through the Bandipur forest is still green in my memory.

Life in Ooty centered around the magnificent Government House and its lovely gardens — a riot of flowers amidst the most luscious foliage. The highlights of the 'season' were the flower show and the races, and the tea parties and dinners at the Government House. There was an order and precedence and my father as the Advocate General was high in that order as a knighted citizen.

The shifting of the Government to the Hills during the summer was a practice initiated by Lord Amherst in 1827. He made Simla the summer resort of the Viceroy. It was a demonstration of the prestige and power of the British Government and the solicitude for the health and well being of the British ladies who had to be sheltered from the rigours of an Indian summer. There was a sense of permanence and supremacy of the British Empire, enhanced by the submissive attitude of the meek Indian intellectuals, and the ultra loyal officials. It was a time when the proud official and his humble chaprasi, the jeweled Maharajah and his liveried attendant, the venerable judge, the astute lawyer, the shrewd businessman, the meek clerk, the dedicated teacher and the learned professor — all believed that the Almighty had placed in the British Government the sacred trust of infusing 'a sense of manliness and moral dignity, a spring of patriotism, a dawn of intellectual enlightenment, a stirring sense of duty' among the vast congeries of races in India. The ruling classes had therefore the right and privilege of enjoying their leisure in the hill resorts away from the burning heat of the scorching plains.

'Silverton' where we stayed was a lovely house with a terraced garden on Fernhill, a charming suburb of Ooty. It was time when the Maharajahs of the native states held an exalted position with their fabulous wealth and prestige in a society so well supported by the British Government. Each native state had its own palace or summer residence in Ootacamund and as boys we watched with interest the colourful parade of Rolls-Royces and the turbaned retinue of the Maharajahs of Mysore, Travancore, Hyderabad, and even the smaller states of North India. All of them paid their homage to the British Governor by calling on him or inviting him for 'hunts' or garden parties and dinners.

In 1936 we rented a house in Ooty called 'Ranga Vilas', not half as good as Silverton, but equally expensive. The most enjoyable part of our stay in Ooty were the long walks. Our private teacher was with us. He considered himself a sturdy walking enthusiast and used to take us to Dodabetta, the highest point on the hills of Ootacamind, where sometimes stray tigers may be prowling in the dark. One of our companions was my father's official typist who was very proud of his Havana suits of shining cotten from the Buckingham and Carnatic (B & C) mills.

The elite of Ooty met at the Lawley Institute, a club maintained in the strict English tradition — clean and comfortable, combining discipline with freedom, with plenty of facilities for sport — like billiards and tennis. Opposite the Lawley Institute was the Assembly Rooms, a film theatre for the elite. We took our father to "House of Rothschild', the only film he must have ever seen! Among other films I remember were '39 steps' with Robert Donat, and the 'Raven' with Bela Lugosi and Boris Karloff. Ooty was a fantasy, yet real, a little England in the vast sub-continent of India, a vision of peace, security, prosperity and tranquility. The reality of India was visible only when we descended from the Hills. It was a world of the well-to-do middle class, paying their homage to the rich, the powerful, and the privileged, with their servants and retinue. The butler was as happy taking care of the master's table as the master was solicitous about his welfare. It looked like a page out of Vicar of Wakefield or Pride and Prejudice, of the pleasing vanities and foibles of human nature, little acts of friendship and politeness, life as in a pleasant summer afternoon in the British countryside in Victorian England. It was just too good for too few to last!

In 1937 we stayed at 'Ambal Vilas' on the other side of Fernhill. It was then that we used to meet the Sarabhai family taking their walks. Ambalal Sarabhai enjoyed a reputation similar to that of the Tatas as the entrepreneur of textiles. One of his sons Vikram was to become a physicist with a Cambridge degree and later the Chairman of India's Atomic Energy Commission of India. Among the Indian elite, my father's friend, T. R. V. Sastriar, tall, gaunt and dignified, used to go on long walks dressed in grey flannels tucked under the woolen socks, looking like a proud British Colonel who had won many battles in his times. My father admired his physical stature, his sense of discipline, his decorum, and his loyalty to his master Rt. Hon. Srinivasa Sastri, a silver tongued orator.

From 1938 we shifted our summer venue to Bangalore and later to Kodaikanal. Meanwhile the world was smitten and blasted by the scourge of war. What emerged from the flame and flood of steel and blood was very different from the halcyon world before the holocaust. Ootacamund of my boyhood days had passed into a fleeting dream.

Notes

1) A *dhothi* or veshti, is usually a plain while cotton cloth with a thin border that is worn by men, and tied around the waist. Veshtis made of silk are often worn on traditional occasions and during festivals.

2) A lakh is one hundred thousand. Large numbers such as population, or even sizeable sums of money, are stated in India quite often in lakhs even today. A crore is one hundred lakhs. In India, we use lakhs and crores instead of millions and billions. The terms lakh and crore are English versions of the Sanskrit words *laksha* and *koti.*

3) In recognition of his great services to the state, Sir Alladi was honoured with Knighthood in 1932. He had other honours previously such as the Kaiser-i-Hind Medal in 1926 and the title of Dewan Bahadur in 1930.

After being knighted, Sir Alladi went to England in the summer of 1933 in connection with the Gollaprole appeal then pending before the Privy Council. There he met leading figures of the legal world including the Lord Chancellor and the Attorney General. When he visited the Old Bailey, he was treated with great deference by the presiding judge, Justice Charles (see Seshachalapathi's article in [1] for more details).

4) On the trip to England, Sir Alladi took with him Mr. Subbaroya Iyer — a close friend who by then was an established lawyer, Mr. Sambasiva Rao — also a lawyer, and Krishna Iyer, to address his dietary needs and to attend on him. Krishna Iyer could not speak English but managed remarkably well when travelling around London on his own to make purchases for Sir Alladi. The joke is that he managed to get what was required by pointing to it and just saying the words "Thank You"! He acquired the name "London Krishna Iyer" after his trip to England with Sir Alladi.

5) Mofussil is a term applied to regions well outside large metropolitan areas. In addition to having a large clientele in Madras, Sir Alladi had a significant clientele in the mofussil (see "My Story" in [1]).

6) A *Dickie* is what is referred to as the trunk of the car in the USA or the boot in England, in the rear for luggage. Not only did this luxury car have two dickies, it had two stepneys as well, one of each side with full size tires. This limited edition luxury Buick is highly regarded by collectors. In 2017, one very well maintained 1933 Buick Model 90L of the type Sir Alladi had, sold for $180,000 in the USA.

Our life moved around our two great parents.

Father in Delhi
(At the Constituent Assembly)

He could not move about without help and Subbanna was always there all eyes and ears.

Sir Alladi had a 1933 Buick 90L exactly like this one. - Editor

Chapter 3

School to College — From Unreal Peace to Reality of War (1936–39)

In Madras, when C.R. assumed the Chief Ministership of the State, it was impossible to describe the enthusiasm and excitement that followed his installation into power. My father was the Advocate General[1] of Madras when there were only three such posts in the whole of India, for Madras, Bombay and Calcutta. Under the new dispensation, the prestige of the Advocate-General's office had in a sense been lessened due to the increase in the number of States. Also the original appointments were made by the Viceroy but from 1935, the Chief of the democratic local government had to choose the Advocate-General and C.R.'s choice was obvious — my father.

The friendship between C.R. and my father dated back to earlier years when C.R. was just a 'soldier' in the Gandhian freedom struggle. No two persons could have been more alike in innate human qualities and more unlike in outward temperament. They loved and admired each other animated by the same ideals (for a comparative study of C.R. and Alladi by Ramakrishnan, see [1]). I can never forget the gleam in my father's eye describing C.R.'s incisive ingenuity in thought, and ruthless directness of expression. During the years 1937 and 1938, C.R. used to visit Ekamra Nivas very frequently at night after dinner. My father, though much younger, would be relaxing in his easy-chair while Rajaji (as C.R. was also known) in his immaculate khadhar dress would sit in front of him on the opposite side of the table. As boys, my brothers and I were excited hearing the voice of the Chief Minister inside our family home. C.R. was a constructive statesman. He wanted to prove that Indian intellectuals were capable of high administration, and not just as civil servants working under the direction of a supreme authority. He wanted to demonstrate that we can take initiative in legislation.

The most curious feature of that period was the total absence of interest of administrators and intellectuals in creative science and original research.

The great technological and scientific advances in the West barely touched our lives. We admired lawyers, civil servants, men of power, and above all the British elite. Our imagination was just filled with politics, tirades against the British administration, discussions about the prospects of war in Europe, biographical tales of the heroes of freedom. We spoke more of Tilak, Subash Bose, the Patel brothers, and of course Gandhi and young Nehru, father and darling of resurgent India. Our Bible was the writings of Nehru who represented the future of our country. It was a pride to subscribe for the 'Harijan' — a newspaper published from Wardha with Gandhi as its soul, spirit and substance. Simplicity became the fashion of the hour and the stoic spirit of Khaddar[2] filled the air. I remember clearly approaching the great Sir S. Radhakrishnan and inviting him to speak at our Student's Union. It was sheer impertinence to do so but I was unaware of it. He very pleasantly replied that we should get young politicians to talk to us — some future Anthony Eden in Madras! Side by side with the desire for freedom, there was a sneaky admiration for everything British and for the English language. For example, our great veneration for Radhakrishnan, the philosopher, was not so much for his erudition and philosophic thought, but for his mastery over the English language and the fluency of speech. C.R.'s popularity was essentially because he could charge his speech with wit and sarcasm when he spoke either in Tamil or English. The peculiar timbre of his voice and the native accent of his speech added an extra sharpness to his incisive humour.

It was in such an atmosphere I completed my school education with an active interest in mathematics generated in the geometry classes of a dedicated teacher. I stood first in my school on the S.S.L.C. Examination (public exam for the School Final)[3] and thus qualified myself for entry into Loyola College, one of the most prestigious colleges in Madras.

I joined the Intermediate[3] in the Loyola College in June 1938. At that time Loyola had a very high reputation for maintenance of discipline and high standards of examination performance. The two years in the Intermediate were happy beyond description. We were very proud of our institution, of the devoted missionary teachers of that day and the beautiful campus which compared favourably with any in the Western world. Every subject was taught so well that study at college was both pleasant and serious. The classes in English prose were unforgettable. Macaulay's History of England was taught by one of the greatest teachers of all time — Mr. K. Subramaniam. He was extraordinarily facile in speech and Macaulay could not have chosen a better exponent of his prose. He wanted to match the text by explanations in chaste English and I remember in one of his lectures his reference to the pathetic fate of a once beautiful woman, 'whose faded beauty is her only stock-in-trade'. The classes in English prose were a treat to attend, almost like watching a cricket match on a sunny day in the English country side. Shakespeare was taught to us by A.L. Krishnan whose sing-song tone added an extra zest to Shakespearean diction. A.L. Krishnan was always immaculately dressed and I envied the crease in his trousers, the smoothness of his shave and the delicateness of his curly hair. Subramaniam was always clad in traditional style, spotless white khaddar dhothi with a 'jibba', a symbol of stoic nationalism, shapeless but comfortable.

In Shakespeare we had as our text 'The midsummer night's dream', not a favourite of mine but which created in me a desire to read Shakespeare, a desire satisfied in later years by reading Macbeth and Hamlet, 'Richard the II and III, King Lear and Othello', 'Henry the Fourth and Fifth', 'Julius Ceasar and Romeo and Juliet and my favourite of favourites — 'Antony and Cleopatra', over and over again. Conversation with two of my brilliant colleagues Thammu Achaya and Vepa Ramakrishna was an exhilarating experience since they had reached such maturity in English style at the age of sixteen that could hardly be improved upon through age and experience!

The Chemistry classes were enjoyable beyond bound. I consider Lakshminarasimhan as one of the most inspiring teachers of all time. His baby-face and rotund figure added new life to an already lively chemistry class. We competed keenly to get the maximum number of marks in each test, but it was always Achaya or Vepa or Rangachari, who won the first place. I could never reproduce the description of experiments as we were taught in the classes; for Lakshminarasimhan would not be satisfied with just an 'yellow' precipitate but only a 'canary yellow one' if the marks were to be fully awarded. His treatment of the Halogens had a dramatic flavour, as he described the intensity of reaction of Fluorine, Chlorine, Bromine and Iodine in corresponding intensity in intonation!

In the physics classes, there was equal emphasis on accuracy; the classes though not so lively were equally serious and instructive. Raghavendracharya who in dress matched A.L. Krishnan, was one of the most serious minded teachers I met in my life. But there was nothing exciting or new or modern in the instruction. It was just steady, uneventful but purposeful. On the other hand the mathematics classes took pride of place in quality and standard. Loyola college at that time stood high above and the rest nowhere.

There was a general reputation that our teacher Adivarahan was a hard taskmaster, a slave driver and teacher whose success depended upon the rigour with which he enforced frequent tests on his students. For reasons inexplicable, he was so different in his attitude towards me. The interest in doing new problems first generated in my school days had become a passion and I had an irrepressible desire to provide a new proof for every geometry problem. Adivarahan (Adi as he was called affectionately by his students) was extraordinarily pleased with this and exempted me from the routine theorems classes. He held a two hour competition for two hundred students in geometry and trigonometry where every conventional geometrical proof has to be replaced by a trigonometrical one and vice versa. Almost all the students returned the question paper without answering any. I was the only person who cleared the entire paper and a Parker pen then costing Rs. 14 was presented to me as a personal gift from Adivarahan. Father Racine was asked to present this at a function held at the college. I consider it one of the proudest days of my life and the intellectual exhilaration I had could only be matched by that when I got the result on 'product densities' ten years later and on 'L-matrix theory' thirteen years thereafter. Since I had the notorious habit of losing things, I gave this pen to my mother. I never could blame myself adequately for having taken it from her some time later and losing it. My mother felt the loss much more than I did and there was no point in replacing it by an identical pen and imagining it was the gift. In fact I quoted this as an example on various occasions to show that sometimes rationalism and sentiment are mutually exclusive! You cannot compensate for the loss of a gift by acquiring another of the same type!

I remember very well how proud and happy I was to go to the Loyola College in our new car which was a Ford Tourer, that is, with a removable top. That was a time when eating out at hotels was not popular among students and it was considered quite normal for us to carry our lunch with us. We used to eat our lunch under the shade of trees just behind the Chemistry hall and our discussions ranged from weekly tests to high politics.

We admired our Principal Murphy for his delectable English which reminded me of Hazlitt's prose. On one occasion when Sir A.L. Mudaliar was a guest on a college day, Murphy sportingly dismissed the criticism about the Loyola College giving too much emphasis to examinations as a 'tribute which jealousy pays to success'. At the end of two years, we became fanatically loyal to our college and we carried it to an illogical extent for sometime even when some of us shifted to the Presidency College for the Physics Honours.

During the period 1936–39 in the transition from school to college, I developed an interest in studying international affairs. My father was a voracious reader of books on world affairs and used to mark in his characteristic style important sentences and paragraphs which impressed him. He read all the time when he was not engaged in legal work — in the car, in bed and in his easy-chair and would read aloud to me drawing my attention to events and opinions. With the confidence of an ignoramus and the brashness of youth, I used to express my opinions casually on the English monarchy, the League of Nations, the Bolshevik revolution and the American New Deal.

The most important events that were discussed were the rise of the dictators, Hitler in Germany, Stalin in Russia, and Mussolini in Italy and the formation of the Rome-Berlin axis. As school boys we heard of the emergence of Soviet Russia with awe and admiration as people spoke of the great success of the Soviet Revolution leading to a rapid rise in the standard of living of the masses. We recoiled with horror at the mass extermination of people who were considered opponents of the revolution. But it was only a distant view of communism for there was no steady source of information reaching us from Soviet Russia. We heard of the rise of Marshal Stalin as a dictator and Party Chief and how he came to power wading through a sea of blood. There was some general interest in the economic doctrines of Karl Marx on which communism was based, but on the whole it was just either admiration or fear born out of ignorance or inadequate knowledge. We also watched the events in Germany and the emergence of Hitler to despotic power claiming that his idealogies were essentially different from Soviet dictatorship. The story of how a house painter of Vienna became the absolute Feuhrer of the 'Vaterland' was read by the younger generation in India with varying emotions. There were some who tried to obtain the translation of Hitler's Meinkampf to understand his real motives. Some thought of him an avenger of Versailles, some reckoned him a bulwark against communism and some hated his racial doctrines and his lust for power. That was the time we heard of Goering and Goebbles and the triumphs of the doctrine of propaganda — the secret of inspiring belief lies in the size of the lie. The masses are taken in by a big lie rather than a small one. Terrorism was made into professional art leading to definite results. In India we could not understand the ideological differences between the two dictatorships, except that big business was banished in Soviet Russia and protected in Nazi Germany. The rise of Mussolini in Italy and the formation of the Rome-Berlin Axis, gave a new dimension to the Nazi threat. There was always the talk of war and the helplessness of the League of Nations.

My father read a variety of books on these international problems with such keen interest that I felt a certain familiarity with world affairs though I had not travelled beyond Trivandrum in the South and Calcutta in the north. What today intrigues me is how my father being in such a caste ridden, class conscious and custom bred society with religious practices dating back to time immemorial, could be excited about the momentous changes in the European scene, the economic and social upheavals, the rise and fall of world leaders. I remember his talking to me of Masaryk of Czechoslovakia, De Valera, the idol of Ireland, Kemal Ataturk who westernised Turkey, the rise of Japan adopting western methods.

When I entered the Intermediate class in July 1938, our interest in international affairs rivalled that in Indian politics and personalities. How would India react in the event of war? The sovereignty of England was still total and complete, the prestige of the Crown still exalted and the entire Indian army and the Indian Civil Service loyal to the British Government. The Congress under Nehru and Gandhi stood for complete independence and the millions of India worshipped them as the messiahs of freedom to be won by peaceful and non-violent struggle. Statesmen like C.R. attempted to bridge the widening gulf between the Congress and the Crown! In six out of eleven provinces, there was Congress rule under the watchful eye and wakeful ear of the British Viceroy in Delhi.

During the first year of my Intermediate (1938–39), just before the daily commencement of classes, we discussed the threat of war in Europe or the various facets of Indian politics in the struggle for freedom. The world was on the brink of war after the Munich pact and Hitler's occupation of Austria in the spring, when the British Prime Minister agreed to sacrifice that little far off land Czechoslovakia to satisfy the appetite of Hitler. There were those who felt that Prime Minister Chamberlain did not have the tough fibre and the stout heart of the British race. Others believed that he represented the caution and circumspection characteristic of a country which would not resort to arms except as a last recourse when negotiations absolutely fail.

Everyone knew that it was an uneasy and unreal peace. War clouds were over the European skies. Throughout the world, the rumble of thunder, the streaks of lightning, the ominous spells of calm before the storm filled the hearts of men with fear of war that chilled their spines, curdled their blood, and froze their minds.

The warlords of Berlin, Rome, and Tokyo waited and watched for the moment to strike as the Wehrmacht, the Luftwaffe, and the panzer divisions bristled and burned in impatient readiness to blast the earth with fire and steel.

Meanwhile the world held its breath. Where would Hitler strike? Would it be across the Maginot line against France, its traditional enemy, or across the channel to invade England with the striking power of the airforce denied to Napoleonic ambition? Would he choose a lesser, weaker victim for an opening triumph?

And then the moment came. On the 23rd of August 1939 a non-aggression pact was signed between Germany and Russia. Poland was to be the first victim, the campaign to erase it was to be the baptism of fire for the German Wehrmacht.

In the darkness before dawn on September 1, a fateful day in human history and a fatal hour for Poland, Hitler struck with the flaming fury of the fiery Hun. He ordered the simultaneous attack on Poland by the Panzer divisions and the Luftwaffe in a lightning campaign which crushed the life out of the hapless country in seventeen days. It was the first successful demonstration of highspeed armoured warfare with aerial support adopted due to the initiative of General Guderain, the young exponent of the Blitzkreig tactics.

Britain reacted true to its history and tradition so well expressed by one of its statesmen: "We were there when the Spanish galleons made for Plymouth. We were on those bloody fields of Netherlands when Louis XIV aimed at the domination of Europe. We were on duty when Napoleon bestrode the world like a demi-god and we answered the roll call in August 1914."

Britain, with France on its side, declared war on September 3, after its unheeded ultimatum. Thus started the convulsion from which mankind was to emerge in six years after an ordeal of blood and tears into a new world invested with a new power and responsibility unprecedented since the dawn of the human race — the use of nuclear power.

The news of war reached us on the balmy, sunny afternoon of September 3, when we were playing cricket on the grounds near my house. Even in the placid atmosphere of Madras, our hearts throbbed in fear and tension. We knew it was going to be a total war sparing neither sovereign nor peasant — death for millions and horrors without a name, affecting every aspect of human life. From that moment till the end of the war six years later, half of each day was spent in discussions of the news of war received through press and radio. Even as boys we felt deeply involved since the course of war would determine the course of our lives.

The outbreak of war found India in a divided mind. To some, Britain's travail looked like India's opportunity. But Indian nationalists were democrats and were strongly antifascist. They knew that the war would lead to mortal peril not only to Britain but to India as well.

On the question of prestige, the Congress ministries resigned and India was ruled by its civil service under the Governors and the Viceroy. The deadlock between the nationalist sentiment and the British lasted till the end of the war.

On the 17th September Russian troops crossed the Polish frontier and that was the end of the luckless off-spring of the Versailles treaty. Britain and France now at war with Germany watched helplessly Hitler's triumphant march into Warsaw.

Notes

1) Sir Alladi was appointed Advocate General in 1928, a position he held with distinction until 1944. He should have finished his term in 1943 at the age of 60, but by the personal request of Sir Arthur Hope (second Baron Rankeillour), the Governor of the Madras Presidency, he served one more year. Speaking about Sir Alladi's term as Advocate General, Suresh Balakrishnan ([4], p. 412) says: "From December 1928, Alladi occupied the position of Advocate General for nearly 16 years until 1944 when he retired. To him belongs the distinction of being the second longest serving Advocate General until this day. The length of his tenure was next only to George Norton, who a hundred years back served as Advocate General for more than 20 years. But conditions vastly differed between the two eras. In George Norton's time, there was less crowding at the Bar and less competition. Alladi resigned from Advocate Generalship precisely for the reason of giving chance to younger members of the Bar to be appointed to that post." Sir Alladi in reply to Justice Hidayattullah's letter regarding his resignation said "There is nothing behind the resignation than the fact that I had been too long at it and do not want to clog other people's progress."

As Advocate General, Sir Alladi represented the Madras Presidency and therefore appeared in the Federal Court in Delhi several times for cases involving Madras. Also as Advocate General, he was the one to deliver felicitation addresses when judges or senior lawyers completed their terms, or when they died. His felicitation addresses are models of English diction, and a good selection of his felicitation speeches could be found in the special volume [7] brought out in 1943 for his 60th birthday.

2) *Khaddar* is home spun cotton cloth made famous by Mahatma Gandhi who said Indians should wear only Khaddar and not fine cotton from British mills. Even after India got its independence, several persons, most notably politicians, wore Khaddar.

3) For standing First in the S.S.L.C. Examination (highest aggregate total in the P.S. High School), Ramakrishnan received as a gift the book "Poems and Essays" by Oscar Wilde. High schools in India in those days were only up to the tenth grade when students took the S.S.L.C. Public Exam (Secondary School Leaving Certificate). Following that there were two years of Intermediate in College before one started the courses for the Bachelors Degree. Thus one always had 12 years of study before enrolling in the Bachelors program.

Chapter 4

Social and Religious Environment of Ekamra Nivas

In the placid, leisurely life of the Hindu middle class family during the period before the world war, the main excitements at home were the religious, social and cultural functions. Hinduism is a way of life and includes in its scope, conformists, non-conformists and partial conformists. It is there like the mother earth and each is involved in some part relevant to his or her life. It means the Vedas, the Upanishads, the Epics, the Ramayana and the Mahabaratha, all the associated literature grown over thousands of years in Sanskrit and the fifty and odd languages of our country. Just as one is a human being, with or without human qualities, good or evil, exalted or bestial, merciful or cruel, so is one a Hindu — he is just born to a great inheritance, a tradition and an attitude to life. Ultimately, theist or atheist, believer or unbeliever, Hinduism determines his attitude to his family, to his wife and children, to his relatives and to his society. Just as we define temperature for an assembly of particles without any significance to a particular one, so a Hindu way of life prevails in a community though each may deviate considerably from it.

I was just inducted into the Ramayana as a child and to me Hinduism is coextensive with it. Everything else is just an amplification, interpretation and modification of the fundamental values of life contained therein. In describing these values we can start anywhere since all of them hang together and are consistent with one another.

The most important message of the Ramayana is the conformity of speech and deed, of profession and action, of feeling and expression, and this is the life which conforms to truth. By definition it eschews hypocrisy, duplicity, cheating and misrepresentation. Equally important is to treat others as you would like to be treated yourself. Since one does not like to be inflicted with pain, mental or physical, it implies one should not do anything that would hurt others. It forbids violence, harshness of speech and unfair judgement. Since one values his own life, one should have a respect for the lives of others. Whether animals are included within the scope of this moral law is a matter of conscience since they have no laws to defend them. Some ancient saints have carried this doctrine of 'Ahimsa' (non-violence) to the extreme that they subsisted on air, on fallen fruit and parts of plant itself. Vegetarianism is just what I can sincerely follow in practice. Any way, it is universally agreed that kindness to animals when they are allowed to live, is a desirable quality. Under no circumstances should a human being derive pleasure by the action of killing. There is no greater sinner than such a one, "Prani Vadhe Rathaha".

Hinduism has been criticised and extolled due to its emphasis on "Karma" and rebirth. "Karma" has been misinterpreted as fatalism, that everything that happens now has been predetermined. "Karma" on the contrary is a natural consequence of free will. A person is responsible for an act only if he has the free will to do or not to do it. He must assume responsibility for the choice and take the consequences. Since it is not a 'physical law', it is naturally postulated that some inexorable power is taking note of balancing acts and consequences. We do find those that perform good or commit atrocities do not reap the good or bad consequences of these acts in the same life. So it was natural to extend the period of time to unite the past, future and the present in which acts and consequences are related. The doctrine of rebirth is a natural consequence of such an assumption. It explains free will and inevitable consequence in one scheme.

The doctrine of karma and rebirth seems to be consistent with all scientific principles. Equally so is the worship of 'idols'. The 'idol' symbolises everything which the worshipper attributes to it — ideals, hopes, dreams, aspirations. It is a means to concentrate his thoughts. The Hindus had realised this from time immemorial, and worshipped only consecrated idols. Consecration is done through 'manthras' or sacred verses. Such a consecration is an attitude of the mind and a matter of personal conviction. There seems to be as much rationality in attributing all possible qualities to just one symbol and affirming faith in one God as attributing them to many symbols representing many Gods. No one has ever proved or disproved that the universe is built up of only one or many basic forms of matter. How many elementary particles or quarks are there? Are they bound or repelled by single or distinct forces? Newton or Maxwell, Einstein or Bohr, Gell-Mann or Feynman, have not found the answer. God and the laws of nature applicable to matter and life, seem to be coextensive. They represent the yearnings of the human spirit despite the inadequacy of our understanding and the finiteness of our existence.

The meaning of the 'sense of awareness', the definition of 'I' and 'You', the dichotomy if it exists, the unity if it has to be realised, engaged the primary attention of our thinkers and savants like Sankara, and Madhavacharya. Unlike a stone, a human body which is an assemblage of atoms and molecules is aware of itself when it is alive. Does awareness imply a collection of memories of past events as located in the human brain? The sense of pain and pleasure can be feelings of the moment even if we had no memory but do they constitute awareness? If a person woke up from sleep having lost his memory and has been transported to another place with his limbs removed or replaced by surgery under anaesthesia, what would he be aware of — he will also speak of an 'I'. He will be aware of himself though not of his past and will start life anew. Does this amount to rebirth or not? These are not idle questions outside our experience. I was confronted with them when I had a terrible accident in 1970. In fact I was told that I had an accident when I found myself in a room in a nursing home in Bangalore without any memory of the past. I could not place the time of my life and slowly with a tremendous struggle things became gradually clear — half dream, half reality, as events ordered themselves in time till memory was totally restored. What if my memory had not returned? Would it not have been equivalent to a rebirth?

These were the questions that held the hearts and minds of our saints, philosophers, of seers and scholars. Their power of intuition and analysis, of logic and reason, were expended in trying to answer these questions relating to the mind, spirit, soul and life. They were not concerned with what are now the primary problems of science, the structure of matter and the nature of the physical forces governing them. Thus our tradition, though highly intellectual, was on a philosophical and moral plane than on scientific analysis, characteristic of the present civilisation.

India is a land of hallowed saints who have thought about these problems since the dawn of the human mind. It was only last year that at the Cambridge University there was a conference on 'consciousness' inspired by the leadership of a Nobel Laureate in Physics. Over thousands of years through thought and experience, through precept and example, a way of life was evolved in our ancient land. Ritual was an essential part of it in just the same way as a signature to a document. Some external symbols were necessary to direct human actions. Without them it will be impossible to communicate love and affection and esteem which make life worth living.

I was born to such a Hindu way of living; I found no reason, no pressure, no impulse to change. On the contrary I felt grateful to God I was blessed with such an attitude to life.

As children we had a great respect and love for our parents, and for our elders and teachers. Violence was taboo in any form and even as children we detested any person using violence. Of course, we had our own quarrels, jealousies and disputes, but it was unnatural to exhibit them through violence. At school I kept away from such company and my chauffeur was always there to scare away the 'bullies' who dared to lay their hands on me.

We looked forward to and enjoyed religious and cultural functions. These were determined according to the lunar and solar calendars, the religious ones by the lunar, and cultural ones by the solar. The English calender year started with the Pongal or harvest festival when my mother used to give presents or bonus to the servants and buy them new clothes and sarees. There was a day (January 16) set apart as a festival for cows and we had three or four in our house with their calves.

After Pongal the marriage season in South India begins, for it is a general convention that Uttarayana (the half year when the sun is above the equator), is a propitious time. My parents felt it a social and religious obligation to attend marriage ceremonies and bless the couples.

Marriages are performed during auspicious periods of the year at auspicious moments (Muhurtham) of the day which are prescribed by the 'Panchangam' or the Hindu calendar. This belief is so deeply ingrained in the Hindu mind that no couple, however 'Westernised or rationalised', would dare to risk their personal happiness in an 'uncharted journey through life' by starting it at an 'illstarred' hour (like Rahukalam) just for proving or disproving the irrationality of such faith! Perhaps it is because marriage is beyond the bounds of rationality — not irrational but transcendental — a sacrament, indissoluble in traditional Hindu society.

On Maha Sivarathri at the end of February, prayers are offered to Lord Siva. The more orthodox and austere used to fast, a practice slowly yielding to easier forms of conformity like 'phala aharam' — light fruit, which by liberal interpretation is implying various types of non-rice meals 'pala aharam'! More austers persons used to wake up all night offering prayers in temples and at home.

In August, on the full moon day, we perform 'Avani Avittam' or the sacred thread ceremony — of great importance to the Brahmin community. Many of our relatives and cousins used to come to our house to change their sacred threads since we had our family priest conducting the ritual. The ceremony according to tradition marks the beginning of the course in Vedic learning for a Brahmin who is being initiated. The importance attached to clear speech, intonation, clarity — emphasis that is characteristic of Vedic culture which has been handed down from our ancestors by spoken word and not by writing.

On Vinayaka Chathurthi we pray to Lord Vinayaka (Ganesha) the elder son of Lord Shiva. In fact every effort, every enterprise starts with a prayer to Ganesa to ward off all obstacles — even the Lords of Lords, Shiva, Vishnu and Brahma propitiate him before embarking on any enterprise. 'Vighnam' or an obstacle is not like an opposition or an opponent; it may be a small incident or occurrence which may thwart a major venture — a delay by the taxi when catching an airplane, a sprain in the ankle to a fast bowler, a wrist injury to a tennis player before a Davis-cup match, unpredictable, unforeseeable impediments which may affect the success of the whole enterprise. The most momentous events can be traced to a human decision and these are swayed or delayed by apparently trivial reasons.

The two great festivals in our social and cultural life are the Navarathri (or Dasara) and the Deepavali. Navarathri means nine nights and if we add Vijaya Dasami we have the ten days of Dasara. During this period every year Vedic scholars are invited to our house to recite the Yajurveda[1] — to which our family belonged. It takes approximately eighty hours at eight hours a day. What an amazing tradition to learn the Vedas by heart in the correct 'Swara' or musical intonation just as they did thousands of years ago! The scholars had learnt it in special schools or institutions 'Veda Patasalas'[2] which are becoming increasingly rare due to lack of financial support. What a tragic irony that our country, in attempting to take to Western education and science, is neglecting the great cultural traditions of our ancient land to the point of wiping it out. Little do we realise that if we cut the roots with the past, the future becomes empty and purposeless. We will drift like leaves in the storm without a sense of direction, with no ideals and principles, with just a desire to increase the standard of life — a mirage which takes us deeper and deeper into an economic desert and the miasma of discontent and frustration.

At Ekamra Nivas before the War, during Dasara, a large Pandal used to be erected under which over a hundred guests had their multi-course lunches and dinners served in the traditional Hindu style on plantain leaves. Though they were not provided elegant cutlery and crockery, they enjoyed the luxury of being attended by a team of fast moving waiters trained as experts under the reticent, efficient chief cook Lakshmana Iyer in the South Indian art of serving lunch within fifteen minutes!

In the evenings, Dasara is a festival for women, young and old, of the display of exquisite colour and beauty. Every house is enlightened by smiles and enlivened by music.[4] The women are dressed in their best, call on friends and offer 'kunkum' (red coloured turmeric powder) and 'vetha-laipakku' (betel leaves and nut) with fruit or coconut, to their guests. It is also the time when young girls make their 'debut' into society so that they are noticed and become candidates for marriage proposals. Prayers are offered to the Trinity of Goddesses[3] — Saraswathi for learning and the arts, Lakshmi for wealth and prosperity, Parvathi for chastity and power. On the ninth day there is a special worship to Saraswathi when we place our books in the Altar and offer prayers to them as representing the Goddess.

Vijayadasami or the tenth day of Dasara is the day of victory. Anything begun then is assured of success and so children are initiated into reading and writing — *Aksharabhyasam* — by the Guru, usually the father,[4] through the Vedic priest.

The most important festival that dominates the heart and spirit of the entire population is Deepavali or the festival of lights. It means various things to various communities and age groups but it is the most important festival throughout the country. It has become a convention to 'fire crackers' especially during the previous evening and early morning before the crack of dawn. Everyone has his sacred bath or 'ganga snanam' smearing himself with oil and washing it off with hot water and cleansing soap or soap nut power. It is a 'round robin' of every one calling at houses of friends and relatives and offering 'sweets and snacks' as gestures of hospitality. Every one wears new clothes and the atmosphere of gaiety, happiness and enthusiasm has only to be experienced or watched. It is really V-day of the triumph of good (in this case Lord Krishna) over evil (the demon Narakasura) — the same spirit in which Times Square was lit up on May 6 when Germany capitulated, and on August 14 when the World War ended with the surrender of Japan.

And then comes Christmas which is a relaxed vacation, with the music season at its peak in Madras. A Carnatic music concert is quite unlike a Western concert in spirit and atmosphere. The audience participates in it by expressing their appreciation in no uncertain terms by looks, gestures and exclamations. It is a social pastime, a religious and spiritual experience, a show of cultural refinement, all rolled into one. There is an air of informality, even of obvious hauteur and it is not considered improper to move around during the concert or move in or out at leisure.

My first interest in Carnatic music was roused in my boyhood by a concert of Semmangudi Srinivasa Iyer, a rising young Vidwan (musical scholar) of that time. We also heard of the names of Ariyakudi Ramanuja Iyengar, Chembai Vaidyanatha Bhagavathar and the violinists, Chowdiah and Rajamanickam Pillai. In the thirties the Madras Music Academy was passing through its adolescence with the support of the elite at Mylapore. A huge pandal was erected every year in the P. S. High School grounds. My own interest for Carnatic music became a passion only after my marriage and after my return from England. In fact I should be considered a late entrant for it was customary even for children of a South Indian family to be interested in Carnatic classical music. But still I remember two incidents which in a sense are the independent starting points of my interest in Carnatic music. My mother took me to a Bharathanatya recital by a young danseuse at the University Senate Hall and the music in the raga Vasantha of 'Natanamadinar' still rings in my years. It was only years later that I understood the meaning of Gell-Mann's reaction when he said that 'Bharatanatyam' is certainly a clear proof that we are 'not an inhibited race'!

In Semmangudi's concert in St. Mary's hall near the beach, I heard Ethavunara in the raga Kalyani and Thunga Tharange Gange in Kunthalavarali. The seeds of desire for Carnatic music were planted then to sprout years later into a soul-stirring, heart-warming passion of my adult life.

Besides these religious and cultural festivals, there were other events which caused excitement and enthusiasm at Ekamra Nivas. My parents were generous hosts and at every possible occasion invited friends and relatives for dinner. 'Westernised guests' were invited for tea parties in the spacious garden. There was a special social circle of judges, lawyers and citizens around my father. Subbaroya Iyer[5] was his most intimate friend in life, closely followed by N.C. (N. Chandrasekhara Iyer). Among the younger associates were Mani Iyer and his brother Chandrasekharan (sons of V. Krishnaswami Iyer venerated by my father), Seshachalapathi (son-in-law of Sir S. Radhakrishnan) and S. Narayanaswamy a well-known industrialist, whose opinions he invited on law, literature, politics and public affairs. As professional associates he had his two remarkably gifted assistants, Rajagopala Iyengar and Ramamurthi Iyer. Among the 'greats' were Sir S. Radhakrishnan and Sir C.P. Ramaswami Iyer. These comprised the 'inner circle' of my father's social life. Of course he was held in great esteem by the leaders of the press like K. Srinivasan of the Hindu, Littlehailes of the Mail (evening daily newspaper) and the editors of the Indian Express and all leading men of business like Rajah Sir Annamalai Chettiar. At that time the Dewans of the Native States held a very exalted position in society — they actually governed the native states in the name of Maharajahs. They were the Wolseys when the kings were not as zealous of their powers as Henry VIII. The unrestrained management of the vast finances of the State invested them with a power and influence beyond the reach of envy and of law. The only constraint was absolute homage to the British Government, paid with obvious ceremony. Among the Dewans I heard of in my boyhood were Sir R.K. Shanmugam Chettiar of Cochin, Sir Mirza Ismail of Mysore, Sir V.T. Krishnamachari of Baroda, Sir C.P. Ramaswami Iyer of Travancore and later Sir N. Gopalaswami Iyengar of Kashmir. Among the elite of Madras feeling prevailed as in Cambridge or Oxford — beyond Trinity College all is desert, beyond Madras and Mylapore, nothing existed of great intellectual value!

The feeling was strengthened by the peace and tranquility that reigned among the intellectual classes under the British Government. Freedom had to be won only by negotiation or by non-violent or passive resistance usually confined to a small percentage of tried and devoted followers of Mahatma Gandhi with the intellectual and moral sympathy of the entire population. It was customary for those in the vanguard of the freedom movement to meet and negotiate with the representatives of British power and prestige. There was so much of freedom of speech that it was almost a social pastime to discuss the statements, rejoinders and debates of various leaders. This is perhaps the highest tribute to the British democratic tradition which persists in India even after the British left in 1947.

It was therefore no wonder that great excitement, exultation and exaltation swept over Mylapore and Madras when Sir S. Radhakrishnan was invited to take up the Spalding Professorship of Eastern Ethics and Philosophy at Oxford University. Imagine a turbaned Hindu, a vegetarian, steeped in our tradition and learned in Hindu and comparative philosophy, being invited as a professor at Oxford where the walls are charged with the learning of its savants and the wisdom of its thinkers. I attended the party given in honour of Sir S. when the cream of Madras society was present. I still remember the sparkling wit and humour of Littlehailes of the Mail. The oratory of Sir Radhakrishnan has to be heard to believe in its possibility! It was like the flow of the Niagara — faultless English without a falter, uninterrupted sequence without repetition, a strange mixture of natural and cultivated accents, Sanskritic phonemics and British diction fused into a harmonious whole.

Sir C.P.'s was a different style of oratory — flamboyant, colourful, charmingly repetitive in phrase, with a generous splash of humour and a versatility rare in these days of specialisation. Satyamurthi, the politician was facile and fluent with a debator's flair for repartee. The Rt. Hon'ble Srinivasa Sastri was silver-tongued, slow, measured, accurate — it was just like setting diamond in jewellery. All this so different from that of Winston Churchill — whose eloquence saved freedom for mankind and directed the course of human history.

Well, orators have different styles — no two gifted minds are alike. Just as they exalt our reason, orators feast our ears and rouse our spirits.

And then there were speakers like C.R. and Nehru, unlike in temperament, unlike in style of speech, yet both devoted to the same ideals — the freedom of India and the desire to emancipate our country and provide a democratic government. C.R.'s sarcasm had no trace of malice in it. It had the bitterness of good coffee and many lovers of the vintage like it black without sugar. Nehru's speeches were charged with emotion, there was repetition of ideas and of phrase but the subject was so exalted that the occasion and the event justified it. He was like a lover sighing like a furnace, with that exaggeration in phrase justifiable by noble passion and exalted purpose.

Among my father's friends Subbaroya Iyer[5] was his closest, a bundle of energy — like a photon he could never be localised. Short in stature, mercurial in movement, he was lively as a squirrel. He enjoyed imitating his masters — Sir C.P. in surprise, my father in legal subtleties, Radhakrishnan in speech, but he was always himself. He never complained against weaknesses in human nature, he exploited them, pampered them and turned them to advantage. He earned with unconcealed zeal and spent generously on deserving causes with greater zeal. He was an ideal companion to my father who admired such qualities without the slightest desire of possessing them. In his presence it was impossible to lapse into depression. Act, act in the living present with an eye on the future and remembrance of things past. He was not content to think in sentimental terms of a departed friend. He had to honour his memory by instituting a prize or writing a memoir and erecting a statue or unveiling a portrait. He had a desire to buy anything which would enhance in value whether it was useful to him or not. He would buy an African tusker or a Siberian tiger if sold well below the market price! I remember his astounding proposition to buy a 'few aircraft hangars' because they were being sold for a song! And he did. These hangars ultimately housed the Madras Institute of Technology in its early years. 'Buy now, you may use it later.'

He was fascinated by the applications of science to technology. He wanted to create something new, an organisation where science and technology are fostered at all levels. He was the architect of three such institutions, Vidya Mandir, an English Medium school maintaining the highest possible standards, the Vivekananda College with emphasis on the highest quality instruction and attracting the best of talent, and the M.I.T. — there was a gleam in his eye and his smile had an additional sparkle as he emphasised that the M.I.T. of Madras would aim at the standards of the M.I.T. in the U.S.

He was a wizard with figures and no wonder he was a successful incometax lawyer. He frequently visited Ekamra Nivas even after he became an affluent lawyer. He enjoyed conversing as he wrote, and the topics of his conversation and writing were different! He was so attached to my father that he would prepare my father's turbans by winding them on his head which was about the same size and shape. My father admired his agility, self-reliance, love of life, and faith in the future. He had no time to discuss politics, but focused on making money, spending it wisely, creating something, and conversing with friends. These filled his mind and his time.

N.C. or N. Chandrasekhara Iyer, equally warm, was of a different sort. He was intensely loyal to his master Sir C.P. He was demonstrative in his affections and enjoyed humour and anecdotes. He had no children and so enjoyed the company of all young children. The best time we had with him was after my S. S. L. C. examination when my brother and I went to stay with him in Kodaikanal in his lovely house. The long walks, the picnics and the boating on the Kodai lake, it all looks a dream now.

Later that summer I accompanied my father to Bangalore where a Vedic scholar Gopala Sastri came to our house everyday to teach me Sandhyavandanam[6] or prayers of a Brahmachari true to the 'Swara' or intonation of many Vedic hymns. That such scholars should languish in penury is the deepest unspoken tragedy of our ancient country which prides in its tradition but ignores such scholarship.

Such was the world in which we lived in the thirties, where literary, cultural and religious pursuits and domestic happiness were all we yearned and craved for in our lives. No one talked of big business and cartels, of factories and joint stock companies, of banking and industry, of armaments and military careers and of science and technology. It was a tranquil, leisure-loving world of a euphoric middle class, secure under an alien government, zealous of their caste, class, custom and faith, complacent in their cultural pursuits and religious rituals and conscious of their appointed place in a time honoured social hierarchy.

The shock of war and the burst of freedom tore the ancient nation from its roots as it quaked and swayed in the tidal gusts of hope and despair, gladness and grief, achievement and frustration in the fateful years following the blazing victory which emerged from the raging flames of Hiroshima and Nagasaki.

Notes

1) The four Vedas (primary Hindu holy scriptures) are the Rigveda, the Yajurveda, the Samaveda, and the Atharvanaveda. Every Brahmin family has a sage whom they consider as their primary mentor, and they associate themselves strongly with one of the four vedas even though they may recite other vedas. Most Brahmins are associated with Yajurveda. This association becomes significant in the performance of religious rites, because the method of performing the rites varies slightly from veda to veda.

2) In the old days, the role of Brahmin men was to learn the vedas. They used to go to special schools and learn the vedas from a guru. These schools called Veda Patasalas were common even up to the first half of the twentieth century but have become rare.

3) The three Goddesses, Saraswathi, Lakshmi, and Parvathi (also known popularly as Durga) are the consorts of the three main male Gods of the Hindu religion — Brahma (The Creator), Vishnu (The Protector), and Shiva (The God of Death or After Life). It is interesting that physical power, wealth, and intellect — three things that we as human beings seek the most — all reside in female deities in the Hindu religion, not male.

4) In Hindu tradition, the mother is the first God to any person, and the father, the first Guru (teacher). So in an Aksharabhyasam, the child gets its initiation into the world of knowledge with the father whispering holy mantras in the child's ears, and teaching the child how to write. The writing is usually done on a bed of rice using fingers. The first words written are usually the names of Gods.

5) M. Subbaroya Iyer, also spelt Subbaraya Iyer (1885–1963), was a very successful income tax lawyer. He and his family were so close to Sir Alladi and his family, that the two families shared a residence in Pelathope in Mylapore for a few years before Sir Alladi built Ekamra Nivas. Subbaroya Iyer was a student in Sir Alladi's history classes at Madras Christian College. When Sir Alladi went to England for a case pending in the Privy Council, his choice for a companion was Subbaroya Iyer.

6) Sandhayavandanam is a prayer done by Brahmin boys and men (who have had their *Upanayanam*), three times a day — at dawn, at noon, and at dusk. All sacred Hindu verses have to be recited with proper intonation and the Sandhyavandanam is no exception to this rule. Ramakrishnan performed Sandhyavandanam daily (even when travelling abroad) until his last day.

Chapter 5

Physics at College, The World at War (1940–41)

After the partition of Poland there followed a period of six months of 'phony war' when everyone was wondering where Hitler would strike next. Meanwhile the Russians pursued a forward security policy by moving forces into Latvia, Estonia and Lithuania 'with their consent', making similar demands on Finland. On the 30th of November, 1939, Soviet forces invaded Finland but we did not read the accounts of those battles in great detail for the campaign did not compare with the dash and daring of the German onslaughts. In fact there was a general surprise how a small country (in population) like Finland could resist the armed strength of the mighty USSR. It was interpreted as the demonstration of the incapacity of the Red Army — a misinterpretation which encouraged Hitler to make his fatal mistake nineteen months later! By March 1940 an uneasy peace was established with Finland's acceptance of the Soviet terms and the Russians halting near the Mannerheim line. So for six months there was a world war without action, with tension mounting everywhere as the belligerents prepared for the moment for decision.

It was in such an atmosphere of world-wide tension that I appeared for the intermediate examination. I spent too many hours at the radio and indulged in discussions of a war beyond our horizon! Yet I did my examinations very well, particularly in mathematics to qualify for a seat in the Physics Honours at Presidency College.[1] With greater concentration I could have secured a coveted rank like my colleagues, Achaya and Vepa.

That summer we were staying in Bangalore in Basavangudi — a favourite residential locality of retired officials who found a haven in that equable climate. We used to take regular walks to Lalbagh and Cubbon park,[2] the lovely gardens, which were the pride of the city, and play tennis with old-timers in a modest and inexpensive club nearby.

It was then we heard on the 10th of May the shattering news that the German army, supported by the blasting fury of the Luftwaffe, invaded Holland and Belgium in a blazing attack of such suddenness and surprise that dazzled the minds and dismayed the hearts of a gaping world. The Belgium armies reeled and gasped before the German onslaught. King Leopold surrendered preferring humiliating survival to total annihilation. In Britain, Chamberlain resigned and true to her history, England found her man of the hour — Winston Churchill — who was destined to send Hitler to his doom and save the world from the deep damnation of total dictatorship. We heard of the gruesome bombing of Rotterdam which horrified the conscience even of a warring world. Within two weeks the German armies under Bock, von-Rundstedt and Leeb, swept like a tidal wave across the low countries and poured their fiendish fury over the fair land of France. Each day brought terrifying news and we wondered what had happened to the much vaunted Maginot line and the famed French army.

Meanwhile the British expeditionary force was swept toward the Channel and it looked as if it would be annihilated unless it withdrew from Dunkirk. This successful withdrawal, Operation 'Dynamo', became one of the greatest legends of human courage and endurance in the face of impossible circumstances. The sheer magnitude and speed of the operation staggered imagination. Over two hundred thousand soldiers were evacuated in less than a week deploying over forty destroyers under an air cover of screaming Spitfires as the German army advanced on all fronts, capturing over a million prisoners in the Netherlands and France by bold and rapid encircling movements.

We heard with terror and admiration of the all conquering surge of the German Wehrmacht and the pathetic helplessness of French generals like Weygand and Marshal Petain. The Germans entered Paris on the 14th of June, a humiliating hour for the arbiters of Versailles, unimagined even by the most incurable pessimist.

France capitulated on the advice of its generals on the 22nd of June. It was divided into two zones, one under German military rule from Paris and the other with full sovereignty, with Vichy as the capital. There were many who condemned the French generals for woeful lack of competence and courage. Among them was a young French soldier, Charles De Gaulle, who worked for the liberation of France from outside the country, destined to become its greatest idol since Napoleon Buonaparte.

Italy entered the war on the 10th of June 1940 against a falling France and dismayed Britain. It achieved little except participating in the victory of Germany. But Britain knew that this would mean war in the Mediterranean and North Africa.

Hitler now held Europe in his iron grip, so tight and deadly that the victim could not squeak or squeal. In triumphant delight he turned his mad-bred fury towards Great Britain. The major question was, would he attempt an actual invasion of Britain or was it going to be relentless bombing to terrorise the country into submission? In considering himself the avenger of Versailles, he had not read the lessons of history too well. The island race that over the centuries had produced the victors of Agincourt and Blenheim, Trafalgar and Waterloo, was not going to be beaten by roaring bombers, bursting shells and blazing fires. The bombing of Britain raged for two months, the so called blitz at its highest intensity and continued till the middle of 1941. The question on the lips of every one outside Britain was, can Britain survive? Only the British and Churchill knew they could and they would.

It was then that Roosevelt declared that the U.S. would extend the material resources of his nation to fight the Axis powers. Churchill conveyed the British predicament to the U.S. in his most famous message on December 8th 1940. America answered the appeal with the 'lease lend' act to aid the British Commonwealth.

As the battle of Britain raged, among the allies and their foes all over the world there arose only one question: When would the U.S. enter the war on the side of the allies'? With the operation of the 'lease-lend' programme, the Atlantic became a scene of naval battles and the Germans intercepted the lifeline of supplies from the U.S. to Britain. The U-boat was Germany's striking weapon, the earlier version of the submarine, which played an important role in the first world war. What a miracle that today the Lufthansa and the British Airways are the aerial armadas of peaceful commerce carrying millions of passengers from both countries across the Atlantic skies?

It was in June 1940 just before the battle of Britain burst in all its fury, that I entered the third honours in Physics at the Presidency College. The three years at that college 1940–43 were a preparation for life and an experiment in independence and experience in friendly cooperation. We were just about a dozen chosen students enjoying unrestrained freedom to work in the laboratories of the college. Everyone from the professor to the laboratory attender allowed us freedom to work, think and plan our own programmes.

During the Third Honours, we primarily studied 'properties of matter' as in Newman and Searle, a textbook remarkable for the problems given at the end of each chapter. Since the examination curriculum did not require the students to solve problems, nobody ever attempted to do them. My pleasant experience at the Loyola College offered me this challenge and I solved every problem in the first few exercise sets in this book.

I seem to remember every day, every hour of the three years' study at the Presidency College. Work was exhilarating for we were an intimate group, one equal temper of friendly hearts.

I remember the classes on Laplace's theory on capillarity, a subject considered 'tough' by students due to total unfamiliarity with three dimensional geometry. I took great pleasure in demonstrating to my colleagues using 'raw mangoes' the concept of the tangent plane and the curvature of three dimensional objects. The sessions concluded with the eating of the raw mangoes with salt and spices, a delectable experience only to practised tongues.

The long mathematical treatment on Laplace's theory was 'out of the syllabus' but it intrigued me. I struggled very hard to understand the basic concept of the principal radii of curvature in arbitrary directions on a three dimensional surface. Actually the proof suggested itself to me during our long walks in Kodaikanal, when I realised (that I had to use the fact that) the drop of elevation in height with distance was independent of the path.

It was at that time during my Third Honours I bought that magnificent introduction to theoretical physics by Joos. The chapter on Special Relativity fascinated me due to the simplicity and profundity, the Lorentz Transformation I wanted one day to solve the historical puzzle: If Einstein's theory of special relativity is based essentially on the Lorentz Transformation, how did Lorentz miss the discovery? Fifty seven years later in 1998 I solved this puzzle by postulating that Lorentz contraction of length should precede Lorentz transformation. It was a 'revelation' that with this assumption, a rod of length $x - vt$ moving with velocity v unlocks Space-Time unity. All my earlier papers on Relativity became natural corollaries of this fundamental 'revelation' (see Appendices in Part II).

There was annual competition for the best essay at that time and with enthusiasm I submitted the solution of problems in Newman and Searle which were never attempted by other students. I also included a simple but rigorous proof of the expression for group velocity, in terms of the wave parameters based on the simple concept of a runner overtaking another moving with nearly the same speed round a circle. But the prize went to a person who reproduced more diagrams from standard textbooks and pasted good photographs! So, next year I did much the same buying a new book on 'Raman and infra red spectroscopy' and almost reproducing it to win the essay prize! This is of course a sad commentary on the apathetic attitude in colleges towards originality.

In the III Honours we heard the first murmurs of modern physics through our senior colleagues, who were studying Richtmyer's splendid text (the 1932 edition). It was based on the vector model of the atom and one of my friends was proud of his facility in deriving almost blind-folded the expression for the splitting of the spectral lines in the Zeeman effect. I wanted to prove myself one better and so studied the chapter in the famous book 'Atomic structure and Spectral-lines' by Arnold Sommerfeld,[3] that famous Sire of Nobel Laureates. The moving ellipse was a fascinating concept which suited my palate for coordinate geometry. But due to conventions, which even today are the bane of Indian education, everyone had to answer according to the prescribed text book and that was the 1932 edition of Richtmeyer.

In fact we ignored the note in Richtmyer's latest (1943) edition about nuclear fission and the possibility of large scale release of energy as being 'out of syllabus'! Just about that time in the sun-drenched deserts of New Mexico, Oppenheimer and his chosen band were engaged in an awesome enterprise based on that same phenomenon of fission, which not only was to deliver the coup-de-grace of the Second World War, but ushered in a new era in human civilisation! Meanwhile our college students yearned for the notes from St. Joseph's College as if they were smuggled watches and considered themselves lucky if they could borrow them from generous rivals in Trichinopoly.

It was during my III honours that I joined my father in a memorable trip to Kerala. In the native states, the Maharajahs in India still enjoyed power, glamour and prestige and Sir C.P. Ramaswamy Iyer reigned Travancore as its illustrious Dewan[4] under the aegis of the Maharaja of Travancore. There was a legal dispute between the Travancore and Madras Governments over the use of the waters of the Periyar. As advocate-general, my father was the counsel for Madras while C.P. himself appeared for the Travancore Government and an arbitrator was chosen from outside. I watched the battle of the giants with the interest of a Roman spectator at the Coliseum — C.P., the master of drama, surprise, rhetoric and repartee, eloquent and versatile, pitted against my father, peerless in the knowledge of law and in the use of it with unerring precision and incisive logic, unperturbed by a show of force, who always knew that the opponents armour is vulnerable at its weakest chink. He smiled when Sir C.P. objected with dramatic emphasis "In the course of my arguments which need not be interrupted" piqued at the well timed but inconvenient interruption by my father, perhaps unwarranted by convention but certainly warranted by the occasion. The gracious standards of hospitality of the ruling families and the established conventions of legal profession were such that outside the court, Sir C.P. and my father were intimate friends.[4] My father was treated as a 'State Guest' in one of the palace guest houses in Trivandrum.

In the first quarter of 1941, we had a visitor for one of our evening seminars in the physics association of the Presidency College. We did not realise that he was destined to transform the scientific scene in India as no scientist or administrator had done before. We the members of the Physics Association had invited a thirty two year old handsome scientist, fresh from his laurels at Cambridge, Dr. Homi Jamshedji Bhabha.[5] He spoke on meson theory and he was quite unlike the conventional Indian University lecturer. He was precise, soft spoken and looked obviously prosperous. His lecture cast a hypnotic spell over me. I did not understand much of what he said, but was thrilled, excited and exhilarated by his voice and presence. Soon we learned that he was a Fellow of the Royal Society and he carried with him the great surges of modern physical thought having worked in world famous centres with men like Bohr, Kramers and Heisenberg. We watched his mannerisms which made him charm more enchanting. My admiration soon turned to a passion to work under his guidance. We heard he had just started a Cosmic Ray Research unit at the Indian Institute of Science, Bangalore.

Beyond the class-room and our home it was the news of war that filled our minds and troubled our hearts. It was clear by the spring of 1941 that Hitler had lost the Battle of Britain on the English skies. Of course the cities of England and London in particular were battered and German might was blunted, bruised and broken by the British fighter force. England emerged stronger in spirit and resolution. It was a bloody but heroic period in her history and the lights of glory shone on the 'blood toil and tear-wrung millions' of that island race. It inspired Churchill's immortal tribute "Never in the course of human history was so much owed by so many to so few". Likewise nowhere in recorded literature, a greater story of human courage has been expressed so well in so short an epigram with such sublime emotion.

The battle of the Atlantic was slowly turning in favour of the allies. The U-boat warfare was intensified but the wonderful exertions by the British Navy and Air Force met the challenge in victorious measure. The might and craft of Hitler's Wehrmacht had failed to crush the dauntless spirit of the defiant island. Britain waited and watched for something to happen.

And in time it happened. Britain held out long enough and there came the moment when the tyrant made a ghastly mistake which altered the balance of the struggle to quote Churchill: "On June 22, Hitler, master as he thought himself of all Europe — nay indeed seem to be master of the world, so he thought, — treacherously, without warning, without the slightest provocation, hurled himself on Russia and came face to face with Marshall Stalin and the numberless millions of the Russian people." The German eagle had entered into a death-grapple with the Russian bear from which it never recovered. It of course took three years, with millions dead or wounded across a ravaged Europe before the end came in June 1944.

In May 1941 after the III honours examination, we went to the lovely hill resort of Kodaikanal where we stayed in a cottage near the Solar Physics Observatory. I used to go frequently to the library there and peruse magazines and books in physics and in particular I read the theory of the seismograph in great detail. During our long walks, picnics and boating, we discussed the war in Europe, admired the courage of the Londoners against the ferocity and vaunted omnipotence of the German air power. While resting the oars as we banked our punts in cosy nooks of the Kodai lake under the shade of overhanging trees, we discussed the war in Europe expressing our opinions on Hitler's folly in Russia when the Luftwaffe was so badly bruised after the battle of Britain. The war looked to us a distant convulsion and we found justification for our noninvolvement in our struggle for freedom against imperial Britain. In India, the only effects of war seemed to be the rationing of petrol and decreasing availability of foreign goods.

Meanwhile two things were happening which were to have a permanent, harmful effect on the Indian economy — the falling value of the currency due to inflation, and the accumulation of black money. The major industries were devoted to the war-effort, supported by demands and contracts resulting in the sudden rise in values of stocks and shares. Contracts and sub-contracts, brokerage for such deals, and sale of hoarded consumer goods generated an uncontrollable abundance of black money. By the end of the war black economy was running parallel to the normal economy. Since such money could not be invested in banks it was expended with gusto by the nouveau riche on consumer goods, the prices of which soared erratically out of control. I remember very well how a 'respectable gentleman' boasted that he had enough stocks of 'Ovaltine' and 'Blue Gillette' at home to outlast not only the war in Europe but in the Pacific, even if it broke out through the vaulting ambition of an impatient Japan!

Notes

1) As in the case of universities in England, universities in India are composed of a number of affiliated colleges. Among the colleges of the University of Madras, the Presidency College is one of the oldest and most prestigious, owing to the excellence of its faculty, and a long list of illustrious alumni. During British rule, India was divided into presidencies, and two Presidency Colleges were created — one in Madras and another in Calcutta.

2) Bangalore used be known as the garden city of India. It had a number of lovely parks throughout the city, of which Lal Bagh and Cubbon Park were, and still are, the most famous. With uncontrolled metropolitan boom, brought about most recently with the rapid rise of Bangalore as the IT capital of India, the city is polluted and congested and definitely not the garden city it once was. Lal Bagh and Cubbon Park still exist but are surrounded by concrete jungles and the deafening noise of automobile honking.

3) Arnold Sommerfeld (1868–1951) did pioneering work in atomic physics and is credited for the introduction of the spin quantum number and the azimuthal quantum number. But he was never awarded the Nobel Prize. He was an outstanding mentor and several of his students won Nobel Prizes — Wolfgang Pauli, Hans Bethe and Linus Pauling — to name three.

4) "Dewan", a Hindi word means ruler, or lord, of a territory, not of a country. Travancore was a princely state in South India that existed until India received its independence from British rule. Sir C.P., an eminent lawyer, a great statesman, and an outstanding orator, was chosen by the Maharajah of Travancore to be his Dewan and assist the Maharajah in the rule of Travancore.

Sir Alladi and Sir C.P. had the highest regard for each other as lawyers, and great mutual affection as friends, whether they appeared on opposing sides in a legal case or not. The Periyar waters dispute was perhaps the best illustration of their true professionalism. Outside of the court, they were together as always as friends, and Sir C.P. hosted Sir Alladi magnificently in Travancore.

5) Homi Bhabha belonged to a wealthy *parsi* family in Bombay and was related to the famous Tata family (also parsis). He received his PhD in 1933 from Cambridge University under the guidance of Nobel Laureate P. A. M. Dirac. His thesis topic was on cosmic radiation, and area in which he and Ramakrishnan were soon to work together. Bhabha was elected Fellow of the Royal Society (FRS) in March 1941.

Chapter 6

War Rages on — The Nightmare is Real (1941–42)

June 22nd, 1941 — the very mention of that date sends a shudder through my heart as if smitten by a nightmare — yet it was real when we heard of three million Germans on the march, with their bayonets and rifles breathing fire and slaughter, on a two thousand mile front, as one blaze of fire lit up the skies behind the fiendish fury of the Luftwaffe and the thundering roll of the Panzer corps across the endless plains of Russia. It was not a dawn that broke, but Hell let loose as Hitler flung his flaming legions which within two years were to recoil, bleed and burn the avenging fury of Stalin and his Red Army. Incredible, that after all the triumphs in Western battles, Hitler should have chosen to attack Russia in a fit of frenzy, hoping for a rapid conquest in six months! It was like waging a war against the swelling oceans, the raging tempests and the consuming fires of Hell.

Operation Barbarossa, the biggest military operation ever mounted in human history as would make Alexander, Genghiz Khan, Tamerlane and Napoleon leap from their graves! It started an hour before dawn when the skies blazed and the earth quaked as six thousand German guns boomed and belched forth an avalanche of fire and steel in synchronous, deafening roar and blinding flame amidst the shriek of bombers, the scream of mortars and the rumble of tanks. The victorious commanders of the West were leading the armies — Leeb, Bock and von-Runstedt, with Kleist's and Guderain's Panzers spearheading the assault.

We heard the awesome news on our return from our holiday in Kodaikanal, bronzed by boating in the noon day sun on the placid lake, speculating on Hitler's next move after his failure to daunt the British by relentless bombing for over nine months. I entered the fourth honours in Physics at the Presidency College in June 1941. War had not touched our lives except through shortages of imported consumer goods. There were complaints that tennis was becoming more expensive since Slazenger balls had to be purchased from dubious sources at arbitrary prices. We compromised by agreeing to play with two balls instead of three, which was hard on the 'picker boys'.[1] For the first time there was talk of rationing of petrol. So four of us, Raghavan, Ramanujachari, Natarajan and I decided to go to college by bicycle or in one car (mine, a small Hillman Minx) with petrol contributed by all of us. This gave us a sense of participation in the world struggle, a share in universal human suffering and a right to discuss the events as they happened, or predict the outcome by a critical appraisal of the present situation.

In the fourth honours we got deeper into physics — Light as in Houston, Preston and Wood's Optics, Heat as in Saha and Srivastava (a massive treatise), Electricity and Magnetism as in Starling, and Modern Physics as in Richtmyer (1932 edition). Two teachers played a special role in our science education — S.B. Bondade who came in as a chief professor, soft spoken, lean and lanky but formally dressed, who had his graduate education under Andrade in London, and T.N. Seshadri, well-built, broad-fronted and impressive, full of confidence in his mastery of fundamental concepts, friendly and eager to teach, easily accessible. Contact with these two continued well into my adult life during my decade long efforts to build a theoretical physics group in Madras.

We enjoyed as usual the experiments we planned and performed. In the second year they were more advanced in physical optics and in electricity and magnetism. It was great excitement to compare notes of 'readings', to argue over the limits of accuracy, or fuss over minor details of arrangement. Above all, friendly cooperation made laboratory work an intellectual picnic.

We knew very little about the geography of Russia except that it spread over almost half the globe from the borders of Poland in the West in Europe to Vladivostok in the East in Asia and that half the year the land was buried in snow. Some said the fall of Moscow would mean the end of war for the vast giant would collapse when his heart is squeezed. Some were of opinion that the Russian armies would withdraw to the Urals and behind to Siberia and wait till the enemy collapsed by losing his breath. It was clear that Hitler should reach Moscow before December or else the long night of Russian winter will freeze his armies to death and destruction.

The German flood poured across the whole front spanning the Baltic and the Black Sea as the Russians reeled under its relentless fury. The German army reached the gates of Leningrad and Moscow and captured Kiev the capital of Ukraine and the granary of Russia, taking millions of prisoners.

And then came winter, just when victory seemed within the grasp with the walls of the Kremlin within the striking distance. Everything froze including the German hearts. It was worse than hell fire — vast interminable expanse of snow, another name for death that unknown land from whose bourn no invading soldier returns, the icy winds and chilling, killing frost. The question was now reversed, can the Germans survive the Russian winter? The German generals pleaded for food and winter clothing for their armies. The spectre of Napoleon's armies rose from the snow bound earth of the Russian plains. It was then we heard of Marshall Zhukov, of the heroic defence of Leningrad and Moscow. What a Christmas to wait for — death standing at attention sparing neither child, soldier or the infirm.

The German conquests startled the imagination of the world. But they turned out to be not the end of Germany's aims but the beginning of its own end. The onset of winter would not only sap the vitality of its army but give a breathing time for the Russians used to such rigours, enough to muster their forces to strike back at a weakened foe. There was also the possibility of an allied attack from the Mediterranean. The Germans had stretched themselves too far and wide. The Eagle had flown too far from its nest and was gasping for breath.

Meanwhile in the Far East, Japan was growing impatient. Its warlords had decided that its future lay with the Axis powers. When and how should it strike? America had its great stakes in the Pacific for it felt its frontiers were to be safeguarded from the Philippines, Wake Island, Guam and the South Pacific isles. Japan was intoxicated with its victories in Northern China. The tiger had tasted blood but the United States was a different matter. Yet Japan felt too strong and too powerful to let the world around live in peace. It looked with envy at the world-wide British Empire, at the European colonies, at the 'good life' of the white man on the African veldt, at the Viceregal splendour of imperial Delhi, at Hong Kong, the glittering jewel of the East, and the East Indies ruled by a little European country under the level of the Atlantic ocean, and above all at the United States with its incomparable wealth, resources, power and affluence. But Japan failed to realise that all these had been built over centuries. The British acquired India slowly and steadily along a tortuous road from Clive to Wellesley, from Hastings to Curzon. Even in Africa the European colonists secured their envied place by systematic conquests steadily over the years. South Africa's Kimberley wealth and adamantine empire needed the vision and leadership of Cecil Rhodes and Jan Christian Smuts followed by decades of exploitation, consolidation and assimilation. Japan wanted to 'telescope' its time of growth into a few months by the apparently obvious expedient of a lightning war. Its warlords and its premier Hideki Tojo did not realise that Alexander, Tamurlane and Napoleon were out of date in a twentieth century world.

The Americans, true to their democratic traditions, were still willing to negotiate with an impatient Japan. We heard of the mission of Suburu Kurusu in November and his talks in Washington. Then suddenly on the 7th of December 1941, which will be remembered in history as a 'day of infamy', even as these talks were going on, Japan struck, with all the advantages of stealth and surprise, a dastardly and devastating blow at Pearl Harbour in Honolulu. Waves of Japanese planes raided the ship-studded harbour and the Japanese naval fleet under Nagumo moved to striking distance. American planes and ships were shot and torpedoed like sitting ducks. The allied world stood aghast as Japan struck with dazzling surprise and blasting fury all over the Pacific — Wake Island, Guam, the Philippines, Hong Kong, Malaya, Singapore and the East Indies. Never in the history of human conflict was there a military and naval operation on so wide a scale of such ferocity and magnitude, stretching from Singapore to Pearl Harbor, from Hong Kong to the Solomon islands.

In Madras we had not heard of the Philippines except as a distant American colony. Only after the Japanese attack we studied the geography of those islands in great detail. The American Commander was General Douglas MacArthur who was destined to send Japan to its doom and later become a controversial figure in American history.

The Philippines collapsed under the Japanese invasion under Masaharu Homma, though Bataan and Corrigedor held out boldly for two months. The grisly horrors of the 'Death March' of the heroic American garrison after surrender was revealed only after the war when retribution came to Homma who was hanged as a war criminal after the Tokyo trials.

It was a blazing victory — so the Japanese thought at Pearl Harbour. About 350 American aircraft were lost or damaged, two battleships sunk, thousands killed and wounded, all within just two hours risking just over three hundred planes. In fact it was no victory — Japan woke up a sleeping giant who would not only wipe out its armies and navies but burn its bowels by nuclear bombs! America was now an active belligerent in the war and no one could be more pleased about it than Stalin and Churchill.

On the 9th of December 1941, China declared war on Japan to invite American support to its long drawn struggle with its imperialist neighbour. Hong Kong fell in a few days and the Japanese entered the glittering city in glowing triumph.

Hitler sealed his fate and dug his grave by the declaration of war against the U.S. on December 11. With the onset of winter and the entry of America into the war, the Axis powers were doomed despite the triumphs of Japan. The nightmare of war was certain to end and the dawn of hope was to break — but how long the wait, what were the price and the stakes? 1942 was to be the year of trial before the first lights of hope were visible.

Life at college was particularly enjoyable for there were no University examinations for us except in the 'subsidiary' subject mathematics in which the examination was to be held in March 1942. We had so much of leisure that I concentrated on tennis and billiards in our neighbouring club. We used to go to the movies quite frequently for they represented a third world — something unreal but excitingly different from the uneventful leisurely world of South India or the war torn world of Europe.

During that Christmas (1941), news of war filled our minds. Suddenly war had become a frightening reality to us as Japanese struck at Malaya and Burma. South India was utterly unprepared for such an emergency. The Indian army was made up of professionals who had taken to the army as a career. Supporting it and supplying it with arms made such demands on such a poor undeveloped country with an agricultural economy, that there was no question of conscription or even training an army of volunteers or local militia. To create a war psychosis, the government organised the ARP (Air Raid Precautionary Force). Except for the steel helmets and Khaki uniforms, there was nothing 'military' about it. It became a fashion for senior civil servants and the police to supervise and conduct mock air raids and train people to run to the shelters. Many construction contractors made their pile by building shelters of dubious value, stitching uniforms by the thousands and providing helmets for the air raid wardens. In our own house under a sprawling tamarind tree, we built a massive underground concrete shelter which would have survived even the bombing of the Luftwaffe! How would people of South India react to the threat of war and bombing? They were used to no more violence and bloodshed than in stray Hindu-Muslim riots or village brawls. The civil servants were used to being attended by turbaned peons dressed in gowns, symbolic of a bygone era, pampered by ushers opening the doors of their cars and carrying their files.

In the wake of the stunning Japanese conquests, everyone whispered. 'The Japs are coming, the Japs are coming' — a shiver passed through our hearts. It was one thing to discuss about the 'larger Asia co-prosperity sphere' and the Japanese attempts to mobilise Asiatic opinion against the white man's empire. It was another thing to meet the Japanese face to face! We were used to the British rulers who allowed us freedom of speech knowing full well it did not affect their sovereignty, who believed in law and justice, may be under their guardianship, in negotiation around the table, in the gradual evolution of freedom of an independent India by slow constitutional means. But we would not know what to do against a relentless foe who knew no law except the bullet, no language except the bayonet, no way of dealing with opposition except by wiping it out.

Singapore, the island bastion of British imperial power, surrendered on the 15th of February, 1942, the tragic end of a tragic story begun with the sinking of the "Prince of Wales" and the "Repulse" shortly after Pearl Harbour. With General Percival's surrender to Yamashita, the lights of the British Empire went out of the East for ever, never to rise again even after the allied victory in the war.

The Japanese invasion of Burma started on the 11th of December, 1941, and by the end of April, 1942, Rangoon had fallen and the British were expelled from the country. The Indians evacuated Rangoon under very difficult circumstances and it was common to see in Indian cities the 'Burma evacuees' who had to be supported by the Government.

In such an atmosphere under the ominous shadow of Japanese invasion, I completed the Mathematics subsidiary examinations. I had followed Asquith's coordinate geometry[2] which was of higher standard than the prescribed text and so it was possible for me to 'clear' the paper in forty five minutes what was intended for two and half hours! Calculus was more laborious since it involved computing coefficients in various expansions.

That summer there was no question of planning a trip to Kodaikanal or Bangalore. The spectre of war and the threat of bombing or even invasion of India through Burma or by sea from Colombo loomed over the darkening horizon. All the tales we had heard of war were no more newspaper accounts to be discussed over morning coffee or afternoon tea. Here the danger was at our gates affecting our lives.

And then we heard that Colombo and Trincomalee were bombed. That settled the plans of the citizens of Madras. The great evacuation began. Within a few weeks the entire population except a few stragglers or members of the 'essential services' left the city. The High Court judges, advocates, their paraphernalia, students, children and parents, everyone left the city. We had our village sixty miles from Madras. So we shifted to our 'country house' and made the journey by our Master Buick leaving Ekamra Nivas in charge of a few tough and loyal staff. The Government encouraged the evacuation since Madras had not a single anti-aircraft gun, no armed forces, no militia. It was a peaceful town for peaceful commerce basking under British rule, with its leisured intellectual class and the contented masses aspiring for freedom through the non-violent struggle of their political leaders or slow constitutional means and round table discussions. Certainly such a population could not comprehend the savage wrath of the German gauleiters, the reign of terror of the SS and the Gestapo, the blood curdling whine of the Stukas and the awesome roar of the flaming Panzers, the madness and fury of the suicide squads of Japanese soldiers. We could not understand that the land of Riemann and Heisenberg, Goethe and Schiller, Bach and Beethoven, would pay homage to a vile dictator who choked the life out of six million Jews. Of what avail are the developments of science if the soil of Europe is to be soaked by the blood of women and children and infants to slake the lust of a fiend on earth?

We prayed for something to turn up to save the situation. In our village absolute serenity prevailed. The peasants were least concerned with the Japs or Germans or the fate of Europe or the sovereignty of the British Empire or the injured might of the United States. Bound to his paternal acres, the peasant was concerned with his daily chores at home and on the field measuring time by the position of the sun and the length of his shadow. We had our loyal band of village hands who were happy to see us and attended to our comforts without feeling the urge for urban luxuries.

In our spacious village house over forty persons were fed on delicious South Indian fare prepared by our expert cooks, Krishna Iyer and Mani Iyer. We had the best rice from our fields and 'country vegetables' (Nattu Karigai[3]) like Brinjal and Bendi and 'local' tomatoes (sour compared to the Bangalore variety not available there) with the incomparable Sambhar and Rasam of Krishna Iyer. Except for our father who carried his learning and his profession with him wherever he went and my mother who had to take care of adult dependents and their spouses, it was idleness par excellence for all of us, justified by the emergency war 'on our frontiers' which we felt was more the concern of our British rulers than the citizens of India!

My father was at ease even with the village environment for it was the place of his birth and the population held him in godly veneration. But he was a restless intellectual and spent all the time reading current literature in law and politics, made available by a very efficient postal system. He studied the laws of war, of the legal effects of occupation, withdrawal and reconquest, of the rights to property and compensation as the Governments changed. I remember his reading McNair's Laws of War[4] and talking to me about the difficult legal questions arising from successive occupations and restoration of original regimes. This in particular applied to Burma and Malaysia and like members of his generation, he was convinced that Britain could never lose and that once the American giant was wakened from his complacency, no power on earth could defeat him. He had so much faith in democracy and the rule of law that he never believed that dictatorships, in spite of their initial triumphs, could prevail over the allies. He admired the heroism of the Russians and emphasised the differences between the totalitarianism of Soviet Russia and of Nazi Germany. Despite the one party regime and consequent concentration of power, the Soviet people did not recognise racial differences. Nothing could compare with the sadism and bestiality of the SS, the Gestapo and the S.A. which wiped out millions and tortured to death as many in their concentration camps.

To the surprise of everyone, the Japanese halted their march into India. For some inexplicable reason, which was clear only later after the Battle of Midway, they stayed their advance and there was an ominous lull and inactivity over the Indian ocean and the Indo-Burmese border. The 'evacuation period' would end by June and we planned to return to Madras to resume normal life.

It was in May 1942 we heard of the German army resuming its invasion of Russia. Two million men on the march from the Baltic to the Black Sea, knocking at the gates of Leningrad and Moscow. The winter had taken its toll — the German army was not the same — it had new recruits, the nascent youth of Germany brought to be sacrificed to feed the insensate ambition of the Moloch in human shape. Hitler's plan was to clean up the Crimea in the south and Leningrad and Moscow in the north. Manstein's eleventh army overran Crimea and even Sebastopol succumbed after a heroic defence for 250 days.

Then came Stalingrad — the bitterest, bloodiest battle in all human history. For the first time since the war, an incredible thing had happened — the German sixth Army under Paulus was surrounded by flanking strokes of the Red Army. The final capitulation came on the 2nd February 1943. Russia demonstrated that the invincibility of the German Wehrmacht was a myth. Hitler had failed to give orders for Paulus to break through. He insisted on his staying there and defending Stalingrad. A hundred thousand German soldiers died and an equal number were taken prisoners.

Meanwhile in the Pacific the tide of war was slowly turning in favour of the Americans. In May 1942 came the battle of the Coral Sea and then the greatest battle in all naval history — the battle of Midway. This was the first major triumph for Admiral Nimitz since America entered the war. Japan lost all four of its largest aircraft carriers while America lost the Yorktown. So the long run of Japanese victories had come to an end. Japan was now on the defensive. With its navy destroyed, its new won empire could not be defended from the ocean. It was then we understood why the Japanese halted in Burma. The monsoon had set in and the jungle terrain was difficult and the Pacific naval battles were lost at high cost.

We returned to Madras by the end of June, 1942. The fear of Japanese invasion had lifted and life was soon restored to normal in our city. Those who had remained felt proud like the defenders of Stalingrad and had stories to tell how they survived the ordeal of suspense and inaction. Among them there were some who had the boldness and wisdom to buy houses and real estate as prices came tumbling down in a city stricken by fear and panic.

Notes

1) "Picker boys" is the term used in India for "ball boys", since they pick the balls during tennis games. In India where labour is cheap, we have picker boys even in regular play at clubs.

2) "Coordinate geometry" is another term used for "analytic geometry", because of the use of Cartesian coordinates to represent points. In India, coordinate geometry was discussed in great detail in the college science classes, because it is an important pre-requisite for single and multi-variable calculus.

3) Vegetables available in India were classified under two types — those that were native to India like brinjal (= egg plant) and bindi (= okra), and those that came to India from the West like cauliflower and green beans. For the annual ceremonies in memory of of our departed ancestors, only native vegetables are used and not those that came from the West. The vegetables that are not native were called "English vegetables". They required a cooler climate and the British grew them in the hill stations like Ooty and Kodaikanal in South India. The English vegetables were available in cities like Madras, but were rare in the villages.

4) The book that Sir Alladi read and mastered was Arnold Duncan McNair's "Legal Effects of War", the First Edition of which appeared in 1920 as a set of lectures. This stood him in good stead in the post independence period, when zamindars (= rich land owners) approached him for advice on ownership when the lands in question had been under the jurisdiction of different rulers.

Chapter 7

War Tide Turns — Crisis in India and in my Career (1942–43)

With the entry of Italy into the war on the 10th June 1940 after the fall of France, the whole of North Africa lay open to the ambitions of the Axis powers. Fallen France was now an ally under the Vichy Government and so the French possessions in North Africa came under Axis domination. The Italians were already in Libya and if only they could drive the British from Egypt, the Italian (or Roman) dream of a Mediterranean empire would become true. With half a million troops under Marshal Graziani, felt that it should not be difficult to oust the numerically inferior British army in Egypt. Thus started the North African battle which became strangely an active theatre of war between European powers and even more so with America's entry. It culminated in the total rout of Italian and German armies and the invasion of the Europe from Italy.

In early February 1941 Hitler had to despatch German forces under a young General, Erwin Rommel, who was to become a Napolean of desert warfare. Meanwhile Japan had declared war in the East and was on the high tide of conquest over the entire Pacific. Singapore had fallen, Malaya taken, and the Japanese were swarming towards Burma, reaching the gates of India by the middle of 1942.

By August 1942 Germany and Japan had stretched their resources and strength, though still on the crest (or edge?) of high tide of victory. Germany was engaged in a titanic struggle with Russia from the Baltic to the Black Sea. In the Pacific the Japanese had suffered a crushing defeat at Midway, which ended their series of surprising victories.

During these months India was passing through a crisis in conscience under the leadership of Gandhiji and Nehru. Gandhiji, the apostle of non-violence, had come to the limit of his patience in attempting to convince British rulers that it was just the time to grant India freedom and gain its voluntary and wholehearted support for the war. The people were already feeling the strain of war, though not a shot had been fired in India. Gandhiji knew that they would not join in any open struggle against Britain, whatever be their sympathies and grievances. So he launched the 'Quit India' movement on August 9, a new campaign in which India expressed its disapproval of the policies of the British Government by a few chosen persons courting arrest — 'Individual Satyagraha'. Among Gandhi's associates there was a conscientious objector to his campaign — C.R., who felt it was unwise to harass the British during the war which would confuse the Indian population, weaken our defence, and encourage the Axis powers. He ran the risk of extreme unpopularity but he was great enough to do so and was hailed as a patriot along with Nehru and Gandhiji by posterity. Gandhiji and Nehru and other leaders courted imprisonment and the strange situation continued till the end of the war — a resolve to fight the Axis powers, a sympathy for the national leaders in their struggle against the British which was engaged in a war against Germany and Japan, an incongruous situation baffling conscience, logic and human reason.

1942–43 was the crucial year of my educational career — the final year of the Honours course. How could we be idle participants to this struggle? Reason told us that British should win the war or else the world would be enslaved by the dictatorships without hope of redemption. Our emotions were with the Mahatma, the uncrowned leader of the people, who had no thought except India's freedom from foreign rule. We discussed the problem between lectures and during experiments in the laboratory. In such an atmosphere I plunged into preparation for the final Honours examination. While we read the massive treatise of Saha and Srivastava, we were aware that our competitors from St. Joseph's College[1] relied on valuable notes, more useful for examinations. I had enjoyed the chapters on radiation taught by Professor Bondade with absolute thoroughness and clarity. I mastered Newman and Searle for 'Properties of Matter' and I referred to Poynting Thompson for some special topics. I read Richtmyer's Modern Physics (1932 edition) from cover to cover, though we were frequently told that the St. Joseph's College was armed with Father Rajam's notes. For our special subject "Wireless", we followed the Admiralty Handbook with the widest (and wildest!) variety of diagrams. We chuckled with pride that the St. Joseph's College, due to lack of equipment, had to choose Kinetic Theory as a special subject though it gained advantage in scoring marks in the examination. With my colleagues Raghavan and Ramanujachari, I planned our strategy to meet the challenge of the St. Joseph's College.

Meanwhile the turbulent situation in India had reached the stage when teenage school and college students were expected to join the turmoil. The student community resorted to strikes and this fever reached the Presidency College. I remembered the wise warning of C.R. that the youth of India should not use the fire of patriotism to burn themselves up like cowdung cakes which used otherwise will make valuable manure for good crops. I did not like the analogy but agreed with the spirit of it. We were willing to express our loyalty to Gandhian ideals but this would not be served by abstaining from classes — rather we should qualify to be better citizens in a modern world. Though the uneasy situation lasted till February, we survived it and were able to take the honours examination in March 1943.

In October 1942 we heard of the battle of El Alamein, the Waterloo on the African desert. Churchill flew to Cairo earlier in August and effected a change in army leadership. Wavell and Auchinleck were despatched to India and considered good enough to be the Viceroy and Commander-in-Chief of a dependent Asiatic country. Sir Bernard Montgomery was put in command of the Eighth Army, and Alexander over all the forces. The air strength was reinforced under Sir Arthur Tedder. The Battle of El Alamein ended the tide of Rommel's victories and he began his retreat which within a few months was to lead to his exit from Africa.

India was taken by surprise when we heard that the Americans had landed in Algiers in North Africa on the 8th of November under the command of a young general Dwight Eisenhower, with Casablanca as their first objective. This was not intended to be the second front which Stalin demanded but it marked an important stage in the war when the Axis powers were thrown on the defensive and had to fear an invasion from the South. So started the race for Tunis, on the west by the Americans and in the east by the British who were hot on Rommel's chase.

We were so much engaged preparing for the final Honours examination that we just read the headlines in the newspapers. The allies failed to capture Tunis by December and Germans poured in reinforcements in desperation. It was to be the last show of force by Rommel before he vacated North Africa, thus closing a bitter and bloody chapter in the course of war.

In the Pacific after the victory at Midway, the Americans moved to the offensive under the direction of General MacArthur and Admiral Nimitz. In November 1942 occurred the battle of Guadalcanal, an American victory at great cost in men and planes. In Burma the British under Wavell started their counter offensive.

It was during the final phase of the African campaign that we wrote our final Honours examination. I had prepared well but the only uncertain part was the 'practicals' which carried 400 marks out of a total of 1400. This was subject to considerable manipulation in a rivalry between colleges resulting in exchange of first classes for ranks! What I feared actually happened in the experiment on 'properties of matter'. I got the 'Oscillating Disc' and just before the experiment, I dropped the metal disc which made such a noise that it caught the attentive ear of Mr. Mandalam Iyer, the examiner from St. Joseph's College! I knew 'my goose was cooked' and indeed it was, since the examiner drew my blood and reduced my marks for not properly handling the apparatus though I did my experiment very well. In spite of this I scored a very high first class with 1009/1400 while the first rank student obtained 1038. But for the 'noise incident' I would have been awarded 80% in the practical and been a contender for the first rank.

But another significant incident generated great confidence in me to seek a research career. In answering a question, I gave a simpler theoretical derivation of a result in heat conduction, though our 'Bible' Saha and Srivastava gave a more cumbersome derivation. I would have been penalised for my initiative but for the timely intervention of one of the examiners who understood the proof.

So my desire to take to research was keen as ever though I did not have any clear preference in any particular area of work. Of course my leanings were mathematical and at that time I did not realise that theoretical physics by itself was a total field of scientific activity. My teacher Professor Bondade advised me to work under Dr. Bhabha in Bangalore but a legend had grown round him that he was distant and inaccessible and that his field of work demanded advanced mathematics. He was one of the youngest Fellows of the Royal Society and was connected with the house of the Tatas. Fresh from his laurels at Cambridge, he started a separate cosmic ray unit at the Indian Institute of Science on a nominal salary of Rs. 500 (which he never needed) and occupied a small rectangular room outside the library. But he was never really there. He stayed most of the time at the Westend Hotel in Bangalore or the Savoy in Ooty or in his Malabar hill home in Bombay. He was the only person who could talk on equal terms with Sir C.V. Raman. I was ready to take up the challenge and meet him in Bangalore. But my father was hoping that I would take to law and practise with my brother but he did not want to impose his will on me. He took me to Bangalore that summer along with him since he was engaged in a important law suit called the Ramalingam Will case. Another veteran lawyer Raja Iyer was appearing on the opposite side. I remember the discussion on the definition of drunkenness, how a few pegs of whiskey which would unhinge a teetotaller would amount to nothing to a habitual drinker. My father was trying to prove that the will disinheriting the wife and children was invalid since it was written under duress or undue influence and that the person was not in full possession of his mental faculties. Newsam, the district judge, did not share my father's aversion to drink, but he could appreciate that the testator could not have been normal.

My father tried to get me interested in the intricacies of the case but finding me obdurate he suggested that I should meet Sir C.V. Raman. It turned out that Sir C.V. was in one of his irascible moods and was very critical about the scientific situation in the country.[2] To me it was a bitter lesson in human life that consultation regarding a career is a contradiction in terms. No one can teach another the path to fame and glory for each successful story is different from another: Sir C.V. did not realise the fire within me which if suppressed would burst sometime, as it did later. His comments were so strong that I had little inclination to work in his laboratory but I was still anxious to pursue a life of science. Alternatively I wanted to do aerodynamics and I met Professor Ghatage who just took charge of the Aeronautics Department in the Indian Institute of Science. The great reputation of Raman, the magnificent library, the spacious grounds with magnificent trees and lovely flowering plants, cast a spell over me. Something went wrong and there was no possibility of pursuing research there but it ultimately turned out to be the greatest boon in my life for I was to get a great opportunity years later to create an institute of advanced learning in the city of my heart, Madras.

At that moment I made a cautious withdrawal and agreed to the suggestion of my father to take to the law course. Of course I was aware of the famous saying of Winston Churchill that wars are not won by withdrawals which can only be a temporary measure to gain strength and confidence before making a major plunge. So I took the law course, which was the resort of members of the affluent generation who wanted a respectable period of waiting before choosing their career.

For the First Year (FL) in the Law College, I studied Salmon's Jurisprudence and Hamid Ali's Muslim Law. The Law College was situated near the High Court. Not a single lecture interested me even though the lecturers were able speakers. I had so much of leisure that I spent hours at the Mylapore Club playing billiards and four sets of tennis a day. Now and then my father used to call me to impress how interesting legal problems could be even for a mathematically inclined person. I was hoping against hope that I would soon get an opportunity to plunge into a scientific career.

On May 11th, 1943, my father was completing sixty years of age, a significant event in a Hindu's life. The Sashtiyabdhapurthi is of religious and social significance, a day of dedication and thanks giving to God for having graced the person with His blessings, and prayers for tranquillity in later life. My father was simple and austere in his personal life and did not want a celebration but only a quiet religious function to which his immediate circle of friends were to be invited. He was gracious enough to phone them personally and the number swelled to two hundred on the eve of the function so that we had to erect overnight a pandal on the terrace of our house to accommodate the guests at lunch. His friends and Chelas, Seshachalapathi and others arranged a function at the Ghokale hall when they released a souvenir[3] and announced an endowment in his name. The souvenir contained an excellent biographical sketch of my father by Seshachalapathi, written in impeccable English, with warmth and feeling equalled only by that of Rajagopala Iyengar in the High Court Centenary volume years later. Another account was by a young enthusiastic advocate Krishna Reddy on the occasion of the unveiling of father's portrait by the Chief Justice of India in 1968 at the Andhra Pradesh High Court.

On the 12th of May, 1943, the Allies entered Tunis and this marked the end of the long-drawn war in North Africa. The stage was set for the invasion of Italy which was to take place a month later.

Notes

1) Madras had several outstanding colleges like the Loyola and Presidency Colleges. But outside of Madras within the state, the real competition was from St. Joseph's College in Trichinopoly. The Jesuit colleges trained their students extremely well and provided them comprehensive notes to help their performance in the university exams.

2) Sir C.V. Raman (1888–1970) was a legend in the world of physics. He earned a Masters Degree from the Presidency College, Madras, in 1907. He was appointed as Professor of Physics at the University of Calcutta in 1917, and there he conducted his famous experiments on light scattering, leading to the discovery of "The Raman Effect" in 1928 for which he won the Nobel Prize in 1930. He was fiercely outspoken and often very critical of the scientific scene in India, as he was when Ramakrishnan met him for the first time. Later (see Chapter 13), after Ramakrishnan took to a career in physics, Raman was very appreciative of his work and got him elected as Fellow of the Indian Academy of Sciences.

3) The beautiful volume that the Sir Alladi 60th Birthday Committee produced, containes messages and tributes by leading lawyers and judges — like Sir Maurice Gwyer — Chief Justice of the Federal Court, Sir Lionel Leach — Chief Justice of the Madras High Court, Sir C.P. Ramaswami Iyer, and others — referring to his brilliance in his legal arguments, as well as messages of admiration from eminent men in other fields like physics Nobel Laureate Sir C.V. Raman.

Chapter 8

Invasion of Europe — Sicily and Operation Overlord (1943–44)

In September 1943, the invasion of Italy (from Sicily) began with the American forces under the command of General Mark Clark landing in Salerno while Montgomery's forces crossed the straits of Messina. To the surprise of the world there was a coup in Italy and Mussolini was overthrown, an ignominious end to a glittering career. On the 25th of July Marshall Badoglio took command and Italy capitulated. Hitler was taken by surprise and he immediately ordered the German entry into Italy under Marshall Kesselring. The fall of Mussolini had the psychological effect of being the first crack of the impregnable Axis.

Life in Ekamra Nivas continued in its even tenor. The world was changing fast and soon would affect India and our ways of life. It was not going to be the same but for the moment things apparently went on as before. My father was the source and sustenance of our prosperity though the children had grown up into dependent adults. My mother was the lenient and generous host and so no one bothered about the blackmarket prices of many of the commodities. It was still a joint family under one earning head and there was enough to share but soon individual ambitions were to assert themselves seeking new opportunities in a changing world. It was certain that the joint family system, like the British Empire, was becoming an anachronism and was to be washed away by the war. The American spirit of individual initiative and enterprise was to overtake the joint responsibilities in Hindu families.

In the Pacific, MacArthur had regained the initiative which was to lead him to the victory over Japan. In the Solomons, the Bismarks, and the Admiralities, the Americans were on the offensive and so also in the Central Pacific where they captured the Gilbert islands.

At the end of 1943, Britain's South East Asia Command (SEAC) was set up under Admiral Lord Louis Mountbatten who was to play a historic role in India after the war. The Indian command was separated from SEAC with Wavell as the Viceroy and Auchinleck as the Commander-in-Chief.

Now that the war situation had turned in favour of the Allies and the Japanese menace departed, life in India resumed its leisurely tenor even though everything had changed. Indian economy faced a new situation — the vast accumulation of black money and the rise of the invisible rich. The Congress leaders were in jail, while the population did not believe in open struggle and were willing to wait to see what would happen after the allied victory. Meanwhile under the leadership of a new Muslim leader Jinnah, there was talk of a separate Moslem State — self determination for a minority which numbered over hundred million.

In early 1944, the time was getting ripe for the 'real' invasion of Europe. England was naturally the springboard for such an operation. Meanwhile Russia had mounted an offensive after the second winter and overtaking Hitler on the fatal plains of Russia. The Russians moved from Caucusus to Kiev. Leningrad was relieved after thirteen months of encirclement, a saga of heroism and tenacity that rivalled the defence of Tobruk.

After my F.L. Examination in the spring of 1944, we went to Kodaikanal for our summer vacation and stayed in a comfortable house called the Maplehurst. We enjoyed the usual picnics, the long walks and boating on the lake. My father was naturally anxious to know where the law course was leading me. I was hoping against hope that something would turn up when I could take a decisive plunge into the scientific career.

It was while we were in Kodaikanaal that we heard of the greatest amphibious military operation in human history, Operation OVERLORD. This invasion of Europe was more breath-taking, more stupendous than Operation Barbarossa in Hitler's invasion of Russia because this was a case of landing on a coast where the triumphant enemy was entrenched expecting an attack everyday. The allies were aware that the Germans had more than fifty divisions, which would strike a deadly blow to any invading army. It was an operation well planned with American efficiency, pragmatism and power, and with British tenacity, determination and toughness. The day of invasion which was to spell the doom of Hitler was kept a closely guarded secret, and so was the place of landing.

General Dwight D. Eisenhower was in supreme command of the Allied forces, with Air Marshall Sir Arthur Tedder and Bertram Ramsay as his deputies, General Walter Bedell Smith as Chief of Staff, General Bernard Montgomery and General Omar N. Bradley commanding the invading armies of the British and American forces. Eisenhower's message on that eventful day ranks in style and spirit with the Shakespearean exhortation by Henry V at Agincourt: "You are embarking on a great crusade. The eyes of the world are upon you and your enemy is well trained, well equipped, and battle hardened. He will fight savagely."

It was then that we heard of the triumphant exploits of George Patton, a man destined to tear the bowels of the German tiger. His armies under the overall command of Omar Bradley, blasted through the Avranches and his tanks flooded the countryside. He was a soldier more colourful than Montgomery and in the next few months competed with him as Bradley held the leash in the conquest of Germany. By the 25th of August 1944, Paris was liberated and General DeGaulle arrived to take over the Government of France at the Hotel Deville.

When I entered the B.L. class, the law college had shifted to a convent school in Mambalam since its main buildings near the High Court were requisitioned for security reasons. I enjoyed the classes on Hindu Law by Venkatasubramania Iyer whose torrential eloquence was a legend in Mylapore circles. My own leanings towards Hindu joint family life made me read Hindu law with enthusiasm.[1]

It was then that an accident occurred to my father when his leg was fractured (thigh bone) which disabled him till the end of his life. He had gone to Madura for a case and was staying with his friend Vaidyanatha Iyer. He slipped and fell in the bathroom and on examination it was found that there was a fracture in the upper thigh bone. Dr. Ramasubramaniam, the eminent surgeon of Madura, was summoned, and he treated the fracture by suspending a weight from the leg and father had to be kept in that position without movement for two months. It was heart-rending to see him in such a predicament: He had to stay a few months in Madura and my mother shifted with a part of our establishment to the 'open house' of the generous Vaidyanatha Iyer. Friends, eminent lawyers, and relatives, all came to see my father. Since the law course sat lightly on me, I spent most of the time in Madura with my father. He returned to Madras well enough to walk with support and resumed his work in the High Court. The judges held him in such high esteem that he was permitted to argue sitting, unprecedented in the annals of the Madras High Court.

Notes

1) Ramakrishnan was a Gold Medallist in Hindu Law. He did very well in the examinations at the Law College, and Hindu Law was his favourite.

Chapter 9

Total Victory and the Birth of the Atomic Era (1945)

The year 1945 turned out to be the most momentous in human history. It brought total victory for the Allies and the free world on the one hand, and to Soviet Russia on the other. It ushered in the atomic age and invested mankind with a new source of power which may turn out to be its salvation or its doom.

The main Russian offensive was launched in January 1945, the sweep from the Vistula to the Oder. Warsaw fell to the Russians and by the end of January the Russians had reached Germany. It was now going to be the race for Berlin — who was to reach it earlier, the Allies or Marshal Stalin?

On March 7, Patton's US Army broke through the Ardennes and crossed the Rhine. With the fervent approval of Omar Bradley, Patton moved southwards and rolled up the German armies. Meanwhile Montgomery was making a great assault on the Rhine near Wessel.

On 12th of April, President Roosevelt died suddenly and the news reached Hitler. Was this the miracle that would save him? Certainly not! Harry Truman became the President. He not only brought about the doom of Hitler, but gave the sanction for the use of the atomic bomb which burnt and blasted the Japanese into submission thus ending a five year old war.

While these momentous events were happening, I prepared for my B.L. Examination comparing the Dayabhaga and Mitakshara systems of Hindu Law. Though I won the Hindu Law Gold Medal, just as my brother did four years before, I was not attracted to the legal profession and so as an interim measure my father asked me to write the competitive examination for the Civil Services. It was the intervening period when the I.C.S. (Indian Civil service) was terminated and the I.A.S. (Indian Administrative Service) was yet to be initiated. I had to choose two new subjects besides physics and I chose British history which was considered an easier subject by the candidates because of the clear sequence of anointed kings and queens, decisive battles and well defined treaties, while Indian history was a morass of conflicting and confusing details. The second was applied mathematics but my friends tried to dissuade me from taking it and I was told it was impossible to reach the Master's Degree standard in Mathematics without a basic honours training. However I consulted T. R. Raghavasastri,[1] the veteran professor of Loyola College, the Harry Hopman[1] for training in M.Sc. Mathematics and he assured me of success provided I would not relax my concentration and effort. In addition I had with me the valuable Loyola college collection of problems and solutions made available to me by my friend and tennismate Thambu (Venkataramani) who was successful in the previous examination. I enjoyed the challenge so much that it turned out in every sense to be the starting point of a renewed desire to take to research again. I cleared the mathematics paper in the competitive examination and in fact gave a new proof to a problem relating to potentials — the dynamics of a heavy spider moving along an elastic thread! It reawakened in me a confidence first generated by solving riders in the S.S.L.C. and in the Intermediate.

In the English paper I had the pleasant experience of writing an essay I had prepared well on Franklin Delano Roosevelt since the great American died a few months before. I based it on the book by Emil Ludwig which my father read to me so often. I pushed in all the phrases and the new words my father had underlined in the book and concluded the well prepared essay with the following famous poem:

O Captain! my captain! our fearful trip is done,
The ship has weathered every rock, the prize we sought is won,
The port is near, the bells I hear, the people all exulting
O Captain! my Captain!
The ship is anchored safe and sound, its voyage closed and done
From fearful trip the victor ship comes in with object won;
Exult O shores, and ring O bells! But I with mournful tread
Walk the deck my captain lies, fallen cold and dead
O captain! my captain!

The sentiment is as true to Roosevelt as it was of Lincoln.

In British history, I read the books, 'England under the Tudors' by Innes and 'England under the Stuarts' by Trevelyan. I prepared thoroughly a series of essays like 'Henry VIII was not only a tyrant king but a kingly tyrant' and 'the Crimean war was a crime' so that these essays or linear combinations of various parts of them would cover the entire syllabus! I was excited when I found the question paper almost full of questions for which I had prepared answers! I enjoyed writing the essay on Marlborough's campaign, the march to the Danube, with profuse quotations from Trevelyan like "At about the time when the last revellers on the Siene were slinking into their beds, English trumpets woke the sleeping soldiers on distant Danube."

At the interview held at the end of the year in Delhi, the questions were simple and direct, and I answered them with great confidence, but it looked like the examiners did not like to believe my frank and free statements. Besides science, I claimed acquaintance with law and was active in sports playing bridge according to Culberston, tennis with Sedgeman as my hero in style, cricket for my college team, and table tennis and billiards at the Mylapore Club. Despite this, I was successful in the competitive examination since my performance in all the papers was uniformly good but a year had to pass before I could make the decision to take to science under Bhabha as circumstances changed and a new opportunity came in a most unexpected way.

The American, British and Russian armies entered Berlin to find that Hitler had committed suicide. On the May 7, at Eisenhower's headquarters in Rheins, Germany surrendered and the armistice was signed.

The war in Europe came to an end officially on May 8, 1945. We in India could not believe it, any more than we did its outbreak in September 1939. To the millions in India, the war was essentially a newspaper story, albeit a terrifying one, since not a shot was fired in our country, not a single inch of land trodden by invading armies. Of course the economic, social and moral consequences were there, but all over the world it was the end of an ordeal of bloodshed, death and destruction.

Only Japan remained. Would it carry on the struggle against impossible odds? For over a year Japan had been subjected to strategic air offensive and bombardment.

On the August 6, while playing tennis at the Mylapore Club, we heard the heart-stopping, nerve-shattering, blood-chilling news that a new type of bomb was dropped on Hiroshima which blasted and burned out the entire city and annihilated its population. It was an 'atomic bomb,' the energy being released from the disappearance of nuclear matter. Three days later came the news that Nagasaki was bombed and wiped out. Russia declared war on Japan on the 8th of August and immediately occupied Manchuria. The Japanese surrendered on August 14.

On September 2nd 1945, representatives of Japan signed the instrument of surrender on board the Missouri before General MacArthur. Thus the

World War II came to an end, just less than six years after it had begun. It had cost over ten million human lives and affected billions on earth.

What circumstances led to the atomic bomb? It was a strange sequence of events that led to the atomic age, but as regards scientific development, it was a natural consequence of 20th century science.

Ever since Einstein formulated his theory of relativity, scientists were aware that matter and energy were equivalent. No one, not even the architects of modern physics, dreamt that matter in bulk could be converted into energy in the laboratory or factory. It was in 1939 in Germany, Otto Hahn discovered the phenomenon of the fission in which Uranium 239 on being struck by neutrons, splits up into two parts, Barium and Krypton, releasing considerable amount of energy and a spray of neutrons. It was realised that these neutrons in turn could break up the nucleus with the possibility of a chain reaction to release energy from matter.

This discovery was made just before the outbreak of war and it was a fortunate circumstance for the Allies that the scientists engaged in such research were Jewish who became refugees from German persecution. Einstein settled down at that haven of learning, the Institute for Advanced Study in Princeton, and expressed his gratitude to the nation which received him as its most illustrious immigrant, by writing the now famous letter to President Roosevelt on the outbreak of war, warning that Germany may have the scientific potential and technological capacity to create weapons based on the results of nuclear energy. So America launched on this project at the cost of a billion dollars which was still worthwhile since victory had to be won at all costs. Thus started the awesome project amidst the deserts of New Mexico at Los Alamos under the direction of forty-year old Robert Oppenheimer who was assisted by a team of established scientists like Enrico Fermi, Niels Bohr, Otto Frisch and Victor Weisskopf, and gifted young men like Richard Feynman. The project took almost two years for completion and the fateful experiment was performed on 15th of July, 1945 at a site called Trinity in the deserts of New Mexico at 5:30 a.m. That was the exact moment when the atomic era was born. The first impression was one of tremendous light of blinding brilliance and then the now famous fire ball mushrooming into a parasol of billowy white smoke. The intensity of light was such that it could have been seen from another planet. Temperature rose to four times than that at the centre of the sun and ten thousand times that on the sun's surface. Tons of sand was sucked up into a swirling column of orange and red darkening as it rose until it looked like the flames of hell as visualised by man on earth. Half a mile from the crater a steel tower lay crumpled on the ground. The spectre of death loomed over the desert in the vicinity of the explosion. Oppenheimer said his prayers and recalled a famous verse from the Bhaghavat Gita.[2]

The news of the success of the explosion was flashed to Harry Truman at the Potsdam Conference on the 17th of July as the message — "the babes are delivered". The three great leaders, Churchill, Stalin and Truman met to discuss how to reconstruct the world from the ruins of World War II. Churchill was replaced by Attlee within two days since he lost the election at home — a strange and striking example of the implacable decree of democracy prevailing over sentimental sense of gratitude.

Two questions had to be decided by Truman: First — Should the secret be revealed to England and to Russia? It was decided that adequate hint should be thrown to Churchill while the news should be kept away from Stalin though he should be informed that America was in possession of a powerful weapon which would decide the fate of Japan. Everyone knew that Stalin was a hard bargainer and would demand a high price of Japanese possessions in Asia if his cooperation was sought in the Pacific War. America now knew that Stalin's cooperation was not necessary for a victory over Japan. If it were revealed that the atomic bomb would hasten the end, Stalin would lose no time in pouncing on a falling Japan and seize its possessions in the Asian continent as indeed he did even in the small interval of three days between Hiroshima and Nagasaki.

Secondly, the great moral question arose, to use or not to use the bomb. Obviously it was not necessary for victory over Japan, whose cities were already incinerated by intense bombing. But the fighting qualities of Japanese were too well known. They refused to believe that they were vanquished but the atom bomb would be too powerful even for their savage obduracy. Besides, so much of effort had been expended and it would amount to waste of time and resources if it were not used against Japan. There was also the irresistible and inexpressible urge to avenge Pearl Harbour. Conscience could be appeased by generosity and massive aid and this is what actually happened. With massive American aid, Japan and Germany have now taken an honoured place among the nations of the free world.

Half of Europe lay under the iron heel of Marshall Stalin, that is under the protectorship of the communist world. Poland, Hungary, Rumania all became Eastern European satellites of Russia. Germany was divided into two — the West under the Allies and the East under Russia, with Berlin situated in the East, under the triple control of America, England and Russia. Egypt, Libya and Morocco all became free and in the Middle East a nation was born in travail — Israel — which was to survive even greater travails. The Arabs who were hitherto ignored as part of the inconsequential East, suddenly became a potent force in the world. The desert was situated on a bed of oil and what started as a convenient exploitation of world resources by multinational oil companies, turned into a negotiation with hard bargainers. The Arabs became wealthy beyond dreams by playing with the price of oil, directing the course of history, knowing fully well that the world moves on automobile wheels and wheels move on oil. "Men in hunger may eat their leather belts (as they did in the War) but vehicles need oil for their motion." (Patton)

What were the effects of war on India? First would be the dawn of freedom for everyone was convinced that Britain, true to its historic tradition would keep its word. The crisis of conscience for Gandhi was over and he had to negotiate with the British. The Indian National Congress was the most powerful political force and it would be the natural successor to British power. Nehru was the darling of new India and the appointed heir to Mahatma Gandhi but things did not turn out as smooth and logical as all that. Meanwhile there arose an equally powerful force, Mohammed Ali Jinnah, leader of the muslims of India, who constituted a third of the Indian population. He asserted that the Muslims were in a majority in a third of the land, Punjab, Sind, North West Frontier and Eastern Bengal and therefore it was reasonable to ask for a separate state having a majority of Muslims. Its logic was perceived by the astute C.R. who knew that Partition was inevitable and wanted to take the sting out of it by greeting it. His logic did not find favour with Indian National Congress. No one knew it better than the British Government which after all had conquered India by exploiting internal dissensions and feuds. The only thing in India's favour was the Labour government under Attlee which was determined to grant freedom whatever the cost despite Churchill's predictions that the Asiatic millions would be plunged in a blood bath, as indeed they were, two years later.

Perhaps the greatest change in India after the war was not so much the political freedom to evolve our own system of government, as the rise of big business in the Anglo-Saxon or American pattern of Joint Stock Companies. The Joint Family system which was a cooperative enterprise of ancient origin, dependent too much on personal ethics and religious beliefs, became an anachronism under the pressure of individual initiative and ambition, in a scientific age offering varied opportunities to diverse talents. However, in India family ties are strong and so the families which were endowed with capital, initiative and enterprise took to the concept of the joint stock company 'as a duck takes to water'. It had the advantage of inviting capital from the public which became investment conscious. Thus arose the big business houses which in turn inspired 'smaller houses' and a race of entrepreneurs who could attract public funds for private enterprise. They knew very well that even in a socialistic society, it is the agricultural landlord who could be condemned as the Bluebeard of a feudal society and made the victim of confiscatory legislation. For the exploitation by the landlord is obvious and strikes the eye, while industries were considered the imperative need for India's advancement. Private fortune could be amassed by invisible means through ownership of shares in such companies, through managing agencies and subsidiaries owned by a privileged few and deriving profits from parent companies by subtler and more effective methods. Moreover, perquisites or 'perks' could be as substantial as regular income without attracting public attention. This also ensured and insured the appointment of friends and relatives who otherwise would have to go through painful and prolonged process of selection for government service, based on qualification or competitive examination. The occupants of large mansions owned by companies would not excite the envy of those less fortunately placed, and such privileges could be enjoyed without the responsibilities of personal ownership. The lawyers also became an important section of the community and they shifted their attention to company law. This in turn led to settlements out of court and companies employed legal advisors, financial advisors, and internal auditors. Accountancy, auditing, and management science, became new avenues for the professional class.

Obviously this could not be an uninterrupted picnic for the upper levels of management. It had also its pains and strains for workers became very conscious of their role in industry. They resorted to strikes, and very often to violence. This spread to the white collared and clerical classes. The government wanted to step into the shoes of private enterprise and thus arose the Public Sector and India became a mixed economy. Banks were nationalised but the press was left free since it was the vehicle of public opinion run by private families. Oil, atomic energy and steel became state enterprises while cement, chemicals and metals were in private hands.

However, in land legislation, extremely socialistic or communistic ideas gained favour leading to ceilings on rural holdings. But by a strange series of circumstances the plantations were left out of such legislation and were considered as industries. The vast tea and coffee estates were the greatest legacy which British ruling class left the Indian businessmen who could acquire thousands of acres either as part of their private fortunes or as joint enterprises. Thus the princes of the native states were soon replaced by a new and more powerful, affluent, influential class — the big business houses and their satellites. Anyway the main result was that two decades after the war India was considered no longer an underdeveloped nation and it qualified itself to a leading membership of the community of developing nations, attracting capital and investment in joint enterprises with affluent countries. Many foreign and multinational companies were 'Indianised' and became part of the economic framework of a free India.

All these developments followed the freedom of India but before it became free, in 1947 it had to go through an ordeal of negotiations, feuds, internecine struggle, an inevitable 'fission' into two nations followed by a blood bath and a holocaust of violence and destruction. What it was spared during the war, it got in twice the measure as the price (or penalty) for sharing the fruits of a victory for which it made no willing effort.

In international affairs, the three worlds, one under American leadership, another, under Russian protection and the third an agglomeration of nations which had become free after the war, agreed to interact through a common organisation — the United Nations, on the pattern of the League of Nations, but a much bigger and stabler structure with various organs of greater complexity and magnitude.

On April 25, 1945, the United Nations came into being. It was signed on June 26th and entered into force on October 24th, 1945, with its six organs: the General Assembly, the Security Council, the Economic and Social Council, the Trusteeship Council, the International Court and the Secretariat.

For my competitive examination I studied in detail the evolution of the United Nations since we felt it would be 'hot' subject as the war came to an end and the tasks of peace engaged the attention of the world. After the competitive examination I entered the apprentice course under Govindarajachari, a chela[3] of my father. I was to complete it by way of abundant caution. It was then I used to hear the legal arguments of my father in the High Court. I realised his greatness in greater measure only years later. He was a man possessed by law. He studied the judge's mood and mind. He would not flood him with facts and figures. He would just take a decisive point and hammer it into his head. He always watched his opponent to find out the weakest chink in his armour. He was never perturbed by the display of flamboyance or flourish by the opponent. He had an analytical mind and he would trounce his opponent but not by sound or fury. His strength lay in reckoning the strength of the opponent and finding the most vulnerable point to force his way out.[4] I was fortunate to imbibe this spirit which has stood me in good stead throughout my life.

Notes

1) Ramakrishnan consulted T. R. Raghava Sastri, one of the finest college teachers of mathematics. After his tenure at the Loyola College, he moved to the newly formed the Vivekananda College in 1946 and eventually became its principal during 1959–62. Ramakrishnan compares Raghava Sastri to Harry Hopman, a famous Australian tennis player who was even more reputed as a tennis coach.

2) Robert Oppenheimer, the Director of the Manhattan Project to build the Atom Bomb, was affected by the destruction it caused. He read the Bhagavat Gita in the Hindu epic Mahabharatha, and drew solace from the words of Lord Krishna to the warrior prince Arjuna who hesitated in fighting his own kith and kin. On observing the atom bomb test in the desert of New Mexico, Oppenheimer is supposed to have said: *I remember the line from the Hindu scripture the Bhagavat Gita; Vishnu (= Lord Krishna in incarnation) is trying to persuade the Prince (= Arjuna) that he should do his duty and, to impress him, takes on his multi-armed form and says "Now I am become Death, the destroyer of worlds." I suppose we all thought that one way or the other.*

3) *Chela* is a term in Hinduism for a student, but it emphasises a special bond between the teacher and his pupil like between a father and son.

4) Sir Alladi had a legendary reputation for the style and effectiveness of his arguments. People used to come to the High Court just to hear him argue a case. The following famous episode as recounted by Suresh Balakrishnan in his fine book "Famous Judges and Lawyers of Madras", illustrates this marvellously ([4], p. 413): "As a lawyer, Alladi appeared quite often in the Federal Court (in Delhi). One he was arguing a case before the Chief Justice Sir Maurice Gwyer. Alladi's arguments concluded before lunch and post lunch it was the other side's turn to make the submission. Sir John Simon of the Simon Commission fame was in Delhi at that time. He visited the Federal Court after lunch and because of his presence, the Chief Justice requested Alladi to continue so that Sir John could hear his arguments! When Sir John visited Madras, the Chief Justice Sir Lionel Leach called a case that was lower down the list so that Alladi could argue and Sir John could hear him."

Sir Maurice Gwyer was Chief Justice of the Federal Court from 1937 to 1943. Sir Lionel Leach was Chief Justice of the Madras High Court from 1937 to 1947.

Here is what Sir Maurice Gwyer said (see [7]) on the occasion of Sir Alladi's 60th Birthday: "It has been my privilege several times to have had Sir Alladi appearing before me in cases of great importance. His massive arguments, with their ruthless logic and profound learning, are among my happiest moments of the bench. I used to admire the way in which he built up his case, the skill with which he selected his authorities, the firmness with which he rejected every irrelevant point, however superficially attractive. The cases which he argued before the Federal Court were concerned with fundamental constitutional doctrines and principles, and Sir Alladi's arguments owed much to his immense knowledge of law and practice of other countries."

Chapter 10

The Years of Decisions (1946–49)

The year 1946 was the most decisive one in my life — I chose my wife and my career, but at the dawn of the year there was nothing to show that these events were to happen.

I was continuing my apprenticeship in law under the flourishing lawyer, Govindarajachari (my father's pupil who later became a judge) but my heart was not in the profession. I was waiting for the results of the competitive examination for central services for which I appeared in September 1945. Unrealistically, I was dreaming of a research career under Professor Bhabha when I saw a brochure from the newly started Tata Institute in which Pauli's association with it was mentioned. It looked a dream impossible of realisation. I had already strayed away from physics for three years, less qualified now for a career of research under so eminent a scientist.

The year was also a stirring period in India's political history — the pre-dawn year of Indian freedom. At a time when Britain was willing to part with power, Hindus and Muslims were divided as never before in the nation's history over political objectives. The Congress under Nehru and Gandhi stood for a united India claiming that the Muslim masses were behind them, while the Muslim leadership of Jinnah was intransigent and demanded nothing less than a separate state. To Gandhiji, India was a 'living child' or a 'sacred cow' which could not be partitioned to satisfy the demands of fighting claimants.

The postwar-British Government of the Labour party under Atlee sent a three member Cabinet mission to India consisting of Sir Stafford Cripps, Lord Pethick Lawrence and A.V. Alexander to discuss the transfer of power in spite of the inveterate opposition of Winston Churchill who taunted his countrymen by reminding them that Gandhi's answer to the earlier Cripps mission was 'Quit India'. Added to all this came the trial of the members of the Indian National Army (INA) which was created under the aegis of the Japanese and which fought against the allied cause of the war. But it was associated with the name of Subhas Chandra Bose who, before he left India in 1941, was the idol of the Indian masses like Gandhi and Nehru. There was a magic about his name which could not be affected even by his association with Germany or Japan. This is the tragic incongruity of the Indian mind which looks at real problems from a sentimental and sometimes irrational way. The horrors perpetrated by Nazi Germany against humanity would not be supported by the Germans themselves after the war. How then could one justify the association of the INA with the Axis powers?

My father was summoned in March 1946 by the Indian government to study some legal questions arising in the INA trials. I accompanied him on the trip to Delhi. On the journey by train, I prepared for my apprenticeship examination in law. Krishna Iyer prepared delectable lunches and dinners in the four berth compartment reserved for us, while Subbanna puffed and huffed around frequently engaging in conversation with the engine driver and the ticket inspector.

After return from Delhi, I appeared for the legal apprenticeship examination in which I stood first, winning the K.V. Krishnaswami Iyer prize.[1] Then we went for the summer to Bangalore and stayed at Basavangudi. It was then that I expressed to my parents my intention to get married which implied that they should be willing to receive proposals from respectable parents offering their daughters. Dr. H. Subramani Iyer,[2] Principal of the Maharajah's College in Trivandrum had proposed one of his daughters Kamalam for my elder brother a few years before, but that proposal could not be accepted because my brother was unofficially engaged. My father had given Subramani Iyer an assurance that he would take up the proposal later when my marriage would be considered. True to his word, my father now approached Subramani Iyer who replied that Kamalam was now married, but he would like us to consider her younger sister Lalitha. He sent as his emissary, his oldest daughter Annapoorni to Bangalore. He could not have chosen a better representative. It was arranged that Lalitha and I should meet, contrary to accepted conventions, in the boy's house. It was the usual custom that meetings are arranged in the girl's house.

It was of course the most exhilarating moment of my life when I first saw her. She arrived with her father — she was dressed in a green silk saree with red and gold border and as she stepped into my house from the porch I found her strikingly beautiful with a poise and grace unusual in such a young girl. She had a dignity of bearing, an immaculate sense of elegance in dress which has been unaltered to this day. She wore a diamond nose ring which accentuated the colour of her face. I was very much impressed with her modesty and confidence. After conversation with her amidst company, I expressed my approval to my parents.[3] Of course I was wondering what her reactions were when she saw me. Later with womanly discretion she told me that she was impressed with the reputation that had grown around that I was a person of very strong will. Probably she wanted to take up the challenge in life to bend that strong will to a fruitful purpose, a task in which she seems to have succeeded in abundant measure!

It was agreed that the marriage should be performed in Madras on the 21st August 1946, the auspicious day and hour (Muhurtham) being fixed by our family priests. Under normal circumstances, the marriage is at the bride's place of residence (Trivandrum in the case of Lalitha), but my father was very anxious that his aged elder brother should be present and Madras was therefore a more convenient venue. A few weeks prior to the marriage, Dr. H. Subramani Iyer arrived to supervise the preparations. A large 'Pandal' was erected besides Norton Lodge, the spacious old house which we owned next to Ekamra Nivas. What a fortunate stroke of destiny that I should build the Alladi House, later in the same site where I was married! The elite of Madras, all our friends and relatives attended the wedding. I invoked the blessings from the venerable Sir P.S. Sivaswami Iyer. At the reception we had a music concert by the great G. N. Balasubramaniam (GNB) and later by Lalitha's sisters. I remember the song 'Himagirithaneye Hemalathe Sri Lalithe' by GNB for obvious reasons.

Lalitha was in her final year at high school and she had to finish her S.S.L.C. before joining me on the journey for life. No one talked of a honeymoon in those days since it was assumed marriage itself was one long honeymoon! This also implies that little differences of opinion make reconciliation sweeter!

Earlier in May 1946 we had received the news in the papers that I was one of the qualified candidates for the Central Services. In fact I had secured a rank high enough for me to obtain one of the first fifteen coveted Finance Officer's posts which later would be absorbed in the higher Administrative Services. I was not overjoyed at it since my hopes were still on a scientific career. I drew consolation of inspiration from the example of India's most illustrious scientist, Sir C.V. Raman, who first took to the Indian Audit and Accounts service and then resigned when offered a professorial position by the enlightened Vice-Chancellor Sir Asutosh Mukherjee because Raman was even more exceptional as a scientist. Such a chance seemed remote for me since Sir Asutosh was an exceptional Vice-Chancellor and Raman was an even more exceptional scientist. At that time of my marriage everyone was congratulating me at the prospect of my entering Government Service.

When the cabinet mission left in June 1946, the Congress and the Muslim League were haggling over the distribution of members in the Cabinet of twelve of the proposed Interim Government and over the question whether the Muslim nominee of the Congress was really a Muslim in reckoning the relative representation. The Congress first formed the Interim Government in September 1946 and the Muslim League joined on the 26th of October after an uneasy and uncertain agreement of the number of members. It was in such an atmosphere of national tension that I got the news that the Medical Board had disqualified me on the ground of high blood pressure — an alarming announcement for me immediately after my marriage! I was aware that the doctor tested me before and after exercise. But I was playing four sets of tennis everyday and so this decision was surprising. Later I learnt that there was politics between the various ministries of the states on communal grounds and that influenced the selection. Anyway I appealed against the decision and it took a few months before I was declared qualified by a new Medical Board! But it was too late because by that time, owing to a series of circumstances I had plunged into a scientific career under Homi Bhabha.

Anticipating a favourable review, I went to Delhi by air in October 1946 and stayed with Sir S. Varadachariar,[4] then Judge of the Supreme Court and a great friend of my father. He represented the quintessence of the South Indian intellectual class — austere and simple in habits but quite conscious of such qualities, unrelenting in intellectual standards, accurate in speech and writing, circumspect in the use of money but generous and hospitable, animated by a sterling sense of duty and honesty — great virtues in the judicial service — and above all imbued with a deep concern for the well being of those near and dear to him. Those six weeks with him were an education and discipline in themselves. Despite his exalted position, there was no barrier between us. He enjoyed my frankness in my opinions on men and matters! He was a man of such deep sincerity and affection, rarely found in these days of unabashed commercialism.

At that time India was passing through its greatest crisis. Hindu-Muslim tension was at its peak and we heard of the violence and rioting in Calcutta and Eastern Bengal. The Interim Cabinet, *ab-initio* was riven into two independent irreconcilable parts — the Congress ministers Jawaharlal Nehru, Vallabhai Patel, Rajendra Prasad, C.R., Asaf Ali, and Mathai on the one

hand, and Muslim League nominees Liaqat Ali, Chundrigar, Nishtar, Ghaz-
anfar Ali, and Mandal on the other. The vaulting idealism of Nehru, the
sentient sagacity of C.R., the iron will of Patel, the suave composure of
Rajendra Prasad, were of no avail against the resolution Quad -I Azam
and Jinnah who argued for a division into Hindu and Muslim majority regions. Jinnah asserted that communal tension will be relieved only on the
day of deliverance. Gandhi, like a new Christ on earth, preached the ways
of God to men and appealed for unity and for the restoration of law and
order.

Meanwhile the British Government had approved the summoning of the
Constituent Assembly toward the end of 1946, and my father, though not a
politician, was invited to be a member of the Drafting Committee.[5] We were
all excited at the prospect of my father playing such a fundamental role in
Indian history. All the years of study, of thought and experience would now
be projected on a task which will affect not only the lives of millions now
but generations to come. Along with my father were T.T. Krishnamachari
(TTK), B.R. Ambedkar,[6] K.M. Munshi, N. Madhava Rao and Sir Benagal
Narsing Rao.

From Madras to Delhi we went by the Deccan Airways DC-3 Dakota
which made stops at Hyderabad, Nagpur, Bhopal and Gwalior.[7] I had made
the trip to Delhi before in connection with my competitive examination
and so I was consulted by my father about the nature of the journey. He
was making his first trip by air. My father and TTK were VIP's indeed
VVIP's in every sense of the term and the airline staff accompanied us to
the aircraft. I felt a great thrill being among the first in our family to
be travelling by air and that too accompanying my father on so great a
mission. The Buick car was transported by train earlier to Delhi.

It is impossible to describe the excitement over the first session of the
Constituent Assembly on the 9th December, 1946. There was a mood of
anxious expectancy. The Muslim League decided to boycott it but the
Congress wanted to get on with its business. Since my father needed
someone to assist him while walking — he lost the power of walking
independently after the accident in Madura — I was given special permission to enter the hall with him but stay a little distance along with the
administrative and supporting Staff. Subbanna would be hovering around
chatting with all and sundry, gathering information about everyone. On
the first day, with tearfilled eyes, Nehru addressed the gathering referring
to the absence of Gandhi (who was in prison) — "there is one who is absent here, who is in the minds of all of you as he is in mine, but his spirit
hovers over the Assembly". For the first time in the annals of our ancient

country, a Constituent Assembly had gathered to devise a constitution for an independent sovereign democratic republic. Many years ago we made a 'tryst with destiny' and that moment had arrived! The Secretary to the Constituent Assembly was a brilliant young I.C.S. Officer, H.V.R. Iyengar, assisted ably by a young aspirant to a diplomatic career, K.V. Padmanabhan both of whom used to meet my father frequently to discuss the procedure and the drafts that were being prepared.

It was during this trip to Delhi that my father was invited by Sir S.S. Bhatnagar, the Director of the Council of Scientific Research, for dinner, to meet Homi Bhabha. My father suggested that I could go to the dinner in his place for he was too busy with his work and I would be more interested in meeting Dr. Bhabha. Indeed I was, for I wanted to make the last and desperate bid for a scientific career. At that dinner I was seated next to Bhabha, more as a representative of my father than in my own right. With unconcealed eagerness I suggested to Bhabha that I would embark on a scientific career if he were willing to take me as his student. I told him that it was 'love at first sight'[8] a coup de foudre in my adolescence when I first saw him at the Presidency College during his lecture, and later it was only Sir C.V. Raman's homily that scared me away from research. He just smiled and asked me to meet him in Madras at the Connemara Hotel where he would be stopping on his way to Trivandrum. Accordingly I met him with unconcealed excitement at the Connemara Hotel in Madras when he assured me that I will hear from him in due course after his return to Bombay. Meanwhile my father had approached Sardar Vallabhai Patel who was just recruiting the staff for the Indian Administrative Service. The regular IAS by examination was to start the next year but a few were to be selected by Patel on the basis of earlier competitive examinations.

During the first few months of 1947, I was accompanying my father to Delhi for the sessions of the Constituent Assembly waiting for things to move at the Tata Institute. I remember the stirring speech of Nehru at the first Asian relations conference at the Ram Lila Grounds. There was strange feeling of awe, anxiety and expectation as we thought of Asia waking up to a new freedom from its age-long slumber under imperial domination. Though Japan lost the war, its initial conquests had destroyed the myth of invincible empires. The teeming millions of Asia were shaking off the shackles of European imperialism.

To boost my spirits during the period of my self-imposed idleness as an unwilling lawyer, my generous father engaged me as a Junior when he was arguing a monopoly case in Delhi relating to two private airlines the

Dalmia-Jain and the Indian National Airways. This meant that I was to get a good share of his personal fee without detriment to his clients and my role was just to prompt him with the names of the aircraft and the directors of the companies! While I performed my role efficiently, I expressed my preference to an unpaid studentship under Bhabha to an affluent career in law. My loving mother argued successfully in my favour that an affluent lawyer could support a son intent on a life dedicated to science. On hearing this, my generous father approved my preference — so did my father-in-law, himself a mathematician with a leaning towards research.

During this period I also made a few trips to Trivandrum to see Lalitha who was there to take her school final exams. Trivandrum is a very clean city and the lush vegetation of Travancore has always held my fascination. The Kerala cuisine at my father-in-law's house was very attractive and Lalitha's mother Ponnakkal used to provide the softest 'iddlies' and the most delicious 'Pulicheri' that I have ever tasted in all my life. The visits to Cape Comorin where we offered worship to Goddess Kanyakumari at the confluence of the three oceans, to the sacred temple of Padmanabha which inspired the sonorous verse of Swathi Thirunal, to the then unspoiled beaches of Kovalam watching the sunsets while lolling in the limpid waters of the sheltered bay, the leisured journeys by train through the fertile plains of Tamilnadu and the undulating terrain of the Western ghats — are unforgettable experiences, memories which sweeten my leisure to this day.

The year 1947 was the most momentous year in Indian history. Lord Louis Mountbatten arrived as the Viceroy in the place of Lord Wavell. He came with a prestige and aura unknown since the days of Lord Curzon. He was cousin of England's anointed Queen — tall, handsome, with the high reputation of the Supreme commander of the South East Asian operations in the world war. He was determined to solve the Hindu-Muslim problem and see India as a free and independent nation. Every day there was something exciting in the news, negotiations and meetings — Delhi was simmering with the heat of expectation.

What a strange and stupendous task faced the Imperial representative, invested with unrestrained power, unrivalled prestige and unsurpassed glamour — to transfer power to a billion voiced nation in which a third of the population was clamouring for division! To divide or not to divide was the question. Should he be influenced by the apostolic faith of Gandhi in a single nation or should he yield to Jinnah and carve out a Pakistan from the loins of British India? After weeks of painful excitement and exhausting negotiation he made the fateful decision to partition India and handed the carving knife to his loyal adjutant Sir Cyril Radcliffe. It was the only possible solution to the age old feud between Hindus and Muslims but at what price in human blood and tears!

It was in this national mood I accepted the 'offer' from Bhabha, to take me as his student. My father-in-law, a mathematician himself, appreciated my desire for a research career. There were yet no stipends available at the Institute. Even Harish Chandra, a world famous mathematician at Princeton, who was Bhabha's earliest student[9] at Bangalore, had held the J.H. Bhabha fellowship of approximately Rs. 75 a month! Most of the fellowships at the Indian Institute of Science, Bangalore, carried a stipend of about Rs. 60 per month and were awarded after prolonged interviews and severe competition. It looked almost madness to throw away a legal career just waiting for me or the Civil Service for which I was qualified, and embark on the uncharted sea of research. My father warned me that I was burning my boats and that I could not again reverse my decision.

Even in 1947, Bombay was an expensive city where it was difficult to find accommodation by reasonable methods at reasonable prices. Already a *Pagdi*[10] system was in vogue where a considerable sum, out of proportion to the monthly rent, has to be paid to a tenant to induce him to vacate! But these difficulties did not deter me and I decided to go to Bombay. Fortunately for the first month of my stay, a spacious well furnished apartment above the Eros Theatre in Central Bombay was made available to me by a friend of my father who was going out of Bombay for a vacation. I took our cook Krishna Iyer with me knowing full well that my spirits can be sustained at the dining table by the master's culinary creations!

The Tata Institute was located in one half of a spacious house, Kenilworth,[11] much smaller in size compared to Ekamra Nivas, on Peddar Road, Cumbala Hill. It had a small garden and in the rest of the house, Bhabha's aunt, the owner of the building, resided. Bhabha at that time owned an old Hudson car with a dull grey paint. The registrar of the Institute was Godbole, a chain-smoking, pleasant-mannered person who was proud of his association with Bhabha and his own qualifications as an academician and administrator. Bhabha had an accountant, a tall, handsome young man, Puthran, with a genial smile, a silvery voice and a friendly accent. Bhabha was an artist and he knew how to pick his assistants. We were five students in theoretical physics — myself, Suryaprakash, George Abraham, K.K. Gupta and P.C. Vaidya. Bhabha said he would allot us different problems but we should do preparatory study in modern physics. On my part, the first books I read were Margenau and Murphy's

"Methods of Mathematical Physics", Heitler's "Quantum theory of Radiation", Dirac's "Quantum mechanics" and Pauli's famous Handbuch der Physik article. Vaidya, a very friendly and communicative person was a loner in research and was interested mainly in the general theory of relativity. The experimental group was being built up steadily. Among the early workers were Shaiar, Daniel, Vaze, Ghokale, Seetharam, and Sreekantan who later became the Director of the Tata Institute.

I flew to Madras for my sister Rajeswari's marriage in July 1947 and returned to Bombay to stay with a friend of mine, who had a comfortable apartment behind the Marine drive. We used to have discussion on politics and science in India on the eve of her emergence into nationhood. We then heard the announcement by the Viceroy that August 15, 1947 was chosen as the day when India would be given its Independence.

Bombay went mad with delight on that fateful night before the dawn of Independence. What an exciting and exalting thought — India, to become free at midnight and we would be the citizens of a sovereign independent nation. All the Indian history I had learnt flashed past my mind. It was impossible for us to think of an India without the British. What about the empire, more magnificent than that of Rome, built by Clive and Hastings, Wellesley and Moira, Dalhousie and Canning, Ripon and Curzon! Unbelievable but true — the triumph of Gandhi's non-violence against the might of an Empire that brought the German Wehrmacht to the dust. The whole of Bombay gathered in the spacious maidan behind the Marine drive. It was a night to remember and the dawn of a new era (Appendix 3).

In grateful admiration, India chose Lord Mountbatten as its Governor-General while Pakistan had its Messiah, Muhammad Ali Jinnah. Nehru announced the Cabinet of a free India — Vallabhai Patel, Rajendra Prasad, John Mathai, Baldev Singh, Jagjivan Ram, C. H. Bhabha, R. A. Kidwai, B. R. Ambedkar, R. K. Shanmugham Chetty, S. P. Mukherjee, and N. V. Gadgil. We felt that the millennium was being ushered in and that all the ills we attributed to foreign rule would just vanish under the light of a new freedom, little realising that freedom implied responsibility and our country had to go through a period of idle hope, harsh reality and stark disillusionment in the years to come.

With the help of some friends, I arranged for a room with kitchen facilities in the Seagreen Hotel, right on the Marine drive. I brought my small Standard car from Madras for my use in Bombay and also our cook Mani Iyer, a staunch puritan intolerant of the mores and manners of a permissive society. I was not getting a stipend when I was maintaining myself in an expensive hotel in an expensive locality in Bombay. Anyone except a generous and lenient father would have scoffed at such prodigality and madness and would have thought I was on a wild goose chase.

My father was busy with the sessions of the Constituent Assembly and I was so hypnotised by the personality of Bhabha that I wanted my father to visit Bombay and meet him to discuss my career. A friend of my father, who had a spacious apartment suggested that father could come and stay for a few days with him. At that time B.G. Kher was the Chief Minister of Bombay under a free India and he called on my father. As usual, due to the excitement of the moment, my father showed signs of nausea which Mr. Kher mistook for a heart attack. A cardiac expert was summoned and he prescribed digitalis, but he warned us that an extreme dose could be fatal. Fortunately my father recovered without the use of any medicine.

It was then that I spent considerable time thinking about the Dirac hole theory.[12] It was years later I was able to do some work extending Feynman's graphical formalism of the positron in greater detail.

For Christmas (1947) I went to Madras since Lalitha was there in Ekamra Nivas. Again the problem of finding accommodation in Bombay became difficult and I stayed for a few days in the Institute itself. It was then that the terrible tragic news of Gandhi's assassination reached us on that day of infamy and shame, January 30, 1948. Within a few hours, life in Bombay came to a standstill. No one knew how to express his grief of anger at the assassin or the fears for a free India without Gandhi. We heard on the radio Nehru's announcement 'The light has gone out of our lives'. India was orphaned and the father of the nation had left us in the hour of trial and triumph. It was as though he was destined to bring freedom for India's millions after years of unrelenting struggle and leave as soon as his work was done. But what a tragic irony that it was the hand of a Hindu fanatic that struck down the greatest Hindu of all who represented the spirit and ideals of this land of sages and saints. By his example and precept he had made the Ramayana and Gita living legends and guides to human action. He woke half a billion people from the torpor of centuries. He opposed the might of the British Empire which had twice defeated the German military power. His will had prevailed against the prejudice and opposition of Winston Churchill who brought Hitler to his doom. India had to pay a heavy ransom in blood and tears in the next few years before it could accept the freedom he had won for it. The Indo-Pakistan conflict brought in its wake the massacre of millions, unprecedented even in its troubled history. There was an orgy of killing and a reign of terror amidst the raging fury of communal passions. The blood of the saint had stained the fair land of the sacred rivers, the Ganges and the Cauvery. In the years to come India was to stray away from the Gandhian ideals. His absence will be felt more and more with the increasing stresses of economic and cultural upheavals.

In May 1948 the last of the British troops left India and there was a colourful ceremony in Bombay. At the Institute, Bhabha had distributed the 'Empire' of theoretical physics to his vassals — Lorentz group to Abraham, relativistic wave equations to Gupta, Meson theory to Suryaprakash, and cascade theory of cosmic ray showers to me. He had the initiative and genius to perceive that cascade multiplication was a stochastic problem — a random process involving probability concepts. I did not know the elements of probability and so I started on the book by Uspensky and first played around with elementary problems. I read Kolmogoroff's "Grundbegriff der Wahrscheinlichkeit theorie". The concept of a random process evolving with time was getting clearer to me, as I contemplated for hours on the meaning of random changes is time. There was no textbook on stochastic processes at that time and in fact my later article in the Handbuch der Physik in 1956 was to be one of the first of its kind in stochastic theory.

Four important and allied concepts became clear to me in the studies in probability theory and their deep understanding turned out to be the basis of all the research I did later in stochastic processes.

1. The most important started with understanding the difference between impossible events and highly improbable events. To an impossible event we ascribe a probability zero, while to a highly improbable event or almost impossible event we ascribe a probability which can be made as small as we please but non-zero. This is infinitesimal but plays a fundamental role when an infinite number of such quantities are added together.

2. There is a fundamental concept of a countable infinite set of entities. The number of integers is countable infinite and the set of entities that can be put into correspondence with them is countable infinite. This leads to the apparently paradoxical statement that the number of even numbers is of the same order of infinity as the total number of integers since, dividing the even numbers by two we get all the positive numbers! But if we take a finite number N and consider even numbers less than N, they constitute half the total as N tends to infinity.

3. However, the set of points in a line is uncountable. We can on the other hand ascribe a 'measure' to such a set of points and the simplest way is to define intervals and the length or 'measure' of such intervals, each containing an uncountable number of points. But one large interval can be

divided into a countable infinite number of small intervals. Though each such infinitesimal interval contains an uncountable number of points, the 'measure' or length of the interval is infinitesimal, and it is this measure that plays a role in the addition of such quantities.

4. The sum of a finite number of infinitesimal quantities is still infinitesimal. However the sum of a countable infinite number of infinitesimal intervals can be of finite measure.

These ideas were extremely relevant to the study of evolutionary processes or processes evolving with time. In such cases we speak of the state of a system at any time which is the result of the accumulated 'events' which happen in the interval 0 to t. The interval can be subdivided into N smaller intervals and as N tends to infinity, each interval becomes infinitesimal. If we consider what happens in this small interval, almost certainly nothing happens, and something happens with very small probability. It is reasonable to suppose that this small probability is proportional to the small interval. Thus the study of a random process is possible if we postulate that the event that can happen in such a small interval has a probability proportional to the interval.

This is to be compared with the method of studying deterministic processes when we are concerned with the change of the dynamical quantities with time. When the changes are infinitesimal and proportional to the infinitesimal time interval, we speak of differential coefficients with respect to time. If however the quantities change in a random manner, we can postulate that no change occurs almost certainly in a small interval of time or a sudden change occurs with infinitesimal probability. But in a finite period of time due to the accumulation of infinitesimal probabilities, finite changes occur with finite probability.

This seems to be the basis of all human actions for every man assumes he is almost certainly going to live during the next small interval of time though he is absolutely certain than he has to die in a finite period of time! Life itself is a stochastic process, the evolution of which depends on the events that happen during small intervals of time — the number of sudden changes or events are infinite in number in a finite period of time though the number of instants of time are uncountable!

Once this concept is clear, we can define complex processes provided we know how to build them up from a basic random process or equivalently break up a complex process into simpler processes.

I was now in a position to study and work on the problem which Bhabha had allotted to me — to consider cascade multiplication in cosmic ray showers as a stochastic process. Only two simple processes were known at that time — the Poisson process and the Furry's basic multiplicative process. In his fundamental paper with Heitler in 1937 which won for him the Fellowship of the Royal Society, Bhabha surmised a Poisson distribution but he knew even then it was not correct and that, since it is a multiplicative process, it had to have some resemblance to the Furry process. Scott and Uhlenbeck had attacked the problem in great detail but some basic mathematical questions had to be settled before formulating the problem rigorously. The infinitesimal probabilities could be postulated only if the process was Markovian and to do this, the state of the system had to be unequivocally defined. How to describe in detail the distribution of a finite number of particles in a continuum of states? This was considered almost pathological and attempting to solve it made or marred the careers of many mathematicians! What was needed was a bold step forward since no adequate literature was available. This was both exciting and frustrating.

At that time I still had not solved the problem of finding suitable accommodation! For a few weeks I stayed with my friend George Abraham in the Y.M.C.A. and then Bhabha arranged for me a single room at the Taj Mahal Hotel, then the best in India — a five-star hotel on international standards. I was awarded a stipend of Rs. 250 per month which was just enough for my transport and pocket expenses, my hotel charges being paid from the funds I received from my generous father. Indian food at the Taj was superb and I enjoyed the delectable breakfasts and lavish lunches. It was at that time I met Professor Vallarta of Mexico who came as a visiting professor to the Institute. His mastery over cosmic ray physics amazed me. He was a charming man with a charming wife and he gave me the secret of domestic harmony — to hold the chair for the wife to sit and then wait for her to serve the food. He also advised me to be generous with compliments particularly in the presence of others. One should not express a difference of opinion with a wife in public! This obviously seems to work — behind every successful marriage there is an assenting husband! Often Vallarta used to accompany me in my small Standard car to the Institute. The most important lesson I learnt from him was that politeness pays not only in domestic life but in public affairs as well! He was a gentleman with a fine sense of humour and conversation with him was by itself an education.

I got interested in the subject of cosmic rays, particularly the study of three dimensional trajectories of particles by considering them as moving on a two dimensional surface which itself is rotating about an axis. Even more interesting was the 'duality' of considering particles leaving the earth and going to infinity and those 'from infinity' reaching the earth. Equally fascinating was the concept of 'asymptotic' trajectories. Indirectly all this had a bearing on a proper understanding of scattering processes in quantum mechanics which start from the 'distant past' ($t = -\infty$) and proceed to the 'distant future' ($t = +\infty$). What then is the meaning finite time ($t = 0$ or finite)? Actually a few years later I found that a quantity which the famous cosmic ray physicist Janossy had plotted as a function of ω graphically was indeed $\cos^3\omega$ which was not noticed in his famous book.

That summer (1948) Bhabha invited me to come to Bangalore where he always liked to spend the vacation, making frequent trips to Ooty. Obviously Bangalore looked cheaper to me than the Taj hotel and I agreed to do so. Moreover my father himself liked Bangalore and so our family and my wife accompanied me. At that time Subbaroya Iyer had rented the upper part of a large house in front of the lovely botanical gardens, Lal Bagh, and he suggested that we could go and spend the summer there. Dr. Bhabha stayed at the West End hotel known for its Western comforts and cleanliness, and Eastern luxury and service. It was a privilege to work so closely with Bhabha and I was able to understand and appreciate him as no other scientist in India could. He was extremely honest in his scientific work, self-critical, at the same time supremely conscious of his powers. He had an extraordinarily analytical mind, with great facility in mathematical calculation, more so in mathematical perception! Though he was not a pure mathematician, he could perceive the mathematical nuances and difficulties in formulating a physical problem in quantitative terms. He was intolerant of vagueness and particularly of pretended knowledge. Though a theoretical physicist, he was able to create and sponsor a school of mathematics in the highly competitive world of creative science.

Our main problem was to describe accurately the probability distribution of a discrete number of particles in continuous energy space. It is too well-known that we cannot put a discrete number in correspondence with a continuum of states. How then could we label the particles with energy? Bhabha's first reaction was to think of the energy-space as made up of a discrete set and make the packing denser and denser and invoke the concept of density of states as in quantum mechanics. In considering correlation, this involves products of such densities at two or more energies. What would happen if they coincide? This led us to some erroneous conclusions and so we obtained at first the obviously wrong result that there was no correlation! Bhabha was in a dilemma — his immaculate intuition suggested that there was correlation, since particles of different energy can be traced to a common parent. The limiting process in mathematics apparently suggested that there was no correlation. So he came to the most logical conclusion: mathematics cannot go wrong, his intuition cannot lead him astray and so we must have made a mistake in the use of mathematics! Like a true scientist he accepted the situation. We were closing in on the problem. Our initial formulation was right, our method was effective but we were applying it in a wrong way.

Our stay in Bangalore was one of the pleasantest holidays we had. Lalitha was with me for such a long period for the first time after our marriage. I had brought the Standard car form Bombay and we went on an excursion to the lovely Shimsa valley and the Sivasamudram falls. We were caught in a cyclonic storm on our way back and we reached home safe with a prayer in our lips.

We invited Bhabha for 'tea' at our house and he graciously accepted our invitation. To me it was a great excitement but to my father it was just one more distinguished visitor to his house. I tried to impress upon my father how different in outlook scientists were from businessmen and with the leniency of a loving father, he smiled at my nascent faith in scientists. Bhabha told my father that we were making good progress in our problem and that I had a capacity to think on my own!

When I returned to Bombay in August 1948, I still did not have any permanent accommodation. My father had spoken to his distinguished friend K.M. Munshi,[13] the famous lawyer in Bombay. He and his wife were kind enough to invite me to stay with them for a few months. The Munshis were gracious hosts and they were particularly kind to me in view of Munshi's great regard and friendship for my father. I stayed in their lovely house in Malabar Hill for a few months. Since I had sent my car to Madras, I was going to the Institute by bus or sometimes by walk. I enjoyed the Gujarathi food in their house, particularly the 'phulkas', almost weightless dry chappathis eaten with delicious salad and vegetables. Their cook was called 'Maharaj' and he was cleanly dressed in spotless white khadi in keeping with his master's standards. Munshi used to talk to me about the role of science in new India, of how useful advanced physics could be in its economic regeneration. He was bouncing with the energy of youth and I loved to hear his frank views on men and matters.

Before my stay with the Munshis, just prior to my visit to Bangalore, when I realised that the Taj hotel serenade could not go on indefinitely, I stayed for a few weeks with hospitable friends at Matunga, the 'Mylapore' of Bombay. The long drive in my Standard car through the crowded highways of Bombay was enough to convince anyone that India had to tackle its population problem before it can hope for a perceptible increase in the average standard of life. Before the end of the year, my long wait for accommodation had borne fruit. I was allotted an apartment right behind the Taj Hotel, a rather spacious one, with one room being given to my friend George Abraham. That December (1948) I went to Madras and I brought to Bombay fine rosewood furniture which I ordered at the annual exhibition in Madras. Lalitha accompanied me to Bombay and so we started 'independent' living on my father's liberal allowance! After a stay of only two months in Bombay, Lalitha returned to Madras. I would see her during my frequent trips to Madras from Bombay.

December–January is the high season of cricket in India. We witnessed the Test Matches between India and the West Indies both at Bombay (November 1948) and in Madras (January 1949). Rae, Stollmeyer, Walcott and above all Everton Weekes drove and glanced, hooked and pulled their way to victory to the thunderous delight of the impartial crowds. While Bombay loses its heart to the king of games, Madras loses both its head and heart, as banker and businessman, professor and student, official and clerk, judge and lawyer, leave their office and factory, home and school, to pay raving homage to the sun-tanned knights of the flashing willows.

It was in early 1949 that I suddenly realised the error we had made in the limiting process which gave us the wrong result that there was no correlation between particles in different energies. It occurred to me that instead of assuming a discrete set of energy states, we should divide the energy range into N intervals making the sub-interval smaller and smaller as N tends to infinity. In doing so if we find that the situation is just like that in continuous time space when we consider events in a small time interval. We assume that in such a case an event occurs with probability proportional to the small interval. Could we not assume similarly that the event here corresponds to the occurrence of a particle in a small energy interval and this is proportional to the interval? The occurrence of two particles in the interval is proportional to the square of this infinitesimal interval and therefore can be neglected. Thus the problem suddenly simplifies itself as the interval becomes smaller and smaller, only one particle can be accommodated with probability proportional to the size of the interval and the no particle case that no event occurs in a small time interval occurs almost certainly i.e. with probability approaching 1. Thus we come to realise that in an infinitesimal energy interval, there can only be one or none. This then is the meaning of density of particles which in a sense is a misnomer since the contribution to the mean density comes from only one particle. This immediately led me to the formulation of product or density correlation functions. These are the functions we should obtain by multiplying the mean values in the discrete case with the density of states but this direct approach obviates the confusion of a limiting process and gives the correlation function correctly. It gives the rule that when two energy intervals coincide there is a contribution coming from the degeneracy.

In my new found enthusiasm and excitement I wrote down my results and submitted them to Professor Bhabha. He himself was quite excited about the problem and was making frantic attempts to make a break-through. During this period he was under the influence of Professor Kosambi[14] who unfortunately had little comprehension of the physical problem and less of its mathematical formulation. Bhabha felt that my method was too simple to be good — but the fact was it was simple enough to be very good! But he arrived very soon at the same conclusions as I did but I was unaware of this development in Bhabha's work. One day in early June 1949, I was having a casual conversation with his private secretary Puthran who in good faith informed me that a nice reference was made about me in a paper which Bhabha had just communicated to the Royal Society. I could not understand the meaning of this statement since I was under the impression that Bhabha was not convinced of the correctness of my method. When I saw the typed version of the paper it was clear to me what had happened — by a long and arduous mathematical process, Bhabha had arrived at the same result as mine, but he was under the impression that his was a rigorous derivation, and the mathematical result was of such importance, independent of its physical application, that it was suitable for the Proceedings of the Royal Society. In that paper he wrote a note that a joint paper with me will follow on the applications of his results. But I was sure that my derivation was correct and if the result was good enough for Bhabha and the Royal Society, my derivation should be really important.

I was now faced with Hamlet's dilemma — should I assert myself and get my result published, perhaps incur Bhabha's displeasure and face all the ensuing consequences in a country where freedom in scientific thought was not an established concept, or should I be satisfied by faint praise, — a reference in Bhabha's work that the applications would follow in a joint paper? I remembered the advice of my father: one can succeed even if one is honest — it may take a longer time but it is more satisfying! Bhabha was a great scientist, but it was a tragic fact that he sincerely believed that my derivation was too elegant to be correct! To me it was a struggle between conscience and discretion and it was conscience that won the battle. I had thrown away a legal career which was open before me, I had given up a civil service position which was obtained by hard work for the competitive examinations. I was therefore willing to risk the displeasure of Bhabha now, hoping to convince him one day that I was right.

Within two hours of my seeing his paper, I was on my way to the Bombay airport to catch the plane to Madras. I knew it would be a very difficult thing to explain to my father that so early in my career, even before I obtained my Ph.D, I was confronted with the almost insuperable difficulty of convincing my guide and hero, Professor Bhabha. It was a strange way to begin an academic life.

My parents felt happy and surprised to see me. It was then I realised the profundity of emotion and sentiment with which our epics were charged. No one except a father would be generous enough to receive a prodigal son — one who had spent his funds to stay in five star hotels in search of a career which brought him to this predicament. I had 'burnt my boats' as my father said. It was as difficult to go back as to go over and so I decided to carry on the struggle. Strange that I should take comfort from the Ramayana on the one hand and Churchill's speeches on the other!

Why should I be worried about the delay in my career? At fifty eight, Marlborough had still his famous victories to win. His illustrious descendant Winston Churchill, at the age of sixty, resigned himself to the position of a tired old man. Six years later he responded to the call of duty — and what a call! For five years he led his country and the world through travail and despair to total victory. And what is more, his greatest achievement came later — the writing of the immortal history of the second World War in four volumes. What a chicken-hearted, pigeon-livered person should I be to yield to doubt and despair when I knew my result was right and I could vindicate myself. I had the financial, moral, and intellectual resources to do so. I decided to seek justice under fair and impartial professors in English Universities,[15] steeped in the tradition of scientific ethics and academic freedom.* I found the men and cleansed my soul (Appendix 4).

In England, my eight page paper, prepared with the advice of D.G. Kendall of Oxford (later professor at Cambridge) was communicated by Professor M.S. Bartlett (University of Manchester) to the Proceedings of the Cambridge Philosophical Society in February 1950 and published in August 1950. Professor Bhabha's twenty two page paper was received at the Royal Society in July 1949 but revised in March 1950 and published in its proceedings in August 1950. Our joint paper was communicated to the Proceedings of the Indian Academy of Sciences in March and published in September 1950 (see Appendices 4A and 4B).

The numerical coefficients in the expansion of the moments in terms of integrals of correlation functions, which were left unidentified by Bhabha, were derived explicitly in my paper. I named the correlation functions "product densities" which has been adopted in later literature. D.G. Kendall who formulated the functions independently in the age distribution problem in population growth, had called them "cumulant densities".

Notes

1) In addition to the Hindu Law Gold Medal, Ramakrishnan won the K. V. Krishnaswami Iyer Prize for standing first in the examination on the Law of Practice and Procedure held by the Bar Council. The Prize was a fine book based on a series of lectures on Professional Conduct and Advocacy delivered in 1939 by K. V. Krishnaswami Iyer to apprentices at law; it was published in 1940 to great acclaim. K. V. Krishnaswami Iyer was not only an eminent lawyer, but also played a role in public life.

2) Ramakrishnan's father-in-law, H. Subramani Iyer, was Professor of Astronomy and Principal of the Maharajah's College in Trivandrum. Starting from 1927, he was the Government Astronomer at the Trivandrum Observatory. In August 1939, the observatory was transferred to be under the charge of the University of Travancore. The Trivandrum Observatory came into the limelight when Subramani Iyer spotted a new comet (1941-C) on the morning of 23 January, 1941.

3) Although Ramakrishnan had enthusiastically agreed to marry Lalitha, Sir Alladi wanted some background information about Lalitha and her family before giving his consent (as was the custom in those days). So he consulted Semmangudi Srinivasa Iyer, a leader in the world of Carnatic music, who at that time was *Aasthana Vidwan* (court music scholar) in the court of the Maharaja of Travancore. Semmangudi was a tenant in Subramani Iyer's home for a few years when he first arrived in Trivandrum, and so he knew Lalitha's family very well. In recommending Lalitha to Sir Alladi, Semmangudi said: "If this girl enters your house, it will be like lighting a wonderful lamp in your mansion!" (This sounds better in Tamil!!) That recommendation of Semmangudi convinced Sir Alladi to proceed with the marriage of Ramakrishnan to Lalitha.

4) Sir S. Varadachariar was a contemporary of Sir Alladi as a lawyer in the Madras High Court and a close friend. He was appointed as a Judge in Madras in 1934 and soon elevated as a Justice of the Federal Court in Delhi in 1939. The Federal Court, along with the Judicial Committee of the Privy Council, was combined in 1950 to form the Supreme Court of India. Sir S. Varadachariar retired as a Justice of the Federal Court in 1946, the same year Ramakrishnan stayed at his residence in Delhi.

When Sir S. Varadachariar left Madras to Delhi to take up his assignment as Justice of the Federal Court, it was Sir Alladi as the Advocate General of Madras, who delivered the felicitation address (see [7]).

5) The original Drafting Committee formed in August 1947 was B. R. Ambedkar (Chair), Alladi Krishnaswami Iyer (Ex-Advocate General, Madras), K. M. Munshi (Ex-Home Minister, Bombay), N. Gopalaswami Iyengar (Member of Nehru Cabinet), Sir B. L. Mitter (Ex-Advocate General, India), Mohammad Saadullah (Ex-Chief Minister of Assam, Muslim League), and D. P. Khaitan (Lawyer), with Sir Benegal

Narsing Rao serving as the Constitutional Advisor. Mitter resigned from the Committee. He was replaced by N. Madhava Rao (Ex-Diwan of Mysore). In 1948, Khaitan passed away. He was replaced by T. T. Krishnamachari, famously known as TTK.

Sir Alladi also served on The Rules Committee, The Advisory Committee, The C. A. Functions Committee, The Union Powers Committee, The Union Constitution Committee, and The Ad-hoc Committee on the Supreme Court (see [3]).

6) B. R. Ambedkar was a famous lawyer of Bombay. He was also a great social reformer who campaigned against the discrimination of the untouchables. In an article entitled "Dr. Ambedkar and Constitution Making", the well-known social essayist and activist Madhu Limaye points out that Ambedkar was not picked for the Constituent Assembly in the first round in December 1946, but only later in 1947. Subsequently, when the Drafting Committee of the Constitution was formed, and Sir Alladi was invited to that Committee, Ambedkar was appointed as the Chairman of the Drafting Committee. Sir Alladi was placed on the Drafting Committee because he was unrivalled in the knowledge of constitutional law. Ambedkar was surprised even being included in the Drafting Committee because as he said in his concluding speech to the Constituent Assembly, his limited aim initially was to serve the interests of the Scheduled Caste (= untouchables). With characteristic humility Ambedkar said that he was even more surprised to be made the Chairman of the Drafting Committee when there were "men bigger, better, and more competent than myself such as my friend Sir Alladi Krishnaswami Iyer". Ambedkar served admirably as the Chairman of the Drafting Committee.

7) Deccan Airways was formed in 1945 by the Nizam (= ruler) of Hyderabad with a small fleet of the reliable DC-3 Dakotas purchased after WW II at throw-away prices. Ramakrishnan was the first in his family to fly in an airplane — a few months earlier on that same Deccan Airways flight from Madras to Delhi. Deccan Airways along with six other private airlines were all absorbed in the government operated Indian Airlines when it was formed in 1953.

8) This "love" of Ramakrishnan for Bhabha, was deep admiration in the academic sense, just as G. H. Hardy of Cambridge University said that his association with Srinivasa Ramanujan was "the one romantic incident" of his life.

9) Harish Chandra is widely regarded as the greatest Indian mathematician after Srinivasa Ramanujan. He started out in physics and was a student of Homi Bhabha before the days of the Tata Institute when Bhabha had the Cosmic Ray Unit at the Indian Institute of Science in Bangalore. While he was with Bhabha, he was introduced to certain questions that the great physicist Paul Dirac at Cambridge University was working on. So he went to Cambridge University to get his PhD under Dirac's supervision. In that process he became interested in certain fundamental mathematical questions and so became a pure mathematician. In the sixties, he was appointed as a Permanent Member (Professor) in the School of Mathematics at the Institute for Advanced Study in Princeton. By the time Ramakrishnan joined Bhabha, Harish-Chandra had left, but they were to interact later (see Part II).

10) Even though India is a democracy, owing to strong socialistic beliefs, and also because political parties rely on votes from the masses who are poor, many laws are against the land owner. Bombay is an extreme example in India where owners of property are helpless in dealing with tenants.

11) B. V. Sreekantan, a former Director of the TIFR, in an essay on the first sixty years of the Tata Institute (1945–2005), talks about the days at Kenilworth; he states that the TATA Institute was given 6000 sq. ft. at Kenilworth. Sreekentan and Ramakrishnan overlapped during the period when the TIFR was at Kenilworth.

12) The Dirac hole theory postulates that in the continuum of energy states filled with electrons (which are negatively charged), there are holes filled by particles with energy of opposite sign, namely positrons. Positrons were experimentally confirmed and Dirac was awarded the Nobel Prize in 1933. In discussing the formalism of Feynman graphs, Ramakrishnan had an alternate interpretation of positrons several years later.

13) In addition to being a famous lawyer, Munshi was also an activist in India's freedom struggle. He founded the *Bharatiya Vidya Bhavan*, an educational trust, in 1938. He had served on the Drafting Committee of the Indian Constitution along with Sir Alladi. Here is what Munshi said of Sir Alladi in his book "Pilgrimage to Freedom", Bharatiya Vidya Bhavan, Bombay (1967), Vol. 1, p. 96:

"Alladi Krishnaswami Iyer, the most eminent lawyer in the Constituent Assembly, scrupulously restricted himself to the specialist's role of a legal and constitutional expert. He was a combative lawyer.... In our endless discussions with him, therefore, we were forced to explore every point of view thoroughly in order to meet his objections. His industry was untiring, his knowledge of law massive, and his subtlety keen as a razor's edge. He had the photographic memory of a Brahmin with a long ancestry of Samhita-Pathis, the reciters of the vedas. His long experience as Advocate General of Madras was also invaluable in what may be called 'applied constitutional law'! In personal relations Alladi was uniformly cordial; but if at any time any of us put forward an erroneous point of law, he would jump upon us with the ferocity of a hungry man-eater!"

14) D. D. Kosambi was a mathematician who worked in the area of differential geometry. He joined the Tata Institute in 1946 after spending time at universities in Aligarh, Benares, and Poona. The School of Mathematics at the Tata Institute began to take shape only from the Autumn of 1949 with the appointment of K. Chandrasekharan (KC) as its head. In September 1949, the TIFR moved out of Kenilworth to the Yacht Club where there was significantly more space available, and there KC started the TIFR mathematics program. Ramakrishnan left TIFR a few months before the move to the Yacht Club. So in early 1949 at Kenilworth, Bhabha consulted Kosambi regarding Ramakrishnan's work.

15) It is very difficult to go against the establishment in any field of activity. In Ramakrishnan's case, his new idea was not appreciated by Homi Bhabha, who was at the top of the Indian academic scene. There was no way that Ramakrishnan could be rescued in India from the difficulty that engulfed him. His only recourse was help from overseas. It was lucky that he got this timely help (see next chapter) and got his work published in the Proceedings of the Cambridge Philosophical Society around the same time that Bhabha's paper appeared in the Proceedings of the Royal Society (see Appendix 4). This dilemma of Ramakrishnan could be compared with that of the great mathematician Paul Erdös who along with the equally eminent mathematician Atle Selberg gave an elementary proof of the Prime Number Theorem in 1949. Erdös and Selberg were supposed to write a joint paper, but they had a misunderstanding and so they wrote separate papers. The Annals of Mathematics accepted Selberg's paper but the Bullettin of the AMS rejected Erdös' submission. Luckily for Erdös, a certain member of the US National Academy of Sciences came to his rescue and communicated his paper to the Proceedings of the National Academy of Sciences. Thus the papers of Erdös and Selberg appeared simultaneously and independently in 1949. In the case of Erdös, he was a famous mathematician who had a lot of contacts, and so he could pull himself out of the hole. In Ramakrishnan's case, he was a mere student going against a leader of Indian Science. Interestingly the drama between Ramakrishnan and Bhabha was being played out around the same period of the Erdös-Selberg controversy (1949)!

It was of course the most exhilarating moment of my life when I first saw her (1946).

Lalitha

My wife, who was very young, accompanied me to England.

Chapter 11

The Manchester Experience
(1949–51)

In the years following the dawn of independence, post-graduate students started going to England for higher studies and research and everyone was ambitious to get a degree from a British University. But the thought of settling abroad never occurred to such students. It was just a desire to acquire a foreign degree even amidst the spartan conditions in England that followed the harrowing years of war, and return to their country seeking better jobs. In my case I had no alternative but to leave for England for I was convinced that my work was right, that it was an original contribution which though not spectacular, would be considered an enduring part of modern probability theory. I had put my heart and soul into it, I wanted to get my work published whatever be the cost in time or money. My ambition was to get it ultimately accepted by Bhabha who was my mentor and hero in every sense of the term. Influenced by some advisers around him, he was not able to perceive the essential simplicity of my derivation in contrast to his long drawn limiting process. It was a challenge which I could not resist though anyone in my position would have been irreparably frustrated. I had given up a career in the civil service by voluntary surrender and turned my back on an affluent legal career which obviously lay before me. I had spent two years in Bombay, rather strenuous for a person used to the leisured comfort of Madras and the pleasures of family life. However, the desire to prove that I was right dominated my mind to the exclusion of every other thought. My loving mother understood the depth of my passion and the strength of my resolve.

My father's generosity was without bound. In his office room he heard my case in great detail and blessed me in my resolve to vindicate myself. It was going to be a very costly vindication for I did not have any prospect of a fellowship in those spartan post-war years in England. The first thing to do was to get a seat under a famous Professor in England and I suggested the name of Professor Bartlett[1] since I had seen his lecture notes as a visiting professor at the University of North Carolina.[2] I was under the impression that he was at Cambridge and therefore I thought I could fulfil my ambition of establishing my work and getting a Ph.D. degree from Cambridge. I wrote a letter directly to Bartlett enclosing the notes of my work.

Whenever any discussion about an English University arose, my father always wanted to consult Sir Radhakrishnan. So he took me to the great professor of philosophy to whom I explained my desire to publish my work. Professor Radhakrishnan was very sympathetic and wrote to the Oxford Institute of Statistics and obtained for me a seat there. If I had accepted it I could have had the privilege of an Oxford education though I would have to do more statistics than stochastic theory. My heart was in mathematical methods and so I was waiting for Bartlett's reply. It came in due course. He mentioned that he was then in Manchester which was certainly not an attractive city but if I was keen on working under him, I could do so. It was one of the happiest moments of my life for I knew I had the chance to prove myself. The pleasure of struggle with the hope of success lay open before me.

For the next three months, in accordance with the custom of 'London going' Indians, I got prepared for the journey abroad. We were well placed in life in the standards of that time for there were not many millionaires and industrialists then, except the established business houses like the Tatas and Birlas besides of course the Maharajahs and Princes of the native states. My father as a lawyer had reached the zenith of his prestige in the early 1930s and stayed there the rest of his life; yet not a member of our family had travelled abroad except my father for a short period of three months. The question of going to England by air did not occur to me for the Air-India had only just started its services and going by boat was still the normal mode. It was decided that my wife should accompany me, a rather unusual thing since a young man of those days would leave his wife in India while going abroad for higher education! I did not consult my wife who was very young, but just informed her that she was accompanying me! She took this news with a calmness that surprised me. Such calmness was natural for her since her own father and uncle had their Ph.D. education in England and the idea of my going there for a science degree did not sound unusual to her. Though my father was very famous, we were all taught to a life of stark simplicity, unaware of the demands of Western life. We were vegetarians both by tradition and conviction and we wished to preserve our way of life. To be otherwise did not occur to us.

We booked our berths through Thos. Cook and Son by the spacious all-tourist P&O liner, "The Maloja" at 150 pounds for a cabin for two persons! It was considered the most luxurious on that ship for most of the singles cost only 50 to 60 pounds. In three months of shopping, we accumulated an enormous quantity of linen which was packed in three large black wooden boxes. My friends advised me to buy the costliest evening hat at Spencers which was the hall-mark of British status in those days. I stitched woollen gaberdine suits in light shades wishing to conform to the picture of the English gentleman strolling in the fashionable streets of metropolitan London. We knew England had suffered irreparable shock and disaster during the war but still in apparel English Tootal fabric was supreme and Scottish worsted wool the pinnacle of fashion. Every young bridegroom insisted that his father-in-law should buy him suits only from expensive shops where the finest English cloth was sold. We went to Bombay by train — the first class coupe was luxurious in those days — and all through we were excited about going to a new country with an uncharted future before us.

At the port of Bombay everything was in a leisured pace — baggage checking, passports and travel formalities. Time was no concern when one was going on a sea voyage. As the 21,000 ton 'Maloja' left the Bombay docks, we felt very proud of being the first among the post-war graduates to leave our family home for study abroad.

We had very good company on the ship; it is a pity that we could not in later years keep track of the friends we made on the 'Maloja'. Every evening was an excitement and experience for we had to order only boiled vegetables and rice and add spices we had brought with us. The little appetisers, nuts, chips, icecreams, served half-an-hour before meals on the deck, were more inviting than the main course at lunch or dinner! Passing through the Suez was an exciting experience. I could not get over the strange feeling of seeing Africa on the left and Asia on the right. At Port Said, we bought colourful bags from peddlars on boats which we still preserve as a memento of our visit. The journey through the Mediterranean was delightful. At Marseilles we got down and made a short excursion into the city. India had been long under British rule and it had just become free. But we valued Western education and adopted Western manners. We felt unduly proud of having had the opportunity of setting foot on European soil. Some of the more impatient of passengers got down at Marseilles and went by train to Calais and then to Dover.

It happened one day when I was musing on the deck, that a mathematical result occurred to me in a flash — the precise expression for the

nth order coefficients in the expansion of moments in terms of integrals of product densities! It turned out to be perhaps one of the finest pieces of work even in comparison with my later efforts of twenty years. Amazing that it should have happened while just gazing on the blue Mediterranean from the Maloja deck! My desire to meet Professor Bartlett was now irrepressible since I knew that these coefficients could not be obtained through the laborious limiting processes of Bhabha.

We reached London after a few days and the passage through Gibraltar and the breathtaking feeling of entering the Atlantic are unforgettable. At London we shipped our luggage straight to Manchester and drove to the old fashioned and comfortable Hotel Waldorf which charged us about 60 shillings a day (as contrasted with 100 pounds for a day for a similar room now). The first few days were spent in looking at the sights of London, travelling in the 'tube' railway and watching the tall dignified policemen speaking with a Londoner's accent, supremely conscious of the imperial prestige of their historic city.

We took the train to Manchester and reached the village of Flixton about twelve miles from Manchester by taxi which cost us only eighteen shillings. The first evening in the countryside around Manchester was unforgettable. I have relived those moments in contemplation many times at leisure. We took advantage of the system of 'Digs' where land-ladies let out a portion of their comfortable houses to paying guests. We were very fortunate in getting almost an entire house for ourselves while the lady stayed in a small room. This was arranged by my cousin and close friend K. B. Subramaniam who had his engineering education in Manchester.

My first meeting with Bartlett was memorable. He conformed to my picture of the English gentleman — reticent, polite, dignified and confident. I gave him the manuscript of the direct method of formulating product densities. He was convinced that I had done something worthwhile but he wanted to verify it himself and so kept the notes with him. The first few weeks at the Manchester University were spent in getting familiar with the environment. I had heard of Rutherford's days in Manchester and I was anxious to meet Blackett at the first opportunity since he was a professor there. The cobbled streets, the old walls made dark by smoke and dust, cast a strange spell, but I felt happy and proud to start my research career in a famous centre of learning. With such a purpose before me, Manchester did not look a dreary city as it was described to me. We found the long walks and leisurely bus rides to small shops and groceries around Flixton quite enjoyable.

Professor Bartlett sent the notes to his young friend and colleague, D.G. Kendall[3] at Oxford and suggested that I should go and meet him (see Appendices 5 and 6). I went to Oxford with my wife and we stayed in a small private hotel, the Isis. We realised why Oxford was considered a great intellectual centre with the tradition of centuries making it almost sacrosanct. Kendall was exceedingly kind to me and was happy to find me so confident about the correctness of my work. He invited me to a high table dinner at Magdalene College and provided special vegetarian food with solicitude for my sentiments. Besides breathing the learned air over the grassy quadrangles of the Oxford Colleges, we also visited the Morris Oxford factory, an American style enterprise in an English University town.

I returned to Manchester invigorated by the support of Kendall who later sent detailed comments on my work suggesting a suitable notation. The paper was written accordingly by me and communicated by Bartlett to the Cambridge Philosophical Society. Its acceptance was the most significant event in my academic career (see Appendix 1).

That winter (1949) I had an opportunity to go to Edinburgh as a participant to the International Conference on Modern Physics with Professor Max Born and Werner Heisenberg as its leading lights. Edinburgh is a beautiful city, its historic forts and castles well preserved. The organisation of the conference was excellent and Max Born himself supervised the arrangements for the delegates. Among the scientists I met were Professor Pryce, young Freeman Dyson who was just becoming famous,[4] and Klaus Fuchs who later created history by being a spy for communist Russia. What a thrill it was to hear Niels Bohr, the father of modern physics, speak on Complementarity in a special lectures outside the conference! It was like Valmiki explaining the message of the Ramayana![5]

In the main session Janossy,[6] to my pleasant surprise, outlined the now familiar G method and tackled the fluctuation problem of Cosmic Radiation by that method avoiding the well-known difficulties in the formulation. But the calculation of moments in his method had the same complexity as that of my product density technique. At the end of his talk, I expressed my desire to speak and with the permission of the chairman I presented briefly my work based on product densities. In the spirit of a true scientist, Professor Jansossy was so interested that he invited me to spend a week in Dublin (see Appendix 4).

Dublin was then connected to Manchester by a modest Dakota service by the Aer Lingus, the Irish Airlines. I was then not convinced of its safety and so we went by a boat from Liverpool to Dublin.

At that time there was working under Janossy an ambitious young Canadian, Harry Messel, extraordinarily competent in mathematical analysis and computation. We became very good friends and later Messel invited me to Australia when he became the Head of the School of Physics at the University of Sydney.

I was exceedingly impressed by the originality of Janossy and by his generating function method. Inspired by his work, I produced a model in which his method yielded very elegant solutions. Quite independently Bartlett had worked on his model by the use of partial differential equations which could not easily be solved. Bartlett found my solutions 'pretty' and I knew it was high tribute (see Appendix 4) being familiar with the English tendency for understatement. These solutions were published in a paper in the Proceedings of the Royal Statistical Society. Soon Bartlett and Kendall found that the Janossy method was anticipated in a different context by Bellman and Harris in their paper in the Proceedings of the National Academy of Sciences in the United States. We drew the attention of Richard Bellman who in turn became interested in the product density methods. Thus it looked at the end of six months after my arrival, the material for my Ph.D. thesis was ready and I had to spend a year and a half more, merely to conform to the University regulations.

In the spring of 1950 I went to Scandinavia but Lalitha stayed behind in Manchester. Arley of Copenhagen had just written his book on the stochastic problem of cosmic radiation unaware of the recent developments. I visited Bohr's Institute in Copenhagen and the Universities of Stockholm, Uppsala and Oslo meeting Oscar Klein, Ivar Waller and Hylleras. I made an excursion to Lillehamer, the ski resort a few hundred miles north of Oslo. In Stockholm I visited Cramer's Institute and I remember the discussions with Oscar Klein on the mathematical foundations of statistical mechanics. My close and critical familiarity with Khinchin's book helped me later to work on the one-dimensional theory of fluids and publish a series of papers in the Philosophical Magazine through Neville F. Mott,[7] Nobel Laureate, its editor.

Lalitha and I became members of a local tennis club in Flixton, the suburb of Manchester and by a series of lucky circumstances obtained two special season tickets for Wimbledon, 'under the Royal Box'. We spent the early summer in Europe, visiting France and Switzerland, one of the finest holidays we had, even in comparison with our later trips round the world. From London we went to Paris by train — 'the Golden Arrow' — and it was a strange experience when the entire train was ferried across the English channel!

Paris was the city of our dreams and fulfilled all our expectations. We joined the organised tours and enjoyed strolling on the Champs-Elysees and shopping at random and at leisure in the small perfume shops in the neighbourhood of the Opera house. We attracted considerable attention by asking for water where wine was more plentiful and insisting on peas sauteed in butter in a gourmet's paradise.

We went to every resort in Switzerland, the playground of Europe — Davos, Lausanne, Lucerne, Bern, Zurich, Interlaken, Zermatt and Jungfraujoch along with young American tourists, as part of summer school excursions intended to stimulate interest in European history. It was a starlit night at Jungfrau and as the day broke we heard the tingling of bells on the cows grazing on the hills and dales. In the evening we watched young and old couples dancing according to the tradition of the residents of the mountain village. The music lingers in my ears and the sight of the Matterhorn still haunts my senses.

We came to London in time for the Wimbledon matches and spent a fortnight there. We feasted our eyes and tranced our hearts watching the most exciting display of youth and talent, beauty and rhythm, affluence and dignity, grace and style, suspense and triumph on the lushest, greenest, liveliest piece of earth — the Wimbledon lawn tennis courts. Wimbledon is as English as its monarchy and no wonder it is still the unquestioned capital of world tennis where Americans and Russians aspire to receive the tantalising crown from Royal hands. That was the year when the fast service and volley triumphed over the traditional baseline tactics. Budge Patty, the lanky American, volleyed his way to final victory beating Frank Sedgman, the hope of Australia. Each match was as interesting as the other as we watched Talbert and Mulloy, Drobny and Seixas, Bromwich and Quist, Brown and Sidwell in exciting doubles matches. The match which took my breath away was the doubles, Patty and Trabert, versus Sedgman and McGregor when one of the sets was extended to 31–29![8] I consider Trabert the greatest player in doubles of all time — what speed and power, what grace and rhythm, as he smashed and served delighting the crowds to raving ecstasy. But it was Bromwich and Quist who stole the doubles crown by lobbing and slicing the opponents out of their wits. The women's singles was won by Louis Brough, supreme in grace and style in her ground strokes.

It was indeed a fortunate circumstance that I should have witnessed Wimbledon before I entered the University tennis team during my second year's stay in Manchester.[9] My enthusiasm for the game became a

passion and I enjoyed the daylong league matches on the lawn tennis courts
of various universities like Leeds, Sheffield, Birmingham and Durham. I
became convinced that the main features of British character, toughness
with politeness, discipline with freedom, reticence with friendliness, all emanated from the national love of sport and sportsmanship. The victor of El
Alamien must have played or watched country cricket on the village green
as the shadows lengthened on summer afternoons.

Since I completed the work for my thesis well ahead of time (see
Appendix 5), I had leisure and inclination to attend with Lalitha many
stageshows and plays, usually previews and premieres in Manchester before being presented in London. I remember vividly every scene of 'Antony
and Cleopatra' with Vivien Leigh and Laurence Olivier. We also went on
a week-end excursion to the Lake district, that tranquil haven, untouched
and unchanged since Wordsworth held his communion with Nature. On
that warm summer afternoon, amidst a silence that hit our senses, we felt
the gentleness of heaven brooding over the lakes and woods.

We visited Cambridge and enjoyed the hospitality of Dr. Shanmugadasan, a student of Dirac. We felt the spell of that lovely University town
where the ivy-mantled towers breathe the spirit of Newton and Maxwell.
No wonder Bhabha and Harish-Chandra, Chandrasekhar and Mahalanobis,
sought inspiration from the hallowed haven to stimulate their native genius
and inborn faculties. Even Ramanujan who lisped in numbers had to seek
a sponsor at Trinity to uncover his discoveries.

My thesis was examined by Professors Peierls and Bartlett and as soon
as it was approved we returned to India by air. We flew to Zurich, a city
we loved very much for its beauty and cleanliness. It was at the Hotel
Richmond in Geneva I met Sir A.L. Mudaliar, and expressed to him my
desire to work in the Madras University where he was the Vice-Chancellor.
He received me warmly and informed me that very soon the University of
Madras would be starting a research department in physics and I could
hope for a readership in that department.

Perhaps the most fruitful and enjoyable part of my academic life in
Manchester, besides research and tennis, was my attending lecture courses
in the University. The most valuable were the hundred and odd lectures
on methods of mathematical physics by M. J. Lighthill.[10] He had earned a
reputation similar to that of Dirac in Cambridge. Later he was to succeed
Dirac as the Lucasian Professor at Cambridge. His lectures on complex
variable theory, partial differential equations, asymptotic expansions, were
perfect, as enjoyable as a musical symphony. He was a ball of fire but with

the warmth and comfort of a Christmas hearth, always friendly and helpful, but intolerant of inaccuracy and ineptitude.

Rosenfeld's lectures gave me an insight into the 'inner shells and hard core' of nuclear physics and stimulated an active interest in basic quantum mechanics.[11] He was an associate of Bohr and from the Great Master imbibed the essence of humanism, characteristic of a true scientist.

Equally enjoyable were the conversations over lunch almost every working day with two Indian friends G. Krishnan and M. Santappa who were working for their Ph.D. in zoology and physical chemistry. The subsidised lunch plate cost only a shilling and six pence, a practical demonstration of plain living and high thinking in British Universities. It is a curious coincidence that all the three of us joined the Madras University with great hopes and small salaries, soon to find ourselves 'exiled' to Madura on promotion as professors. Frequently on weekends, Lalitha enjoyed providing them with South Indian food and Madras style coffee.

We returned to India in April 1951 by the Air India super constellation, that leaping beauty of the pre-jet era.[12] Back to Madras with thankfulness to God, to the life with my parents amidst the comforts of our family home.

Notes

1) M. S. Bartlett, a highly reputed statistician, was at the University of Manchester in England. He was widely known for his work in multivariate analysis, stochastic processes, and applications of statistics to genetics. He wrote a number of influential papers and books. Among his honours are the Guy Silver (1952) and Gold (1969) Medals of the Royal Statistical Society, Fellowship of the Royal Society — FRS (1961), and his election as Foreign Associate to the US National Academy of Sciences (1966). Bartlett was appointed as Professor at the University of Manchester in 1947.

2) The notes that Ramakrishnan read was: M. S. Bartlett, "Stochastic Processes", Notes of a course at the University of North Carolina, Chapel Hill (1946).

3) D. G. Kendall (FRS) was a former PhD student of Professor Bartlett, and perhaps his most distinguished disciple. He was initially appointed as a mathematics tutor at Oxford University in 1946, but later was appointed at Cambridge University as a Professor. Kendall is widely regarded as the Father of British Probability. He served on Ramakrishnan's Phd Committee which was chaired by Bartlett.

4) Freeman Dyson was born in 1923, the same year as Ramakrishnan. He first made a splash in number theory as an undergraduate student at Cambridge University when he gave a combinatorial explanation for two of Srinivasa Ramanujan's famous congruences for the partition function. Dyson published this in the Cambridge undergraduate journal *Eureka* in 1944. But as a graduate student, Dyson shifted to physics and came under the influence of the great Hans Bethe at Cornell University. In describing the 1949 Edinburgh conference, Ramakrishnan talks about "a young Freeman Dyson who was just becoming famous". In 1949 Dyson published a rigorous proof of the equivalence of the two versions of quantum electro dynamics proposed by Feynman, and by Schwinger and Tomonaga. This brought worldwide fame to Dyson. Subsequently, Ramakrishnan found an alternate and simpler way to understand this equivalence (see Appendix 7, Part II).

5) Sage Valmiki, the author of the Ramayana, is often referred to as *Adi Kavi*, which means the very first poet. Many eminent poets of later generations offered salutations to Valmiki who was their inspiration. Likewise, Niels Bohr is considered the founding father of atomic physics having proposed a model of the atom. So to Ramakrishnan, listening to Bohr was like listening to Valmiki.

6) Lajos Janossy (1912–78) was a Hungarian physicist who made important contributions to several areas such as astrophysics, nuclear physics, and quantum mechanics. He applied statistical methods very effectively. He worked on cosmic ray showers and that aspect of his work which he presented at Edinburgh conference intersected with Ramakrishnan's research. At that time Janossy was at the Dublin Institute for Advanced Studies as the group leader of the cosmic ray research laboratory.

7) The *Philosophical Magazine* founded in 1798 is one of the oldest English scientific journals. It has a hallowed tradition with legendary physicists like Michael Faraday, James Clerk Maxwell and Lord Rutherford publishing some of their work there. Sir Neville Mott (FRS), as Editor of the journal, accepted Ramakrishnan's papers for Phil. Mag. Mott won the Physics Nobel Prize in 1977 (along with Philip Anderson and J. H. van Vleck) for work on the electronic structure of magnetic and disordered systems.

8) Those were the days before the tie breaker was introduced in tennis. The precursor to the tiebreaker was VASSS, the Van Alen Streamlined Scoring System (introduced by Jimmy Van Alen) to avoid very long tennis matches. In a refined form, VASSS became the present day tie breaker that introduced in the US Open in 1970. Even though Wimbledon nowadays uses the tie breaker, in the fifth set in men's tennis at Wimblesdon, the tiebreaker is not allowed.

9) Ramakrishnan was a fine tennis player and was especially strong in doubles. He was on the University of Manchester tennis team as a doubles player. In Madras, Ramakrishnan played tennis regularly at the Mylapore Club close to his home.

10) Sir James Lighthill (FRS) was one of the most eminent and productive applied mathematicians in England. He held the Beyer Chair at the University of Manchester during 1946–49 when Ramakrishnan was there. Lighthill appreciated Ramakrishnan's work on product densities. He was an acknowledged authority on aeroacoustics and fluid mechanics. His recognitions include the Royal Medal (1964) and the Copley Medal (1998). In 1964 he was appointed Royal Society Resident Professor at Imperial College, London. Later Lighthill served as Lucasian Professor at Cambridge University. Ramakrishnan and Lighthill continued to interact during the fifties and sixties and hosted each other in Madras and in England.

11) Leon Rosenfeld, a Belgian physicist, was a collaborator of Niels Bohr. He was appointed Professor of Theoretical Physics at Manchester in 1947, and so Ramakrishnan was fortunate to attend his lectures there. After serving in Manchester, Rosenfeld moved to Copenhagen to work with Bohr.

12) Although Air India was founded in 1932 by J. R. D. Tata, its international operations began only in 1948 with the acquisition of a Lockheed L-749 Constellation named the Malabar Princess. Unfortunately, the Malabar Princess was destroyed when it crashed into Mont Blanc in November 1950. In 1951 when Ramakrishnan and Lalitha flew back from England by Air India, it was a very young international airline. Given the fact that Air India lost the Malabar Princess a few months earlier, it was a bold choice to fly back on Air India instead of taking a steam ship — a choice due to Lalitha.

My Mentors

H. J. Bhabha
Source: Image by Konrad Jacobs, Erlangen - Oberwolfach Photo Collection,
https://commons.wikimedia.org/wiki/Category:Homi_Jehangir_Bhabha#/media/
File:Homi_Jehangir_Bhabha.jpg

M. S. Bartlett
Source: The Making of Statisticians
(J. Gani, Ed.). Springer, New York
(1982), p. 41.
Permission granted by Springer.

D. G. Kendall
Source: Image by Mathematisches
Forschungsinstitut Oberwolfach,
https://commons.wikimedia.org/wiki/
File:David_Kendall.jpg

Chapter 12

The Beautiful Years (1951–53)

I consider the two years following my arrival from England, the most beautiful years of my life, for I was closest to my father, listening to his views on men and matters, hearing him express his deepest thoughts and emotions, while my mother stood like a ministering angel beside him taking care of every comfort, treating every pain and pleasure of his as her own. If I were to live my life again I would like this period to have stretched longer with one difference that my father should have been physically healthier.

On our journey back to India in April 1951, we stayed a few weeks in Switzerland since we loved its scenic beauty and tranquillity, so zealously preserved by its citizens, and so jealously enjoyed by foreign visitors.

Lalitha and I arrived in Madras in May and the first months were spent in recounting our experiences to our friends and relatives. I had met the Vice-Chancellor of the University of Madras, Sir A.L. Mudaliar,[1] at the Hotel Richmond in Geneva and he had informed me that the University was starting a research department in physics and that I could be accommodated as a theoretical physicist though the primary objective was to have an experimental department on Crystallography. It was expected that with the recommendation of Sir C.V. Raman, G.N. Ramachandran[2] was to be appointed as a professor but he could take up the position only in late 1952. I had to go through the normal channels of application which meant a long wait and a vexatious interview before a duly appointed committee. Meanwhile I was expected to wait and watch like a 'chakravaka' bird which eases its thirst from the gentle drops of rain as they fall from the heavens.

My father was naturally very anxious that I should take up a position in Madras for he wanted me to be with him since his health was failing and he needed someone dear to him to be near him all the time. Among his children I was the only one who could discuss with him general topics on men and affairs. He knew that the job as a reader carried a very meagre salary (Rs. 400 a month and no allowances!) just enough to meet the petrol expenses of my car, and barely half a day fee for a moderate lawyer.

I brought from England a new Black Hillman[3] car (the Minx 10) which I used for over twenty years till I shifted my loyalties to the native Ambassador. I bought the Hillman upon reaching England but kept it in a garage there since I felt too shy to use it when my professor was coming by bus and Fellows of the Royal Society by bicycle to the University. The spartan attitude of life during the war continued for many years till Britain tried to emulate the standard of life of its affluent cousin across the Atlantic. But in Madras I enjoyed taking my guests from abroad on frequent trips to Mahabalipuram in the Hillman and my chauffeur Govindan zealously guarded the shine of its paint and sheen of its chrome, emulating the example of the devotion of Soundarraj to father's 'Master Buick'.

Perhaps the greatest mistake, almost an unpardonable 'crime' we committed was in disposing of the priceless Buick after my father passed away in 1953 in view of the high cost of petrol. Had I known that I was one day going to write the Alladi Diary, I would have prevented such an act of obvious folly. For it was the symbol of the glorious era of my father's eminence, his dignity, his rectitude, his generosity, his scholarship and above all his wide humanity. It carried his juniors who were seniors in the profession, it provided him the cushioned comfort for his extra legal reading. Sunderraj, our driver was a broken man, a Cleopatra without her Antony, and we could see the obvious indifference with which he drove the new Chevrolet, though from the same factory, not a match in form, size, beauty and dignity to the Master Buick.

I accompanied my father to Delhi since he had meetings to attend even after the main work of the Constituent Assembly was over. Lalitha came with me and our stay in Delhi could not have been pleasanter. Winter in Delhi is a wonderful season, sunny and cold — made more enjoyable with Krishna Iyer's cuisine and luxury rides in our magnificent Buick.

Delhi was simmering with hope and aspirations after the exit of the British. Nehru was the uncrowned king of India. The largest democracy in the world willingly yielded its power to the glamour and prestige of a single man. I had the opportunity to meet the Prime Minister many times with my father who could not move about without help. Subbanna was always there, all eyes and ears attentive to every need of my father and every syllable of conversation of friends and bystanders.

The idea of having diplomatic missions and embassies was new, and it excited the imagination of the elite and shrewd office seekers. Everyone who had the slightest access to Nehru aspired to be an ambassador with visions of luxurious life in Paris, New York and London. There was also the desire to participate in the counsels of the U.N. and obviously the first chance for such an assignment came to Mrs. Vijayalakshmi Pandit,[4] endowed with all talents and graces for success. I heard her lecture at the Constitution house on the characteristics of a good ambassador.

Politics also became an attractive profession. No longer did it mean arrests, trials and imprisonment. Elections, the imagination of the populace, intoxicated by universal suffrage which gave to the poorest peasant, the unemployed graduate, and unskilled worker, a sense of participation in the high affairs of state. Everyone aspired for a seat in the legislature to make speeches in the language of his choice on any subject he liked from rural sanitation to nuclear arms limitation.

Anyway it was a glorious hour for India and my father was right in the heart of it in the country's capital. I continued my work on stochastic processes since I had just completed a short paper simplifying the derivation of an equation of Chandrasekhar and Munch. It was then I started playing with the problem of random fragmentation of mass, whether it should be treated as happening in a sequence albeit in a short interval of time, or all at once. This led me to the formulation of the problem of 'Random points in a line'. I got so many results that I wrote them up as a paper and sent it to professor Bartlett to be communicated to the Proceedings of the Cambridge Philosophical Society for publication. I was naturally proud and happy when I received a formal letter from the editors that it was accepted for publication — a paper based on work done at home between newspaper gossip and social engagements.

During this period my father was invited to deliver the convocation address of the Delhi University.[5] He prepared the address with his usual diligence and analytic precision but included a few references to science and scientists just to please me. He enjoyed teasing me by referring to 'your country', which meant England and 'your subject' which meant 'science.'

On returning to Madras my father became weaker and could not walk even a few steps without help. He was under the care of Dr. Guruswami Mudaliar[6] who visited him everyday. He would just feel his pulse and discuss politics in an inaudible whisper and my father would be restored to good spirits on such an assurance.

My father was always dependent on doctors and felt thankful to them for the personal attention they gave to him with respect and concern. The first doctor I remember was Rangachari in the thirties who had a legendary reputation as a surgeon in Madras — fair, swift footed and deft-fingered, endowed with a charmed hand and an enchanting smile, exuberant, effusive, exuding confidence and a love of life. He came in his Rolls Royce car which symbolised his excellence, the spotless white cotton covers for the seats revealing a surgeon's cleanliness. I remember how he used to take the cricket bat from my hands and ask me to bowl and how even before I did, rushed to his Rolls Royce, parked in our spacious portico. Madras lost an eminent citizen when he died of typhoid in 1940. It was then Dr. Guruswamy Mudaliar was summoned to take charge of my father. He was quite unlike Rangachari in temperament but inspired by the same high ideals. He was a physician of the first magnitude and was endowed with a sixth sense and just by feeling the pulse, examining the eyes and expression on the face, he could deduce analytically the disease of the patient. He was as erect in his professional standards as in his physical frame.

For many years we had a family doctor who, in particular, took care of my mother. That was Dr. Ramakrishna Iyer, the father of my classmate Natarajan. His most striking quality was cleanliness — the white collar, the well ironed silk suit, and the tie with a perfect knot, the handkerchief in the appointed place, polished shoes clean even on the heels. He propagated cleanliness and hygiene by example. He spoke in a low, clear, halting voice, with a gentle smile. He was available night and day and would never drug his patients with strong medicine unless absolutely necessary. In later life he made it a point to attend all the marriages to which he was invited and shared the joys of happily united families.

We later relied on the service of a younger man, a close assistant of Rangachari, Dr. Narayanaswami, who by sheer honest effort to live up to his master's reputation, rose to success in his profession. He had an orthodox tuft of hair and therefore used a turban in cosmopolitan society. He died in the prime of his career and later we had his assistant K. Srinivasan who for many years took care of my mother till she left us.

Our family has been very fortunate in its family doctors, physicians and surgeons and I developed a great respect for the medical profession. It is not given to everyone to enjoy perfect health and 'the thousand ills which the flesh is heir to' can only be cured or alleviated by doctors.

Every morning I used to read The Hindu and the Indian Express to my father. He enjoyed my style of reading the topics he liked — first, the headlines, then domestic politics, about Nehru and his all powerful popular team, of the complex relations between Pakistan and India, of local speeches, and finally the headlines on the financial page. Subbanna was always there with his comments on men and affairs. Any illness looked lighter in his presence which was so reassuring. He would confirm the diagnosis of the doctors and even alter the dosage of the medicine at his discretion!

Three things engaged my attention till I got the job in the University — travel with my father to Delhi, lecturing at various institutions propagating theoretical physics, playing tennis at the Mylapore Club and in local matches. Since I played for the Manchester team, I became a doubles player with the volley as my main stroke. I gave up the forehand top spin and changed from the Eastern forehand to the Continental grip, more favourable for a back hand volley. Sedgman was my hero with his flexible midcourt volley, the stroke next to the serve or a forcing drive. Savitt won the Wimbledon in 1951 and Sedgman had to wait a year before winning the crown. I used to play four sets a day at the Mylapore Club admiring the elegant court craft of the ageless wizard Balagopal.

In March 1952, I was appointed as a Reader in the University. I was then indifferent to the meagreness of the emoluments and was convinced of the dignity of academic position in English tradition where even an F.R.S. could be a Reader in Cambridge and Oxford. I felt I should act as a self appointed reformer emphasising the dignity of a scientist, trust in colleagues, pride in work, and indifference to material gain as the characteristics of true academic life.

On the first day I went to the University in my new Hanava (not Havana!) suit with cotton manufactured by the B and C Mills, specially stitched for the first day of my work. I was of opinion that trust should be the basis of life and so I left my purse in my coat which was hung on the stand in my room and moved around, only to find that the entire money was stolen with the purse intact! The British tradition of keeping newspapers in a stand where buyers would just place the change and take the newspapers would not work in Madras. The newspapers and the cash will both be missing.

I started lecturing at every college in the city like an inspired missionary. At the University I gave a hundred lectures based on Lighthill's course on 'Methods of Mathematical Physics' and gathered my own audience by discreet persuasion — elderly teachers to young students. In the beginning there were a hundred eager listeners. Slowly as I got deeper into the subject, particularly using complex variable theory, the audience reduced to a few by the principle of natural selection, and among them were P. M. Mathews and S. K. Srinivasan who later became my students.

In 1952, we heard of the first jet passenger service by deHavilland Comet, London to Johannesburg, on the 2nd May, when 6724 miles were covered in less than twenty four hours.

Among those who frequently called on my father, and whose company he enjoyed, was Seshachalapathi, or 'Seshu' as he was lovingly called, the son-in-law of Sir S. Radhakrishnan. Seshu was an ardent admirer of my father's legal genius, his natural simplicity and his regard for the higher values of life. Father adored Seshu for his love and mastery of the English language, his sparkling wit and humour, his caricature of the foibles of the high and the mighty. He was a master at mimicry and could imitate the trenchant wit of C.R., the measured diction of Srinivasa Sastri, the Homeric eloquence of Churchill, the Oxford accent of Anthony Eden, the emotional excess of Nehru, and the casual confidence of Krishnamachari. My father and Seshu seem to have been made for each other enjoying the same brand of humour, the same power of expression, with the same estimate of men and affairs. I remember Seshu, with characteristic irony saying that the British Government had to blame itself for introducing English in India which instilled through its literature a desire for democracy and freedom in Gandhi and Nehru. Otherwise India would have been involved inextricably in religious dissensions, caste prejudices and language controversies that British rule would have been ensured for centuries to come. Father used to tease Seshu that like Sir S. Radhakrishnan, he was a *Niyogi*,[7] too conscious of decorum and dignity that he would not readily accept a cup of coffee unless offered thrice like Caesar's crown while the Alladis were just simple *Vaidiks*[7] easily pleased and easier humoured.

The other frequent visitor was N. Rajagopala Iyengar, my father's able assistant in law, known for his incisive reasoning, unrelenting accuracy and mastery of constitutional law. Father used to speak of him so highly to everyone and to C.R. every time he met him, so that when the occasion came, he was elevated to the bench at the Supreme Court. N. Rajagopala Iyengar repaid his master's generosity by writing a most touching, accurate and analytical, biographical account of my father, one of the finest of its kind ever written, in the High Court Centenary volume (see [3]).

Subbaroya Iyer, my father's dearest friend, supplied him a hospital cot which could be adjusted so that he could rest or recline in any angle. He would visit him at all hours, without notice, discharge some exciting news or message, and feel my father's pulse with one hand and write rapid notes with the other while consulting him on a difficult question of law. He would vanish before Krishna Iyer brought him coffee, leaving father with his spirits boosted, but longing for his company at his next unappointed visit. My father lay in the drawing room with its four doors wide open overlooking the spacious garden and providing a glimpse of the Luz Church Road. He loved people from the most humble servant and poor relation to the exalted Maharaja or the Prime Minister. He was lying there like Bhishma in full possession of his mental faculties, but infirm in body, giving legal opinions on important matters to his established clients. He had to get his exercise in bed only by moving his limbs and so we summoned one Krishnamachari who taught Yogasanas and breathing exercises. I remember his telling my father that 'Arogyam' or good health can be acquired by human effort while 'Aayas' or the span of life, is determined by the will of God and the two need not be closely correlated. It was incredible how my father kept to the exercise schedule with enthusiastic rigour.

Every evening my father used to hear the Ramayana which was read to him by Subramanya Sastrigal, a venerable Pandit from the Sanskrit College. He would call me frequently and impress upon me how the Ramayana is relevant to the modern age. I understood the depth of his statements years later, more so under the stress and strain of modern life.

Since he knew he could not move about, he was anxious to be reminded about those who played a role in his career and in his life. We arranged a 'portrait gallery' with pictures of his great associates, V. Krishnaswami Iyer, P.R. Sundaram Iyer, Sir S. Subramania Iyer and K. Srinivasa Iyengar and of Devar[8] who helped him with his school fees, Professor Kellet of the Christian College, C. Rajagopalachari and Sir Radhakrishnan, his intellectual peers and personal friends, and Dr. Guruswami Mudaliar, the doctor who felt his pulse everyday.

It was then that Nehru, the Prime Minister of India, visited Madras and we received a phone call that he wished to call on my father in the evening. My father suddenly realised that we did not have a portrait of Nehru in our 'gallery' and so I immediately rushed to get one, have it framed and hung on the wall.[9] It was a portrait of the great leader in a pensive pose 'carrying half a billion problems' of the Indian population in his handsome head. Nehru called on father and unconsciously lapsed into a similar mood while talking to him. On seeing his portrait he smiled as he saw the image of his present self in it!

On December 9, 1952, on U.N. day, the All India Radio wanted my father to give a talk. This was recorded at home. A savant who should have been participating in the U.N. debates was now confined to bed, physically infirm, and had to be content with a recorded speech. A voice which had won a thousand legal battles was now enfeebled by time and fate and had to be amplified by an electronic device.

He wanted to write his memoirs but he had not kept a diary and not one of his near relatives who had lived on his largesse, had the ability or inclination to provide the incentive. So he dictated the recollections of his early life to me and to a young junior who wanted to get inspiration for entry into the profession by contact with him. It was then that I realised how many valuable years I had missed in getting his advice and imbibing his ideals when he was at the height of his powers, how his frail frame carried such a mighty brain, how he earned also in an absent minded fashion, enough to keep all of us in great comfort throughout our lives. He built magnificent houses for all his four daughters. He acquired enough landed property which was later eroded by socialistic legislation. He had supported the education of hundreds of relatives from school to college who now are flourishing in various professions. He read voraciously and his learning in law astounded even the leaders of the profession. He did not use any adventitious aids or attempt to establish legal contacts for practice. He could not help being a legal genius and he could not control the prosperity that resulted therefrom. He was generous in his support of many deserving causes. There was no flourish or fanfare in his munificence. It was personal, warm, and friendly, with a deep concern for the well being of the recipient. Ekamra Nivas was an open house and everyone had his share in his bounty though he never expected anything in return.

There was one who matched him in generosity and that was my mother. She never lost her temper though aware of the faults and foibles in others which she ignored with natural grace. She knew no pleasure other than serving him night and day with a smile on her lips.

Perhaps the only effective manner in which I could express my affection and gratitude to my father was by informing him of international events through the newspapers. The year 1953 was a significant year in world affairs. On the 5th of March occurred the death of Stalin — incredible that a person who saved and claimed, controlled and directed, the lives of millions on earth and guided the destiny of the world through the greatest war in human history, should breathe his last. It is the inscrutable, inexorable law of nature, that no Alexander or Napoleon, Newton or Einstein, prince or pauper, tyrant or saint, can escape.

In May came the news of the conquest of Mt. Everest by the New Zealander Edmund Hillary and the Indian Sherpa Tenzing Norgay in an expedition under John Hunt. With characteristic promptitude, Britain honoured its citizens by conferring knighthood on Hillary and Hunt. Tenzing became a national hero and the idol of the mountaineering world. My father listened to the news with considerable interest and I remembered how years before he read to me the classic work of Sir Francis Younghusband entitled 'Everest — the challenge'. Weak in health, he could admire the spirit of the mountaineers, so well expressed by Edmund Hillary who, when asked why he wanted to scale the Everest, replied 'just because it is there'.

On the 30th May, Eisenhower turned over the European command to General Mathew Ridgway. He became a candidate for the Presidency and on November 4th that year was elected president with Nixon as Vice-President, defeating Adlai Stevenson, the democratic candidate.

That year the Nobel Prize for literature was awarded to Sir Winston Churchill and I was one of those who felt that he was a man of the century though I was aware of his unreasoning, obdurate antipathy to India. Next to Shakespeare, he was synonymous with the power, vitality, flexibility, resilience and above all the durability of the English language. He was bred so much in the British tradition of Imperial glory, that he could not see a future for India beyond its place in the British empire. To him a Hindu was a bundle of superstitions and there was no way in which our great literature, the Ramayana and Mahabharatha, the Vedas and the Upanishads, could reach him, since he was by choice and circumstance insulated from the rest of the world by an implacable faith in the primary role of the Anglo-Saxon or English speaking peoples in guiding the destiny of the world. Even men like Rutherford felt that Cambridge was the 'centre of things' though a few decades later the centre of gravity of the scientific world shifted to the United States. The same insularity which was the strength of Britain and its pride also proved to be the source of its weakness and consequently Britain not only lost its Empire, but yielded its pre-eminent place to the English speaking nation across the Atlantic, which, as a haven for immigrants was willing to receive and foster talent and initiative from all over the world.

On his seventieth birthday May 11, 1953, my father dictated a message which is the starting point of this diary.

What a pleasant surprise when I received an invitation from my great friend Harry Messel[10] of Sydney, inviting me to spend a few months as a visiting scientist in his rapidly growing department of theoretical physics there! This was in pursuance of the 'Dublin pact' between us which we entered into during our first meeting in the winter of 1949 as research students working for the Ph.D. degree when we had agreed to invite one another at the earliest opportunity! My father was overwhelmed with joy for it was the first evidence I had given him that the career of a scientist had its own rewards though not financially comparable with those of the legal profession. He did not live to see me fulfil the academic assignment, which was unprecedented in the career of an Indian scientist at that time.

On the second of October father complained of uneasiness in his stomach and Guruswami Mudaliar came in and assured there was nothing to worry about. My father spent a sleepless night and the uneasiness continued next day but we thought it was still the stomach trouble. I played tennis in the evening and returned home early. We were discussing loudly in the 'koodam' over coffee when our servant Bakka uttered an unearthly cry that father was no more. We summoned the doctor who confirmed the sad news. I burst into tears and could not control my grief. All his friends and relatives thronged to Ekamra Nivas to pay their last respects.

It is impossible to think of Mylapore and Luz Church Road without my father in Ekamra Nivas. He was the unquestioned leader of the Madras Bar for forty years, the most eminent jurist of his time in India. His name had become was synonymous with legal learning and forensic genius, and he had gained household currency. What would my mother do without him? It was impossible to think of so tender and gracious a woman without her husband. It was God's will and no one can resist His summons.

To me he is a living force and there is not a moment of action when I do not think of him as guide to my will, not a moment of repose which is not enlightened by sweet memories of his voice and his presence. The rest of my life will be a trek along the road he had shown us in pursuance of principles he stood for. He had demonstrated that the pursuit of wealth and prosperity is consistent with honesty and rectitude, that wealth should be shared with those whom you love and spared to those who are in need of it. Generosity and friendship must be part of our nature, frankness and candour the spirit behind our actions. It is criminal to waste time, it is a life of action, of ambition and achievement. We serve our country best by exalting our spirits, cleansing our thoughts, sharpening our talents and enlightening our minds. He never preached any of these qualities but practised them, and was not conscious he was doing so.

To India he left a constitution, to the legal world his reputation, to his children an inheritance, to his friends an honoured memory, and to me the driving force of my life.

Notes

1) Sir Arcot Lakhmanaswami Mudaliar was a leading obstetrician and gynaecologist. He was the longest serving Vice-Chancellor of the University of Madras (for 27 years). He played a leading role in the World Health Organization (WHO) and therefore often visited Geneva, where Ramakrishnan met him in 1951.

2) G. N. Ramachandran (1922–2001) — popularly known as GNR — was an internationally reputed physicist who worked primarily in crystal physics and crystal optics. He was appointed as Professor and Head of the newly formed Department of Physics at Madras University in 1952; Ramakrishnan joined the Physics Department in March 1952 as a Reader a few months before GNR arrived. GNR's main interest was to start a program in crystallography. Ramakrishnan's interests in physics were quite different and so there was not much interaction between them as physicists.

3) The Hillman Minx was a mid-sized British made family car manufactured from 1931 to 1970. The car Ramakrishnan bought was a 1949 model soon after he arrived in England. The Hillman Minx was a rugged and popular car and enjoyed sales worldwide.

4) Vijayalakshmi Pandit (1900–90) was the younger sister of Jawaharlal Nehru. She too was a very successful diplomat and politician. She became the first female President of the United Nations General Assembly in 1953.

5) During the Convocation of the University of Delhi in November 1951, two persons were awarded Honorary Doctorates: The President of India (Babu) Rajendra Prasad, and Sir Alladi for his contribution to the drafting of the Indian Constitution. Sir Alladi was also invited to deliver the Convocation Address on that occasion (for some excerpts, see Appendix 3B; for the full speech, go to [1]).

6) Guruswami Mudaliar (1880–1958), was one of the most revered medical doctors of Madras. He was a master of diagnosis and considered an expert on the technique of *percussion diagnosis*, where by tapping the body one could diagnose the problem. He has been honoured by having a statue of him installed on the campus of the Madras Medical College. He was a close friend of Sir Alladi and regularly attended on him. A portrait of Dr. Guruswami Mudaliar was put up in the drawing room of Ekamra Nivas, by Sir Alladi in gratitude to the doctor.

7) Smarthas are Brahmins who worship all Gods of the Hindu religion with equanimity. The Smarthas of Andhra Pradesh can be broadly classified as *Niyogis* and *Vaidiks*. The former group were very successful in administration whereas the latter primarily pursued the priestly profession, like Sir Alladi's father.

8) Venkanna Devar was a prosperous landlord who was a neighbour when Sir Alladi lived in the George Town section of Madras. Sir Alladi's father was a priest and was not wealthy. Once in school, when Alladi could not pay his school fees to continue his education, Mr. Devar immediately gave young Alladi the money he needed. Sir Alladi was grateful for the timely help and kindness of Devar. He had an enlarged portrait of Devar put on the wall of the Drawing Room at Ekamra Nivas. In "My Story" (see [1]) Sir Alladi acknowledges Devar's help more than once.

9) Ramakrishnan rushed to the Luz Corner, a busy shopping junction near Ekamra Nivas to get a photograph of Nehru. When he could not find a photograph of Nehru, he grabbed a calendar carrying a picture of Nehru in a store and had that calendar picture framed.

10) Ramakrishnan first met Harry Messel in Dublin in the winter of 1949 when he went there at the invitation of Professor Janossy to deliver a lecture on stochastic processes. Messel at that time was working under Janossy. Messel and Ramakrishnan became very good friends and had common research interests on cosmic rays.

Chapter 13

The Hallowed Years (1953–56): Australia and a Gift Therefrom

Our primary duty after the loss of our father was to take care of our mother who lost the will to live in the absence of her husband. It was heart-rending to see her in such a predicament for, true to her name Venkalakshmi, the sacred Goddess of happiness and wealth, she was the light of Ekamra Nivas. We felt as Arjuna did at the Nirvana of Krishna, that with the demise of my father, we had lost the charioteer, guide, minister, philosopher, friend, relative lord, guru, protector and source of our wealth and prosperity. I was drowned with grief but my duty was clear — to be at my mother's side and carry on the ideals my father stood for.

The years that followed were hallowed by her saintly presence and I had the privilege of spending considerable time with her. She was such an enlightened woman endowed with a sweet understanding of human nature and an awareness of human affairs that I could talk science, politics, and international affairs freely with her. I had to dispel her sense of loneliness and unconcern for life. She was herself frail and weak though a mother of many children. All her physical ailments came to the fore after the loss of my father. It was as though her prayers held them back so that she could be of greater service to him.

I used to inform her of the news in the newspapers and explain to her the scientific and technological progress that was altering the pace and even the texture of our lives. Ritual did not have any special hold on her. Religion enlightened her life and she was totally free from pride and prejudice. Gentleness was part of her nature, tolerance her way of life though she did distinguish actively but politely, between right and wrong. I remember her telling me once that a relative of ours had such a vicious tongue that we have to blame ourselves to get into conversation with him!

She felt very proud and happy that I was invited to Australia and on my part I felt it a pleasure to tell her the nature of the lectures I was going to deliver there at the invitation of my friend Harry Messel. She had always been on my side in any discussion with my father about my career. Once she felt bold enough to tell him that he should not worry too much that a scientific career in India implied meagre salary since he had earned enough as a lawyer so that I could enjoy the luxury of a research career! Certainly an irrefutable argument, since my standard of life had been sustained not by my earnings as a scientist but my father's bounty.

The two years when I was close to my mother turned out also to be the most fruitful in my research career. Whatever I touched became gold and I was able to complete a paper every month, perhaps not exciting research but certainly satisfying to an ardent votary of a scientific career. With my first student Mathews I had completed a comprehensive paper on Cascade theory which was published in the Japanese journal *Progress in Theoretical Physics*. In March 1954 I followed this up by a paper entitled "Simple models on cascade multiplication". On the basis of these papers I had the privilege of being offered the contract to write a book on Cascade Theory for the Pergamon Press[3] by Professor Heitler,[2] the joint architect of Cascade Theory[1] with Professor Bhabha. I accepted this with enthusiasm and started preparing the material for the monograph.

I also started working on stochastic problems in astrophysics and completed a series of papers which was published in the Astrophysical Journal of which Professor Chandrasekar was the editor (Appendix 9). A curious situation arose once when the referee insisted that the infinite series solution in my paper should be proved equivalent to that of Professor Chandrasekhar and Munch.[4] I was aware that there can be no two distinct infinite series expansions for the same function and therefore it should be possible to derive one expansion from the other by algebraic manipulation. So when a new student, S.K. Srinivasan expressed his desire to work with me, I allotted him this task which he did satisfactorily and therefore I was able to achieve two things at one stroke, the acceptance of a paper and the admission of a new student. I was elated that Chandrasekhar accepted paper for the Astrophysical Journal, since there was a reputation that he was uncompromising in standards even to his countrymen!

My mother was so kind and generous that she wanted me to go to Australia in response to Messel's invitation despite her weak health when she needed my constant presence. She was also particular that Lalitha should accompany me for she felt that a wife should be with her husband sharing his thoughts, feelings and emotions, as she did in her own life.

Our voyage from Bombay to Australia was made in the P&O[5] liner 'Stratheden'. We reserved a comfortable first class cabin and my brother and his wife from Poona came to Bombay to wish us bon voyage. The ship stopped at Colombo for one day where we did some sight-seeing and enjoyed a delightful lunch at the Mount Lavinia Hotel.[6] Our next stop was the beautiful city of Perth in West Australia where every house looked pretty with its exquisitely manicured lawn and sparkling sprays from swirling sprinklers. The journey throughout was made in good weather and we enjoyed relaxing on the deck under moonlight watching the restless waves and listening to lilting music. I was a runner-up in a table-tennis tournament, beaten in the finals by a young businessman from Sydney. We shook hands and became friends like Robin Hood and Little John after a doughty duel. At Adelaide we made a short excursion by bus and it was a great excitement to go through the residential area where the legendary Don Bradman lived. We disembarked at Melbourne and travelled by the fast train 'The spirit of progress' to Sydney. We were received at the station by Harry Messel himself and photographs were taken by the newsmen. We were taken to Kings Cross where a furnished apartment was reserved for us.

I had quite a busy time at Sydney lecturing on stochastic processes and the notes I prepared at that time became the basis of my later article in the Handbuch der Physik. It was at that time I arrived at the interpretation of the integrals of random functions which had applicability to various fields. Professor Messel was interested more in Cosmic Radiation than in the theory of stochastic processes. He developed the Monte Carlo technique and later wrote the monumental work based on the computer calculations.

For Lalitha the stay in Sydney turned out to be even more significant than to me. It was eight years since we were married and we were anxious to have children but for inexplicable reasons, Lalitha was having repeatedly a miscarriage during the third month of pregnancy. It happened again in Sydney when she was under the care of one of the greatest surgeons in Australia — Sir Angus Murray.[7] A surgical operation was performed to remove a defect causing the miscarriages and he assured us that we would soon have a child and a year later Krishna was born.

After her period of convalescence, Lalitha and I attended tennis matches at the White City Stadium. That was the time when Ken Rosewall was eighteen and Lewis Hoad was nineteen, making a bid for world championship. There were also Mervyn Rose and Ashley Cooper and we attended whole day matches for a week which convinced us that the game of the big serve and the bold volley had come to stay. I played plenty of tennis on the excellent courts of the Sydney University. On many evenings we watched the sail boats on Rushcutter's bay and sea planes land on water and take off from the sea for New Zealand. The Sydney skyline is dominated by the famous harbour bridge and we never got tired of strolling near it. The young businessman with whom we made friends on our journey by ship, took us out on excursions to various beaches for which Sydney is so famous.

We went to Canberra, the lovely capital of the island continent. There we were treated to the most generous hospitality by the Indian High Commissioner General Cariappa.[8] I lectured in the Statistics department headed by Professor Moran, a friend of Professor Bartlett. I also spent time with Sir Mark Oliphant,[9] an old associate of Rutherford. Since he was planning to visit India under the auspices of the Royal Society, I invited him to Madras to deliver the Rutherford Memorial Lecture.

Prior to our return to India we went to Melbourne at the invitation of Professor Belz of the Statistics department. Our plane could not land at Melbourne airport which was blanketed by fog. It landed in a small airport called Mangalore and our slow journey by bus through dense fog still stays in my memory. We had an excellent vegetarian dinner waiting for us at the house of Professor Belz. It is considered bad manners to ask for soup twice but the onion soup tasted so good — like Sambhar — that we almost made a meal of it! My lecture in the Statistics department was well received. I addressed a gathering of Indian students one evening and found that many students from Kerala with medical degrees had come to Australia.

We took the magnificent P&O liner,[5] S.S. Himalaya (28,000 tons) from Melbourne which called on Adelaide and Perth, and so we took the opportunity to go on short excursions round the cities. We disembarked at Colombo where we spent two days staying at the Mount Lavinia hotel[6] admiring the luscious loveliness of that tropical island. We took the opportunity to watch an elephants tea party and enjoyed a relaxed evening in the botanical gardens. We returned to Madras and I was happy to be back at my mother's side with tales to tell of our pleasant experiences in Australia.

I enjoyed my tennis at the Mylapore club with added zest after watching the Australian quartet — Hoad and Rosewall, Rose and Cooper — in Sydney. At Madras we had the opportunity to watch some first rate tennis — Lennart Bergelin, the blonde Swede with an immaculate style, played with a Japanese Davis cup player Nakano at the Egmore tennis courts. Later Bergelin became famous as a coach of that tennis phenomenon Bjorn Borg who won Wimbledon five times successively. We had also the good fortune to watch Frank Sedgman in Madras after his Wimbledon triumph and before he became part of Kramer's professional circuit.

What a pleasant surprise when I received a letter from Sir C.V. Raman that he was anxious to have me elected as a Fellow of the Indian Academy of Sciences* (Appendices 7C & D) and that he would ask G.N. Ramachandran to propose my name! The pleasure was enhanced since a series of papers by me appeared in the Philosophical Magazine in which the great Lord Raleigh had published some of his important contributions.

In December 1954 we had the privilege of receiving (Nobel Laureate) Professor Dirac[10] and his wife the sister of (Nobel Laureate) Professor Eugene Wigner, at Madras. He graciously agreed to give a general lecture at the Senate Hall and I sent round a circular that he was the architect of the only valid relativistic wave equation of an elementary particle, the electron, imbedding with unqualified success, relativity into quantum mechanics. The hall overflowed with eager listeners and many heard the lecture from their cars as in a drive-in theatre.

He was a legend even in Cambridge which was the womb Nobel Prize winners and Fellows of the Royal Society. Many stories had grown around him of his reticence, that it was an incredible surprise to me when he was quite expressive about his sympathy towards our efforts at theoretical physics and about his desire to appreciate the cultural attractions of our city. When I took him to an excursion to Mahabalipuram, he insisted on walking up the hill in Thirukalikundram, leaving his tired wife behind under the care of my chauffeur. He had read about the eagles that come at the appointed hour to be fed by the priests and indeed they came. Professor Dirac watched with unconcealed delight this legendary scene, now familiar even in tourist folders. On our way back he satisfied his curiosity by watching a village potter exhibit his skill at the spinning wheel, unconscious of the presence of the father of the spinning electron! He equally amused himself watching a barber at work on the roadside, conversing with his customer who was enjoying the open air ritual but obviously incapable of participating in the conversation.

I gave a dinner in honour of our distinguished guest at Ekamra Nivas when Krishna Iyer and Subramania Iyer exerted their skills to satisfy his Cambridge palate with South Indian delicacies. We had prepared 'parallel dishes' with and without Karam (hot chillies) and it was amusing to find Dirac choosing the hot stuff, a preference not shared by his partner in life.[10] I invited a sprinkling of Mylapore elite — a judge, a professor, a businessman, to convince Professor Dirac that Madras was the cultural centre of South India.

I took him to the Presidency College and when out of my uncontrolled enthusiasm I introduced every member of the staff, he politely drew me aside and told me that I should not expect him to remember so many names and carry so many details in his head. After his lecture a young student thoughtlessly asked him a provocative question. The positron, when explained as a hole in a negative energy sea, seems so simple an idea, why is it that it had to be formulated by a Nobel prizeman? To this he patiently explained the nature of evolution of physical theories. A physical phenomenon is explained by a new mathematical formulation leading to additional results which in turn should be interpreted physically. These may look simple but one should have boldness or madness to assert them and what is more, the truth has to be borne out by experiment. In turn I explained to my mother this incident in great detail. She had a keen appreciation of the spirit of science and the desire for achievement. Moreover she had a great admiration for the true British tradition as exemplified by the Cambridge scientist — habits of reserve, devotion to duty, shyness of publicity, punctiliousness in work, modulation of aesthetic inclinations, art providing pleasure, not distraction from work.

My contact with Dirac created in me an irrepressible desire to understand the transition from Pauli to Dirac matrices in a logical manner; this I did twenty years later leading me to higher dimensional anti-commuting matrices and even ω-commutation, the Clifford Algebra and its generalisation, comprising my book 'L-Matrix Theory or the Grammar of Dirac Matrices' published by Tata-McGraw Hill in 1972 (see Part II).

A natural consequence of my visit to Australia was Professor Oliphant's acceptance of our invitation to deliver the Rutherford Memorial lecture in Madras in early 1955. To our pleasant surprise we found he was a strict vegetarian and there was no problem in meeting his needs through delicious south Indian dishes. Professor Oliphant had his lecture written up and was planning to deliver the same in Madras and Bangalore. I was unaware of this and spoke to the Hindu representative with such emphasis about the importance of the lecturer that he called on Professor Oliphant who inadvertently handed him a transcript of the lecture. The next morning when Professor Oliphant was in Bangalore he found it disconcerting that the Hindu, as a tribute to the greatness of the scientist, had published the entire lecture verbatim in four full columns, and what is more Professor Raman phoned to him that he enjoyed reading his speech in the morning newspaper! He had obviously to prepare another version for the Bangalore lecture since Sir Raman was to preside over the meeting!

During this period I worked with my students P.M. Mathews and S.K. Srinivasan whom I used to call the Mellin and Laplace Transforms of the department since they used these techniques so well in our joint work on stochastic processes. They later became professors at the University and the I.I.T. continuing their research effectively with their own groups of students.

In the spring of 1955 we consulted Mrs. John, our family doctor, who informed us that Lalitha will soon be having a child, but she advised her to take great care in her movements. Lalitha took the advice so seriously that she rarely moved out of Ekamra Nivas. By the middle of 1955 Lalitha left for Trivandrum to stay with her parents as was the custom in a Hindu family before the birth of the first child. The happiest moment in my life was when on the 5th of October 1955 my father-in-law informed me by phone that 'little Alladi was born'. In a sense I considered Krishna a gift from Australia since Lalitha could bear a child only after she underwent expert surgery under Sir Angus Murray — an excellent example that the grace of God is made available through human agency.

Among my friends Harry Messel was the first to be informed of the happiest event in my life. His reply was characteristic of his affection and friendship for me. I rushed to Trivandrum to share with Lalitha and her parents the joy of Namakaranam ceremony when we named the child after our father and decided to call him Krishna.[11]

It was during December 1955 when my mother was very weak and ailing that we had a distinguished visitor to Madras, Professor C.F. Powell.[12] She knew of the importance of the man and my esteem for him and asked me to take care of the guest without bothering about her. I took him for dinner at the roof garden of the Dasaprakash when during our conversation I told him of my love of Shakespeare, he immediately burst forth reciting the sonnet — "Since brass nor stone nor earth nor boundless sea, but sad morality oversways their power, how with this rage shall beauty hold a plea whose action is no stronger than a flower?" I mused sadly how true it was of my mother who was then too frail and weak to fight against the unrelenting hand of fate. Within a few days she became unconscious and left us on the 5th of January 1956. I have never known a woman more womanly than my mother, so gentle in speech, so soft at heart, endowed with all the graces that a woman could wish for, beauty, a famous and devoted husband, a large family, never conscious of her earthly possessions or her social status, ever conscious of the immanence of God and of His will that directs our lives.

The happiest day of my life was 5th October, 1955 when Krishna was born.

Notes

1) Cascade Theory here refers to the theory of electronic showers — originally proposed in 1937 simultaneously and independently by Bhabha-Heitler and Carlson-Oppenheimer. Both treatments assume certain approximations without full justification. Subsequently, in 1938, two Russian physicists L. Landau and G. Rumer provided a rigorous treatment.

2) Walter Heitler (1904–81), a German physicist, made fundamental contributions to quantum electrodynamics and quantum field theory. Through his theory of valence bonding, he brought chemistry under the fold of quantum mechanics. With Homi Bhabha, he proposed the theory of electron cascades. He held the Directorship of the School of Theoretical Physics at the University of Zurich from 1949 until his retirement in 1974. Ramakrishnan interacted with Heitler while he was in Zurich. Heitler invited and encouraged Ramakrishnan to write a book on Cascade Theory for the Pergamon Press.

3) Pergamon Press, based in Oxford, England, was founded in 1948 by Paul Rosbaud. It initially had close ties with Springer. Robert Maxwell (also known as Captain Maxwell) acquired the company in 1951 as the major share holder and changed the name to Pergamon Press. At the suggestion of Heitler, Rosbaud and Maxwell offered the contract to Ramakrishnan to write the book on Cascade Theory for the Pergamon Press. After Maxwell died in 1991, Pergamon Press was acquired by Elsevier.

4) Guido Munch worked with the great astrophysicist Subrahmanyam Chandrasekhar at the famous Yerkes Observatory outside Chicago in the 1940s. Munch hailed from Mexico. He later became a faculty member at the prestigious California Institute of Technology (Caltech) and was associated with the Mt. Wilson and Palomar observatories. Munch invited Ramakrishnan to Mt. Wilson in 1956 and also in 1962.

5) The Peninsular and Oriental Steam Navigation Company (P&O) was a British shipping and logistics company formed in 1835. After World War II, the travel market to Australia boomed and P&O had 15 large ships built for operations to Australia. Subsequently P&O started operating cruises. P&O Cruises are now operated by the Carnival Corporation.

6) Mt. Lavinia Hotel, is one of the oldest and most famous in Ceylon (now Sri Lanka). It opened in 1947, and so was quite new when Ramakrishnan and Lalitha stopped there enroute to Australia. The hotel has featured in several Hollywood movies.

7) Dr. Sir Angus Johnston Murray (1896–1968), an obstetrician and gynaecologist, was an outstanding figure in the Australian medical profession. He was one of the founders of the Australian Medical Association of which he was the President from 1964 to 67. He worked for many years at the North Shore Hospital in Sydney.

8) Field Marshall K. M. Cariappa, was the first Indian Commander and Chief of the Indian Army, the post being given to him in 1949. After his retirement in 1953, he served as the Indian High Commissioner to Australia and New Zealand until 1956.

9) Sir Mark Oliphant (1939–2000) was an Australian physicist of international repute. He did his doctoral work at Cavendish Laboratory of Cambridge University under the direction of Lord Ernest Rutherford, and investigated heavy hydrogen nuclei and discovered the nucleus of helium-3. In 1937 he was elected Fellow of the Royal Society and in the same year appointed as Poynting Professor at the University of Birmingham. He returned to Australia after World War II to be the Director of the newly formed Research School of Physical Sciences and Engineering at the Australian National University in Canberra, which is where Ramakrishnan met him for the first time. He later was Governor of South Australia (1971–76) and had Ramakrishnan was his personal guest in the Governor's Residence in 1973 (see Part II).

10) P. A. M. Dirac (1902–84) was a legend in the world of science. He won the 1933 Nobel Prize in physics along with Erwin Schrodinger. He was the first international

visitor to the newly formed department of physics at Madras University. Dirac and his wife were on a round-the-world tour in 1954–55 and they spent four months in India (13 Oct, 1954 to 21 Feb, 1955). Mrs. Dirac detested spicy food but Dirac loved it. So it was prudent of Ramakrishnan and Lalitha to offer the Diracs both spicy and non-spicy Indian food when they dined at Ekamra Nivas.

11) While it is common in Hindu families to name a child after the grandfather or grandmother, it is rare to do so when the grandparent is alive. If a child is named after a grandparent who happens to be alive, then that name is the official name in the records, but the child will be called by a different name. Krishna was the first paternal grandchild born after the demise of Sir Alladi and so was named after him.

12) C. F. Powell (FRS), a British physicist, won the Nobel Prize in 1950 for developing the photographic method to study nuclear processes, which resulted in the discovery of the sub-atomic particle pion. He received his PhD from Cambridge University in 1927. From 1928 he was at the University of Bristol, where he subsequently was Melville Wills Professor of Physics.

My wife and I with Krishna, Mathura, and his daughters Lalitha and Amritha in Florida. My grand-daughters lovingly presented me a book on Einstein on Father's Day.

Chapter 14

Yukawa Hall and Round the World (1956)

On February 16, I left Madras for Calcutta by the night Skymaster DC-4 service. The DC-4 was the 'elder brother' of the DC-3, the workhorse of the airways, and the Indian Airlines had bought a few such sturdy, well-proven, weather-beaten, survivors of WW II for carrying the night airmail.

On arrival at the Dum Dum (Calcutta) airport after a noisy but uneventful journey, I found that Pan Am had made arrangements for my accommodation at the Grand Hotel, a symbol of colonial affluence and Victorian luxury amidst the squalor and poverty of the teeming city. In the evening I boarded the Pan Am DC-6B,[1] the intercontinental of the Douglas series. A journey through the Far East brings an excitement which has only to be experienced but cannot be adequately described. After a short halt in Bangkok, we landed in Hong Kong Kaitak airport and were taken for lunch to the opulent Peninsula Hotel[2] with its characteristic British colonial atmosphere of leisure and dignity.

Hong Kong was breathtakingly beautiful — the shimmering sea, the sunkist mountains, the sprawling bay and the teeming city. No wonder it was the setting for one of the greatest love stories, 'Love is a many splendoured thing' portrayed so well on the screen (1955) by Jennifer Jones and William Holden.

In the evening I arrived at Haneda International Airport, Tokyo by Pan Am DC-6B to be received by Dr. Kotani, a young physicist and a son of a Supreme Court judge. We were taken in a Buick limousine to Kokusai Kaikan, a new hotel which was the last word in creature comforts. The girls clad in Kimonos served coffee and it was an interesting exercise explaining to them my vegetarian needs. Tokyo is in a sense more exciting than European cities. The hustle and bustle have only to be seen to be realised.

I left for Kyoto by a comfortable train watching the over-cultivated countryside with interest. I was taken to the Rakuyu Kaikan or the faculty club, the main attractions of which were the wholesome vegetarian food prepared specially for me and the warmth of the Japanese baths.

My seminar on 'Probability aspects of Cascade theory' was well attended and it was a thrilling moment when Professor Yukawa,[3] the Nobel Prize winner, entered the hall and sat in front to listen to my talk on Cascade theory along with other professors Hayakawa and Nishimura. Since I was one of the first visitors to the Yukawa Hall supported by the Asia Foundation, I met Mr. Sheldon, its representative in Kyoto, and I remember clearly the Japanese style vegetarian lunch with him at Professor Kobayashi's house. Among the friends I made in Kyoto was Dr. Ueno who was interested in astro-physical problems. He had read my papers in the astrophysical journal so closely that he had an exaggerated idea of my knowledge in that subject! He was soon surprised at my ignorance of various experimental devices when he took me round the observatory! I remember the pleasant evening when Ueno took me out and provided an excellent vegetarian dinner at the restaurant Alaska. From the sixth floor, Kyoto at night looked like a fairy land.

My first contact with field theory started in Kyoto where I noticed among Japanese scientists a great admiration for Feynman who had visited that country a year before. Ziro Koba was an expert on quantum mechanics but he so frequently described Feynman's formalism as 'difficult' to understand that I was nervous to start the study of the subject till I heard Feynman himself talk about it at Caltech during my visit a month later. I was soon to learn that the Japanese use the words 'difficult', 'complicated' and 'ingenious' in a weaker sense than in modern English usage.

I had many exalting experiences during my stay at the Yukawa Hall. Professor Powell who had won the Nobel prize in 1950 for his work on cosmic radiation, visited Kyoto briefly at that time and gave a magnificent lecture on 'new particles'. In his honour was arranged a seminar in which Sakata announced his now famous model. We did not realise at that time that it was going to be the primitive source of Gell-Mann's formulation of unitary symmetry which dominates physical thought today.

Since I was supported by the Asia Foundation, there was a formal function in which a cheque towards my fellowship was presented to Professor Yukawa by the Kyoto representative of the Asia foundation and photographs of this appeared in the Kyoto newspapers.

On April 22, I bade good-bye to Professor Yukawa and left for Tokyo the next day by the express train 'Tsubame'. I stayed with the Kotanis and enjoyed their generous hospitality in their impeccably clean Japanese style house.

I boarded the Pan American Clipper 'Golden Gate', a big bellied Stratocruiser,[4] for Honolulu. The flight was very comfortable at twenty three thousand feet above sea level. I enjoyed the dinner at midnight over the Pacific but suddenly early morning I woke up to see the sign 'Fasten Seat Belts' and became exceedingly nervous since we were only half way to Honolulu. The plane landed in Wake Island on an airstrip beside a beautiful lagoon. This island stands sentinel as the symbol of American power and prestige in the Pacific. Giant Globe Masters were parked like taxi cabs. We crossed the International Date Line and landed in Honolulu earlier than when we left Tokyo! Hawaii was then an American possession, soon to become the fiftieth state of the union in 1959. The Honolulu airport was full of American tourists and I had to take the flight to Los Angeles which was delayed by an hour. The microphone was not working in the room where I sat and I almost missed my flight.

After a comfortable flight by Pan American, I arrived in Los Angeles to be greeted by a tall, straight-backed handsome young man, Dick Bellman.[5] He drove me to Santa Monica in his new Packard car. Thus started a friendship which grew deep in substance and in strength in years to come. I stayed at the Broadmoor Hotel, modest but very comfortable, where I was provided a room with bath for six dollars per day! Next day I visited the Rand Corporation and Dick asked me what I wanted to do most in Los Angles. Instead of asking him to arrange a visit to Hollywood or Disneyland, I desired an interview with Feynman! Dick took me out to dinner at a fabulous restaurant where mounds of fruit and cream were served in exotic bowls.

The most interesting day of my entire academic career was when I met Feynman[6] at Caltech. Dick took me to Pasadena through the famous Californian freeways. Feynman is an amazing personality and I took courage to talk to him about his version of quantum electro-dynamics. He spent over three hours with me and gave a really first hand account of his interpretation of the positron as an electron travelling back in time. To hear Feynman talk on his work was like listening to Shakespeare explaining the soliloquy of Hamlet, the chastity of Desdemona, the jealousy of Othello, the passion of King Lear, the charm of Cleopatra or the oration of Antony. However, while I admired Feynman's genius, at the end of the discussion

I was excited about the possibility of reinterpreting his concept of 'travelling back in time' as really 'tracing back in time' in the light of my own familiarity with stochastic processes and inverse probability.

Years later I was able to complete some comparative studies on such concepts in evolutionary, stochastic and quantum-mechanical processes. Moreover, it was clear to me that in spite of the total triumph of Feynman's graphical formalism, three questions still remained to be studied — the structure of the Dirac matrices, the meaning of virtual states, and imbedding spin in a relativistic equation. On these also I was able to make some contributions in later years by taking an entirely new view point.

I met Guido Munch, a collaborator of Chandrasekhar in a work which inspired me to write a series of papers in the Astrophysical journal. He took me to Mount Wilson one day, where I met Otto Struve[7] observing through the telescope. It was like watching Don Bradman[7] at the nets! At Rand, I met Ted Harris[8] who invited me to his home where his wife asked so many questions about Hinduism and Hindu habits that I found it difficult to keep pace with her interrogation. During the weekend I relaxed by taking the Greyhound to Santa Barbara along the Pacific Highway. This lovely little town was soon to become the venue of the most beautiful campuses in the world with a location exciting the envy of even established Universities.

I left Los Angeles for Chicago by the T.W.A. Super G Constellation flight. On the route the plane did not fly very high and since we had clear weather, I was able to see every mile of the American west being covered — The Grand Canyon, the sprawling desert lands sculptured by timeless winds, the towering Rockies dividing the continent near the mile-high city of Denver. At Chicago I changed planes and went to Rochester where I attended the High Energy Physics Conference for four days. The entire panorama of High Energy Physics was unfolded in a flash. The conference had a terrific impact on me who till then was a complacent votary of British institutions. I found that the centre of gravity of creative science had shifted from Europe to the United States. The affluence and the academic freedom of the American university campus were attracting even the most nationalistic scientists from Europe for permanent residence in the United States. Besides these, a new brand of American born, American trained, physicists had gained leadership in the mathematical sciences. Gell-Mann, Goldberger, Feynman, Chew and Low were all there at the conference organised by Marshak, the architect of the 'Rochester conferences' which were held later by rotation at various centres of the world.[9] The young Chinese physicists Yang and Lee were there preparing their minds for their great discovery of parity nonconservation as they heard Dalitz formulate the famous tau theta puzzle. Feynman almost took time by the forelock when he posed the question: what if parity is not conserved in weak interactions? Oppenheimer with his omniscient wisdom called the assembly of elementary particles the sub-nuclear zoo.

It was during lunch time that a fortuitous incident occurred which altered the course of my scientific career. Since I was a stranger to the High Energy group, I sat alone at a separate table in the cafeteria when Oppenheimer walked in and with a politeness, characteristic of true greatness, asked me whether he could join me for lunch at the table. I took the opportunity to express to him my desire to spend a year at the Institute of Advanced Study. Though I was a 'seasoned probabilist', I was only a novice in relativistic quantum mechanics. I wanted to learn from high-powered seminars rather than through the grinding mill of graduate courses! My ambition was to be realised a year later when I received from him a gracious invitation.

At the last day of the conference it snowed so heavily that the flight to Boston was cancelled and so I went there by train. My brother Prabhu and my nephew Babu greeted me at the station and a friend of theirs, Abernathy who later became a Professor at Harvard, drove me to the faculty club. Though I formally registered at the Faculty Club, I spent all my time in the dorm room with Prabhu and Babu. They were grappling with the rigorous graduate courses at Harvard. Watching them at work I felt what a salutary effect, insistence on such high standards of lecture courses will have, on scientific education in India. I also met two students of Schwinger — Baker and Glashow, who later became professors at Seattle and Harvard.

I gave a series of four lectures on Stochastic Processes at M.I.T. at the Norbert Wiener seminar arranged by my friendly host Bayard Rankin. I was invited for dinner by the Italian Physicist Bruno Rossi. He was disappointed that I did not take advantage of the variety of wines from Italy.

From Boston I went to Chicago at the kind invitation of Professor Chandrasekhar. At Chicago Mid-Way Airport I was asked to choose between a limousine and taxi cab service. Not knowing the American usage, I imagined that the limousine service was an expensive luxury and I asked for a taxi which proved to be much more expensive! I stayed at the International Club which was visited by Chandrasekhar frequently. My German friend Friedman was also staying there. I made a 'pilgrimage' to the Institute of Nuclear Study where the first atomic pile was started by Enrico Fermi.

When I met Chandrasekhar or Chandra as friends called him, in his room, I watched his smile and flashing eyes. I heard his lecture on radiation theory and was impressed by his thoroughness in preparing even regular lectures.[10] It was a stormy night when he took me to Williamsbay in his new Buick. The observatory was situated sixty miles from Chicago. He had to pick up his wife at the Drake hotel and he had to drive round the hotel many times since there was no parking space, a characteristic feature of an advanced automobile civilisation.

I stayed in the lovely house where Mrs. Van Biesbrock was running a boarding establishment. I remember the breakfast room clearly because it was so similar to the dining area in my Madras home, Ekamra Nivas. I remember the intimate discussions I had with Chandra and his students, Backus and Prendergast and his colleague Kuiper. It was an exciting intellectual experience giving a seminar on stochastic problems in astronomy before Professors Kuiper, Chandrasekhar and Stromgren.[11] I was to renew my acquaintance with Professor Stromgren next year at Princeton.

Another impressive incident related to the lecture at Chicago by Eugene Parker, then a young Ph.D. student. The controversial discussion on the magnetism of the earth convinced me that we know less about the earth under our feet than of the heavens above. I also had discussions on Cosmic Radiation with the celebrated Simpson's group.

I had a curious experience while walking back towards the International House. I was about to be accosted by threatening characters. Discretion proved to be the better part of valour and I ran toward the building and escaped unhurt. My wife was certainly thankful for my timely cowardice!

From Chicago I went to Buffalo by train where I saw the fabulous Niagara falls and then travelled by bus to Washington. I stayed in the comfortable home of Peter Chiarulli of the National Bureau of Standards in Chevychase, a lovely suburb of Washington. I lectured at the Naval Research Laboratory at the kind invitation of Maurice Shapiro.[12] Thus started a friendship which soon brought beneficial consequences to Indian science.

I then left for New York where I was met by Peter Lax[13] of the Courant Institute. I visited my good friend Viswanathan in his office at the U.N. building and later went to his home in Jamaica. I saw all the sights of New York — the Empire State building, the Radio City Music Hall, the planetarium, the natural history museum, Fifth Avenue and Times Square. Then I returned to Boston to be greeted again by Prabhu and Babu.

I boarded the BOAC stratocruiser bound for Glasgow. The plane wobbled so badly in turbulent weather that I was really frightened. After a overnight halt in a comfortable hotel inside the city of Glasgow I arrived in Manchester by the BEA to be received by my teacher Professor Bartlett. After my seminar in the University, he invited me for dinner at an Indian restaurant in Oxford Street and there I met Joe Gani, a handsome young man from Australia who later became the editor of 'Applied Probability'.

I left Manchester for Zurich by the BEA Viscount. It was one of the best flights of my life for the weather was so good that I saw every square inch of the fair land of France. I met Professor Heitler several times and we had pleasant discussions over coffee in the restaurants overlooking the lake of Zurich. It was during that visit that I met S. R. Ranganathan,[14] the veteran scholar who created and fostered Library Science in India.

The journey from Zurich to Stuttgart by train was through beautiful countryside. As I enjoyed the 'gemischte gemuse platte' in the 'Speisewagon' watching the moving panorama of the 'Vaterland', I wondered what had happened to the scars left by Hitler's legions and Patton's tanks.

My participation in the GAMM conference turned out to be one of the most memorable experiences in my academic life. I read a paper on the 'Ergodic properties of some simple stochastic processes' in which I introduced new concepts like "the probability distribution of the time spent in a particular physical state". I spent the evenings at the lovely Killesberg gardens, a riot of flowers and foliage. I also attended a series of delightful operas in the open air festival. I spent pleasant evenings at the roof top restaurant in the television tower and cultivated an insatiable taste for 'Schwarzen Johannisbeer Sussmost' (Black currant juice).

One of the best excursions in my life was the trip through the 'Schwarzwald', the Black Forest. I went to Heidelberg at the kind invitation of Professor Hans Maass.[15] I stayed at the comfortable Reich Post Hotel and walked through the castle grounds and the shopping center of Bismarkplatz. I gave a talk at the Max Planck Institute and listened to a lecture by Hopf from America.

My visit to Gottingen became a very significant event in my academic life. It was a comfortable journey from Frankfurt to Gottingen via Kassel. I stayed in the well-furnished guest room of the Max Planck Institute. I liked the idea to have accommodation for visiting seminar speakers, adopted later at Matscience. I spoke at the Heisenberg seminar on the stochastic problem of Cosmic radiation. The favourable remarks of Professor Heisenberg[16] who attended the lecture naturally encouraged me to pursue the subject in greater detail. Both Dr. and Mrs. Gottstein were very hospitable and so was young Mr. Fay who explained his experimental work on electromagnetic cascades to me. He and his wife took me out on an excursion on a boat trip on the river Weser which gave me an idea of the extraordinary beautiful countryside of Germany. Professor Flugge who attended my lecture invited me to come to Marburg, a University town in some respects lovelier than Cambridge because of its wooded environs. He offered me a contract to write the article on stochastic process for the Handbuch der Physik[17] which I accepted with readiness particularly after Heisenberg's support. I later took the express train to Zurich and enjoyed the lunch in the Spiese wagon for which the German trains were so famous.

At Zurich I met Professor Heitler again and discussed the contents of my book. I was so elated that I took a ten mile walk around Uetliberg with a song in my heart and a smile for each passer-by. That weekend I went on a trip to Lake Lugano the beauty of which has to be enjoyed at least once in a lifetime. It was a wonderful journey through Bellinzona where suddenly Lake Lugano burst into view amongst the mountains. I then understood the preference of the fabled Aga Khan for the demi-paradise of lakes and mountains. I returned to Zurich by boat via the Lake Lucerne.

I lay in bed thinking of the Cascade problem described by Fay. A new approach to Cascade theory suggested itself to me and this implied that I could give my student Srinivasan enough work to complete his Ph.D. thesis.

I returned to Madras by the Air India super-constellation after a most fruitful trip to Europe. As I entered Ekamra Nivas I felt the void created by the absence of my parents. But a new light had entered into our lives — Krishna just a few months old, learning to meddle with the battery-driven model of the stratocruiser I brought for him from Japan.

Within a few hours of my arrival, I summoned Srinivasan and told him of the new approach to Cascade theory and of the 'revelation' of the Feynman formalism. The excitement of introducing the four vector for the first time into South Indian physics was too great to resist.

Notes

1) Among all the piston engined airlines, the DC-6B was considered to be the "thoroughbred airliner" in view of its reliability, operational efficiency, and economics. The airlines loved the DC-6B so much, that they used it well into the late sixties, a decade after the commercial jets took to the skies.

2) Those were the great days of flying when airlines treated passengers both in Economy Class and in First Class with dignity and comfort. If a passenger had a stop over of at least six hours, then a hotel would be provided. Pan Am, "The World's Most Experienced Airline", was especially generous to its passengers.

3) Hideki Yukawa won the 1949 Nobel Prize in physics for predicting the existence of the pion. He was the first Japanese citizen to receive the Nobel Prize. In his honour,

the President of Kyoto University created Yukawa Hall in 1952. This then was expanded in 1953 to the Research Institute of Fundamental Physics and Yukawa was appointed as its first Director. Yukawa was a great inspiration to the young Japanese physicists when Japan was recovering from the ravages of WWII.

4) The Boeing Stratocruiser was the first double decker commercial aircraft, but had its origins as a military aircraft. Although only 55 were produced for the airlines, it enjoyed great popularity because it was spacious and had a lounge in the lower deck where passengers could relax on long flights. This motivated Boeing to introduce an upper deck lounge years later in the 747 Jumbo Jet. The Stratocruiser "Golden Gate" that Ramakrishnan flew in was the first that was delivered to Pan Am.

5) Richard Bellman, a very well known applied mathematician, was a senior scientist at RAND (acronym for Research and Development) Corporation, located in Santa Monica, a lovely suburb of Los Angeles. In 1949, at Manchester, Ramakrishnan became aware of the fundamental work of Bellman and Ted Harris, who was a colleague of Bellman at RAND. A correspondence followed. Bellman became interested in Ramakrishnan's work on product densities and invited Ramakrishnan to visit RAND in 1956, and that was the beginning of a great friendship. As a token of his appreciation, Bellman presented Ramakrishnan with the book "Science and the Method" by Henri Poincare, with the inscription "To my friend Alladi, in memory of a pleasant but all too brief, visit."

6) Richard Feynman was one of the most brilliant physicists of the 20th century. His penetrating insight was admired by all. He won the Nobel Prize in 1965, but ho was a legend two decades earlier. Inspired by the meeting with Feynman in Caltech, Ramakrishnan subsequently obtained a new and simple proof of the equivalence of the Feynman and field theoretic formalism of quantum electrodynamics by splitting the Feynman propagator into its real and imaginary parts. Ramakrishnan's paper appeared in the Journal of Mathematical Analysis and Applications of which Bellman was the Founder and Editor-in-Chief.

7) Otto Struve was one of the most distinguished and prolific astrophysicists of the 20th century. It was Struve who brought the Yerkes Observatory to world prominence, and it was he who appointed the great Indian astrophysicist Subrahmanyan Chandrasekar at the University of Chicago. Struve was a leader in the field of astrophysics, serving as the Editor-in-Chief of the Astrophysical Journal and as President of the American Astronomical Society. The Mt. Wilson telescope that Struve was peering through when Ramakrishnan visited there, was at its time, the largest aperture telescope in the world. The comparison of Struve to Bradman is quite appropriate and interesting in view of Ramakrishnan's passion for cricket.

8) Theodore (Ted) Harris was a noted American mathematician who made significant contributions to stochastic processes, including some joint work with Dick Bellman. He was head of the mathematics division at RAND from 1959 to 65.

9) Robert Marshak, an eminent physicist, was a great statesman for the discipline. He launched this successful series of conferences in high energy physics, which are called the Rochester conferences because they were first held at the University of Rochester where Marshak was. Subsequently these high energy physics conferences were held in different parts of the globe, and Marshak continued to be a key component in the conferences. Ramakrishnan attended several of these conferences. It was at the 1956 Rochester conference that Ramakrishnan first met Marshak and they became close friends. They had a mutual admiration for their efforts to encourage the development of physics in their home countries. Marshak invited Ramakrishnan many times to Rochester in the sixties. In return, Ramakrishnan invited Marshak to MATSCIENCE in 1963 as the First Niels Bohr Visiting Professor (see Part II).

10) Subrahmanyam Chandrasekhar, a nephew of Nobel Laureate Sir C. V. Raman, was one of the most eminent astrophysicists of the 20th century. He was Morton D. Hull Distinguished Service Professor at the University of Chicago, and often went to Yerkes Observatory located in Willams Bay in the neighbouring state of Wisconsin for stellar observations. Willams Bay is a two hour drive from Chicago.

Chandrasekhar was very meticulous in whatever he did, including preparations for his lectures, and in his commitment to his students. There is the story that for one class with just two students (both very talented), he would drive several miles to teach them. The students were T. D. Lee and C. N. Yang who later won the physics Nobel prize!

Ramakrishnan and Chandrasekhar had close interaction, both from the point of view of research, and on issues relating to the developments of science in India (see chapter on the inauguration of MATSCIENCE). After Otto Struve, Chandrasekhar served as Editor of the Astrophysical Journal from 1952 to 71, and during this period accepted seven of Ramakrishnan's papers (some in collaboration with students) for publication in the journal (see Appendix 9).

11) Kuiper, Stromgren, and Chandrasekhar, were are all recruited by Otto Struve.

12) Maurice Shapiro was a leading authority on Cosmic Rays and that is how he got interested in Ramakrishnan's work in cascade theory of cosmic ray showers. He joined the Naval Research Laboratory (NRL) in Maryland in 1949. Ramakrishnan and he became close friends. Shapiro invited Ramakrishnan to lecture at the NRL on almost every one of his trips to the USA. Shapiro's positive comments to Education Minister Mr. C. Subramaniamon observing Ramakrishnan's Theoretical Physics Seminar, were crucial for Mr. Subramaniam to write to Prime Minister Jawaharlal Nehru to start a new research institute (see Chapter 21).

13) Peter Lax was born in Budapest, Hungary in 1926, but grew up in the United States. He received his PhD in 1948 at the Courant Institute of NYU, and has been there ever since. He is arguably the world's most eminent applied mathematician.

14) S. R. Ranganathan (1892–1972) was the father of Library Science in India. After a brief stint as an Assistant Professor of Mathematics at the Presidency College in Madras, he accepted the position as the University Librarian in Madras in 1924, a post he held for 21 years until his voluntary retirement in 1945. In 1931 he developed his famous *Colon Classification*, a system that is used throughout the world, and his *Five Laws of Library Science*. Between 1955 and 57, he was in Zurich, and Ramakrishnan met him there in 1956. In his personal annual diaries, Ramakrishnan says in an entry dated 20 May, 1956 that "Ranganathan is one of the finest men I have met."

15) Hans Maass (1911–92) was a German mathematician who is most known for introducing *Maass wave forms* which have now become crucial in understanding the relationship between Ramanujan's mock theta functions and the theory of modular forms. Even though Ramakrishnan was a physicist, he maintained good contact with the mathematics community as is indicated by his visit to Heidelberg at the invitation of Maass.

16) Werner Heisenberg (1901–76) was one of the pioneers of quantum mechanics. He won the physics Nobel Prize in 1932. He was a leader in the development of physics in Germany before and after WW II. Ramakrishnan first met Heisenberg in the Winter of 1949 at the Conference on Modern Physics in Edinburgh that he attended, while doing his PhD at Manchester. Heisenberg invited Ramakrishnan to lecture in his Seminar. He was impressed that the theoretical work of Ramakrishnan concurred with the experiments his group was conducting.

17) The Handbuch der Physik begun in 1926 was a 24 volume project of Springer-Verlag, the renowned German publishing house. When Seigfried Flugge took over as the Editor in 1955, the series was expanded to 55 volumes. On observing Heisenberg's

favourable comments about Ramakrishnan's lecture, Flugge offered Ramakrishnan a contract to write an article on probability for the Handbuch. Ramakrishnan's 125 page article on Probability and Stochastic Processes appeared in 1959 in the volume devoted to the Principles of Thermodynamics and Statistics.

Richard Bellman of the RAND Corporation.
Source: Permission granted by IEEE to reprint this Image #2679 of the IEEE History Center Image Archive.

Chapter 15

Prelude to Princeton (1956–57)

The round the world journey through Japan, America and Germany had a tremendous impact on my academic life. I plunged into work, for it was the only relief from the absence of my parents. Of course the presence of little Krishna enlightened Ekamra Nivas and watching him play with Japanese airplane models I could foresee that air travel was soon to become his insatiable passion.

I started in right earnest writing the article on Stochastic Processes for the Handbuch series almost like a man possessed, as if it was the chance of a lifetime. I had the hearty cooperation of my students, Srinivasan, Ranganathan and Vasudevan. I felt like a pioneer for there was no text book on the subject and I could feel happy not only on introducing new concepts but even fundamental ideas of probability theory and the physical meaning of frequency interpretation.

Since Oppenheimer reacted favourably at Rochester to my desire to spend a year at Princeton, I was hopeful that he would send an official invitation, which he did a few months later. So I prepared myself by studying Quantum Electrodynamics according to Feynman and found that the best way to learn was to lecture on the subject to eager and critical audiences. Here again I felt the spirit of a pioneer in talking on relativistic quantum mechanics, Dirac matrices and invariance under Lorentz transformation to large audiences in the city colleges which were complacent under the tyranny of an outdated syllabus that would not easily admit changes in structure. That made the task more demanding and challenging and I never enjoyed anything more than reading of Feynman's lectures line by line and page by page to my students whose willingness to learn was as keen as mine.

I thought it was also an opportune moment to write an article of a general nature on the physical approach to stochastic processes and at the suggestion of Sir C.V. Raman, I gave a talk on the meeting of the Indian Academy of Sciences in December 1956.

I received a formal offer of a professorship at the Texas A & M University by Professor J.G. Potter. That was the period when American institutions were seeking experts from many countries for senior academic positions. It was not possible for me to settle down in the U.S. especially when I reckoned myself as a trustee to the traditions of Ekamra Nivas.

We had the pleasure of receiving Professor Harry Messel in early January on his way to Australia after attending a conference on Theoretical Physics in Seattle. We took him to the best sights and sounds of Madras — the cleaner parts of the Madras beach, the rock sculptures of Mahabalipuram, the staid dignity of Fort St. George, the sombre exhibits of the Madras museum and the sedate strains of Carnatic music — rather sober fare for a tireless globetrotter used to the five star comforts of Acapulco and Pago Pago, a zealous head of a Foundation sponsored by million dollar donors and an ardent wildlife conservationist interested in Alaskan bears, Siberian tigers, Canadian seal pups and Australian crocodiles.

Harry and his wife are such warm friends who enjoyed their stay in Ekamra Nivas with genuine enthusiasm that their visit put me and Lalitha in the right mood to think of Princeton, that haven of learning which offered Einstein and von Neumann not merely an asylum but a home away from their homelands.

Chapter 16

The Princeton Experience (1957–58)

Our stay in Princeton fulfilled a desire I nursed since I heard of Oppenheimer as its Director, Einstein and Neumann as its luminaries, Dirac and Pauli as its distinguished visitors. Matscience in turn was the realisation of my Princeton dreams.

It was in February 1957 that I received a short and friendly letter from Oppenheimer inviting me to spend a year at the Institute as a visiting member (Appendix 11). I applied for a travel grant to the Asia Foundation by an informal air letter and it was both flattering and pleasing to receive a generous cheque by return of post.

We decided to leave two year old Krishna with my elder brother Kuppuswami at Hyderabad. He came to Madras to take delivery of his toddling ward and wish us bon voyage to the United States. My younger brother also arrived for we had to perform our father's annual ceremony at Ekamra Nivas the day before our departure.

We left Madras on the 20th September afternoon by the Indian Airlines flight to Bombay and took the Air India Super Constellation "Rani of Jhansi" for Rome. Then occurred one of the most frightening experiences in our flying career. As the engines roared and the plane sped on the runway, we actually saw one of the left engines catch fire. The pilot applied air brakes and the plane came to a screeching halt at the end of the runway. The fire in the engine was automatically extinguished. The pilot tried to take off once again — we do not know why — and with the same result! The plane was delayed for four hours and the defect rectified. On behalf of the passengers I made a complaint that the pilot should not have risked our lives and his by making a second trial without locating the defect. The Air India manager told us that I could accept his assurances that the defect was rectified since his son was travelling on the same flight!

The plane then took off with the leaping grace of a 'super connie' and after halting in Beirut we arrived in Rome the next morning. We indulged in the luxury of engaging a large Fiat taxi for six hours (for eight thousand lire) and motored round Rome seeing the Vatican and St. Peters and later the Catacombs. The visit to the spiritual capital of the Catholic world was a thrilling experience because the evolution of Europe was as much influenced by Popes as by potentates, as much by religion as by diplomacy.

We caught the luxury Rapido Express to Venice and arrived at the astonishingly beautiful city, the Queen of the Adriatic, with its exotic situation on seagirt canals. We stayed in the deluxe Park Hotel, very expensive, but had the satisfaction that Professors T.D. Lee, Emilio Segre and Robert Marshak also stayed there. The next day I went by the autostrada to Padua to attend the conference on high energy physics.[1] I heard the opening talk by Steinberger, with his characteristic American humour, consciously casual when really implying hot business.

Lali accompanied me to Padua and as she waited outside the conference hall she had the opportunity to watch Professor Pauli in action, discussing with his colleagues. We attended a delightful Venetian ballet that night in the fabulous opera house.

The next day the conference shifted to the beautiful St. George's Island in Venice. Yang and Lee had just done their famous work on the non-conservation of parity in weak interactions which was experimentally confirmed and led to the universal Fermi interaction postulated by Feynman and Gell-Mann and independently by Marshak and Sudarshan. We also heard Professor Powell's lecture[1] — a magnificent summary of experimental data reminding me of his lecture I heard earlier at Kyoto.

We returned to Rome and were told that the TWA plane to Zurich was six hours late. We went around on an exciting half a day excursion by motor bus to the Villa Borghese with its treasures of painting and sculpture — Bernini's Rape of Persephone and Raphael's Descent from the Cross — and the Capitoline Hill designed by Michelangelo, the grandest sight in Rome. We left that night by the TWA constellation and arrived in Zurich past midnight and drove by taxi to a comfortable hotel.

We had a comfortable flight by Swiss Air DC-6B from Zurich to Frankfurt. We hired a Mercedes Benz for a three day motor excursion to Marburg and Gottingen, through the famed autobahns and the lovely countryside. The handsome student engaged as a driver imagined that we were members of a princely family from India who could afford a chauffeured car! We visited the Max Planck Institute and then went to Marburg, where we stayed at the lovely Kur Hotel. We had a delightful time in Marburg with our friends the Flugges who provided us delicious vegetarian food. Returning to Frankfurt, at the BEA office we received the comforting news by cable that little Krishna was happy in Hyderabad with my elder brother.

We left by a BEA Viscount 'Sir Francis Young Husband' for London where we spent the afternoon strolling inside the city. We boarded the BOAC DC-7C[2] at London Heathrow Airport and after a superb twelve hour non-stop flight over the Atlantic arrived at Idlewild Airport, New York where we were received by my nephews, Babu and Prasad, and our friend Viswanathan of the United Nations. We spent the whole day with our gracious hosts the Viswanathans and left for Princeton by train. We entered our house 196-A Springdale Road and the first night we slept under our overcoats since we did not have sheets and blankets.

Housing for the members at the Institute was well designed, to ensure privacy but provide easy communication amongst them. The apartments were modern (at that time!) with a generous expanse of lawns and plenty of trees, a typical American small town atmosphere of elegance and comfort.

Fuld Hall, the main building of the Institute, was very impressive with its colonial architecture — red brick, with white wood work, high ceilings and large doors and windows. Members were allotted rooms in two additional buildings and I shared my Room 100 with Mike Cohen, a recent student of Feynman. This turned out to be fortuitous since I was able to recommend one of my M.Sc. students Deshpande for a fellowship under him when he later took up a faculty position in Pennsylvania.

It was autumn when the trees were swathed in resplendent colours against the luscious green of the sprawling lawns. The first few days at the institute we naturally spent walking round its beautiful grounds inhaling the air sanctified by the breath of Einstein. He died in 1955 but his spirit was everywhere in that university town where even the cabman or the store clerk claimed to have seen or talked to him.

My first meeting with Oppenheimer fulfilled my expectations about this legendary figure who dominated not only American science but influenced the destiny of the world as the architect of the atomic bomb. Lean and of medium height, he had an oval head with prominent cheek bones and piercing eyes. He could pick his men while lighting his pipe, each for the appointed task according to his talent and inclination, from a Nobel prizeman to a truck driver. He was magnanimous in providing opportunities for young scientists and enjoyed discussions at every seminar where his very presence stimulated creative thought and invited impartial criticism.[3]

The Oppenheimer legend is just a record of incredible facts. Born to prosperity in 1904, he was educated at Harvard under Whitehead and Bridgman and took his Ph.D. at twenty three in Gottingen after a preliminary stay at the Cavendish. His intellectual interests ranged from theoretical physics to Hindu philosophy, and in Rabi's estimate "he understood the whole structure of physics with absolute clarity" that one wonders why his creative work was not of that seminal quality of Dirac's or Heisenberg's. It was said he had two passions — physics and the desert! He found the one in the other when at forty he was called upon to undertake at Los Alamos a task unprecedented in its objective, undefined in its scope, unpredictable in its consequences — the creation of the atomic bomb. It was a leap into the uncharted future of mankind and he achieved it with the pragmatism of an American and the vision of a universalist. Tormented by the moral ambiguities of twentieth century science, he found his haven at the Institute for Advanced Study, the environment to which he belonged, contemplating on the nature of the physical universe from the innermost recesses of Gell-Mann's quarks to the outermost bounds of Einstein's universe.

The eight months at the Institute were exceedingly pleasant and fruitful in the lovely town of Princeton. I attended over a hundred seminars at the Institute and the University, the first being on the CPT theorem by Jost and the next by Trieman on the Venice conference.

It was going to be the transition period in my scientific career for I was moving into Elementary Particle Physics having completed my Handbuch article on stochastic processes. Princeton was the 'centre of things' and I could not ask for a better place being a late entrant into theoretical physics. Nor could I have chosen a more propitious time to be in Princeton. Yang and Lee[4] were in residence and everyone was excited about their discovery of non-conservation of Parity. Would they win the Nobel Prize? If they did, it would be the quickest award since its inception. And so it happened and even in Princeton where genius could be seen rambling in the streets, there was unconcealed excitement about the discovery and the award!

I enjoyed watching Lee and Yang in various moods and what impressed me most was their involvement in understanding the laws governing the universe. It was obvious that their main purpose was to understand Nature, its symmetry and deviations therefrom and that the Nobel award was 'incidental', though a natural consequence of their efforts. How we wish we could infuse this spirit in our country where so much time and effort of scientists are wasted in scrambling for official status and privileges?

It was in October that the world was stunned by the news that Russia had launched the Sputnik into space, initiating the new space era in human history so soon after the birth of the atomic age. It shook America out of its complacency in its claims for leadership in space technology. Everywhere and in particular in Princeton the question was asked: What were the favourable factors in Russian scientific effort which made this achievement possible? The frank assessment of the situation in the U.S. was the most striking demonstration of the vitality of American democracy resulting in an 'explosive' support for science and creative talent. The decade of the sixties was to become the glorious period of Man's conquest of space which he marked by setting his footprints on the moon.

In Princeton I thought about the distinction between multiplicative and additive quantum numbers. When speaking of conservation of energy and momentum, we are thinking of additive quantities but parity was a multiplicative quantum number and conservation had to be understood in a corresponding manner. The second revelation was the structure and algebra of the Dirac matrices. While in the Dirac equation there occurred only four anti-commuting matrices, there was a fifth one, 'the gamma five' which was ignored hitherto but had to be invoked to explain weak interactions. It took me ten years to understand this mystery and it was totally resolved when I obtained the hierarchy of matrices of the Clifford algebra and their generalisation in 1967. I remember how with friendly interest Oppenheimer gave me the preprint of Feynman and Gell-Mann asking me to understand the role of 'gamma five'!

Life in Princeton during the radiant months of autumn had a fragrant tranquil charm. We enjoyed walking in the spacious wooded environs of the Institute, noting with tireless interest Einstein's home, 112, Mercer Street, strolling around Palmer square, riding in the comfortable Institute station wagon to the supermarket, or browsing in the University Library. What impressed me most was the peaceful coexistence of the Princeton University and the Institute, with no bounds marked between their territories, no barriers psychological or physical for communication between the members of the two Institutions. It was just one Princeton community sharing the common excitement of seminars and informal discussions, zealous and conscious of its pre-eminent place in American, nay, international science.

There was an interesting evening party at the house of Rodnicks, a couple who had just returned from a tour of India. It was there I heard that a party in Princeton was considered a success if it was attended by a Permanent Member of the Institute preferably a Nobel Prizeman and the host was supposed to have 'made it' if he was able to get Oppenheimer! One can imagine the situation when Einstein was living when he imparted to a social gathering the prestige and aura comparable to the presence of a Rockefeller or a Roosevelt.

The first snowfall occurred on November 22 and Princeton lay under a white blanket for the next few months. In early December, we started on our journey to Canada in response to the kind invitation from Shanmugadasan of the National Research Council to lecture on stochastic processes. Enroute we went to Boston to spend a few days with my nephews, Babu and Prasad who were graduate students at Harvard. While in Boston, we had dinner with professor and Mrs. Buechner of M.I.T. at the faculty club. They recalled their visit to Madras and Ekamra Nivas with flattering warmth and friendship.

In Ottawa we stayed at the comfortable Lord Elgin Hotel. Herzberg, the Director of the Laboratory of the National Research Council, Professor Wu and Dr. Shanmugadasan were very kind to me and my lectures on Cascade Theory were well received. Shanmugadasan was an old student of Dirac and we had become friends during a short visit to Cambridge when we were in England in 1951. He was a conversationalist of sparkling wit which also meant he was a friendly listener with a natural concern for other's feelings. We had dinner at the house of K. Srinivasan, a senior official in the Indian High Commission. I also lectured at the University of Ottawa and we spent our leisure with Gopinath Kartha, a young bio-physicist from Madras, and Mathews, my old student, both of whom were working as post-doctoral fellows at the N.R.C.

We left for Toronto where we stayed at the luxurious Lord Simcoe hotel and went to the Niagara falls in midwinter. We stayed at the Honeymoon Motel, twelve years after our marriage! It turned out to be a beautiful and sunny day and instead of seeing a frozen Niagara we saw the falls in the best of moods. The floodlit waters at night had a strange garish beauty. We caught the 'Maple leaf', the express train to New York on our way back to Princeton.

The next two months in Princeton were very interesting from an academic point of view. I worked on a paper on multiple product densities in collaboration through correspondence with my student S.K. Srinivasan who was then at Sydney with Harry Messel. It was also engaged in interpreting limiting operations in stochastic processes.

It was an exalting experience to listen to Niels Bohr on superconductivity when the hall was so crowded that Lee and Yang sat on the floor watching the 'master'. I also heard Moller on relativity and Pawsey on radio astronomy.

At the end of January we went to New York to attend the Annual Physical Society meeting. Feynman's lecture on Beta decay was such a spectacular performance, that his resonant voice and flashing smile inspired me whenever I lectured on quantum mechanics at home and abroad. My old student Mathews came from Ottawa for the meeting and I introduced and recommended him to Professor Falkoff of the Brandeis University who offered him a post doctoral fellowship for the ensuing year.

Back in Princeton, I attended seminars Stromgren on astrophysics, by Goldberger on dispersion theory, Haag on Scattering, Pais and Trieman on weak interactions, Gold on magneto-hydrodynamics and on the solar origin of cosmic rays. Stromgren[5] started his lectures with an introduction to the cosmic magnitudes so that particle physicists used to subnuclear scales of space and time could appreciate the size of galaxies, the age of stars and the range of the expanding universe! Goldberger represented the explosive brilliance of the new American generation of physicists while Pais provided the charm and elegance of European culture and refinement.

Among other seminars of 'educative value' were those of Oskar Klein on gravitation. I was pleased when he recalled my conversation with him on stochastic theory during the spring of 1950. Another illuminating talk was by Treiman on 'Poor man's party' to explain the concept of parity to a layman. To me the simplest way of transmitting the idea is to state that if the law of motion of an airplane is that it moves forward if the propeller rotates clockwise as seen by the pilot then in the reflected world (as in a mirror) the airplane moves forward if the propeller rotates anticlockwise as seen by the pilot inside the mirror. Thus the law of motion is different for the airplane inside the mirror and parity or reflection symmetry is not conserved!

An interesting incident occurred at a seminar at the Princeton University on 'Spin and Statistics'. There was an embarrassing silence when the too attentive audience did not realise that the learned lecturer had completed the proof in twenty minutes while they were under the impression that he was just getting started after going through the preliminaries! A clear case of a proof being too simple to be understood! No wonder that my derivation of product densities was considered too elegant to be correct by a mathematical adviser to Professor Bhabha!

In Princeton we had a pleasant companion in Indersingh Luthar, a young mathematician who took his PhD from Urbana under Professor Day. I enjoyed conversations with him since he went into raptures over excellence in mathematics as revealed by the work of Ramanujan and Hardy, Serre and

Weil, Artin and Harish-Chandra. Through him I learnt about the activities of the pure mathematicians at the Institute.[6] Selberg, a world famous mathematician at the Institute, associated with the elementary proof of the prime number theorem like Erdös, was one of the leading lights, and Serre one of the invited professors. There was a well founded rumour going round that one had to prove a theorem which enlightens a whole domain in mathematics to be invited to a professorship at the Institute. At that time I could hardly imagine that one day Krishna would become a number theorist aspiring to study the works of Selberg and Erdös.

In February we had a most enjoyable journey to Washington where we stayed with the Shapiros. We went by the fast train 'Washingtonian' and called on our Indian Ambassador[7] and perused Indian newspapers at the Embassy with eagerness. My lecture at the Naval Research Laboratory arranged by Shapiro was well attended and received.

The next day we had an unforgettably unusual experience. Normally in Washington it does not snow even in winter. We missed hearing a radio announcement that in Washington there was to be a snow storm in the evening. We went into the city, lunched at an Italian restaurant and walked round the Capitol. When coming out of the Lincoln memorial we found there was a peculiar concentration of cars towards the highways as though there was some warning for evacuation of the city! We joined the race and it suddenly became dark and started snowing heavily. Within a few miles of the city we found there was an impassable traffic jam and we tried to take a detour. By that time it was so dark and the snow fall became so heavy that there was no question of travelling by car. We left the new Pontiac in the snow and climbed a hill imagining ourselves to be the mountaineers scaling the Everest! We lost our way at night and asked for shelter and were received by a kind couple into a beautiful house of such striking elegance that when later I built my new house in Madras I copied the design of the staircase and the general arrangement of the bookshelves. Even an accident had its own pleasant consequences! We left Washington with vivid memories of American hospitality and the vagaries of a Washington winter.

In late March we were eagerly expecting Sir Radhakrishnan who was to lecture at Princeton. But due to heavy snow fall, Princeton was completely isolated and his trip was cancelled. Even electricity failed and we had to burn candles to light our apartment.

We went to Cleveland by the 'Chicagoan' and were received by Professor Bayard Rankin[8] who took us to the Wade Park Manor hotel where we indulged in the luxury of asking for room service for lunch. Later that evening we had a delectable vegetarian dinner at Rankin's house. At the Case Institute of Technology, I lectured on 'the physical interpretation of some limiting stochastic operations'. In the evening we had a wonderful dinner with Professor McKuskey who prepared a most delicious vegetarian fare reminding us of the wonderful dinner at Belz's house at Melbourne. Among the guests were Professor Foldy well-known for his work on the transformation of the Dirac equation.

We left for Chicago by the 'Prairee State' and were received by Peter Chiarulli who took us to his modern apartment. But he warned us that it was not safe to walk in the neighbourhood after sunset.

We met Chandrasekhar who invited us for lunch to the Faculty Club and as usual we talked of the state of science in India. My lecture at the Illinois Institute of Technology was well attended and we joined the lunch in honour of Sir Radhakrishnan the next day. So high in esteem was he held, that everyone went vegetarian even in Chicago, just a few miles from its bleeding stockyards!

Back in Princeton, we started preparing for our return to India. One evening Professor Wolfe of the History Faculty of the Institute invited us for coffee at his house where we saw his colour slides of Amsterdam. They looked so good that we decided to travel by K.L.M. and visit Amsterdam on our way back to India.

It was a pleasant surprise to meet Professor T.G. Room of Australia who was spending a few months at Princeton. When he invited us to his house he expressed his desire to record Lalitha singing the famous song Bhajagovindam with my English commentary. The song is an exhortation to a person to concentrate his thoughts on God and not waste the evening of his life in the idle subtleties of grammar!

We left Princeton on the 4th of April for New York and stayed with the Viswanathans of the United Nations before taking the KLM DC 7C for Amsterdam from the Idlewild airport. The thirteen hour flight in wonderful weather was enjoyable followed by a relaxed stay in a comfortable hotel in Amsterdam. The excursion on the canal was interesting, more so was the visit to the villages and the cheese factories nearby where women dressed themselves in ancient Dutch costumes to satisfy the curiosity of American tourists. We took a Viscount flight to Zurich and after two delightful days there, we left for Bombay by the Air India super constellation.

We returned to Madras after visiting Hyderabad where we picked up Krishna who was kept in my brother's care during our stay in Princeton. He was just three years old, lisping Sanskrit verses of Suprabhatham which he learnt by listening to my uncle at prayer every morning.

I was full of visions of creating a centre of advanced learning in Madras. It was too much to expect our university to respond readily to my personal dreams and aspirations but it was certainly possible to introduce a flavour of Princeton into the intellectual environment of Ekamra Nivas.

Notes

1) The specific title of the conference was "Mesons and recently discovered particles". C. F. Powell, a principal speaker at this conference, played a key role in the discovery of the pion (= pi meson) predicted by Hideki Yukawa.

2) The Douglas DC-7C, affectionately called the "Seven Seas", was the largest and the last of the great piston engined aircraft produced by the Douglas Aircraft Company. It genuinely had a range long enough to cross the Atlantic without a refuelling stop in Gander, New Foundland, or Shannon, Ireland. To stay competitive, the British Overseas Airways Corporation (BOAC) bought the American made DC-7C before the introduction of the British made Bristol Brittannia.

3) The period when Oppenheimer was the Director of the Institute (1947–66), saw the most intense activity in theoretical physics. At that time, the faculties of mathematics and physics were both under the School of Mathematics but functioned independently.

4) T. D. Lee was a Visiting Member in 1957–58. C. N. Yang was a Visiting Member at the Institute from 1949 until 1952 when he made Permanent Member, and then promoted to the rank of Professor at the Institute in 1955. The announcement of their (joint) Nobel Prize in physics came in the winter of 1957.

5) The Institute for Advanced Study first tried to get Subrahmanyam Chandrasekhar to begin their program in Astrophysics. But Chandrasekhar was not inclined to leave Chicago, due to his work at Yerkes observatory. Then the Institute offered the position to Stromgren who accepted it.

Bengt Stromgren (1908–87) was an eminent Danish astronomer and astrophysicist. He was appointed in 1957 at the Institute for Advanced Study as the first professor of astrophysics and occupied Einstein's office. So the lectures of Stromgren that Ramakrishnan heard were during Stromgren's first year at the Institute.

6) The mathematics program at the Institute has always been of the highest level year after year. The Norwegian Atle Selberg, considered to be one of the most eminent mathematicians of the 20th century was a Professor of Mathematics, with stalwarts like Armand Borel, Kurt Godel as his colleagues. The world class visitors in mathematics during 1957–58 included Jean Pierre Serre, Georges DeRham, Hans Grauert, Paul Halmos, Kunihiko Kodaira, Jean Leray, and Louis Nirenberg. There was a lot of excitement and expectation that Andre Weil was to join the mathematics faculty the next year.

7) The Indian Ambassador in Washington D.C. was G. L. Mehta. He was briefly a member of the Constituent Assembly. In a letter to Ramakrishnan dated 3 January, 1958, Mehta said "I had the privilege of knowing your late and distinguished father when I was a member of the Constituent Assembly for a short time."

8) Bayard Rankin was a well-known probabilist who received his PhD in physics in 1955 from UC Berkeley. His thesis was on stochastic processes and its uses in cascade theory, an area where Ramakrishnan had made notable contributions. Rankin went to MIT soon after his PhD and invited Ramakrishnan to speak in the Norbert Weiner Seminar there in 1956. Subsequently, Rankin moved to Case Institute of Technology in Cleveland and invited Ramakrishnan for a seminar there in 1957–58.

The Institute for Advanced Study, Princeton.

Robert Oppenheimer
Source: This image is in Public Domain and is a work of a United States Department of Energy (or predecessor organisation) employee, taken or made as part of that person's official duties, https://commons.wikimedia.org/wiki/File:JROppenheimer-LosAlamos.jpg

Albert Einstein
Source: The image is in Public Domain, https://commons.wikimedia.org/wiki/File:Einstein1921_by_F_Schmutzer_2.jpg

I was full of visions in creating a centre for advanced learning in the city of Madras (1958).

Chapter 17

The Year of Exile (1958–59)

Back from Princeton to the supreme comfort of Ekamra Nivas, I dreamt of creating in Madras an environment similar to that at the Institute for Advanced Study. During my stay there I learned that the seminar was the essence of intellectual activity, where there was as much a desire to impart as to imbibe, where opportunities are provided for the clash of intellects which would produce creative ideas.

I started a vigorous series of seminars at Ekamra Nivas based on the lectures I heard at Princeton, in which my senior research students took an active part. Vasudevan and Ranganathan were well set on their road to a Ph.D on stochastic problems in physics and astronomy while Devanathan had started his research work. During his M.Sc. Devanathan had impressed me with his clear analysis of angular momentum in quantum mechanics from Schiff's classic text and I therefore assigned him to low energy nuclear physics involving detailed calculations using the concept of angular momentum. I had just brought with me from Princeton Edmond's latest book on angular momentum which he mastered with great diligence. He justified my estimate of him by later becoming Professor and head of the new department of Nuclear Physics in the University of Madras.

During the first month of my arrival, a group of eager lecturers from St. Joseph College, Trichy, with their venerable leader Ananthakrishnan, approached me for lectures on complex variable theory so that they can prepare themselves to teach according to the new syllabus which was being introduced that academic year. During the month of April when the heat in Madras was intolerable even to its proud citizens, I lectured with fanatical zeal in the spacious hall of Ekamra Nivas. It was not merely a labour of love but a love of labour to teach the group how to teach, from the Argand diagram to the Residue theorem. There was adequate reward for such an effort in the warm hearted response from the academic community of that famous college and even more in the lasting friendship with Ananthakrishnan whom I loved to call the Bhishma of South Indian physics.

Meanwhile a new group of enthusiastic and talented students joined the new theoretical physics M.Sc. course which I initiated at the University. Among them was A.P. Balachandran whom I recommended to Dalitz for a post-doctoral fellowship after he took his Ph.D. under me two years later. There was also in the group two dedicated young men, V.K. Viswanathan and V. Radhakrishnan who gave up the security of an ill-paid lecturership and a permanent clerkship respectively to seek their fortune in the 'open domain' of theoretical physics. I was able to send Viswanathan to Rochester after his MSc at Madras. Later he worked with the Ferrand Optical Company in Manhattan before settling down at Los Alamos as an active scientist in Fusion Research. Radhakrishnan joined the permanent staff of Matscience after his PhD degree and post-doctoral experience in Canada and the United States.

By July we had the seminar programme in full swing when we discussed page by page and line by line Yang and Lee's elegant brochure on weak interactions and Chew's work on strong interactions. The new M.Sc. students were fed on Feynman kernels as the main course with the Dirac algebra as an appetiser and field theory as the dessert!

It was almost a shock and disillusionment when I found that the University would not entertain any plans for the expansion of the physics department to include theoretical physics within its scope. Instead of strengthening and expanding existing departments, the University decided to have an extension centre in Madurai and advertised for professorships there. It was impossible for me or any one to understand the wisdom or necessity of establishing an extension centre far from the metropolitan area, like a distant colony being administered from the Imperial capital! This would obviously result in creating a caste system among the professors, those in Madras being distinguished from those in the mofussil centre. This would in turn lead to intrigues for transfers, dissension over such transfers and unscrupulous abuse of power by dubious representatives in the name of the Vice-Chancellor. Actually this resulted in the inevitable creation of a separate independent Madurai University a few years later.[1]

What really pained me was that the University took no cognisance of the valuable experience I gained at Princeton. I had heard over hundred and fifty seminars by world famous physicists and I wanted to initiate work in high energy physics in a University which had among its alumni,

Ramanujan, Raman and Chandrasekhar. The only reward for my enthusiasm was the modest compliment of banishing me as a professor to the
extension centre. I found no alternative except to apply for and secure the
professorship since otherwise someone else would be appointed who would
later 'wangle' a transfer to Madras 'above my head'. I remembered at that
time the example of C.R. on the one hand and Winston Churchill on the
other. C.R. accepted the Chief Ministership in 1937 not out of a desire for
office, but to convince the British Government that the Indians were competent to govern themselves. He used it as a waiting period for the great
struggle to come a few years later. Churchill ordered the British Expeditionary forces to withdraw from Dunkirk and waited for four years before
Normandy. It was in that spirit that I accepted the Madurai assignment
and it was indeed a coincidence that the order of appointment was issued
on my birthday, the 9th of August! I was actually willing to accept even
an honorary professorship in Madras if adequate opportunities were given
to do research and initiate young scholars in modern physics. It was really
tragic that the university should have ignored my idealism on the ground of
'no vacancy' and exiled me without the slightest compunction while finding
funds for wasteful and purposeless expansion.

On my birthday my father-in-law presented me with five volumes of
the Valmiki Ramayana, knowing full well it was a living inspiration for
me, keeping the memories of my parents vivid and sacrosanct. It turned
out to be most appropriate for I was to draw comfort by reading about
Rama's Vanavasam during my period of stay at Madurai. As I travelled to
Madurai I thought of the immortal verses in the Ramayana when Sri Rama
told his aunt Kaikeyi, the architect of his banishment, "I am not bothered
about going to the forest but what hurts me is that my father did not give
this injunction himself under the impression that I would be unhappy and
disappointed. Gladly would I give the entire empire and all my possession
to my brother Bharatha".

Krishna was just three years old and I thought that it was not fair on my
part to take him and Lalitha away from the comforts of Ekamra Nivas and
subject them to the rigours of life in Madurai. I therefore planned to stay
in Madurai alone making frequent visits to Madras at every possible opportunity. I also wanted to spare the research scholars and the newly admitted
MSc students the inconvenience of life in Madurai without accommodation
and elementary facilities.

The University Extension Centre was a mere name. It had neither
buildings nor library. The ill-furnished and crowded rooms with poor toilet

facilities at the local colleges where I was asked to teach had a depressing effect on a person who had just spent a year at the Princeton Institute with its
world renowned professors and magnificent campus on sprawling lawns, and
who was used to the comforts of Ekamra Nivas, with its spacious halls and
ancient trees. What a far cry were the ill equipped, noisy and over-crowded
rooms at Madurai from the seminar hall at Princeton where I heard Yang
and Lee, Goldberger and Stromgren, Pais and Trieman! Yet I took the
Madurai assignment as a challenge and gave lectures on modern quantum
electrodynamics based on Feynman's work. The only relief from boredom
besides work was the conversations with my colleagues Drs. M. Santappa[2]
and G. Krishnan, both highly qualified with PhD degrees from Manchester and in the same predicament as myself. We compared ourselves with
Nehru and C.R. who spent the best years of their lives in prison. Our only
prayer was that the ordeal would not be long, and the redemption would
come without travail. My friends had no plans except to wait on the mercy
and generosity of the authorities but I was not willing to do so.

The first opportunity came for me when I had a curious offer from the
Ministry of Education inviting me to be a member of the expert committee
for Hindi terms in the physical sciences. I did not certainly have any faith
in this project for I knew no Hindi but I felt it was wisdom to accept
it since I could spend a day or two in Madras on my way to and from
Delhi. I knew that money was available in India for such committee work
though funds were denied for symposia and lectures! This would give me
greater freedom of movement and possibility of contact with the 'higher
authorities' in Delhi to draw their attention to the gross injustice done to
me. The greatest attraction to me of course was to see Lalitha and Krishna
and spend a weekend at Ekamra Nivas and work with my students there
on problems in high energy physics.

There was a slight relief when the University permitted me to conduct
the MSc course in theoretical physics at Madras during specified periods
of the academic year 1958–59. I was therefore able to spend Navarathri in
October at Madras. I held rigorous tests for the new MSc students insisting on very high standards of performance based on texts used at leading
American Universities. The students on their part shared 'the Frontier
Spirit' with a justifiable pride of their being the first batch to be examined
on Feynman graphs, strange particles, and invariance principles.

It was during this period we heard the news that the BOAC resumed
the Comet jet flight across the Atlantic and Sir Goeffrey De Havilland was
alive to receive the news of the triumphant resurgence of 'his creation' which

had unfortunate accidents in 1954. On October 27th, 1958, Pan American began the daily Jet Clipper (Boeing 707) service from Paris to New York[3] with justifiable fanfare but these made no particular impression on us at that time since we were mostly concerned with train reservations which required a strange combination of versatile skills, of exercising influence in the face of widespread malpractices relating to 'ghost reservations', 'VIP quotas', etc. The social status of a person or official was enhanced by his ability to get a 'coupe' without advance reservation and to announce his achievement to an envious and admiring group of friends and relatives. As a University professor in an 'extension centre' I had to be content with small mercies like meetings of the Academic Council and the Board of Studies which enabled me to visit Madras at frequent intervals when I could spend some time with Lalitha and Krishna.

By November I arranged a house in Thirunagar, a clean and pleasant suburb of Madurai. I bought a few items of furniture and enjoyed the experience of simple living and hoping that the period of austerity would soon come to an end.

I invited Lalitha and Krishna to spend a week at Thirunagar. to share the pleasures of Madura. We made a delightful trip by car to Cape Comorin and to Trivandrum to visit Lalitha's parents. The scenic drive to Kanyakumari at sunset was of breath taking beauty. The crimson glow over the hills still haunts my senses even as the voice of little Krishna standing in the front seat of our car reciting certain verses of the Ramayana. In his innocence he did not realise that these verses were relevant to my situation! Through them the sage Valmiki had described how Hanuman on his first meeting Rama and Lakshmana in the forest asked them 'Why is it that you, who look like princes, are walking through a forest, infested with snakes and wild animals?', to which with the grace and dignity of a gentleman of royal blood, Lakshmana replied 'A wicked stroke of fate has brought us here!'

The drive through Travancore brought back thrilling memories of my first trip with my father in 1941 and my later visit with Lalitha immediately after my marriage. In every sense Travancore was Krishna's 'Mother State' for he was born there and his early months of infancy were spent under the fostering care of his grand parents.

In late December 1958, the meeting of the Indian Academy of Sciences was held in Baroda and with the permission of the University and without its financial support, I was able to take my research students there and participate in a special symposium arranged at my request by Sir C.V. Raman, who insisted in his frank and forthright manner that the talks should be of general interest and adequate references be made to the experimental background in the subject. It was a most memorable trip which gave a boost to the morale of my group which was kept high by frequent seminars and publication of research papers in established journals.

January is not only the radiant month of the harvest (Pongal) festival but the period when Madras pays it homage to the King of games — cricket. The West Indies team made its second visit to India since 1948 and our city simmered with excitement weeks before its arrival. In our boyhood we had heard of George Headley, the Black Bradman and Sir Learie Constantine, names which conjured up visions of cricket in the Carribean isles. The Madras test, the fourth in the series, was held from the 21st to 26th January. Fortunately, a meeting at Delhi was arranged on the 27th and I was able to watch the West Indies pile up 500 runs during the first three days before I left for Delhi on the 25th. Garfield Sobers[4] had just entered the scene like a Comet in the cricket skies. I had the opportunity to watch the West Indian bowling at Hyderabad on my way back from Delhi. Wesley Hall, soon to become Krishna's hero, bowled with extraordinary grace and power, running to the crease like a cheetah on its chase.

On the 25th of January I went to Delhi in time to witness the Republic day parade the next day, before attending the Expert Committee meeting on Hindi terms. At the end of January my distinguished friend Janossy arrived in Madras from Hungary and I had an opportunity to reciprocate the hospitality he showed me during the winter of 1949 in Dublin. He had an exceedingly original mind and talking with him was both an inspiration and an education. He was engaged in the probabilistic interpretation of quantum mechanics, a rock on which great minds have foundered.

I made frequent appeals to the Vice-Chancellor before leaving Madurai and returning to Madras to teach the MSc Classes. It was then that Sir K.S. Krishnan arrived and evinced considerable interest in the theoretical physics group. His polite sympathy was of great moral support to us though by temperament he was not a person to take a positive action or render active assistance in furtherance of our cause. On the 7th of March I left for Bombay to participate in a symposium at the Tata Institute where I gave a lecture on Cascade theory. It was attended by Professor Bhabha and I had great satisfaction at this generous gesture by my teacher.[5]

I held the university MSc classes in a room in the Senate House, and informal seminars at home. This became a regular feature that by early 1959 we decided to formalise the informal group. Since I was regularly reading the Ramayana at that time, and since I was under a sentence of exile, we thought it would be appropriate to start an association called the Theoretical Physics Seminar in Madras on Rama Navami day[6] in April 1959. Devanathan, the most pleasant-mannered of my associates, was unanimously selected as the Secretary to take charge of all the academic and social arrangements. We decide to take advantage of the presence of eminent scientists who may visit the city for short periods, often times on their way to Australia or Japan or back to Europe and the United States. We had a monthly subscription of Rs. 10 which was used for dinners in honour of the visiting scientists and under favourable circumstances for their travel expenses from nearby places, like Bangalore! Of course it was expected they would stay with me at Ekamra Nivas.

It was then that I received a pleasant letter from Brueckner[7] (Appendix 12) at LaJolla that he was willing to award a postdoctoral fellowship to any of my PhD students whom I would recommend. This was in response to a conversation we had in U.S. after we met in a seminar at Princeton University. I named Vasudevan as the most suitable and he accepted my recommendation spontaneously. Vasu justified Brueckner's interest in him by staying in California for four years with single minded devotion to many-body problems, away from his family in Madras.

During May and June there was an opportunity to participate in a Summer School in Mussorie organised by the Ministry of Scientific Research. It was the first and last summer school of its kind where scientists and physicists from all over India were invited and everyone was asked to give a lecture on his research work. The moving spirit of the conference was Satyen Bose,[8] the doyen of theoretical physics in India very sympathetic to the aspirations of the young scientific community. He was more interested in stimulating creative thought than in expository talks on a single topic at the conference. This attitude was not relished by the physics group at the Tata Institute of Fundamental Research which did not participate in the conference. However there were enough representatives from various institutes to impart an all India character and in my own speech I exhorted that the Mussoorie spirit should pervade Indian sciences in due course. On my part I presented some work on a new approach to perturbation theory discussing the meaning of a kernel function, forward and backward in time, and deducing the Feynman formalism from a new point of view. Professor Bose was quite enthusiastic about my effort for he was more concerned about originality than the fashionableness of a topic!

The conference was held in a British style hotel which was going out of business but later became the venue of the Indian Administrative Service

School. Professor Mazumdar of Delhi took charge of all the arrangements from the dining table to the blackboard, yet many participants took the liberty of complaining about the monotony of the menu or the quality of coffee and he answered them with great patience. The deputy Secretary Mustafy became almost a member of the academic community and participated in the academic discussions. When I told him that Vasudevan's presence would be very useful at the conference he agreed to summon him by air, a gesture of surprising informality, unusual in rule-ridden officialdom in India. Vasudevan joined us a few days later and he was in a great mood since he was proceeding to La Jolla to work with Breuckner as post-doctoral fellow. Mussoorie is one of the most beautiful hill stations in India, with an atmosphere slightly different from that of Ooty and Kodaikanal. The actual town does not have such lush vegetation as Ooty but the Himalayas are visible at a distance and against such a background the sunsets over the surrounding hills are of exquisite beauty. I found satisfaction in long walks to the wooded hills, sometimes doing fifteen miles a day. Vasudevan would judicially avoid me just at the time when I started on my walk.

Vasudevan and I took the opportunity to make a pilgrimage to Haridwar and Lakshmanjhoola where we had an unforgettable experience. Deeply impressed by the scenic beauty, we crossed the jhoo la (bridge) on the Ganges by walk and rambled such a distance away from the bridge that on return we decided to cross the river in a boat which we found in a Jetty with a solicitous boatman inviting us to get in for a fare for Rs. 2 per person! We thought the ride was too cheap to be safe and so it proved to be! The boatman had no intention to move when we got in and we had to wait one full hour when suddenly out of nowhere two hundred persons appeared and entered the boat with such gusto that we thought it would topple over even before it started on its hazardous journey. Finally two other boatmen joined and started rowing while the water came up to the very rim as our hearts throbbed with inexpressible fear. As the boat crossed the river at oblique angles to adjust itself to the whirling current, Vasu and I said our prayers, which were reinforced by the loud voices of the passengers crying 'Ganga Mathaki Jai', which meant 'victory for Mother Ganges', which of course did not mean that we will be taken as its victims! Anyway we reached the other side (or else I would not be writing this diary now) and returned by taxi via Dehra Dhun to Mussourie.

There was another excursion by bus in which the old contraption swung round hair-pin bends at such speed that we feared it would go off the hill any time. When I complained to the driver he replied in friendly Hindi with the spirit of comradeship implied by the common national language "Yes, it is dangerous, but everything lies in the hands of God." Certainly not a comfortable thought even for a person with abiding faith in Divine dispensation!

Hazards apart, it was impossible even for the dullest soul not to be impressed by the scenic grandeur of the Ganges against the background of the Himalayan foothills. We also visited an ashram in Rishikesh where to my pleasant surprise I found there was a tablet acknowledging the assistance given by my mother to the Swamiji of the ashram when he visited Madras!

I resumed work at Madurai having arranged a more spacious house at Thirunagar. I took with me a cook who provided the only luxury available there — excellent South Indian food using fresh vegetables plucked from nearby farms, and fresh milk drawn from the cow's udder before our eyes. I also took with me a team of research students, Vasudevan who was completing his Ph.D., Ranganathan and Balachandran. The M.Sc. theoretical physics course was shifted to Madurai and so we had new M.Sc. students.

Since my house contained just enough furniture for my use, all the students sat down on the floor in true gurukulam style as I lectured to them on Feynman graphs asking them to imagine that it was a room at Harvard or Princeton! There was a strange but exhilarating feeling in teaching positron-electron scattering in Thirunagar, enjoying the luxury of eating raw guava fruit freshly plucked from the tree. We reminded ourselves of Oppenheimer at Los Alamos though of course his was a billion dollar effort to make the world safe for freedom by a single blast!

In early August 1959, I received an invitation from the University Grants Commission to serve on an Expert Committee to recommend proposals for the advancement of mathematics in India. What a strange irony when I was struggling to find a place in the sun for our theoretical physics group against implacable odds! Yet I accepted the assignment hoping to get acquainted with the mathematics scene in India, which I did during the 'inspection' visits to various centres before the committee made its final recommendations.

The first visitor to the theoretical physics seminar at Madras was Professor Ziro Koba,[9] one of the most active physicists in Japan, whom I had the pleasure of meeting in 1956 during my stay at the Yukawa Hall and later at the Rochester conference in the U.S. when I shared my room with him at the Seneca Hotel. This was followed by the visit of Dr. and Mrs. Kotani from Tokyo, who were my hosts in Japan in 1956 and who later visited me in Princeton. Their visit to Madras gave us an opportunity to reciprocate their generous hospitality and it was gratifying that they enjoyed their stay in Ekamra Nivas.

It was a particular pleasure to receive Andre Mercier[9] of the University of Bern, Switzerland. He was a person with a built-in smile in a friendly face, an excellent conversationalist who could tell charming stories of his association with the 'grand masters' of European science.[10]

Meanwhile I made repeated appeals to the University and received no response. In this predicament while invoking inspiration from the Ramayana I also drew comfort from human sources! I read Churchill's speeches to my students with irrepressible faith in his 'dictum': *And if you hold out alone long enough, there always comes a time when the tyrant makes some ghastly mistake which alters the whole balance of the struggle.* And indeed the balance was altered in October not by a tyrant's mistake, but by the visitation of friends from nowhere, an act of God through human agencies!

Notes

1) In 1958 the University of Madras first started an extension centre in Madurai (referred to also as Madura), the second largest city in the state of Tamil Nadu, about 200 miles south of Madras. This extension centre was formally made into Madurai University in 1966 with its own administrative set up.

2) In order to start a program in science at any university, one must first begin with the departments of mathematics, physics, and chemistry. Ramakrishnan was asked to start the program in physics in Madurai. Professor Santappa was asked to create the program in Chemistry. Finally, Professor M. Venkataraman was asked to launch the program in mathematics. Both Santappa and Ramakrishnan returned to Madras a few years later. Whereas Ramakrishnan became the Director of MATSCIENCE, Santappa stayed on in the Department of Chemistry of the University of Madras and eventually became its Vice-Chancellor. Venkataraman stayed in Madurai and became the head of the mathematics department when Madurai University was formed, and he shaped its development of the mathematics department with dedication.

3) Juan Trippe, the CEO of Pan Am, never wanted to be second in anything in the aviation industry. BOAC beat Pan Am by inaugurating the first trans-Atlantic jet service. The Boeing 707 that Pan Am used and favoured, was not only larger, but faster than the Comet. A few months after the launch of BOAC's Comet service from London to New York, Pan Am operated its first Boeing 707 service. The Pan Am aircraft deliberately left London about half an hour after the Comet but arrived at New York Idlewild airport about 15 minutes earlier, just to demonstrate its superiority!

4) Just as Donald Bradman was considered as the greatest batsman ever, Sir Garfield Sobers is widely regarded as the greatest all rounder in cricket history. He excelled as a batsman, but also was very impressive as a bowler, fielder, wicket-keeper, and captain. In 1958–59, the West Indies dominated India in all aspects of the game, but Sobers was the most consistent and forceful in batting.

5) Even though Ramakrishnan left the Tata Institute in 1947 because Bhabha did not have him a co-author in an important paper, he felt only disappointment at this action of Bhabha, but never had anger or resentment. Throughout his career, Ramakrishnan spoke about Bhabha very highly.

6) Rama Navami is the birthday of Lord Rama, just as Krishna Janmashtami is the birthday of Lord Krishna. Ashtami and Navami are the eighth and ninth days in the lunar calendar and occur once each between every new moon and full moon, and vice-versa. Since both Rama and Krishna faced several obstacles in their lives, Ashtami and Navami are considered by Hindus to be inauspicious days to start anything. Ramakrishnan's devotion to Lord Rama was so great, that he did not let that superstition stand in the way of formally launching the Theoretical Physics Seminar on Rama Navami day. Even after the creation of MATSCIENCE, Ramakrishnan always gave a special seminar at his Institute on Rama Navami each year to commemorate the Seminar.

7) K. A. Brueckner (1924–2014) was a eminent theoretical physicist who did fundamental work in the difficult area of many-body problems. In 1959, Brueckner was recruited by the University of California at San Diego (UCSD) at their lovely campus in La Jolla as one of the founders of the Department of Physics. He was instrumental in recruiting several excellent researchers to that department. As soon as he knew he was moving to UCSD, he wrote to Ramakrishnan offering a post-doctoral fellowship to ANY student working under him (see Brueckner's letter in Appendix 12), and Ramakrishnan selected Vasudevan to go to UCSD.

8) Satyendranath Bose (1894–1974) was a physicist of world renown. He laid the foundation of the Bose-Einstein statistics for thermodynamic equilibrium. The great physicist Dirac named the particles that obeyed the Bose-Einstein statistics as *Bosons*. For his seminal work, he was elected Fellow of the Royal Society (FRS).

9) Ziro Koba, an eminent Japanese physicist, was the first visitor to Ekamra Nivas after Ramakrishnan's seminar was formally named the Theoretical Physics Seminar. Koba received his PhD under the direction of Tomonaga who won the 1965 Nobel Prize Prize in Physics jointly with Feynman and Schwinger. Koba did very fundamental work on the foundations of string theory.

10) Andre Mercier was a Swiss scientist who first started in geology but then moved into mathematics, physics, and philosophy. He studied mathematics under the great Elie Cartan in Paris, and physics under Nobel Laureate Niels Bohr in Copenhagen. Thus he had moved closely with the grand masters of physics and mathematics. After holding temporary positions at ETH in Zurich and the University of Geneva, he became a professor of physics at the University of Berne. He was so much impressed with the Theoretical Physics Seminar, that he invited Ramakrishnan to the University of Bern for a series of lectures (see Chapter 19 on the Swiss Interlude).

The seminar was formally named the Theoretical Physics Seminar (TPS) in 1959. All the members of the TPS with Professor Kotani (1959).

Chapter 18

The Track of Destiny (1959–60)

The destiny of an individual is the result of interaction between his free will and external circumstance. Such circumstance, fortunate or unfortunate, is called an act of God over which the individual has no control. In many cases God acts through human agency, 'Daivam Manusha Roopena'. In October occurred an incident which in every sense was the starting point of an incredible series of events which led not only to the redemption from the Madurai assignment but the creation of Matscience.

I came to Madras just before Dasara and Krishna's Aksharabyasam (initiation into reading and writing) which was fixed on the 11th October, 1959, Vijayadasami Day.[1] On the one hand I was in a happy mood on the eve of the pleasant function. On the other, I had a feeling of frustration that there was no response to my repeated appeals to the University. It was then I received an invitation to an international gathering of African and Asian students at the Woodlands Hotel with C. Subramaniam as their Chief Guest. He was then the minister for Education and Finance at the same time.[2] I knew of him and had met him in Delhi when he was a handsome black-haired young Congressman during the momentous days of the Constituent Assembly. But I did not know him personally. I did not feel enthusiastic about going to the meeting since I felt that he may not have a real interest in higher eduction or creative science. So, that evening I decided to go to and relax on the beach with Lalitha. As the car was about to turn towards the Marina beach, she suggested that we should respond to the invitation and attend the meeting at least for half an hour, to which I consented most unwillingly.

At the meeting, to the embarrassment of everyone, an African student who was a member of an International group, mentioned in his welcome speech that there was racial prejudice in Madras though not of the rabid type as in South Africa! He complained that no African student was invited to any home during his stay in Madras. Naturally our Chief Guest was embarrassed and in his speech gave an explanation, not too convincing, that the social and domestic habits were such that we hesitated to invite foreigners and that such 'shyness' was not due to any racial prejudice. I was then asked to give a speech and I readily agreed making the point that prejudice gets strengthened by acknowledging its existence and sometimes has to be 'actively' ignored. Such snobbery existed even in the realm of science and in many professions. I recalled how my great father, when he started on his legal profession, was warned that it was difficult to succeed unless one had tasted the waters of the Cauvery! Within seven years he was the unrivalled leader of the Madras Bar, with many 'juniors' who hailed from the banks of the famous river! A similar situation arose in modern physics where it was generally agreed that Europe was the centre of science and culture. But through Oppenheimer and his new generation, the centre of gravity after the war shifted to the United States. Similarly there was the rise of the Californian Universities rivalling in prestige the Eastern Quintet, Harvard, Princeton, Yale, Cornell and Columbia. This speech of mine impressed Subramaniam so much that he made kind inquiries and asked me to meet him at his official residence, the Cooum House near the Marina. I did not waste any time and emphasised to him the need for providing suitable opportunities to the band of theoretical physicists working with me. I invited him for dinner to my house to which he readily responded[2] and it was at that dinner he suddenly asked me to explain what I meant by 'suitable opportunities' for creative science! I told him plainly that I meant something like the Institute for Advanced Study at Princeton! (see Appendices 13 and 14).

During Dasara we had a distinguished guest, Professor Dallaporta[3] from Italy. He was one of the most enlightened men of science I have met, a gentleman to the manner born, quite firm in his scientific opinions but with a deep concern for the feelings of others. I invited him for dinner to Ekamra Nivas and he enjoyed sitting in traditional Hindu style to hear Vedic recitations from our family priests.

Krishna's Aksharabyasam was celebrated in true orthodox style. Whenever I hear Vedic incantations, I feel that the Vedic tradition is the primary source of the intellectual vitality of our nation despite its economic and social disabilities.

As a member of the UGC committee I had the opportunity to visit the mathematics departments of the I.I.T., Kharagpur and the Calcutta

University. It was clear that the Queen of Sciences should find a more honoured place in our education. It was being assigned a secondary role in our education as an ill paid handmaid to science and technology which in turn could not meet the exacting demands and standards of a competitive world.

In early December we had two distinguished visitors who evinced keen interest in our efforts at theoretical physics. A. M. Lane, a young nuclear physicist from Harwell and his wife stayed at Ekamra Nivas and we took them on the inevitable excursion to Mahabalipuram before introducing him to our students. He was deeply impressed by our serious efforts in nuclear physics the down to earth approach with emphasis on the numerical calculations and comparison with empirical data. The other visitor was George Gamow, a true American phenomenon — a brilliant nuclear physicist who had influenced American scientific life not only through his famous research but through his interpretations of relativity and quantum mechanics to the world at large. His very presence was an inspiration — his scintillating wit, his exuberant friendliness, sent a wave of ecstasy through our group. The 'Gamow effect' was matched only by the 'Gell-Mannic impact' two years later.

Then came an opportunity for me to organise a symposium on elementary particle physics at the Indian Science Congress session in Bombay. I was told that Professor Abdus Salam was coming to India as the guest of the Congress and since our group was very well acquainted with his work, I took the liberty of inviting him to participate in the symposium. With the characteristic generosity of a true scientist, he readily agreed. I also invited him to spend a few days at Ekamra Nivas after the Congress — an invitation he readily accepted.

It was a privilege and pleasure to receive Salam as our house guest and it was gratifying that he enjoyed the vegetarian food prepared by Mani Iyer under the direction of Lalitha. He gave a magnificent two hour lecture in the seminar hall at Ekamra Nivas on his work on weak interactions and was thrilled to find such an eager and enthusiastic audience. We presented him with a portrait of Lord Krishna which he accepted with grace.[4] I expressed my appreciation of his gesture by accompanying him to a function where he performed *Namaas* (prayer) with his Moslem brethren.

What an incredible coincidence that Professor Niels Bohr (see Appendix 16) should visit Madras at the same time as the personal guest of the Prime Minister of India! He was accompanied by a deputy secretary of the Ministry of Scientific Research who travelled with him as his personal aide.

I had written to him earlier and he graciously consented to visit my house and meet my students. Though he stayed at the Government House, he accepted our invitation for dinner at the Woodlands Hotel and later at my residence. It was very flattering when he and his wife stayed on till midnight after an open air dinner on the lawns of Ekamra Nivas under the luscious foliage of the mango tree near the portico in spite of repeated reminders from the official aide to get back to the Government House. Mrs. Bohr, a gracious woman of great elegance and charm, kept a watchful eye to prevent her husband from putting his burning cigar into his coat pocket! Krishna watched with wide eyed interest and cherubic wonder the gentle face and genial smile of the father of modern physics. It was the best introduction he could have to twentieth century science even before he learnt to read or write!

We could not believe that such a great physicist,[5] engrossed in his own thoughts, who along with Einstein set in motion the atomic age, could evince interest in a group of 'stripling physicists' in a far off country! He did so in generous measure at a press interview at the airport which was reported in The Hindu[6] in good detail as follows:

Dr. Bohr said that the Atomic Energy Establishment was a mighty endeavour where research was being conducted in the very best way under the leadership of Dr. H.J. Bhabha, a scientist and at the same time a very great administrator. Asked about the place mathematics should occupy in the pursuit of Theoretical Physics, the professor said that in Bombay and Madras energetic efforts were being made for the promotion of knowledge of physics which demanded new mathematical methods and education of young people to be able fruitfully to contribute to such work. Wonderful work was being done in the field of theoretical physics by Professor Alladi Ramakrishnan of the Madras University.

He also seems to have mentioned this to the Prime Minister when he met him before leaving India. We received a communication from the Prime Minister's secretary to supply further information about our efforts in theoretical physics. A period of gestation had to follow before these inquiries were transmuted into action but it looked as if we were driven on the track of destiny or how else could one explain the triple interest of the Finance and Education Minister of the State Government, of scientists of international fame like Bohr and Salam and in turn of our esteemed Prime Minister! It is said in the Ramayana that the anger or pleasure of great men cannot go without a consequence — "amogha krodha harshascha". Obviously something salutary must happen if the heads of Indian administration on the one hand and world renowned scientists on the other, took simultaneous interest in our group. We had just to sail on the track of destiny, wait and watch for the goal to which it would lead us in God's good time. I therefore took courage to write to the Vice-Chancellor that it would be impossible for me to continue the sterile exile in Madurai, that I should be given two months leave to enable me to go to Switzerland in response to a kind invitation from Andre Mercier, and then transferred to Madras. The University reluctantly agreed and after my stay in Switzerland I was to return to Madras as a 'professor without portfolio' till suitable arrangements were made.

The UGC committee In March visited Madras and as one of its members I invited the entire group to an open air dinner on the grounds of Ekamra Nivas to meet my students and colleagues. My distinguished friend C.R. Rao, India's foremost statistician, evinced considerable interest in our aspirations.

In March there was the visit by Professor Philip Morrison[7] of M.I.T. who stayed with his wife at Ekamra Nivas. He could not control his excitement at watching an active group of theoretical physicists, trained in quantum mechanics and relativity in true 'ivy league' style and standard.

I participated with some of my students in the nuclear physics conference held under the auspices of the Atomic Energy Commission at Waltair. It was a splendid opportunity to meet scientists from various centres of India. In particular I enjoyed meeting the physicists from my alma mater, the Tata Institute of Fundamental Research, Bombay.

Meanwhile I got busy with the preparations for my journey to Bern. Lalitha decided to stay in Ekamra Nivas since Krishna who had his Aksharabyasam had to be admitted to school. Anyway he was too young to desire a Swiss holiday and would enjoy romping about Ekamra Nivas than being confined to a little apartment in Bern. Since I was travelling alone, I decided to 'try' the BOAC flight by the new comet jet from Delhi with a 'pioneer spirit' even as I did the first Dakota flight in 1946 on my trip to Delhi.

There would be enough time to dream and hope while rambling in the wooded environs of Bern and Zurich, or gazing at the snow-clad splendour of the Matterhorn and the Alpine panorama at Zermatt or watching the silvery trails of the motor boats on the limpid blue of the lakes of Geneva and Lucerne. The dreams and hopes were soon to be realised through the will of Man and the grace of God.

Notes

1) Vijayadasami, the last day of the Dasara festival, is the most auspicious day in the Hindu calendar. Hindus believe that anything started of Vijayadasami will be a success. Thus Ramakrishnan and Lalitha decided to perform Krishna's Aksharabhyasam on Vijayadasami.

2) Mr. C. Subramaniam, was one of the most well-known politicians in India. He held several portfolios both in the State Government of Madras as well as in the Central Government. He was even tipped to be President of India at one time. Mr. Subramanian was not the typical politician. He was very passionate about whatever he undertook and never stood on formalities. When Ramakrishnan invited him to Ekamra Nivas to meet the students of the Theoretical Physics Seminar, he readily agreed. That speaks volumes about his genuine interest in encouraging young talent and also about his warmth and informality.

3) Professor N. Dallaporta (1910–2003) was one of the leaders in the rebuilding of Theoretical Physics in post-war Italy. His early work on cosmic rays was very important, and is is due to common interest in cosmic rays, that he and Ramakrishnan came into contact. Dallaporta, who was Professor in Padua, was also responsible for the development of astrophysics in Italy. In astrophysics, his work ranged from stellar evolution to cosmology. A gentleman to the core, he was interested in understanding other religions and cultures. He was a man of profound religious convictions, and in his personal life he attained a harmony between religion and science.

4) Abdus Salam was not only an eminent scientist, but a great human being as well. He was a statesman for the discipline. Although he was not a Nobel Laureate in 1960, he was very famous in the scientific world. He readily accepted Ramakrishnan's invitation to visit Ekamra Nivas and mingle with the students. He had a very tolerant view of all religions, and happily accepted the portrait of Lord Krishna that Ramakrishnan presented him. He appreciated Ramakrishnan's efforts to infuse the spirit of Princeton in Madras because he too wanted something similar in his native Pakistan. Whereas Ramakrishnan succeeded in creating MATSCIENCE in Madras, Salam's International Centre for Theoretical Physics was founded in Trieste, Italy, in 1964 and not in Pakistan (see Part II).

5) The entire atomic theory in physics rested on Bohr's model of the atom. Bohr matched Einstein in the universal respect he commanded over all of science, and so it was natural that he was invited to India as the personal guest of Prime Minister Nehru who was passionate about science.

6) *The Hindu* reports news about science and education more frequently and in greater detail than any other newspaper in India. Although based in Madras, The Hindu is a national newspaper and is read throughout India.

7) Professor Philip Morrison (1915–2005) was on the faculty of MIT. His fundamental paper of 1958 on cosmic rays is considered be the beginning of gamma ray astronomy. Here too, common interest in cosmic rays was what brought Ramakrishnan into contact with Philip Morrison. In addition to doing fundamental research, Morrison believed in reaching out to the public by writing popular books, and being involved in radio and television shows on science.

Professor Abdus Salam lecturing at the Theoretical Physics Seminar (Jan 1960).

Lalitha and me with Professor Abdus Salam at Ekamra Nivas (Jan 1960).

Dinner for Nobel Laureate Niels Bohr (Center) at Ekamra Nivas (Jan 1960).

The author and his wife Lalitha with Professor and Mrs. Niels Bohr at Ekamra Nivas (Jan 1960).

Chapter 19

The Swiss Interlude (1960)

On the 22nd April 1960, I left by the Viscount service for Delhi enroute to Switzerland while Lalitha and Krishna stayed behind in Madras. I boarded for the first time in my travels, a jet aircraft — a BOAC comet. Naturally, I was excited about the take-off which was fast, smooth and steep unlike that of the conventional propeller aircraft. After stops at Bahrain and Beirut, it flew direct to Zurich. Dawn broke over the Alps and I watched the sunrise above the clouds, a sight denied to Shelley and Keats before the era of aviation. The jet-eye view of the Alpine panorama was of fantastic beauty — snow-capped mountains bathed in the light of the gleaming dawn, awakening thoughts of sublime ecstasy and sweet melancholy.

Zurich was in a holiday mood at the early spring festival, a time when daffodils come before the swallow dares and take the winds of March with beauty! After a pleasant and fruitful meeting with Heitler, I left for Bern where I was received by Professor Mercier and his student Gorge. I stayed in the modest and elegant hotel Regina which served wholesome vegetarian food and I started attending the Institute the very next day. I gave my first series of lectures on stochastic processes. That weekend I went on a wonderful excursion by boat to Lucerne, Interlaken and then through Lake Thun to Geneva. The scenery was spectacularly beautiful and particularly enjoyable over a delectable lunch in the spotlessly clean dining room of the Swiss motor boat.

Mercier drove me to Winterthur for the meeting of the Swiss Physical Society, where I had interesting conversation with Stueckelberg,[1] known to physicists as a forerunner of Feynman. We talked about vegetarianism and he caught me on the wrong foot and questioned me about my leather shoes!

I gave my second series of lectures on scattering theory and the new interpretation of the Feynman formalism. Professor Wouthuysen[2] was intrigued by this approach and invited me to give a lecture at CERN.

Professor Mercier drove me in his smooth English Rover to Geneva and we drank the delightful beauty of the Swiss countryside as we sped along the 'petit' highways through pretty little towns set amidst the exquisite loveliness of winding lakes and verdant valleys. Professor Wouthuysen received me at CERN where I lectured on the Feynman graphical formalism. I met Dr. A.L. Mudaliar at the hotel du Rhone and talked about initiating a visiting scientists programme at Madras, which, despite his support turned out to be a futile effort. That weekend I went out on an excursion to the Chamonix valley in France abut eighty kilometers from Geneva. The French countryside was lush and beautiful, a little wilder than in Switzerland but the houses did not look as trim and elegant. We saw the snowbound Mount Blanc and went to Pic-Du-Midi by cable car. It was a thrilling experience made more so because the cable car was an old structure compared to its counterpart in Switzerland. I made up my mind to bring Lalitha and Krishna to such a lovely spot on their next European trip.

In the evening I had dinner in Professor Wouthuysen's house where he provided delicious mushrooms and tomatoes on toast to suit my vegetarian palate. I met Professor Casimir,[3] the Director of Philips, one of the wittiest conversationalists who reminded me of Professor Gamow. He gave an entertaining account of how to distinguish between the upper and lower classes in England through table manners — those that do not and do divide their coat tails before sitting down, those that add milk to tea or tea to milk, those that move the spoon forward or backward while drinking soup. He raised the legal question whether a cable autobus was an 'automobile' in the strict legal sense since it drew power from an external source.

An interesting incident occurred one evening when Professor Yang asked Wouthuysen for a 'lift' in his car. He took him to the town and Yang got down and caught a tram quite casually — something unthinkable in India where Nobel prizemen are lionised. In fact I had heard a story earlier that one of the reasons why Einstein did not visit India was his fear of being idolised and smothered by adulation. I returned to Bern and after a most enjoyable stay there, I went to Zurich where I called on Heitler and discussed his probable visit to Madras. I left for London by a Viscount and had the pleasant surprise of meeting my teacher Professor Bhabha at the airport. I stayed at the De Vere hotel next to the Kensington Palace hotel where my father stayed thirty years before! I was the guest of Professor Salam at the Imperial College and that evening he took me out to a London coffee house. I had discussions with Salam, Mathews and Feldman on my viewpoint regarding Feynman graphs.

It was really difficult to stimulate interest in theoretical physicists on a new viewpoint in Feynman graphs. No one dared to retouch the Master! It was believed that the Feynman graph, like Shakespearean verse, was too perfect to admit interpolation or interference! I also met Captain Maxwell of the Pergamon Press and he was very lenient about the date of submission of my manuscript. I visited Oxford with Professor P.T. Mathews where I attended a seminar and spent a bright sunny afternoon watching dull cricket. It was eleven years since I visited Oxford and I found considerable American influence on the social life of students. In the evening I was the guest at a high table[4] dinner of Professor D.G. Kendall, kind and generous as a friend but demanding uncompromising standards in mathematics. Then I left for Harwell and stayed with the Lanes in their well furnished apartment. I lectured on 'Hyper-fragments' and had a discussion on inverse probability with Skyrme who later visited our Institute in Madras. At dinner there were interesting discussions on Indian marriages and the Hindu way of life.

I returned to London to see a musical and a ballet before taking the Scandinavian Airlines (SAS) Caravelle to Copenhagen. It was a delightful flight and on arrival I went to an elegant hotel arranged by Professor Bohr. I called on Professor Bohr at his residence and was impressed by the lovely garden and the beautiful mansion.[5] I attended a symposium on Nuclear Structure to which 'old friends' of the Copenhagen school were invited.

The most memorable experience was the dinner at Bohr's residence to which Professor Jensen the Nobel Laureate was also invited. A superb vegetarian dinner was provided in the main dining hall under glowing chandeliers. There was also a wonderful party at Aage Bohr's[6] house where I met young short-cropped Americans, riotous in conversation but serious in the business of nuclear physics. I took photographs at Bohr's house which I treasure with pride and delight.

I left for Paris by SAS Caravelle and arrived at Le Bourget airport where I was received by Kichenasamy, a colleague of Madame Tonelat who had invited me to lecture in Paris. That evening I spent walking along the Champs-Elysee and the Concorde where Paris lays bare its architectural opulence and the splendour of the Napoleonic era. Later I went to the opera La Traviata at the Opera house. At Madame Tonelat's house I met Papapetrou, the well-known gravitationist and another professor from the Moscow University. The next two days I rambled in the Louvre museum which treasured human achievement in art too magnificent for the human eye to comprehend.

I went to Frankfurt by Lufthansa Boeing 707 and after a short stay there I visited Marburg where I spent a week in the style which would have satisfied a 'nature addict' like Robert Louis Stevenson, The weather was so good that my lectures were arranged at night so that the afternoons could be spent on walking tours enjoying the ambrosial fragrance of wild flowers amidst wooded hills of stateliest view. I used to walkabout fifteen miles a day through the pleasant environs of Marburg, lunching casually at wayside restaurants enjoying 'frucht milch' and shopping in the small towns. I frequently made motor trips through the cattle country. At dinner with my host Professor Flugge we had the usual discussions on the place of Sanskrit in India, the meaning of Gotra and Sakha and the comparison of Rama and Krishna as Avatars of Vishnu. My evening lectures on cascade theory and Feynman graphs were well-attended and well-received. It is hard to believe that amidst the peaceful beauty of Marburg, that there was a time when Hitler's armies and panzer divisions swelled the countryside. Nobody now liked talking of War. They enjoyed watching the cows grazing on the meadows in the mellow light of a tawny sunset.

I left Frankfurt by Olympic Airways DC-6B for Zurich. There was a telegram from Professor Dallaporta inviting me to a conference in Trieste. So from Zurich I flew by Swissair DC-6B and arrived in Milan. Then went to Trieste by train and enjoyed watching the beautiful countryside. I stayed in the lovely hotel Riviera in Trieste overlooking the Adriatic. It was a wonderful conference at the Castle Miramare. It was there that the idea of an International Centre for Theoretical Physics (ICTP) at Trieste was conceived and soon plans for the new Institute engaged the mind of Professor Salam.[7] I lectured for one and half hours on my views on Feynman formalism before full-blooded theoretical physicists like Fubini. I reached Rome after a train journey through Milan and took the Air India to Madras.

Notes

1) Ernst Stueckelberg was a noted Swiss physicist who did fundamental work but unfortunately did not get the recognition he deserved. In 1935 he gave an explanation of nuclear interactions as being due to an exchange of vector bosons. But he did not publish this work since the great Wolfgang Pauli told him that it was ridiculous. Hideki Yukawa received the 1949 Nobel Prize for giving a similar explanation of nuclear interactions.

One of the major advances in theoretical physics in the 1940s was the renormalisation program in quantum field theory. In the early 1940s, Stueckelberg wrote a long paper providing a complete and correct description the renormalising procedure for quantum electrodynamics. His paper was rejected by the *Physical Review*. In 1965, Richard Feynman, Julian Schwinger, and S. Tomonaga were jointly awarded the Nobel Prize in physics for work in quantum electrodynamics.

2) Siegfried Wouthuysen was a Dutch physicist who received his PhD in 1948 at Berkeley under the direction of Robert Oppenheimer. Wouthuysen became famous in 1949 for his joint work with Leslie Lawrence Foldy and Robert Marshak at the University

of Rochester, with subsequent improvements by Wouthuysen and Foldy. Wouthuysen had a long association with CERN where he hosted Ramakrishnan in 1960.

3) H. B. G. Casimir (1909–2000) was a very eminent Dutch physicist. Several things in physics are named after him, such as the 'Casimir effect'. He was Director of the Phillips Research Lab in Eindhoven, Holland.

4) British Universities are composed of a collection of affiliated colleges, a system that is followed by universities in India as well. Each student and faculty member in a British University is associated with one of the colleges, and these members meet in the college "hall" for dinner. In the dining hall, one of the tables, or one row of the tables, is on a raised platform — hence the term "high table" — and only Fellows of that college or their invited guests, sit there for dinner. It is an honour to be invited to sit at the high table, especially at Cambridge and Oxford Universities which are steeped in tradition and known for their excellence.

5) Niels Bohr had a magnificent mansion in Copenhagen, like Ramakrishnan's Ekamra Nivas in Madras. And like Ekamra Nivas, Bohr's mansion had a lovely statue in the middle of the garden (see photograph below).

6) Aage Bohr (1992–2009) was the distinguished son of a distinguished father. Aage Bohr was at the University of Copenhagen. He won the Nobel Prize in physics in 1975.

7) A small group of physicists gathered at the lovely Castle Miramar near Trieste, Italy, for a Symposium on Elementary Particles Interactions 22–26, June 1960. Salam was then a Professor at Imperial College, London, and attended this Symposium. It was there that he first proposed to this group of scientists that included Ramakrishnan, the idea of creating an International Centre for Theoretical Physics (ICTP).

The author with Nobel Laureate Niels Bohr at the latter's home in Copenhagen (1960).

Chapter 20

Back to Ekamra Nivas (1960)

After my return from Switzerland, my task was just to wait for the official communication from the University transferring me from Madurai to Madras. I received it on the 9th of July when I promptly wrote a letter to the Vice Chancellor thanking him for his solicitous interest in my career. No one could be happier than Lalitha and little Krishna about my return to Ekamra Nivas after the two year exile. It was the flamelike purity of my parents' lives that sustained me through all trials and tribulations. I felt thankful to God for giving me the courage to strive, through honest means, without yielding to unkind circumstance. I therefore plunged into the academic task of writing the book on elementary particles with cooperation my enthusiastic students. I wrote the preface with the dedication,

To my great father

ALLADI

(Sir Alladi Krishnaswami Iyer)

One of the principal architects of the Indian Constitution.

His was not a life just lived,
but a life charged with humanity and intellect.

The inscription was taken from a touching letter from my good friend Tammu Achaya in 1953. I sent a copy of it to Sir S. Radhakrishnan, then Vice-President of India, who had a paternal interest in my scientific career. I received a delightful letter from him which encouraged me in my efforts to build a theoretical physics group in Madras.

The next few months I made frequent trips to Delhi using the visits to contact the University Grants Commission through Dr. Laroia and the Department of Scientific Research headed by Minister Humayun Kabir. I also attempted to initiate a visiting scientists programme at the Madras

University starting with a post-doctoral fellow from Switzerland but this did not fructify due to interference from some academicians in high places.

I was considered a *Professor without portfolio* and so had no office at the University. I was permitted to lecture in the German class room in the University Senate building and the University addressed me as Professor of Physics, care the German Class Room! This did not deter or discourage me for it was indeed a blessing in disguise since Ekamra Nivas became a defacto Institute of Advanced Study where I conducted seminars and lecture courses. It was then we had the privilege of receiving a mathematician, Professor A.H. Copeland[1] who expressed sincere appreciation of our efforts.

I gave a radio talk on "Atoms for tomorrow" and another at the Rotary Club on "Incentive to research" at the Hotel Oceanic, rather ironic when I had still to find a place under the sun for theoretical physics at Madras!

On Vijayadasami day I started gathering some material on my father's life and work starting with what he dictated to me as 'My Story.' Little did I realise that in a few years it was to be transformed into the Alladi Diary with a more comprehensive aim and purpose.

At the end of December 1960, I became deeply concerned when Lalitha developed some trouble in her vocal chords which was affecting her voice. On consulting Dr. Sathyanarayana the ENT expert, he advised a delicate operation for the removal of nodules on the vocal chords. Lalitha was admitted to the Wellington Nursing Home where she underwent the operation. With the grace of God and the skill of the surgeon, she recovered her voice and was able to practise Carnatic music a few years later, first with the encouragement of Mannargudi Sambasiva Bhagavathar[2] who became her teacher, and later with the blessings of the great Chembai Vaidyanatha Bhagavatar, the ardent votary of Lord of Guruvayoor.

Notes

1) Arthur Copeland (1898–1970) was an American mathematician who worked on the foundations of probability. He evinced a lot of interest not only in the idea of Ramakrishnan conducting the seminar at home, but on the work of Ramakrishnan and his group in stochastic processes. Copeland was from the University of Michigan where he introduced new courses in such as on game theory. So he could appreciate Ramakrishnan's desire to change the outdated syllabus of Madras University.

2) Mannargudi Sambasiva Bhagavatar was a renowned in *Harikatha*, namely the art of telling religious and cultural stories interspersed with classical (Carnatic) music. He was a disciple of Maharajapuram Viswanatha Iyer, one of the greatest Carnatic vocalists of the 20th century. Sambasiva Bhagavatar was also an outstanding music teacher, and Lalitha was his student starting from the mid-sixties.

Chapter 21

A Web of Mingled Yarn (1961)

The year 1961 was eventful, a period of fluctuating fortunes, domestic tragedies, tensions, excitement, anxious waiting and ultimate triumph.

The first few months I spent in correcting the proofs of my book on 'Elementary particles and cosmic rays'. In early January I completed the paper 'On stochastic methods in quantum mechanics' with my student Ranganathan and communicated it for publication. With my students I spent many hours at my home discussing the Mandelstam representation and the analyticity of scattering amplitudes. I never could understand why this problem never engaged the attention of Feynman who solved greater problems in quantum electro-dynamics.

It was then I received the amusing offer from the Education Ministry to be a part-time member of the Hindi language commission! I accepted it to continue my contact with the 'pulse' of Delhi. But destiny had willed that our hopes were to be answered not from such efforts but from incredible sponsors before the year was out.

A trivial domestic incident occurred which made a great impression on my future life. Six year old Krishna was getting interested in cricket, in fact crazy about it and when I once showed no interest in joining him at play he just remarked that I was getting old! And he was right! In life there is a time to play, a time to love, to work, to achieve and then to rest and fade away. Anyway I joined Krishna and his friends in the cricket 'Test Matches' played with a tennis ball on the vacant ground next to Ekamra Nivas. It was great fun watching Krishna imitate his hero, Wesley Hall of the West Indies. I enjoyed this excitement till we started building 'the Alladi House' on that adjacent compound in 1967.

One of the most interesting lectures in early 1961 was by Kampe de Feriet of France with whom I discussed random functions. I also got very much interested in strong interactions hoping to discern analytical properties of scattering amplitudes even in the Feynman formalism.

In early February Professor Heitler of Zurich visited Madras and gave
two lectures at Ekamra Nivas on 'The finite size of elementary particles'
during my absence since I had to be away in Trivandrum to see my father-
in-law who suddenly took ill. After he improved I returned to be with
Heitler in Madras leaving Lalitha and Krishna behind in Trivandrum. The
journey by Dakota was memorable since there were only two passengers
and I stayed all the time in the cockpit and watched the landing in Madras
by night seated just behind the pilot!

The calm and placid atmosphere of Madras life was suddenly enlightened by the golden flash of a Royal visit. On the 8th February, the BOAC
Britannia landed at Meenambakkam Airport bringing the Queen of England. There was a reception at the University and we had an opportunity to see the Queen. How fortunate is Britain in having such a queenly
queen whose grace, dignity and bearing have inspired the adoration and
loyalty of a country which had preserved its traditions for a thousand years
from the Norman invasion of England to the British landing at Normandy!
Throughout my boyhood I had heard of the Viceregal splendour, the glitter
and pomp which dazzled the Indian prince and pauper alike into worshipful submission. It was incredible that the Queen, before whom the Viceroy
would kneel, was right there before us in a simple ceremony, acknowledging
the greetings with a gracious smile.

The next two months were a period of deep domestic tragedy. Lalitha
lost both her parents and since she had no brother, I performed the obse-
quies in accordance with Hindu rites. The tragedy was even more poignant
since her parents were looking forward to see Krishna grow and my dreams
for an Institute come true.

In March, between the domestic tragedies, I participated in two symposia, one in Bombay and another in Chandigarh. I flew to Bombay from
Trivandrum by a Dakota and the flight by the noisy but reliable aircraft
had its own pleasures and attractions. What a beautiful coastline from
Kerala to Bombay fringed by coconut palms beside golden sands! Amidst
the profusion of Nature's delights why should man be subject to the numb-
ing sadness of losing those he loves? It is the Maker's plan that pleasure
and pain should be the dual states of man?

At Chandigarh I spoke on the possibilities of studying Feynman Kernels
in greater detail. I visited the Bakra dam and remembered the statement of
Nehru about the economic revolution due to such projects though my own
faith and Hindu sentiment would not allow me to think of them as temples
as Nehru did. On my way back to Madras I lectured at the Delhi University

on perturbation theory. I returned to Madras by the night service via Nagpur by a Dakota which was really treated as working horse!

For a month I was busy with the proofs of my book on elementary particles and cosmic rays. I heard that Professor Bartlett was made a Fellow of the Royal Society, a belated recognition to so famous a statistician. I congratulated him by cablegram.

The Mathematics Review Committee of the University Grants Commission, of which I was a member, submitted its report on the development of Mathematics. While other members of the committee recommended advanced centres within the University, I insisted that there should be provision for creating institutions outside the University. Though my dissent was not formally included in the report, it turned out to be prophetic by the creation of Matscience at the end of the year.

During April we received the great American mathematicians, Marshall Stone[1] and had many long conversations with him on the development of mathematics in India and the U.S. When I asked the professor, a resident of Chicago, what he wished to see during his stay with us, he expressed his desire to sleep one night in the Ashram of Ramana Maharishi[2] in Tiruvannamalai! I then understood the depth of wisdom in Thyagaraja's immortal song that true happiness lies in sublime tranquillity. The din and bustle of O'Hare inspire a longing for a saint's serenity!

We held many seminars at Ekamra Nivas, sometimes six hour sessions! At that time I was working on a paper on 'The physical basis of quantum field theory' which was a departure from convention. I derived the free field representations by using perturbation theory!

At the end of May I went to Dalhousie for a summer school where I lectured on this new work on the physical basis of quantum field theory which I communicated for publication soon after. The most exciting part of the stay in Dalhousie was the sight of the snow-capped Himalayas, the beauty and magnificence of which have to be gazed at least once in a lifetime. What a divine dispensation that a Royal Knight from New Zealand and a hardy Sherpa from the Indian hills should, in joint endeavour, pay homage by human presence to those smokeless altars of sun and snow!

On the 7th of June I called on C.S. and he showed considerable interest in starting an Institute of Mathematical Sciences. He was leaving for the United States next day.

A summer school was being arranged at Bangalore by the Tata Institute with Gell-Mann and Dalitz as the lecturers. With Krishna and Lalitha I started by car in the morning for Bangalore and went by the southern route through Vellore picnicking under shady groves of mango and coconut on the way. The first lectures were by Gell-Mann and the impact was Gell-Mannic! It was like seeing Don Bradman when he was breaking records, or Frank Sedgman volleying his way to the Wimbledon crown. I had the opportunity to watch Bhabha, Dalitz and Gell-Mann in close discussion. In his lectures Gell-Mann introduced the now famous unitary symmetry which won for him the Nobel prize. It is impossible to find lectures more thorough than those of Dalitz and it was no wonder that the Dalitz analysis of the famous tau-theta puzzle prepared the ground for the Nobel Prize conquest of the Chinese physicists Yang and Lee.

We took Gell-Mann by car to Nandi Hills and conversations with him convinced me that genius was the offspring of enthusiasm and talent. We had a wonderful excursion to Belur, Halebid and Somnathpur — marvels of sculpture and carving amidst the lush profusion of nature. All these are creations of anonymous artisans — what religious fervour, what aesthetic genius must have brought forth such intricate and lovely friezes in stone! At Sravanabelagola we saw the monolithic idol of Gomateswara to which millions pay worship once in twelve years. We returned to Bangalore after visiting the exquisitely lovely waterfalls at Sivasamudram.

At the end of the conference I invited Gell-Mann[3] and Dalitz[4] to spend a few days as my house guests in Ekamra Nivas. The impact of their stay for those few days on my scientific career was greater than that due to years of conventional education! The admiration of Dalitz for Gell-Mann was so great that sometimes he wondered whether it was worth pursuing research when Gell-Mann was striding the stage like a colossus. I did not share the view of Dalitz and I was hopeful of doing something on the fundamental problems that engaged the attention of Feynman and Gell-Mann. After all, there were votaries of Sri Rama who were not content with reading the epic of Valmiki. Thyagaraja and even his 'chelas' composed verses to express, in their mortal lives, their immortal longings for the lotus feet of the Lord. Likewise the study of Nature is not reserved only to a chosen few.

In Madras I took Gell-Mann to a Bharatanatyam performance. He was making good humoured fun of me that I was an orthodox Guru, too much weighed down by Hindu tradition. After watching the movements of the danseuse and hearing the lilting music, he conceded that Hindu society was not so inhibited as he thought!

I took Gell-Mann to meet our Vice-Chancellor, having informed him earlier that Gell-Mann was a legend in a country which does not easily accept legends. The interview between a venerable Vice-Chancellor whose power over the University was so great and young Gell-Mann whose influence on science was universal was watched by me with anxious interest.

Life looked very dull after Gell-Mann and Dalitz left but fortunately our students received part of the enthusiasm and we resumed seminars at Ekamra Nivas. We tried to reproduce Gell-Mann's lectures in the course of twelve seminars.

It was then I received a letter from Vikram Sarabhai asking me to take care of Professor Sandstrom, a cosmic ray physicist of Sweden, during his visit to Madras. I was well aware of Vikram's interest in the time variation of cosmic radiation. Sandstorm was entertained at dinner in Ekamra Nivas when I introduced all my students to him with a teacher's pride.

On August 2, 1961, the University Centenary building was opened by Sir S. Radhakrishnan, then Acting President of India. There was a festive air of sparkling humour and glowing tributes when so many distinguished men of science, politics and administration met in a joyous mood of jubilant celebration. Sir A.R. Mudaliar received the honorary doctorate from his twin brother Sir A.L. Mudaliar, the Vice-Chancellor who in the citation expressed that he knew the recipient from the time of his birth!

I despatched the final proofs of my book and felt a younger man. I relaxed by reading a thought-provoking essay on causality and dispersion relations by Aage Bohr who won the Nobel Prize a few years later in the tradition of his famous father.

In August I was able to complete some work on the concept of virtual states in the Feynman formalism. I also did some work on density correlations based on Weisskopf's paper and included this as a part of my paper on quantum field theory.

At that time there was a visit to Ekamra Nivas by Professor Glaser,[5] the Nobel Prizeman who evinced great interest in the work of our theoretical physics group. A few days later C.S. returned from the United States and related his experiences in the Rand Corporation where he met Vasu and Bellman. About the same time I got a letter from Maurice Shapiro that he would visit Madras on his way from a conference in Kyoto.

Shapiro arrived from Kyoto and we took him to the dance drama 'Ramayana' which thrilled him to distraction. He gave a two hour talk in Ekamra Nivas on the Kyoto conference and later met our Vice-Chancellor. The same evening he had an interview with C. Subramaniam when I made a great gamble for a theoretical physics institute. Shapiro told Subramaniam that watching the students at work in Ekamra Nivas reminded him of the manner in which scientists gathered round Oppenheimer at Los Alamos!

That was high and generous tribute which made a great impression on Subramaniam.[6] Shapiro went on to suggest that the students should meet the Prime Minister of India. On the 30th of September, I received a telegram from C.S. after he went to Delhi that the Prime Minister was willing to meet the members of the Seminar on October 8th.[7] That day turned out to be one of the happiest days in my life. After hearing a splendid concert of Madura Mani Iyer, I went to the Raj Bhavan for a dinner in honour of the Prime Minister.

In spite of a crowded series of official engagements for the Prime Minister, it was found possible by the Hon'ble Subramaniam to arrange an interview at the Raj Bhavan at 9:30 p.m. after an official dinner, to which my wife and I were invited to have a preliminary occasion to meet the Prime Minister. I need hardly describe my trepidation and anxiety, for so much was to depend upon a few minutes of conversation — almost the dreams and aspirations of a whole scientific community hung on the smiling lips of our Prime Minister! Ten minutes before the close of the dinner, he summoned me and all the students to have a personal interview with him. It was a strange feeling for me to do all the talking with a man who held almost unquestioned sway over the destinies of our country. At the end he asked me only one question: "Are you really convinced that we should have an Institute of the kind you are insisting upon?"[7] I naturally said 'yes' with all the emphasis at my command and he just smiled (see Appendix 17).

The Navarathri was therefore celebrated with pleasure and expectation. Then followed two months of teasing suspense. When I went to Delhi I met Kothari who was always obsessed with the idea of centres of advanced study within the University, obviously an unhelpful attitude when we were planning a centre independent of the University! I had a rather hectic schedule in Delhi, meeting Chandrasekhar who was there at that time, Professor Humayun Kabir, the Cabinet Minister for scientific research and Laroia of the University Grants Commission.

Later Chandrasekhar visited Madras and gave a series of lectures in the university and took an active interest in our theoretical physics group. We arranged a plantain leaf dinner in traditional Hindu style at Ekamra Nivas and he answered the gesture by wearing a dhoti in Madras style.[8] Then followed the visits of Hazlett,[9] the Vice President of the University of Rochester, and Lighthill, then Director of the Royal Aircraft Establishment. It was as though emissaries were sent by Divine dispensation to prepare the ground for the creation of our Institute. On the 31st November I made a final appeal to Subramaniam to start the Institute of Mathematical Sciences (see Appendix 18). There was a dinner at Ekamra Nivas in honour of Professor Lighthill and Subramaniam. Meanwhile things happened favourably within the Government of Madras and the Chief Secretary asked me to formulate the aims and objects of the proposed institute.[10]

December 7 was a significant day when I met C. Subramaniam and the Chief Secretary both of whom now agreed to the creation of the Institute. Meanwhile seminars were going on with unabated vigour in Ekamra Nivas and then followed a drama which I describe in some detail to convey the spirit of the hour.

The annual symposium on Cosmic Rays under the auspices of the Department of the Atomic Energy was arranged in Madras, and inaugurated by Sir A.L. Mudaliar. I was in charge of the arrangements for the symposium but my mind was set on the new Institute. On the 20th of December I met C. Subramaniam at the Madras Airport on his return from Delhi. What a thrilling moment it was when he informed me that the Prime Minister agreed to be the Patron of the Institute!

December 22nd was a day of dedication in my career. At 1:30 p.m. I was called by the Education Secretary, Mr. K. Srinivasan,[11] who issued an order of appointment to me as the Director of the Institute (see Appendix 19). What a providential coincidence that it should be the birthday of Srinivasa Ramanujan! I met Professor Chandrasekhar the same evening to inquire whether he could inaugurate the Institute on the 3rd of January 1962. He agreed to do so provided we obtained the suitable air reservations for U.S. the next day. What a simple request from so great a man!

That evening we had a music performance in Ekamra Nivas by Lalitha's sister Kamalam,[12] concert originally intended for the delegates of the conference. As the music rose with sonorous swell, our hearts throbbed in resonant unison. It was an evening of prayerful thanksgiving and cheerful dedication for a dream realised. For a fleeting moment I wondered — was it just a dream that a dream had come true!

My loyal secretary Nambi Iyengar[13] typed in one evening over seventy letters to scientists all over the world informing them of the inauguration of the Institute.

The next day I tendered my resignation to the University of Madras, an event for which I was well prepared all these years.[14] I was reminded of the great verses of Thyagaraja and Syama Sastri that there is no greater folly than attempting to please someone who refuses to be pleased. Even Gell-Mann's genius and charm which had enlightened physics had failed to impress the University of Madras of the necessity of an advanced centre of learning in theoretical physics.

The first cable of greetings arrived on the 27th of December from my dear friend Messel of the Sydney University.[15] Within twenty four hours over a hundred cables arrived from all parts of the world (see Appendix 20D).

Thus ended a year of tragedy and triumph, tension and relief, despair and hope, suspense and success. The web of life is of mingled yarn, good and ill, spun in equal measure.

Notes

1) Marshall Stone was one of the most influential mathematicians of the 20th century. The *Stone-Weierstrass theorem* is so fundamental, that everyone going through graduate school sees it in a course on real analysis. Stone was more than just a great mathematician. He believed in making contributions to the profession. Under his dynamic leadership as Chairman, the mathematics department at the University of Chicago grew to great heights in the 1950s. The period when Stone was chairman in Chicago has been referred to as *The Stone Age*! Stone was also President of the American Mathematical Society. Thus with his own desire to mould the shape of mathematics education and research in America, he could understand and appreciate Ramakrishnan's interest in creating a stimulating atmosphere for scientific research in Madras. Also Marshall Stone's father was a Justice of the US Supreme Court. Thus Professor Stone could appreciate and understand Ramakrishnan's background as well.

2) Ramana Maharishi (1979–1950) was a great sage. In Sanskrit, *Maha* means great, and *rishi* means sage. *Ashram* means the residence of a sage which is usually in a forest setting that is peaceful for meditation. Ramana's ashram is in Thiruvannamalai, which is an hour outside of Madras.

3) Murray Gell-Mann is one of the most influential theoretical physicists. His quark theory of the atom won him the Nobel Prize in 1969. But even before that he was a legend in physics, for his fundamental work and for his brilliant and charismatic lecturing style. Gell-Mann was one of two lead speakers at the 1961 Tata Institute Summer School in Bangalore that Ramakrishnan attended with about half a dozen of his students. Ramakrishnan was on the Organizing Committee of this Summer School.

It was gracious of Gell-Mann to accept Ramakrishnan's invitation to stay in Ekamra Nivas after the summer school and interact with the students of the Theoretical Physics Seminar. Gell-Mann was very much interested in foreign languages. He understood how Indian names like Ramakrishnan are formed by combining the names of Gods Rama and Krishna. However in such combinations, the order is chosen to sound pleasing to the ear. Krishnarama for instance is never used. Each morning at Ekamra Nivas, at breakfast, Gell-Mann would tell Ramakrishnan that he had come up with a few names such as Sivaraman, and he was always correct in the combinations.

4) Richard Dalitz (1925–2006) was a well known Australian physicist who was a professor at the Enrico Fermi Institute in Chicago when he visited India in 1961. He and Gell-Mann were the two lead speakers of the Tata Institute Summer School in Theoretical Physics in 1961. And he, along with Gell-Mann, stayed in Ekamra Nivas after that summer school. Dalitz visited India soon after being elected Fellow of the Royal Society in 1960. After his visit Dalitz wrote a fine letter to Ramakrishnan dated Aug. 29, 1961, in which he said "Your courage and persistence in setting up such an active theoretical physics group at Madras won my admiration and I confess to leaving Madras with the question on my conscience, why I was not trying to do the same thing in my own country, Australia".

5) Donald Glaser won the Nobel Prize in physics in 1960 for the invention of the *bubble chamber*, which plays a crucial role in observing elementary particles in high energy accelerator experiments. Glaser joined the University of California, Berkeley in 1959 and received the Nobel Prize when he was there.

6) Education Minister C. Subramaniam was not only impressed with Ramakrishnan's Theoretical Physics Seminar, but he was also convinced that a new Institute as proposed by Ramakrishnan should be created. Shapiro's high praise of the Seminar and the comparison of it with the group that gathered around Oppenheimer during the Manhattan project, was just what was required for Mr. Subramaniam to request the Prime Minister to meet my father and his students.

7) There is perhaps no parallel in the history of science anywhere in the world, where a Prime Minister agrees to meet a professor and all his students, discusses with them, and asks what they really need with a genuine desire to do something for them. Ramakrishnan has often said that when a Prime Minister asks you what you need, you do not request something meagre — you ask for something big. Ramakrishnan asked for a new institute, and he got one!

8) When Ramakrishnan invited Chandrasekhar to address the students of the Theoretical Physics Seminar, he jocularly said that would agree if Ramakrishnan and Lalitha would provide him a South Indian style dinner served on a banana leaf! That was the easiest thing to do at Ekamra Nivas. So after his lecture, Chandrasekhar and his wife and all the students of the seminar were treated to a magnificent dinner at Ekamra Nivas.

Chandrasekhar gave a two hour lecture at Ekamra Nivas as was his practice. He was dressed in his customary full suit for his lecture. But he wore a dhothi to enjoy the South Indian style dinner for which one has to be seated on the floor.

9) McCrea Hazlett visited Madras a few months after he was appointed as Provost (= Vice-President for Faculty Affairs) at the University of Rochester. Hazzlett's visit was triggered by the suggestion of Robert Marshak of the University of Rochester who had formed a favourable impression of Ramakrishnan's efforts in Madras. Hazzlett had a strong interest in, and ties to, India and even served for a few years as the First Director of the American Institute of Indian Studies in Pune, India. When Ramakrishnan visited the University of Rochester in 1963, Hazzlett as Provost hosted him graciously.

10) Even though Jawaharlal Nehru was much impressed by the meeting he had with Ramakrishnan and his students, he consulted Homi Bhabha, Director of the Tata Institute and Director of the Atomic Energy Commission, concerning Ramakrishnan's proposal for a new Institute. Bhabha was not in favour of the idea of a new Institute and so Nehru did not act rightaway. It was Mr. C. Subramaniam who convinced Prime Minister Nehru that Ramakrishnan's proposal needs to be approved. I will quote from Mr. Subramaniam's autobiography *The Hand of Destiny*:

"Ramakrishnan mentioned to me that for the purpose of encouraging young talent in theoretical physics and mathematics, a new institution was necessary. At that time the entire research work in mathematics and theoretical physics was done only in the Tata Institute of Fundamental Research (TIFR) in Bombay under the auspices of the Atomic Energy Commission (both headed by Dr. Homi Bhabha). Ramakrishnan emphasised the need for another institute so that there might be some competition... instead of one institute monopolising the entire research work. But another institute could be started only with the concurrence of Atomic Energy Commission and the Government of India....

Jawaharlalji was greatly impressed by the enthusiasm shown by the students (of Professor Ramakrishnan)... and in particular to see four girls among the students. When the students told him that they needed an institution for the development of theoretical physics and mathematics, he asked me to examine the proposal and put up a note for his consideration. Ramakrishnan prepared a note for the purpose and I sent it to the Prime Minister."

It took two months before the Prime Minister gave the nod, and the events during these two months are described by Mr. C. Subramaniam:

"The Prime Minister referred the matter to Dr. Homi Bhabha for his advice. Unfortunately Dr. Bhabha was not very enthusiastic. His contention was that the available limited resources would have to be utilised for the existing institution — the Tata Institute of Fundamental Research.

When the Prime Minister passed on the opinion of Dr. Bhabha to me, I requested him to arrange a meeting between me, Dr. Bhabha and himself to discuss this matter. The meeting was arranged and I argued my case for a separate institution in the South, particularly when talented students in the South were not getting opportunities for pursuing their interests because of the limited number of students admitted to TIFR. I also emphasised that mathematical sciences did not require heavy investment. Panditji also showed his inclination to accept my point of view. So Dr. Bhabha also gave his consent. Thereafter steps were taken to establish what is now well-known as the Institute of Mathematical Sciences.... For the purpose of emphasising the importance of this Institute, and for its proper funding, I thought we should have Jawaharlal Nehru himself as the Patron of the Institute. When I mentioned this to him, he gladly agreed. I requested Dr. Bhabha to be a member of the first Governing Body of the Institute (and he agreed)."

11) British rule in India had a lot of effects in our social life and in the Government. Civil Service was handed to India by the British. The advisors to the Government and to the various ministers were IAS officials, namely individuals who had passed the tough Indian Administrative Service (IAS) exams. Having the support of the Secretary of Education (an IAS official) is absolutely crucial for the development of any educational or academic institution. Ramakrishnan was fortunate that K. Srinivasan, IAS, the Education Secretary at the time, was totally supportive.

12) Mrs. Kamala Kailasanathan was Lalitha's elder sister by two years. She was a professional musician who gave concerts all over South India. The concert she gave at Ekamra Nivas for the delegates of the Cosmic Rays Conference, turned out also to be a celebration for the creation of MATSCIENCE.

13) Nambi Iyengar lived with his family at Ekamra Nivas and was there to assist Ramakrishnan any time of the day or night. Those were days before xerox machines and other copying facilities. So each of the one hundred or more letters were typed by him one by one. He shared Ramakrishnan's excitement and did his work cheerfully and with total dedication. After MATSCIENCE was created, he became Ramakrishnan's personal assistant (Director's secretary) at the Institute.

14) Ramakrishnan's letter of resignation to the Vice-Chancellor was written in very polite terms. Ramakrishnan thanked Dr. Lakshmanaswami Mudaliar for the "parental interest" he showed in his career.

15) It was only on December 22 and 23 that Ramakrishnan sent the letters to the international community of scientists announcing the creation of MATSCIENCE. That the congratulatory telegrams arrived within a week proves that international mail was delivered in just four to five days. There used to be a Mobile Post Office — that is a postal van — that would be parked in front of the Mylapore Club close to Ekamra Nivas, in the evenings. Our staff would take the letters and post them there. The Mobile Post Office would proceed at night straight to the airport, and the letters would be despatched that same night. These Mobile Post Offices do not exist anymore.

L to R: Richard Dalitz, Lalitha Ramakrishnan, Murray Gell-Mann at Ekamra Nivas (1961).

Astrophysicist S. Chandrasekhar (second from left) enjoying a South Indian dinner at Ekamra Nivas with the members of the Theoretical Physics Seminar (1961).

Prime Minister Jawaharlal Nehru meeting the author and the students of the Theoretical Physics Seminar at the Raj Bhavan (Governor's Residence), Madras, in October 1961.

Professor M. J. Lighthill (seated) and Mr. C. Subramaniam (standing) at Ekamra Nivas (1961).

Chapter 22

The Birth of Matscience (1962)

What a thrill and excitement it was to greet the new year, for the Institute was to be inaugurated on the 3rd of January (see Appendix 21) by Professor S. Chandrasekhar at a public function presided over by its primary sponsor, C. Subramaniam. I could not sleep the whole night as cablegrams of greetings were being delivered every hour from various parts of the world (for scans of a good collection of these telegrams, see [2]). I was too excited to prepare my inaugural speech and I decided to deliver it extempore under the inspiration of the exalted moment and Nambi Iyengar knew that I would do so!

The institute was inaugurated at 9 a.m. on the 3rd of January in the main English Lecture Hall of the Presidency College, where I had listened to many lectures during my study in the honours course.[1] I gave my speech almost in tears of joy and gratitude to providence for a dream fulfilled. I consider it perhaps the best speech I made in my life, for it was a reflection of the past, an estimate of the present, and a peep into the future of science in India.

It was a day of prayerful dedication when I thought of my parents who should have been there with us in 'our finest hour'. Among the happiest persons were Lalitha who had shared my hopes and sustained by efforts, and little Krishna who understood the spirit of the hour in equal measure. Commenting later on my speech with sparkling candour, Krishna said that it was an emotional blend of a dash of Churchill and a splash of Shakespeare and recited sentences there from my speech to prove his point, imitating my style and accent.[2]

I was deeply moved when my father's lifelong friend Subbaroya Iyer, who was anxious to attend the function, was carried into the hall in an easy-chair since he was stricken by partial paralysis. He took the place of my father in that glorious hour, as the surviving symbol of a true and fast friendship, which like a happy marriage, was the result of Divine decree.

The academic work of the new born Institute started the same day with a lecture by Professor Chandrasekar on gravitation in the physics lecture room where he and his illustrious uncle and I also had studied as undergraduates.[3] We started work in right earnest and I asked my students to lecture on subjects with which they were not familiar, like complex variables and astro-physics.

It was then that the Test match was held between England and India with young Nawab of Pataudi as the captain of our side. I did not care to attend it since I wanted to get on with the work of the new Institute. I started with a course on elementary particle theory based on the elegant book of Marshak and Sudarshan. We were very fortunate in having a stream of visitors even during the first few months after the inception of the Institute. I was as excited as a bridegroom before a honeymoon, for so much of freedom was given to me to organise the academic programme and direct the affairs of the Institute. What a pleasure it was to receive the first cheque of Rupees Five Thousand from the Government of Madras for incidental expenses to get things going! The present budget of MATSCIENCE is several crores of rupees.

The first lecturer was Professor Skyrme[4] of the Atomic Energy Research establishment of England who talked on non-linear field theory. I invited my good friend A. N. Mitra of Delhi to give a series of lectures on separable potentials, a subject to which he made substantial contributions. Next came Ugo Fano of the National Bureau of Standards and our friendship was strengthened by my visit later to Washington where I stayed with him in his comfortable house. Professor Bloch[5] of Saclay visited us in May but we had no difficulty in making him comfortable even amidst the rigours of a Madras summer. He was one of the most charming men I have met, Parisian to the manner born. He gave a masterly series of lectures on many-body problems. Our association was strengthened by my frequent visits to Paris in the ensuing years.

As our next visitor, we had an eminent mathematician Professor Bourgin[6] from Urbana, Illinois, who gave the lecture on topological features of some physical problems. I had met his son-in-law Baker six years ago when he was a graduate student under Schwinger at Harvard and Bourgin told me that he was set on his career to a professorship at Seattle.

The Government was gracious enough to allow me to contact famous professors in Europe and U.S.A. to implement the visiting professorship scheme of the Institute. My student Balachandran who completed his Ph.D. was granted a Post-Doctoral fellowship in Chicago under Dalitz, a natural consequence of the discussions I had with him when he was my house guest in Madras along with Gell-Mann. Meanwhile Professor Thirring also offered Balachandran a fellowship for six months in Vienna and I obtained special permission from the Vice-Chancellor to allow him to go a few months before submitting his thesis.[7]

I visited Bangalore to lecture at a Summer School organised by the Indian Statistical Institute on stochastic processes. It was then we had the pleasure of receiving the young Yugoslav scientist from CERN, Dr. Bogdan Maglic, the discoverer of Omega Meson. Thus started a friendship which continued in strength and substance in the years to come.

In June I underwent an operation for hernia at Dr. Mohan Rao's Nursing Home. I was deeply touched by the manner in which my father's aging friend Subbaroya Iyer called on me in spite of his crippling ailment. Subbanna was of great assistance to Lalitha in attending on me day and night. Dr. Mohan Rao, true to his name, was a charming personality besides being a surgeon of first magnitude. But for the pains of surgery, the patient felt it was a privilege to be operated by him. I myself felt like that in such measure that I wished him to perform 'in advance' all the operations that I may have to undergo in my life!

During my convalescence, Dr. Frautschi[8] a young associate of Gell-Mann visited us, a lean and lanky American, seemingly casual but extremely serious about his work. He gave a series of lectures on strong interactions, a subject which was engaging the minds of high energy physicists following Chew's work on the 'Bootstrap Philosophy'.

I submitted the papers for registration of the Institute on Friday the 13th July and the Registrar issued the Certificate on the 23rd July (see Appendix 25C). The next few months we had active seminars on various topics in physics. I received a handsome invitation from the RAND Corporation, California, to spend a few months as a consultant in the Mathematics Department, in which Ted Harris and Richard Bellman, were the leading mathematicians.

Notes

1) The Presidency College of the University of Madras has a most enviable location facing the ocean. Presidency College has a moghul style architecture, as do some other other buildings of Madras University. The Old English Lecture Hall is a stately auditorium style room commanding a splendid view of the famous Marina Beach.

2) Ramakrishnan was an orator par-excellence, and he gave his speeches extempore, as he did for the inauguration of MATSCIENCE. The speech at the inauguration was the finest he gave in his entire life (see Appendix 21). The Hindu, India's National Newspaper, gave a full report of the inauguration, quoting parts of the speeches of C. Subramaniam, S. Chandrasekhar, and Ramakrishnan.

3) No time was wasted in starting the activities of the new Institute. The Theoretical Physics Seminar was formally closed and lectures moved to the rooms at the Presidency College. Since Chandrasekhar inaugurated the Institute, it was natural to ask him to give the very first lecture of MATSCIENCE that same day. Chandrasekhar and his Nobel Laureate uncle Sir C. V. Raman, were alumni of the Presidency College, and portraits of them hang on the walls near the entrance.

4) Tony Skyrme (1922–87) was a noted British physicist, who in 1962 was at the Atomic Energy Research Establishment in Harwell, England, as the head of the theoretical nuclear physics group. The particles that he predicted would manifest a certain unusual behaviour, were later named as *skyrmions*. He was recognised for his work in 1985 by the Hughes Medal of the Royal Society. Ramakrishnan first met Skyrme in 1960 in Harwell during his trip to Europe (see Chapter 20).

5) Claude Bloch (1823–71) was a noted French physicist who made significant contributions to quantum field theory and nuclear physics. He studied at the Bohr Institute in Copenhagen and after a stay at Caltech, returned to France. He was the Head of the Theoretical Physics Department at Saclay in France when he visited MATSCIENCE in the summer of 1962. Later in 1962 Ramakrishnan visited Saclay at his gracious invitation.

6) David G. Bourgin was a highly reputed mathematician who received his PhD from Harvard University in 1926. He was for many years at the University of Illinois in Urbana where he guided more than 25 PhD students. There is a Fellowship for PhD students now at the University of Illinois that is named after Bourgin. He was so much involved in the training of PhD students that he was naturally impressed with Ramakrishnan's commitment to provide suitable facilities for research students in Madras.

7) Since Ramakrishnan knew how important it was to work in close contact with eminent scientists, he went out of the way to help his students in getting fellowships abroad, and making sure they get the required leave to accept such fellowships. This involved the risk of losing such students who may decide not to return, but Ramakrishnan was generous and never stood in their way since he believed that science was an international enterprise. Such large heartedness is hard to find among administrators.

8) Steven Frautschi, is a distinguished physicist at Caltech, who is especially known for fundamental work related to Geoffrey Chew's bootstrap theory. In addition to being an eminent researcher, Frautschi is an outstanding teacher at all levels, and has received several awards for teaching; as recently as 2014, he was awarded the Richard Feynman Prize for Excellence in Teaching at Caltech, even though he is now an Emeritus Professor.

(L to R) The author, the great astrophysicist Prof. S. Chandrasekhar, and Education Minister Mr. C. Subramaniam, walking to the dais for the inauguration of MATSCIENCE, 3 Jan, 1962. - Editor

The author giving his speech "The miracle has happened" at the Inauguration of MATSCIENCE, 3 Jan, 1962. On the dais are Mr. C. Subramaniam and Professor Chandrasekhar. - Editor

Appendix 1

The Dedicated Life

Passages from Viscount Haldane's address to the Edinburgh University read to me by my father* when I was a school boy, influenced my life and career.

"I mean by the expression 'Dedicated Life', one that is with all its strength concentrated on a high purpose. The purpose though high, may be restricted. The end may never be reached. Yet the man is great for the quality of striving is great.

The first duty of life is to seek to comprehend clearly what out strength will let us accomplish, then do it with all our might. A life into which our whole strength is thrown, in which we look neither to the right or to the left, such a life is a 'Dedicated Life'. The forms may be manifold. The lives of all great men have been dedicated; singleness of purpose has dominated them throughout. Thus it is with the life of Socrates, a Spinoza or a Newton, thus with lives of men of action like Ceasar, Cromwell and Napoleon."

Notes

Richard Burdon Haldane (1856–1928), also called the Right Honourable Viscount Haldane, was England's Secretary of State for War during 1905–12 when this speech was given and the book was published. Haldane was an influential writer and for this he was elected Fellow of the British Academy in 1914. In 1907 when he was President of the Edinburgh Sir Walter Scott Club, he gave a speech entitled "The Dedicated Life" to a group of students at the Club's annual dinner. Subsequently this was published as set of four lectures on the topic "The Dedicated Life" in his book of 1912 referred below.

Sir Alladi was a prolific reader and his vast library at Ekamra Nivas contained this book by Haldane.

===============
*From "Universities and National Life", Viscount Haldane, John Murray, London (1912), 179 pp.

Appendix 2

MATSCIENCE Calendar (1941–62)

The incredible story of the concept, conception and birth of Matscience

1941: Visit of young Dr. H.J. Bhabha, fresh from his laurels in Cambridge, to the Presidency College, Madras. His lecture on "Meson Theory". Alladi Ramakrishnan (AR) aspires to work under him after his Honours course.

1943: AR meets Sir C.V. Raman who was pessimistic about the future of research in India. Therefore AR postpones meeting Bhabha and decides to wait for a suitable opportunity. Meanwhile AR joins the Law College in Madras.

1946(December): AR accompanies his father to the opening of the Constituent Assembly on December 9th. AR meets Professor Bhabha at Bhatnagar's house in Delhi at dinner and expresses his desire to work with him.

1947(January): Meeting with Professor Bhabha at the Connemara Hotel, Madras, when he agrees to take AR as his student.

1947(July): AR joins the Tata Institute of Fundamental Research as an unpaid scholar, when it was located at Kenilworth, Peddar Road, Bombay.

1948(January): The best gift from a Professor and guide is a good research problem and AR receives one from Bhabha relating to Cascade Theory. He studies Uspensky's 'probability' and Heitler's 'Quantum Theory of Radiation'.

1948(April–June): Active collaboration with Bhabha on Cascade Theory at Bangalore West End Hotel.

1948(December): AR formulates *product densities* directly while Bhabha prefers a long limiting process.

1949(July): AR decides to do his Phd in Stochastic Theory under Professor M.S. Bartlett at Manchester and secures a seat for doctoral study with the help of Sir S. Radhakrishnan.

1949–51: Study at Manchester under Professor Bartlett. Professor D.G. Kendall endorses work on product densities. AR visits to various leading centres of research in in Britain and Europe.

1952–57: AR initiates theoretical physics at Madras University. Teaches quantum mechanics based on Mott and Sneddon. Contact with world famous scientists who come as house guests to Ekamra Nivas.

1954: Visit of Professor P.A.M. Dirac to Madras and dinner at Ekamra Nivas. AR starts lectures on Quantum Mechanics based on Dirac's book.

1954: AR's visit to Australia. Inspired by the School of Physics under his good friend Harry Messel, an examplar of success due to sheer devotion, hard work and initiative.

1954: Invited to become a Fellow of the Indian Academy of Sciences by Sir C.V. Raman.

1956: Round the world on lecture assignments at various centres in Japan, U.S., Britain and Europe — Yukawa Hall, Rochester and Gamm Conferences. First introduction to Feynman Graphs in Kyoto. Meets Feynman who gives a three hour private lecture to him at his Caltech office on Feynman Graphs. AR offers Feynman lecture notes and the concept of four vector as 'gifts' to his students in Madras.

1957–58: The Princeton experience — an education, revelation and inspiration. Close contact with Oppenheimer.

1958–60: The Madura assignment. AR teaches Feynman graphs in "gurukula" style to a band of dedicated students.

1959–61: Theoretical Physics Seminar started in April 1959 at Ekamra Nivas. About fifty visitors from Nobel Laureates to 'Nobel aspirants', lectured at Ekamra Nivas. Frequent appeals to the Vice-Chancellor to start a Theoretical Physics Department at Madras University. Member of expert committee for Mathematics which recommended Advanced Centres in Universities. AR suggested the creation of Institutes outside Universities also, but the suggestion was not included in the report. Suggestions, negotiations, criticisms and replies by Bhabha, representatives of the Madras Government and AR.

1959(October): At the suggestion of Mrs. Lalitha Ramakrishnan, AR attends an international gathering of students at Woodlands. His talk attracts the attention of the Chief Guest Mr. C. Subramaniam (Finance & Education Minister in Madras) who invites him for discussions to Cooum House.

1959(November): Dinner to Mr. C. Subramaniam at Ekamra Nivas. AR suggests creation of an Institute for Advanced Study in Madras.

1960(January): Visit of Niels Bohr and Abdus Salam. Dinners at Woodlands and later at Ekamra Nivas. Bohr's press statement in The Hindu about AR's Theoretical Physics Seminar attracts Prime Minister Nehru's attention.

1961(July): Visit of Gell-Mann and Dalitz as guests at Ekamra Nivas after the Bangalore Summer School (organised by TIFR) in which AR's eight students participated. Gell-Mannic impact on Physics at Madras.

1961: Visit of Dr. Maurice M. Shapiro (Naval Research Laboratory, Washington, D.C.) who suggests to CS that there ought to be a meeting of AR and his students with the Prime Minister.

1961(October): Meeting with Prime Minister Nehru at the Raj Bhavan, Guindy, Madras, arranged by Mr. C. Subramaniam. AR suggests to the Prime Minister the creation of an Institute for Advanced Study in Madras.

1961(December): Dinner at Ekamra Nivas to Sir James Lighthill (Lucasian Professor at Cambridge) who commends the creation of an Institute in the presence of Mr. C. Subramaniam.

1961(December 22): Government of Madras decides to create an Institute and offers Directorship to AR. Prime Minister Nehru agrees to be Patron of the Institute.

1962(January 3): Inauguration of Matscience at the (Old) English Lecture Hall, Presidency College by Professor S. Chandrasekhar, who delivers the first lecture of the new Institute on the subject of Gravitation.

1962(July 23): Registration of Matscience as a society. Bhabha agrees to be a member of the Board of Governors.

1962: Personal interview with Nehru who graciously sanctions a special grant for the visiting membership programme for young scientists in the presence of Mr. C. Subramaniam.

Appendix 3

An Ancient Nation and
a New Republic

(A)

Preface

On the 26th January 1950, India became a sovereign independent republic — an ancient country, tracing its traditions to the dawn of human civilisation and its scriptural heritage to divine origins. The proud British Empire, after a hard-won victory in a world war, bowed out of India with grace and dignity but the democratic way of life held the minds of the half a billion people of our country which adopted a republican constitution drafted by its wisest leaders, its learned jurists and seasoned statesmen.

Lalitha and I were in Manchester, England at that time where I was working for my Ph.D. degree. I was invited to talk to a gathering of international students on that historic occasion, the birth of a republic. My 'speech-writer' was one of the founding fathers of the constitution, a privilege which I could enjoy as a demanding son. The typescript of the speech dictated by my father to my brother in Madras was air mailed early enough for me to peruse and understand what I was to read!

I would like to make available that document describing the baptism of a nascent republic at this time when that constitution, so well conceived and formulated, is being put to test whether it will long endure.

Alladi Ramakrishnan

An Ancient Nation and a new Republic

Alladi Ramakrishnan
(January 26, 1950)

(1) We are assembled here to celebrate the most memorable day in the long and chequered history of ancient India, the day when under a Constitution framed by an assembly representing the will of her people, India is constituted a sovereign independent Republic. It was only in 1947 after thirty years of struggle to become a free and independent nation, adopting the unique weapon of non-violence under the inspiring guidance of Mahatma Gandhi, the more enlightened and liberal minded statesmen of Great Britain now at the helm of affairs in England, realised that the best way to secure amity and friendship between England and India was to cease to govern or rule it against her will. The result of this realisation was the Indian Independence Act constituting the then British India into the two Dominions of India and Pakistan, with a right to frame their own constitutions through the medium of duly elected constituent assemblies. The Constituent Assembly of India began its task of framing the Constitution for India on December 9th, 1946 and the new constitution was finally passed on 26th November 1949 to take effect from today. It is a source of immense satisfaction to note that ever since the 15th of August 1947, when India was declared to be a self-governing dominion, there has been a great metamorphosis in the relationship between England and India. Britain has begun to realise that India's freedom is a great source of strength to the principles of democracy and the erstwhile fighters for freedom who are now guiding the destiny of India have forgotten old feuds and feel that a new friendship between the Sovereign Republic of India and Great Britain is for the well-being of the Indian people and of the British.

(2) Though India is from today an Independent Republic, still it is happily a member of the British Commonwealth. In the London Conference of last year which marked a very important stage in the development of Commonwealth relations, it was decided that India though avowedly republican under her new Constitution, would continue to be a member of the British Commonwealth. Though she owes no allegiance to the British Crown, she recognises the Crown as the symbol of the free association of all the members of the Commonwealth, with the hope that along with the other members of this happy family of nations, she will play an important role in securing world peace and maintaining harmony among the different nations.

(3) Having regard to the vastness of the territory comprising India and the difficulty of governing a subcontinent like India under a unitary form of Government, it was rightly decided by the framers of the Constitution that India should be, if not strictly federal, quasi-federal in its constitutional structure. This Union of India has been considerably strengthened by the several Indian States having joined the Dominion of India and deciding to form part of the Union when so constituted. When the Independence Act was passed and as a result of the Act every one of the hundreds of Indian States varying greatly in size and resources became independent, it was feared that these islands scattered through the length and breadth of the Union of India would stand in the way of the solidarity and progress of the Indian Union. But thanks to the drive and energy of Sardar Patel, the Home Minister and the Deputy Prime Minister, and the breadth of vision of the rulers of the Indian States, we are now in the happy position that there is not a single Indian State, excepting those that have joined the Dominion of Pakistan, which is not now part of the Indian Union.

(4) While it was recognised on all hands that the State or Provincial autonomy should be secured, it was also realised by those responsible for the constitution that in the difficult days ahead it is very essential that India should have a strong centre. With communism gaining ground on the borders of India, in Burma and China, with India along with the rest of the world passing through food crisis, with the necessity to reorganise her industries, agriculture, transport, and education, to suit the needs of the modern age, it was felt that there should be a strong central Government which could unite the various parts of India and lead her to progress. While therefore the units have unrestricted legislative and executive power in their respective fields, in order to meet unforeseen national emergencies and economic situations, special provisions have been inserted for central intervention. In the matter of Legislative power, the centre is invested with residuary power, specific subjects of national and all India importance being expressly mentioned. A large number of subjects has been included in the concurrent list to enable the centre to intervene wherever there is necessity to intervene and over-ride State Legislation.

(5) The executive power of the union is vested in the President, but unlike in the U.S.A., he does not enjoy large powers and except in very exceptional circumstances serves merely as a symbolic head of the State. The actual Government is carried on by means of a Council of Ministers collectively responsible to the Legislatures, aiding and advising the President in the exercise of his functions. In other words, the British system of

Cabinet Government, which has proved a notable success in the Dominions, has been adopted.

(6) The new Constitution also provides for a strong and independent judiciary which is essential to the proper working of the Federal Constitution. As has been pointed out on several occasions during the debates of the Constituent Assembly, the Supreme Court of India has larger powers and wider jurisdiction than the highest court in any other known federation. It is a general court of appeal from all civil cases, from every High Court in the units including the native States. It has original jurisdiction not only in respect of disputes between two or more units, but for the purpose of issuing writs in appropriate cases throughout the length and breadth of India. It is the ultimate arbiter in all matters involving the interpretation of the Constitution. The independence of the judiciary is safeguarded by making the judges irremovable from office except for misbehaviour or incapacity after an elaborate procedure and by providing that their salaries cannot be varied to their disadvantage during their time of office. With these powers and privileges, it can be confidently asserted that the courts will serve as the balance-wheel of the Constitution, seeing that the different organs of Government do not exceed their limits or powers and upholding the rights of the individuals while at the same time safeguarding the interests of the State.

(7) India under the new Constitution is pledged to the principles of democracy. As is set out clearly in the preamble, the Constitution aims at securing to all its citizens social, economic and political equality, opportunity of thought, expression, belief, faith and worship, equality of status and of opportunity, and to promote among them fraternity, assuring the dignity of the individual and the unity of the nation. In consonance with these objects and ideals set out in the preamble, special provisions are made in the Constitution guaranteeing to the individuals certain fundamental rights: Equality before the law and equal protection of laws; equality of opportunity in matters relating to employment freedom of speech and expression; freedom of conscience and free profession, practice and propagation of religion and the freedom to acquire, hold and dispose of property are some of the more important rights guaranteed to the citizen under the new Constitution. No doubt, certain limitations and restrictions are imposed on the exercise of these fundamental rights or Parliament is given the power to impose such restrictions, whenever necessary. But such restrictions are inevitable in the complex conditions of modern society, when a large degree of social control is necessary. To people in England, who have exercised such

rights ever since the date of the Magna Carta and have come to regard them as part of the fundamental law of the land, a specific enumeration of such rights may seem unnecessary. The constitution following the British example does not contain any chapter or provision regarding the fundamental rights. The idea of having a special chapter formulating fundamental rights originated at the time of the Cabinet Mission scheme and the constitution of a committee for the enunciation of such rights was insisted upon as an essential step in the framing of the Constitution. Even after the Independence Act, the formulation of fundamental rights by the Committees constituted for the purpose was proceeded with, as it was thought that during the infancy of Independent India, it is better to safeguard the rights of individuals. In addition to the fundamental rights which are justiciable, the Constitution, following the Irish model, has laid down certain directive principles of social policy. Provisions in Part IV lay down that the State shall direct its policy towards securing, that the citizens have the right to an adequate means of livelihood, that the ownership and control of the material resources of the community are so distributed as best to subserve the common good; that the operation of the economic system does not result in the concentration of wealth and means of production of the common detriment, and so on. These principles, though not justifiable, are nevertheless fundamental in the governance of the country and to a nation like India, which is anxious to advance with the times. These serve as a guide to those at the helm of affairs in steering the ship of State.

(8) No reference to the chapter on Fundamental Rights will be complete without a reference to the clause regarding abolition of untouchability. Every progressive Indian cannot refer without a feeling of regret to this practice of considering a fairly large section of the people as degrading and even untouchable. Mahatma Gandhi fought this scourge throughout his career and succeed in making a large section of India realise the injustice which was being done to their fellowmen. The new Constitution forbids the practice of untouchability in any form and there is no doubt that with the growing consciousness of the people of India, that this is one of the greatest blots in the social landscape of India and with the express prohibition contained in this Constitution, untouchability will cease to exist in the course of a few years. It is not an exaggeration to state that this is the provision in the Constitution which contributes most to the forces of Unity in India.

(9) In this connection two other very important factors which will go a long way to consolidate and unify the nation are worthy of attention. The first is the concept of a single citizenship for the whole of India. Though

India consists of various Provinces each nearly as big as England, it was considered desirable that there should be a common citizenship for India having regard to the common culture of its people in the various Provinces and in the interests of the unity of India. The second important step taken to ensure the unification of India was the introduction of a common language. No other provision in the constitution created a greater controversy than this provision regarding the common language but so great was the desire on the part of the framers to find a solution, that in the end a formula agreeable to all was devised. Hindi which is being spoken by a vast majority of the people of India was chosen as the official language of India. At the same time, the units were permitted to use their regional languages. It was also realised that with science, medicine and industry in the west far ahead of India, she could not do without the English language altogether, if she had to learn all that the West had to teach her in these matters. Therefore it was decided to continue the use of the English language for a period of fifteen years by which time it is hoped that Hindi will make considerable headway and become the medium of communicating scientific thought to the millions of India. The cultured and advanced section in India and those at the helm of affairs are quite alive to the importance of English as a world language, as the proper vehicle of scientific thought and knowledge and as a powerful factor in the moulding of our legal institutions. While therefore provision is made for a common language being developed, the importance of English is not lost sight of.

(10) In order to ensure that the democracy is broadbased and the legislatures are truly representatives of the people of India, the Constitution recognises the principle of adult suffrage. It is only those who are acquainted with the mass illiteracy prevailing in India and the complex problems which face the country at the present day that can realise that this is perhaps the boldest experiment undertaken in any country. The framers of the Constitution were distinctly alive to the possible dangers and difficulties in conferring the right to vote upon a large mass of illiterate citizens. Nevertheless, adult suffrage had to follow as necessary corollary to true democracy and it has been introduced with abundant faith in the common man and in the full belief that introduction of adult suffrage would bring enlightenment and promote the well-being, the standard of life and the comfort of the common man and will hasten to remove mass illiteracy, a problem which has not been tackled during British rule. The introduction of any property or educational qualification for the exercise of their franchise would have been a negation of the principle of democracy and it would be denying the

right to a large number of labouring classes and women of India. It is hoped that the passage of time will prove that this abundant faith in the common man is completely justified.

(11) Closely realising in full that communal electorate and democracy could not co-exist, the Constituent Assembly did away with communal electorate, while making special provision to the scheduled castes and the tribes (Backward Classes) on the basis of joint electorate for a temporary period. It is a noteworthy feature that the Indian Christians, the Parsees and other minority communities have wholeheartedly supported this decision and even advanced Muslim opinion has supported this great change. The perpetuation of communal electorates would have been a negation of the principles of democracy.

(12) A perusal of the preamble and the various provisions in the Constitution clearly shows that the framers have succeeded in the idea of building up a secular State; a State which does not recognise any discrimination in favour of any particular form of religion, community or in any of that kind. It is a State in which every human being will be given the opportunity of rising to his full height and stature irrespective of the religion which he professes.

(13) The provisions regarding the amendment of the Constitution are so designed as to make it sufficiently flexible with a view to facilitating any changes which might be found necessary in the working of the Constitution in its early stages.

(14) On the whole, the Indian people may be justly proud of their new Constitution, which represents the labour of the members of the Constituent Assembly, who have been progressive in their outlook without being revolutionary.

(15) The Constitution has been most favourably reviewed in the leading organs of public opinion and in informed journals both in England and America.

(B)

Sir Alladi's Speech at the Convocation of Delhi University (1951)
Excerpts

"I do not know if I will be in order in touching on a controversial topic — the place of Hindi and regional languages in university education. While I yield to none in recognising the need for having a common language for welding together the people of this country, I would like to sound a note of caution. Scientific knowledge in the world today is progressing at a giddy pace and India cannot afford to stand still when other nations are progressing. The student and the teacher alike must be in touch with the latest discoveries and treatises both in the pure and applied sciences. It would be almost impossible task to attempt translating into Hindi or the vernacular language all the great works that are being produced in the world of science from day to day...While I yield to none in the zeal for the gradual spread of Hindi and the adoption of the Indian regional languages as the media of instruction, I would plead for the students getting a sound working knowledge of English in the high school classes and due importance being attached to English in collegiate and university education. There is no use of ignoring the fact that English is today a world language."

Scientific attitude and early education: Scientific advancement in universities is not possible or feasible unless foundation is laid in the secondary courses and if I may say so, even in the elementary classes. The scientific attitude must be developed and cultivated almost from the beginning of one's studies. The mind, the eye, the ear and intellect, should be trained to observe natural phenomena from an early age and the scientific attitude must be fostered from the very beginning. If a student is not taught to observe, but merely undergoes in his early days purely mechanical instruction in particular subjects and is made to believe that the passing of an examination is the be-all and end-all of study, he is not likely to get or develop a scientific attitude in the later years. This may not apply to an outstanding genius who has an inner urge to observe natural phenomena almost from his infancy."

Notes

In the Constituent Assembly, Sir Alladi vehemently argued against dropping English in favour of Hindi. He urged that the consideration of Hindi as a national language be postponed by at least 15 years.

For the full text of Sir Alladi's speech, see [1].

Appendix 4

Papers of Bhabha and Ramakrishnan (1950)

(A)

Explanatory Note

In the next two pages, the opening pages of Homi Bhabha's paper in the Proceedings of the Royal Society (Mar. 1950) and Alladi Ramakrishnan paper in the Proceedings of the Cambridge Philosophical Society (Feb. 1950) communicated by Professor M. S. Bartlett are displayed. Bhabha used a limiting process to obtain distribution functions in the continuous range. The same functions were named as *product densities* in Ramakrishnan's paper with the approval of Professor D. G. Kendall (Oxford) who had called them cumulant densities in his work on age distribution of populations. Ramakrishnan derived these product densities by a direct formulation in continuous space. This is among Ramakrishnan's best contributions, and these functions have been extensively used in the literature acknowledging the work of Bhabha and Ramakrishnan.

Although submitted in July 1949, Bhabha's paper got slightly delayed in publication because it was revised. Ramakrishnan's paper was accepted rightaway in February 1950 and so it was his luck to have it published around the same time as Bhabha's!

(B)

Reprinted without change of pagination from the
Proceedings of the Royal Society, A, *volume* 202, 1950

On the stochastic theory of continuous parametric systems and its application to electron cascades

By H. J. Bhabha, F.R.S.

Tata Institute of Fundamental Research, Bombay

(*Received* 11 *July* 1949—*Revised* 31 *March* 1950)

A mathematical definition of an assembly of continuous parametric systems is given and its theory developed which makes it correspond precisely to most of the continuous parametric assemblies known in nature. Certain general theorems (formulae (19) and (22)) are deduced which hold for all such assemblies. It is shown that the usual method of treating a continuous parametric assembly by dividing up the domain of the parameter into a number of small segments, treating each segment as belonging to a discrete state of the system and then passing to the limit of making the segments infinitely small does not lead to a continuous parametric assembly of the type described above, but one of much wider generality which does not correspond to any type of physical system met with in nature. The general method and the theorems are of immediate application in calculating the fluctuations of the number of particles in chain reacting systems, and not only for systems in thermodynamic equilibrium.

The general theory is applied, as an illustration, to the stochastic treatment of an electron cascade to derive the differential equations which determine the functions from which the mean number of particles in any energy interval and the mean square deviation of this number can be calculated. It is shown in the appendix how the application of the usual method to this problem leads to the same results only if particular boundary conditions are imposed on the problem.

1. Introduction

When a physical system only possesses a number of discrete states, the stochastic treatment of an assembly of such systems presents no difficulties in principle. For brevity, it is convenient to call an assembly of such systems possessing only a number of discrete states an 'assembly of discrete state systems', or even a 'discrete state assembly'. One obtains all the statistical information that is possible about the assembly as soon as one can calculate the mean of the number of systems in any given state, the mean of the square and higher powers of this number, and the mean of all products such as the number in one state times the number in another state, etc.

In nature, however, there is a whole host of physical systems whose states pass continuously into one another, so that a state is labelled by giving a particular set of values to a set of parameters each of which can vary continuously in some domain. For example, a free material particle is of this type, and any of its states is labelled by giving a particular set of values to the three components of its momentum. We call an assembly of such systems a 'continuous parametric assembly' for brevity. The stochastic treatment of a continuous parametric assembly presents some interesting features. The usual method of treating such an assembly is to divide the domain of every parameter into a very large number of small segments and to assume that with regard to each segment there are only two possibilities for a system: either to lie in the segment, or to lie out of it. No difference is assumed to exist between the states of a system lying in different parts of the same segment. This

(C)

[Extracted from the *Proceedings of the Cambridge Philosophical Society*.
Vol. 46, Pt. 4.]
PRINTED IN GREAT BRITAIN

STOCHASTIC PROCESSES RELATING TO PARTICLES DISTRIBUTED IN A CONTINUOUS INFINITY OF STATES

By ALLADI RAMAKRISHNAN

Communicated by M. S. BARTLETT

Received 2 February 1950

1. INTRODUCTION

Many stochastic problems arise in physics where we have to deal with a stochastic variable representing the number of particles distributed in a continuous infinity of states characterized by a parameter E, and this distribution varies with another parameter t (which may be continuous or discrete; if t represents time or thickness it is of course continuous). This variation occurs because of transitions characteristic of the stochastic process under consideration. If the E-space were discrete and the states represented by $E_1, E_2, ...$, then it would be possible to define a function

$$\pi(\nu_1, E_1; \nu_2, E_2; ... ; t)$$

representing the probability that there are ν_1 particles in E_1, ν_2 particles in E_2, ..., at t. The variation of π with t is governed by the transitions defined for the process; $\nu_1, \nu_2, ...$ are thus stochastic variables, and it is possible to study the moments or the distribution function of the sum of such stochastic variables

$$N = \nu_1 + \nu_2 ...$$

with the help of the π function which yields also the correlation between the stochastic variables ν_i.

But if the E-space is a continuum no such π function can be defined, for we have a continuous infinity of stochastic variables representing the numbers in the elementary ranges dE. The concept of correlation has to be generalized and a consistent formulation is necessary before we deal with such a system.

In quantum mechanics such processes arise since the transitions may occur between continuous sets of variables. Examples of such transitions are collision loss, radiation loss by fast particles or pair creation by high-energy photons.

The method described in this paper is quite general, and the word 'particle' is used to facilitate understanding of the problem from a physical point of view. It is also to be noted that the continuous parameter referred to in the title is E and not t. There is no restriction on t. Whether t is continuous or discrete depends upon the definition of transition probabilities.

2. DESCRIPTION OF THE METHOD

Let $M(E; t)$ represent the stochastic variable denoting the number of particles with parametric values less than E. Then $dM(E; t)$ represents the stochastic variable denoting the number of particles in the elementary range dE. We shall assume that the probability that there occurs one particle in dE is proportional to dE, while the probability that there occurs more than one particle, say n, is of order $(dE)^n$ and hence is vanishingly

(D)

Letter from Prof. L. Janossy

DUBLIN INSTITUTE FOR ADVANCED STUDIES

SCHOOL OF COSMIC PHYSICS

5 MERRION SQUARE

DUBLIN

UOS JÁNOSSY.
NTOP PROFESSOR.

TELEPHONE: 7433
TELEGRAPHIC ADDRESS
DIAS, DUBLIN

4 March 1950

Mr. A. Ramakrishnan,
24 Lawrence Rd.,
Flixton,
Manchester.

Dear Ramakrishnan,

I have communicated your paper to the
Proc. Phys. Soc.,

Yours sincerely,

NOTE: *This is the first published paper by Ramakrishnan, appearing just before the 1950 paper in the Cambridge Philosophical Society. Though the joint work of Bhabha-Ramakrishnan on the stochastic problem of electron-photon cascades was done in India earlier, it was published only later in 1950 in the Proceedings of the Indian Academy of Sciences since Bhabha waited a bit before communicating that paper (see [2], p. 81).*

Appendix 5

Letters from Prof. M. S. Bartlett

(A)

DEPARTMENT OF MATHEMATICS.

PROFESSOR M. H. A. NEWMAN, F.R.S.
PROFESSOR S. GOLDSTEIN, F.R.S.
PROFESSOR M. S. BARTLETT.

THE UNIVERSITY.
MANCHESTER, 13.

ARDwick 2681

6th January 1950.

Dear Ramakrishnan,

I have not finished looking at the notes you gave me yesterday, but your solution of the G,F equations is very pretty. I notice that you did not derive the solution for the moments from the complete solution, and it would seem to me advisable to do this, at least for the mean, as a check on the expression for the complete solution.

The immediate point I would like to make about your solution is that in all cases where you can solve these equations for the marginal distributions you can solve my complete equation, (equation (32) in my symposium paper), as the auxiliary equations for the latter are equivalent to your two equations. I attach such a complete solution, based on your own method of deriving the G,F solution.

Yours sincerely,

M S Bartlett

Alladi Ramakrishnan, Esq.,
Mathematics Department.

(B)

DEPARTMENT OF MATHEMATICS.

PROFESSOR M. H. A. NEWMAN, F.R.S.
PROFESSOR M. J. LIGHTHILL,
PROFESSOR M. S. BARTLETT.

THE UNIVERSITY,
MANCHESTER, 13.

ARDWICK 2681

2nd July, 1951.

      ALLADI RAMAKRISHNAN has been working at the University of Manchester for the past two years in Mathematical Statistics and Mathematical Physics, especially, as the title of his Ph.D. thesis indicates, on "Applications of the Theory of Stochastic Processes to Physical Problems". Part of the work contained in this thesis has been written up and accepted for publication.

M S Bartlett

(Professor of Mathematical
Statistics.)

Appendix 6

D. G. Kendall's Letter to Sir Alladi

From THE CLERK-TO THE COLLEGE,
MAGDALEN COLLEGE, OXFORD.

16. i. 51

Dear Sir,

I am glad to be able to assure you that when I saw them last (January 4th) your son and daughter seemed to be well on the way to recovery. I have since received a letter from your son, dated 8.1.51, in which he makes no mention of his indisposition, so I do not think you need worry.

I have much appreciated getting to know them. I have great respect for A.R's originality and am convinced that he will go far.

Sincerely yours,

David G. Kendall

Appendix 7

Letters from Sir C. V. Raman

(A)

Telegrams:
"RAMAN ACADEMY BANGALORE"

Telephone:
Office: 2646 Link.
Residence: 2023.

Sir C.V.Raman, F.R.S.,N.L.

Director.

Reg : Mc. 6/4-10.1189

RAMAN RESEARCH INSTITUTE
INDIAN ACADEMY OF SCIENCES,
HEBBAL POST, BANGALORE 6.

13th February 1952.

My Dear Ramakrishnan,

      I heard from your father that you had come back to India after getting your Doctorate. I have now received your very kind letter of the 2nd February and the accompanying reprints. It is very pleasing indeed to find that you have done such excellent work.

      Professor Chandrasekhar, I know, is deeply interested in Stochastic processes and problems, and you will certainly find it useful to work under his guidance. I shall be very glad to hear in due course that he has succeeded in getting you a scholarship to proceed for the United States.

      The Government of India have given me a position as National Research Professor and are giving me a small grant which just suffices to allow my Institute to function on a basis of minimum activity.

      With kind regards,

Yours sincerely,

C.V. Raman

Dr. Alladi Ramakrishnan,
27, Luz Church Road,
Mylapore, MADRAS-4.

239

(B)

Telegrams: Telephone:
RAMAN ACADEMY BANGALORE   Office: 2546 Link
                                                 Residence: 2023

Sir C.V.Raman, F.R.S., N.L.,   RAMAN RESEARCH INSTITUTE
        Director.   HEBBAL POST, BANGALORE 6

        Ref:No.P/A-10/ *1365*   22nd March 1952.

My dear Ramakrishnan,

        I have just received your kind letter of the

21st March 1952.

        With Ramachandran and yourself as Professor and

Reader respectively, ~~in~~ the newly established Department

of Physics makes the best possible beginning that it could.

I feel sure that the new Department has a bright future

before it.

                                Yours sincerely,

Dr.Alladi Ramakrishnan,
27, Luz Church Road,
Mylapore, Madras 4.

RP

(C)

Sir C.V.Raman,F.R.S.,N.L..   RAMAN RESEARCH INSTITUTE
            Director.   HEBBAL POST, BANGALORE 6

Ref: No. P/A-10 / 523   10th August, 1954.

My dear Dr. Ramakrishnan,

       I have just received and read with
pleasure your kind letter of the 7th August as also
the reprints of your papers you have been so good as to
send me.

       I can assure you that I am gratified
to see how successfully you are working your way into
the top ranks of scientific men in our country, I would
like very much to see you elected as a fellow of the
Indian Academy of Sciences in the next annual meeting
and I am writing to Dr. G.N.Ramachandran to speak to
you and if you are agreeable , to nominate you for
election.

             Yours sincerely,

             C.V.Raman

Dr. Alladi Ramakrishnan,
Reader,
Department of Physics,
University of Madras,
Madras.

(D)

Sir C.V.Raman,F.R.S.,N.L.,   RAMAN RESEARCH INSTITUTE
          Director.   HEBBAL POST, BANGALORE 6

Ref:No.P/R- /   4th January, 1955.

My dear Dr.Ramakrishnan

    Many thanks for your kind New Year greetings.

    I am very glad that you have become a Life
Fellow of the Academy. I quite agree with your
proposal to send us a series of three papers under
the title " A Physical Approach to Stochastic
Processes, I, II and III " and you may rest assured
that they will be given the promptest possible
publication.   When I was at your age I would have
been overjoyed to receive an invitation to Japan,
but I did not get it.

                Yours sincerely,

                C. V. Raman

Dr.Alladi Ramakrishnan,
Department of Physics,
University of Madras,
A.C.College of Technology,
Guindy, MADRAS 35

Appendix 8

Contract from Springer Verlag

SPRINGER-VERLAG
BERLIN · GÖTTINGEN · HEIDELBERG

Geschäftsinhaber: FERDINAND SPRINGER, Dr. med. h. c. Dr. phil. h. c. · JULIUS SPRINGER, Dr. Ing. e. h. · TÖNJES LANGE, Dr. med. h. c.

Dr. M. Mayer-Kaupp

Air Mail
R e g i s t e r e d (17a) HEIDELBERG, den
Neuenheimer Landstraße 24
Telefon 24 40

August 7, 1956
1b

Dr. Alladi R a m a k r i s h n a n
Department of Physics
University of Madras
A.C. College Buildings
M a d r a s - 25 / INDIA

Re : Handbuch der Physik / Encyclopedia of Physics, Vol. 3/II.

Dear Professor R a m a k r i s h n a n ,

thank you very much for kindly returning the signed articles of agreement. Please find enclosed herewith one copy of the contract, as undersigned by Dr. Ferdinand S p r i n g e r , for your file.

Sincerely yours,

M. Mayer-Kaupp

cc: A.S....

Bankkonten: Süddeutsche Bank, Filiale Heidelberg; Berliner Bank, Berlin-Charlottenburg Kto. Nr. 75 72; Berliner Disconto-Bank, Berlin W 35
Postscheckkonten : Karlsruhe Nr. 477 26; Berlin West Nr. 17 30

Appendix 9

Communications and Letters from Prof. S. Chandrasekhar

(A)

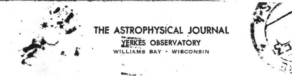

DEAR Dr. Ramakrishnan

 I am glad to inform you that the paper On the Solution of an Integral Equation of Chandrasekhar and   which you Munch. recently communicated has been passed for publication in the

January   issue of THE ASTROPHYSICAL JOURNAL.

            S. CHANDRASEKHAR, *Editor*

THE ASTROPHYSICAL JOURNAL
YERKES OBSERVATORY
WILLIAMS BAY · WISCONSIN

DEAR Mr. Ramakrishnan:

 I am glad to inform you that the paper A Stochastic Mode of a Fluctuating Density Field   which you recently communicated has been passed for publication in the

March   issue of THE ASTROPHYSICAL JOURNAL.

            S. CHANDRASEKHAR, *Editor*

THE ASTROPHYSICAL JOURNAL
YERKES OBSERVATORY
WILLIAMS BAY · WISCONSIN
January 26, 1954.

DEAR Dr. Ramakrishnan

I am glad to inform you that the paper A STOCHASTIC MODEL OF A FLUCTUATING DENSITY FIELD. II.   which you recently communicated has been passed for publication in the   May   issue of THE ASTROPHYSICAL JOURNAL.

S. CHANDRASEKHAR, *Editor*

Note: All in all, Professor Chandrasekhar, as Editor, accepted seven of Ramakrishnan's papers (some co-authored with his students) for publication in the Astrophysical Journal (Ap. J) during the years 1952–57. For the list of papers of Ramakrishnan in the Ap. J., see [2], pp. 81–82.

(B)

1956, February 6

Dear Ramakrishnan,

I was delighted to learn from your letter of January 24 that your plans for visiting this country are now more or less definite.

It is possible that I shall be away from Yerkes during April 19-26. But in all cases I expect to be here during the last week of April and the first week of May; and that should leave us enough time for discussions, etc.,.

Perhaps you would be willing to talk to our colloquium during your stay at Williams Bay. Our colloquia are on Mondays and I have tentatively put you down for the one on April 30: I hope this will be satisfactory to you.

I have arranged that a room be set aside for you at the boarding house run by Mrs. George Van Biesbroeck, Yerkes Observatory, Williams Bay, Wisconsin, for the benefit of the students and visitors at the Observatory. I should recommend that you write to Mrs. Van Biesbroeck directly when you are definite about dates.

From Chicago you can take a train directly to Williams Bay (from the North Western station: trains for Williams Bay leave at 4:42 p.m., 5:20 p.m., and 1:30 p.m. The 1:30 p.m. train leaves on Saturday only.) or to Walworth (by the Milwaukee Road from the Union Station: trains leave 9:30 a.m., 4:40 p.m. and 9:30 p.m.). I shall be glad to meet the train if I should be here and if you will let me know before hand. If I should be away, I do not suppose you will have any difficulty in getting to the Van Biesbroeck's (taxis are available at the stations).

Looking forward to seeing you here, I am,

Yours sincerely,

S Chandrasekhar

* not on Saturday.

(C)

Yerkes Observatory
Williams Bay, Wisconsin
1956, September 11

Dear Ramakrishnan,

On returning to Yerkes after an absence of some ten weeks, I was delighted to find a copy of your father's book on "The Constitution and Fundamental Rights" awaiting me. It was good of you to send this to me and I appreciate your inscription. I believe we told you while you were here that we once heard a lecture by Justice Felix Frankfurter on the Constitution of India and of his happy references to your father. We shall both read the book carefully.

With best wishes,

Yours sincerely,

S. Chandrasekhar

Appendix 10

Letter from Prof. W. Heisenberg

Professor W. Heisenberg

MAX-PLANCK-INSTITUT FÜR PHYSIK

· GÖTTINGEN   BÖTTINGERSTRASSE 4

GÖTTINGEN, 8.10.1956
Tel.: 8653

Dr. Alladi Ramakrishnan
Department of Physics
University of Madras
P.T. College Buildings,

M a d r a s - 25, (Indien)

Dear Dr. Ramakrishnan,

Many thanks for your letter and for your interesting manuscript on cascade theory. Your paper has been studied by the theoreticians of the institute, and we agree with you that your new approach may simplify the comparison between theory and experiment. From the view point of the experiments we would be very interested to have numerical calculations on the fluctuations of cascade showers of very high primary energy (above 10^{12} eV) at small thicknesses (t=0,5 - 2,0).

You would certainly be welcome for a visit of a few months in our institute during the next summer, and the physicists especially of our experimental group would be glad to discuss the problems of electronic cascade with you. One could scarcely speak of a lecturership in this connection because the Max-Planck-Institute is a pure research institution in which lectures to students are not given (it is not connected with the University), but you would be completely free to work in the institute in the same way as the other visiting scientists from abroad.

Yours very sincerely,

W. Heisenberg

248

Appendix 11

Letters from Prof. J. R. Oppenheimer, Sir S. Radhakrishnan, and Sir C. V. Raman

(A)
Letter from J. R. Oppenheimer

THE INSTITUTE FOR ADVANCED STUDY
PRINCETON, NEW JERSEY

OFFICE OF THE DIRECTOR
27 February 1957

Dear Dr. Ramakrishnan:

With the concurrence of my colleagues in Physics, I am pleased formally to offer you a membership in the School of Mathematics of the Institute for Advanced Study for the academic year 1957-1958. We can make available to you a grant-in-aid of $4,500 to defray the expenses of your sojourn in Princeton.

We all look forward with pleasure to having you with us for a visit.

Sincerely yours,

Robert Oppenheimer

Dr. Alladi Ramakrishnan
Department of Physics
University of Madras
A. C. College Buildings
Madras 25
India

enclosure: Exchange-Visitor memorandum

(B)

Letter from Sir S. Radhakrishnan

VICE-PRESIDENT

VICE-PRESIDENT'S LODGE
2, KING EDWARD ROAD
NEW DELHI

March 20, 1957

My dear Balu,

    Thank you for your letter.

    I am delighted to know that you are going to Princeton and will be a visiting member of the Institute there.   I know your father would have been greatly pleased at this recognition.

Yours sincerely,

Shri Alladi Ramakrishnan,
Department of Physics,
University of Madras,
Alagappa Chettia College of Technology,
Guindy, Madras 25

(C)

Letter from Sir C. V. Raman

Sir C.V.Raman,F.R.S.,N.L., RAMAN RESEARCH INSTITUTE
   Director.   HEBBAL POST, BANGALORE 6

Ref:No.P/A- / 1530   18th March, 1957.

Dear Dr.Ramakrishnan,

  The manuscript of the paper referred to in your letter of the 16th March has not yet reached me. It will be promptly dealt with when it comes to hand. || I am not surprised at your having received the call to Princeton. I see you are very pleased about it, but I wonder what the reactions of the Madras University would be at letting you go away for the best part of each year.

   With kind regards,

       Yours sincerely,

       C. V. Raman

Dr.Alladi Ramakrishnan,
Department of Physics,
University of Madras,
A.C.College of Technology,
GUINDY, MADRAS 25.

(D)

Letter from Sir C. V. Raman

Sir C.V.Raman,F.R.S.,N.L.,
   .Director.

RAMAN RESEARCH INSTITUTE
HEBBAL POST, BANGALORE 6

Ref:No. 1076 ·

12th December 1957.

My dear Dr.Ramakrishnan,

    This is just a line to acknowledge your kind letter of the 6th December. I am happy to learn that you are enjoying your stay at Princeton. The recent award of the Nobel Prize to two young Chinese physicists has made a great impression everywhere as an example of how young talent can blaze its way in the field of science. We are all hoping that some day, India will have some more Nobel Prizemen to talk about. What about Chandrasekhar and yourself for an early choice ?

    With kind regards,

Yours sincerely,

C.N.Raman

Dr.Alladi Ramakrishnan,
  Institute for Advanced Study,
  School of Mathematics,
  PRINCETON, N.J., USA.

Appendix 12

Letter from Prof. K. A. Brueckner

UNIVERSITY of PENNSYLVANIA
PHILADELPHIA 4

The College

DEPARTMENT OF PHYSICS

April 7, 1959

Professor Alladi Ramakrishnan
27, Luz Church Road
Mylapore, Madras - 4
India

Dear Professor Ramakrishnan:

Thank you for your letter reminding me of your desire to send one of your senior students to work with me. I would like to offer a position as research fellow to whichever of your students you feel is best qualified. This position would pay approximately $6500 for twelve months with an option for one month vacation. The appointment would be at the University of California at La Jolla where I will be accepting a position next September. The appointment would be for a two year period, although if the recipient so desired, he could of course terminate it earlier.

After I am advised by you as to which student you recommend, I will write to him to confirm this offer. I hope very much that both your student and we are able to benefit from a period of close contact.

Sincerely yours,

Keith A. Brueckner
Professor of Physics

KAB:rjs

Appendix 13

Letter to Mr. C. Subramaniam

Personal   31st October 1959

Dear Hon'ble Minister,

    Perhaps you will remember our conversation when we met casually at the Woodlands a few weeks ago. I should like to meet you sometime tomorrow (Sunday, 1st November, 1959).

    On behalf of the Theoretical Physics group I have great pleasure in inviting you for an informal dinner on the evening of 21st or 22nd November 1959 when I will be in Madras in connection with the Senate meeting of the University.

Yours sincerely,

Alladi Ramakrishnan

(Alladi Ramakrishnan)
Professor of Physics,
University of Madras

The Hon'ble
Sri C.Subramaniam,
'Cooum House',
Government Estate,
Madras.2.

Appendix 14

Need for the Creation of an Institute of Mathematical Sciences in Madras

The most significant feature of scientific development in the twentieth century has been the simultaneous advance of fundamental research especially of a theoretical nature on the one hand and of applied sciences with special emphasis on their relation to technological progress on the other. Before the birth of modern physics as it is understood today, the role of the mathematics in physical sciences has been essentially that of a tool, albeit valuable, for the study and systematisation of observations. When a relativistic and quantum mechanical description of matter was forced upon the physical world by the discoveries of Einstein, Planck, Bohr, Heisenberg and Dirac during the first three decades of this century, the deductive and mathematical study of nature assumed an importance equal to that of experimental investigation. The role of mathematical sciences in general and theoretical physics in particular in the scientific advancement of a country was realised especially by America and Russia. The creation and spectacular rise of great schools of theoretical physics in the U.S. at Harvard, Princeton, Rochester and Berkeley coincided with the corresponding advance in experimental physics. Unfortunately, the situation in India has been quite different. Little or no emphasis has been laid till now on high standards of mathematical discipline even for post-graduate studies of science. There has been an indifference and apathy towards fundamental research of a theoretical nature. The direct consequence of such an attitude has been the falling of standards in every branch of science. It is common knowledge that our students sent for training abroad found the mathematical discipline in America almost "oppressive" and this has naturally affected the prestige of Indian education in the eyes of the world. Particularly unfortunate has been the situation in South India where no effort has been made hitherto to foster fundamental research in mathematical sciences in spite of the obvious fact that mathematical talent is available in abundance and in quality in the State of Madras. The time is now propitious for the starting of an Institute essentially devoted to the pursuit of fundamental sciences with accepted emphasis on original work and research. Even in the United States, where universities have been endowed with extraordinary resources, the need has been felt for such centres like the Institute for Advanced Study where the main objective is pure research of a very advanced theoretical nature. A few years ago, a similar Institute was started in Japan under the leadership of Professor Yukawa for the advancement of theoretical physics, a subject to which the Japanese have made significant contributions in recent years.

The only expenditure that will be incurred by such an Institute is the payment of salaries to the staff and research workers. Almost no capital expenditure is involved. The creation of an Institute of Mathematical Sciences will earn the gratitude and admiration of the young scientific community in India which has been hoping and yearning for opportunities which have hitherto been available only outside our country.

Details of the proposals

1.   The Institute shall be named "The Institute of Mathematical Sciences".

2.   The Institute shall be administered by the Government of Madras through a Governing Body appointed by it.

3.   The aims and objectives of the Institute are:

(i) To conduct research and original investigation in fundamental sciences in general with particular emphasis in mathematics, applied mathematics and theoretical physics.

(ii) To foster a rigorous mathematical discipline, to stimulate a zest for creative work and cultivate a spirit of intellectual collaboration among academic workers in pure and applied branches of sciences.

(iii) To arrange lectures, meetings, seminars and symposia in pursuance of its academic work.

(iv) To invite scientists of established reputation and active research work in and outside India to deliver lectures and participate in its academic activities.

4.   The academic staff of the Institute shall consist of:

(i) A Director: Grade of pay Rs. 1300-100-1800 (Post permanent).

(ii) Professors: Grade of Pay Rs. 800-50-1000 (during the period of probation. If research work is found satisfactory, permanent position on Rs. 1250 per month (Standard University Grants Commission Scale).

(iii) Assistant Professors: Grade of Pay Rs. 500-50-700 (during the period of probation of five years. If research work is found satisfactory, permanent position on Rs. 800 per month (Standard University Grants Commission scale).

(iv) Temporary Members: Appointment for a term of one year Rs. 500 a month. Maximum period of stay — three terms.

(v) Honorary Members: Active research workers in fundamental sciences in other institutions can be admitted on application as honorary members by the Governing Body.

(iv) Research Students: Stipend Rs. 200/month. On probation of one year. On successful completion of probation, tenable for two years. A research student shall work under the direction of a Professor.

5.   The Institute shall assist the Departments of Physics and Mathematics at the Presidency College in the organisation and conduct of post-graduate classes.

6. The professors of Physics and Mathematics of Presidency College shall be ex-officio honorary members of the Institute.

Duties

A. *Director*: The normal administration of the Institute shall be carried on by the Director who should be a scientist of established reputation and should be responsible for the effective pursuit of the aims and objectives of the Institute. The Director shall be an ex-officio member of the Governing Body.

B. *Professors and Assistant Professors*: The duties of the Professors and Assistant Professors shall be to conduct and guide research and assist the Director in the academic work of the Institute.

C. *Temporary Members*: They shall be engaged in original investigation and research either independently or in collaboration.

D. *Members and Honorary Members*: They shall be entitled to participate in the academic programme of the Institute.

Tentative Annual Budget

The Institute can start functioning in the first instance with the following staff:

ANNUAL EXPENDITURE

Director — Rs. 18,000
Six Research Students — Rs. 14,400
One Temporary Member — Rs. 6,000
One Assistant Professor — Rs. 7,200
One Stenographer (PA to the Director) — Rs. 2,000
Administrative stationery and postal expenditure — Rs. 5,000
Honorarium for invited lectures — Rs. 5,000
Library — Rs. 10,000

The recurring expenditure will be less than Rs. 75,000/year. The Institute may be temporarily located in the Presidency College. At present two rooms will be sufficient for the staff. Lectures, Seminars and Symposia can be held in the lecture halls of the Physics and Mathematics Departments of the Presidency College.

Appendix 15

Letter from Prof. Homi Bhabha

GOVERNMENT OF INDIA
Atomic Energy Commission

December 2, 1959

My dear Alladi,

On my return from abroad I have seen your letter of the 2nd August and thank you for it and for the kind sentiments you have expressed. I propose to look through your articles as soon as I have a little time. I am happy to see that you have been doing so well.

Science in India can only be built up by each one of us working with integrity to the best of his ability, and by all good workers co-operating with each other in furthering the cause of science.

I send you my best wishes for greater success in your work.

Yours sincerely,

H. J. Bhabha

Appendix 16

Letters to and from Prof. Niels Bohr

(A)

India

31st October 1959

lladi Ramakrishnan

ly may be sent to:
Luz Church Road,
pore, Madras.4.

Dear Professor Bohr,

It was with pleasure and pride we heard that the Indian Science
Congress has invited you again for the forthcoming session in Bombay
early January 1960. It is the earnest hope of the entire scientific
community in India that it will be possible for you to accept the invitation and spend sometime visiting research centres in our country.

In particular I am expressing the desire of the theoretical physics
group at Madras that you should visit our city to give us an opportunity
to express our warmest admiration and offer our personal tribute to the
father of modern physics whose discoveries ushered in the atomic age
and whose scientific philosophy has had the profoundest influence on
human thought since the age of Newton. Your very presence here will be
a source of inspiration to the young and enthusiastic band of workers
with whom I have the good fortune to be associated. With the pardonable
vanity of a citizen of Madras , I can assure you that we can make your
visit quite interesting since Madras still preserves some strains of
Hindu culture and ways of life despite the rapid onset of industrialisation.

I shall deem it a great privilege if you can stay at my family
residence when you visit Madras and then my young friends here will have

31st October 1959

(2)

the opportunity of listening to you during informal meetings.
I will be in Bombay for the Indian Science Congress session
when we can discuss $\overset{the}{a}$ detailed programme of your visit to Madras.
An assurance now from you that you are accepting our invitation
will naturally be a source of great pleasure to us.

With kindest regards,

Yours sincerely,

(Alladi Ramakrishnan)
Professor of Physics

(B)

ALLADI RAMAKRISHNAN   'EKAMRA NIVAS'   **17th January 1960**
                              LUZ. MYLAPORE.
                              MADRAS-4.

Dear Professor Bohr,

      My wife and I cannot adequately express in words our
gratitude to you and Mrs. Bohr for having accepted our invitation to visit our family residence and have dinner with us.
All the members of the Theoretical Physics Seminar in Madras
feel that it was the realisation of a dream they had cherished
all these years. Personal contact with you has been an inspiration which will sustain them throughout their scientific career.

      Your public expression through the press about the work
of our group was obviously stimulated by your kindness and
generosity which are without measure even as your reputation
as a scientist. May I hope that very soon some of our young
men here will justify in a humble measure atleast the affectionate interest that you have taken in them. It is our earnest
prayer that you should be endowed with health and strength to
continue as the leading light not only of the scientific community but of whole humanity in dispelling the fears of a
distracted world.

      My wife and I look forward to meeting you and Mrs. Bohr
at Copenhagen this summer.

                              With kindest regards,

                              Yours sincerely,

Professor Neils Bohr,
(visiting Scientist),
c/o Prime Minister,
NEW DELHI.

                         Alladi Ramakrishnan

                           (Alladi Ramakrishnan)
                           Professor of Physics
                           University of Madras.

(C)

UNIVERSITETETS INSTITUT
FOR
TEORETISK FYSIK

BLEGDAMSVEJ 17
COPENHAGEN, DENMARK
TELEGRAMS: PHYSICUM, COPENHAGEN
February 19, 1960.

Professor Alladi Ramakrishnan
27, Luz Church Road
Mylapore, Madras-4.

Dear Professor Ramakrishnan,

As my wife has written to Mrs. Ramakrishnan, our whole journey to India was a most inspiring experience and on my return from Geneva I want to thank you most heartily for the kindness you showed us during our stay in Madras where we greatly enjoyed the evening in your beautiful home and the meetings with your students.

I deeply admire the effort you are making in building up such an enthusiastic and competent group of theoretical physicsts, and even if the primary aim is purely scientific, I feel that the effort will most essentially contribute to the background for the endeavours to promote prosperity of your country by help of modern technology.

We look forward with great pleasure to your visit here in a near future and send your family and yourself and common friends in Madras our kindest regards and warmest wishes.

Yours,

Niels Bohr.

P.S.: I have just received your letter of February 13th and am delighted to learn that we can expect you here in May.- My wife has also received Mrs. Ramakrishnan's letter and is very grateful for the trouble she has taken on her behalf. We both hope that she will be able to accompany you on your visit to Copenhagen.

(D)

6 July, 1960

TO: Professor Niels Bohr
Copenhagen, Denmark

FROM: Dr. Alladi Ramakrishnan
27 Luz Church Road
Mylapore, Madras-4

Dear Professor Bohr,

Pardon me for not writing to you earlier. I wanted to wait till I returned home after my trip to Europe which now looks like a beautiful dream. The most exhilarating part of that dream was my stay in Copenhagen where I had the pleasure and privilege of meeting you again. Copenhagen is a city which has enchanted visitors all over the world and especially from across the Atlantic. To me it was twice enchanting since it is the home of the greatest living theoretical physicist of this century and the seat of one of the greatest schools of research in the world.

The hospitality and generosity showered on me by you and your gracious wife were flattering even to my vanity. I do not know whether I deserve all that kindness which is more a measure of your humanity than my worth. It is also a gesture of goodwill between two great nations and of the academic spirit which has brought together in an intimate and almost holy association despite the stresses and strains of war — of scientists and savants, great and small of various races and nationalities. There is no doubt that in the years to come this comradeship will pervade the general atmosphere and make the world so safe and clean that men will have the leisure and opportunity to realise the pleasures of living and thinking together.

When I attended the Symposium at your Institute, I seemed to imbibe the "Copenhagen spirit" which I shall endeavour to transmit to my younger and brighter colleagues in Madras. It was gratifying that this young school has attracted your attention, but my hopes and prayers are that they should also deserve it. Very soon, I should like the most competent among them to spend a year or two at Copenhagen and I was naturally delighted when your son agreed to consider such a proposal favourably.

Perhaps you will permit me to express a sentiment so clear in the Hindu mind and heart. What a thrill it was to see a son justify the greatness of his father by his own efforts. Professor Aage Bohr is carrying on the great traditions of the Copenhagen school and he has in him the strains of humanity and genius worthy of the name he bears.

With warmest regards,

Yours sincerely,

(sd) Alladi Ramakrishnan

PS: My wife was very happy to receive a letter from Mrs. Bohr to whom she will be writing in a few days.

Appendix 17

That Magic Moment

The world is so worshipful of greatness that we rarely wait to think of its true source of strength and sustenance. It lies in the manner in which it touches the individual lives of those who feel the depth and extent of its influence. Such is the quality of the greatness of our beloved Prime Minister whose life has almost directed our way of life.

It was in 1947, during the eventful days of the Constituent Assembly, that I had the privilege of feeling his benign presence and watching that handsome face with pleasure and wonder. Even my great father who was entrusted with the task of drafting our constitution could not hold his attention, for the distant look in his eyes seemed to peer into the uncharted future and was oblivious of the immediate environment. It was unbelievable that an occasion would arise fifteen years later when by a fortuitous circumstance, I would be called upon to place before him, in person, a proposal which is to affect the lives of young scientists in the years to come. I cherish that moment with blatant delight and the best homage I can pay to so noble a soul is to transmit the magic of the momentous interview to my fellow scientists. To estimate the significance of that event I have to describe the background against which it took place.

Ever since the war, there was so much discussion about the migration of Indian scientists abroad and the difficult conditions that inhibited the growth of creative work in India. This looked paradoxical, particularly when the same period saw the establishment of various Governmental organisations to stimulate and support scientific research. While there was basic agreement among the academic community and the organisers of scientific endeavour that something should be done to vivify and vitalise the atmosphere, there was considerable conflict of opinion as to how this could be done. Of course, there was the very conventional view that higher

266

learning should be pursued in the universities since creative work could only be sustained in consonance with a teaching programme. On the other hand, the need for specialised institutes and laboratories was too obvious since the financial resources available had to be conserved and their magnitude demanded direct support from the Central Government.

The mathematical sciences demanded a combination of these two modes of organisation. It is too well-known that the advances in physics in the last decade were comparable with the progress in physical sciences over a century before, for, with the development of giant accelerators and new experimental techniques, the physicist became aware of new particles and new phenomena associated with them, not anticipated even by the creators of quantum mechanics and relativity.

The American Universities, true to the pioneering traditions of that great nation, adjusted themselves to the rapid changes in the scientific scene by creating semi-autonomous and highly specialised institutions within their expanding framework. Unfortunately, nothing like this had happened in India and was likely to happen, in view of the repeated emphasis on insurmountable difficulties and too firmly established conventions. A breakthrough was necessary and therefore it was suggested that an autonomous institution should be created and supported by the Government but which would actively collaborate in academic work with the Universities. It was a miraculous sequence of events that culminated in the decision of our then Finance Minister of the Government of Madras, Mr. C. Subramaniam, to create an institute of this kind. Even 'this man of steel' required the support and assent of our Prime Minister since this idea was considered almost revolutionary in the domain of our scientific education. To whom else could we submit such a proposal than to one who had effected the greatest revolution in our minds — the desire for a free and independent India which was achieved within his own life-time?

Indeed this suggestion of consulting the Prime Minister was put forward by an American physicist, Dr. M.M. Shapiro who visited Madras as a guest of the theoretical physics group here. In the course of a casual conversation when the Finance Minister complained of various difficulties and obstacles, the Professor interposed and said, "Why not let the Prime Minister see the members of the theoretical physics group during his visit to Madras and find out his reactions"?

In spite of a crowded series of official engagements for the Prime Minister, it was found possible by the Hon'ble Subramaniam to arrange an interview at the Raj Bhavan at 9:30 p.m. after an official dinner, to which my wife and I were invited to have a preliminary occasion to meet the Prime Minister. I need hardly describe my trepidation and anxiety, for so much was to depend upon a few minute conversation — almost the dreams and aspirations of a whole scientific community hung on the smiling lips of our Prime Minister. Ten minutes before the close of the dinner, he summoned me and all the students gathered to have a personal interview with him. It was strange feeling for me to do all the talking with a man who held almost unquestioned sway over the destinies of our country. At the end, he asked me only one question: "Are you really convinced that we should have an Institute of the kind you are insisting upon?" I naturally said 'yes' with all the emphasis at my command and he just smiled.

Two months later, the Finance Minister decided to obtain the formal assent of the Prime Minister. We waited with bated breath at the Madras airport as Mr. Subramaniam came across the tarmac and said with his inevitable smile, "It is going to be all right. The Prime Minister has consented to be our patron." Later on, he recounted to us what a miracle it was to succeed in directing the attention of the Prime Minister to this question. It was the day on which the decision for military action in Goa was being taken that the Prime Minister agreed to be the patron of the Institute.

Sixteen months later, I met him again to report on the progress of the Institute. He just asked, "Do you want anything particular to be done now?" It was too valuable an opportunity to miss and I stressed the need for extending the visiting programme to young post-doctoral workers of outstanding promise. Through his aegis has now been set in motion an international collaboration in science which in the words of Niels Bohr, "offers great opportunities for understanding among the peoples of the world" — a cause nearest to Nehru's mind and heart.

The triumph of Nehru's life is the triumph of imagination over prejudice and ignorance — the characteristic feature of all scientific endeavour. To this ideal, aspirant members of the Indian scientific community should dedicate themselves.

Alladi Ramakrishnan

Appendix 18

Letter to Mr. C. Subramaniam

26th September 1961

Dear Hon'ble Minister (C. Subramaniam)

I wish to express my opinion freely and frankly on the present status and the possibility of the future development of fundamental science in our country with the confidence that I am doing so on behalf of the entire scientific community in India.

At the outset I would emphasize that a clear and marked distinct to be kept in mind between:

1) The development of atomic energy,

2) The advancement of fundamental research

when discussing the nature of organisations to foster these efforts. Of course I am aware of the close relationship between the two and how technological development can only be sustained by a corresponding advance in fundamental sciences. The development of atomic energy needs enormous financial resources and is closely linked with problems of defence, agriculture and the economic progress of the country. It is only natural that it has to be under the direct control and sponsorship of the federal government of the nation, as is being done throughout the world. However as regards fundamental research, the situation is completely different since creative achievement is a product of intellectual endeavour demanding the largest measure of freedom and availability of opportunities. The following factors are necessary for the advancement of fundamental science:

1) Teachers and guides of great competence who are able to attract, due to their idealism and devotion to science, research workers of talent and promise.

2) A steady stream of gifted students and senior researchers of established creative ability.

3) Freedom to work in a atmosphere of "enlightened and contemplative leisure".

4) A spirit of competitive endeavour and active collaboration among the various research groups.

It is the unanimous opinion of the scientific groups throughout the world that academic freedom on the one hand and competitive endeavour on the other are the soul and substance of scientific life which every teacher and student hold dear and almost sacrosanct. While we enjoy a certain amount of academic freedom especially in the universities, it is a matter of regret that there is an almost total absence of competitive spirit and collaboration among research groups. What is more of deep concern is that the importance of these ideals has not been realised. On the contrary we are attempting to carry over the principle of concentration of resources which may be necessary to carry out multi-purpose projects for economic development into the domain of fundamental science. Scientific work languishes and even petrifies in the absence of the stimulus of competition. This has been fully realised in all the advanced countries, particularly in the United States of America. From Harvard and Princeton in the East to Berkeley and Pasadena in the West, have grown in recent years various centres of research activity comprising the greatest cooperative and competitive endeavour in the history of science. It amazes one beyond comprehension to see such active collaboration among various groups attached to different organisations and the absolute freedom of a scientist to move from one centre to another.

As a strikingly illustrative example, in the city and environs of Boston there are four institutions: Harvard University, the Massachusetts Institute of Technology, Tufts and the Brandeis Universities — in which research workers meet and discuss with mutual benefit and sometimes hold joint seminars and symposia.

Once the dual principle of competitive endeavour and active collaboration is accepted, the only solution for the development of fundamental sciences in India and the quickening of our intellectual life is through the creation of large number of centres of fundamental research with frequent exchange of staff and personnel.

There cannot be fetters on the mind of man and the creative worker should have the freedom to choose his own field of intellectual activity and a suitable venue for his work. It is only under such conditions that the sensitive plant of intellectual effort will thrive and grow to its full stature. Such an atmosphere can be provided only if a choice of various research centres situated in different parts of the country are available to the worker. A single mammoth institution consuming all the resources of the country cannot provide such an atmosphere since an aspirant to a research career is left with no choice and this in a sense indirectly amounts to a restraint upon his academic freedom.

The only question therefore is the feasibility of finding persons to direct and man such institutes. If enough talent were not available, I could have agreed with the view that such institutions may be saddled with mediocrity. But I feel quite strongly that the conventional arguments that there are not enough persons to direct research is without foundation. It is an unhappy tendency in our country to recognise talent only when it is discovered and honoured by foreign societies like the Royal Society of England or the Universities of the United States. If we were to justify the high intellectual tradition to which our ancient country is heir, we should be bold enough to recognise talent in our country. The postulate that there is not enough talent in the country is not borne out by facts. Professor S. Chandrasekhar, the greatest theoretical astrophysicist in the world today, was constrained to start his research career in Cambridge and has now settled down in Chicago as an American citizen. Professor Harish Chandra, perhaps the most gifted of our mathematicians, made his great reputation in the mathematical world when he was only a research student under Professor Bhabha. Professor Salam's work was recognised as outstanding in fundamental physics four years before he was invited to the Indian Science Congress on his election to the Fellowship of the Royal Society. It is a tragedy too deep for tears that we seem to wait so long before accepting the reputation of Indian scientists that we are not able to attract them to our institutions after they have settled down as permanent members of the staff of leading institutions in the world. It is time we shake ourselves out of an attitude of apathy towards indigenous talent. I have not the least doubt that if we have the willingness to start institutions of fundamental research in various parts of India, we can find leaders of competence inspired by the necessary sense of idealism.

With almost no resources at my disposal except my abiding faith in these ideals, I started on my return from Princeton, a humble experiment just two years ago in our own home city in forming a group of workers in theoretical physics devoted to the pursuit of science in the spirit of their compeers elsewhere in the world. The success of this experiment has gone far beyond my wildest dreams and if today, I am able to claim that this group is as active as any in India and bids fair to be the nucleus of an intellectual renaissance, it is because of the abundance of talent available here which is just waiting for opportunities for expression. During the past twelve months our group had the pleasure of personal contact with some of the distinguished physicists who have come to Madras and who have expressed their warmest appreciation of the nature of the effort of this scientific community.

Far from being an intrepid venture, the creation of an institute for theoretical physics would become an example to be followed in other parts of India. The existence of such institutes will not be inconsistent with an expansion of the physics and mathematics departments in the Universities. In Princeton for instance, there flourish side by side the Institute for Advanced Study and the Princeton University and their collaborative achievement has become part of the scientific tradition of the United States. At present it is a general feature in our Universities that each subject is pursued only under the direction of one professor. Since a University is concerned with a large number of departments, the possibility of having many professors in one department may seem remote under the present circumstances.

Only one point perhaps remains to be clarified. I agree that research institutes should not be divorced from teaching. Even at the outset I have stated that a true scientist must be willing to impart knowledge and it is only through the propagation of ideas from teacher to students that we can ensure a steady stream of fresh talent. The sanctity of the relationship between the teacher and a student has been handed down to us from the Vedic age and constitutes an integral part of our way of life. Lectures, symposia and seminars can be organised frequently in these specialised institutes so that teachers from the various colleges in the neighbourhood and young aspirants to research career can come into close contact with active workers. In the words of Oppenheimer, the teaching of science is at its best when it is most like an apprenticeship and there can be no better venue of apprenticeship than an institute which professes to foster creative work.

In conclusion, let me express my hope that the idea of starting an Institute of Theoretical Physics so well conceived, will be pursued to its realisation in the near future.

(sd) Alladi Ramakrishnan

Appendix 19

Letter from Hon'ble Humayun Kabir

MINISTER,
SCIENTIFIC RESEARCH AND CULTURAL AFFAIRS, INDIA,
NEW DELHI.

29th January 1960.

Dear Prof. Ramakrishnan,

This is just a line to say how delighted I was to hear from Prof. Niels Bohr and Prof. Abdus Salam of the splendid work that you are doing. Both of them said that they were greatly impressed by the enthusiasm and keenness of your colleagues under your able leadership. In fact, what they said has made me confident that we can expect great things from the Madras School of Physics in the course of next four or five years.

I am endorsing a copy of this letter to Dr. Lakshmanaswami Mudaliar.

With kind regards,

Yours sincerely,

(Humayun Kabir)

Prof. A. Ramakrishnan,
University of Madras,
MADRAS.

Appendix 20

Appointment as Director of
MATSCIENCE

(A)

Letter from Mr. K. Srinivasan, IAS

(By Special Messenger)

SRI K.SRINIVASAN, I.A.S.,
SECRETARY TO GOVERNMENT.

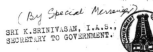

FORT ST. GEORGE
MADRAS
Dated 22nd December 1961.

EDUCATION AND PUBLIC HEALTH DEPARTMENT.

D.O.No.S-134/61-1/EPH.

Dear Professor Ramakrishnan,

      Will you please refer to the correspondence resting with your letter dated the 13th April 1960, and addressed to the Minister for Education and your subsequent discussions with him and myself on the subject of setting up an Institute of Mathematical Sciences for higher research? I am glad to inform you that the State Government have decided to set up such an Institute more or less on the lines suggested by you and that the Prime Minister of India has also agreed to become its patron in due course. It is also proposed to appoint you as the first Director of the Institute on terms and conditions which may be finalised later. The Madras University authorities are being approached for your relief to enable you to take up the appointment of Director. In the meantime, I am to request you to take necessary steps to arrange for your relief and for taking preliminary action for the inauguration of the Institute at an early date.

Yours sincerely,

(K.SRINIVASAN)

To
/Prof.Alladi Ramakrishnan,
Professor of Theoretical Physics,
University of Madras,
M A D R A S--5.

(B)

Letter to the Vice-Chancellor, Madras University

23rd December 1961

Dear Vice-Chancellor,

The time for making the most crucial decision in my academic career has arrived and I wish to convey to you my desire to relinquish my post as Professor in the University.

A new situation has arisen by the grace of God and the will of man which offers me and my colleagues great opportunities for the pursuit of our ideals and aspirations. I have been asked to assume the Directorship of the new Institute of Mathematical Sciences created by the Government of Madras where creative work can be fostered in a spirit and environment as exhilarating as that in the Institute for Advanced Study in Princeton. It is perhaps the most gracious gesture that have ever been made by an administrative authority to an academician in our country and I feel the urge and necessity to accept it without hesitation.

In these circumstances I would request the authorities of the University to accept my resignation and relieve me of the duties forthwith. No inconvenience will be caused to the University for the obvious reason that I am just holding the Madurai Professorship which has been transferred to Madras under the stress and strain of circumstance. My resignation will restore the conditions to normal and I wish to convey to the Syndicate my apologies for the inconvenience that has been caused by this transfer.

On this occasion, I wish to express my deepfelt gratitude to you for your almost parental interest in my career, which sustained me during these difficult years. I shall always cherish with undiminished pride the kindness and consideration you have shown me since we met in Geneva.

Only students registered for their Ph.D. are now working under me and I presume that suitable arrangements for the transfer of these students to the new Institute of Mathematical Sciences can be made ~~immediately~~ as soon as possible

With kindest regards,
Yours sincerely,

Alladi Ramakrishnan

Vice-Chancellor,
University of Madras,
Madras.

(C)

Letter Accepting Directorship of MATSCIENCE

ALLADI RAMAKRISHNAN
  PROFESSOR OF PHYSICS
  UNIVERSITY OF MADRAS.

'EKAMRA NIVAS
27, LUZ, MYLAPORE,
MADRAS.
Phone: 71106

23rd December 1961

Sri K. Srinivasan,
Secretary to Government,
Education and Public Health Department,
Fort St. George,
Madras.

Dear Mr. Srinivasan,   Ref: Your D.O.No. 8-134/61-1/EPH d/22-12-19
           Your letter, Ref. No. 162634/E4/59-13,
           Education, dated 22nd December, 1961
           addressed by you to the Registrar, Uni
           versity of Madras, Madras-5 in re:
           Institute of Mathematical Sciences at

    It was with unconcealed pleasure that I received your
kind invitation to assume the Directorship of the Institute of
Mathematical Sciences to be created by the Government of Madras.

    To me and my colleagues, it is the realisation of a long
cherished dream. The creation of such an Institute sets in motion
an intellectual renaissance in our country of a kind and magnitude
unknown before. I feel it a most pleasant duty to accept this offer
especially as I am aware that it originated from our esteemed Financ
Minister who was a great friend of my great father. It is perhaps
the most gracious gesture ever made by any administrator in India
to an academician. My colleagues and students have offered their
active co-operation and goodwill in this great endeavour.

    I am placing my resignation before the University today
to be operative forthwith. I presume there will be no difficulty
in the University authorities accepting my resignation since I am
just holding the Madurai Professorship which has been transferred
to Madras under the stress and strain of circumstances. My resignation will restore the conditions to normal and I have conveyed to
the Syndicate my apologies for the inconvenience that has been
caused by this transfer.

         Yours sincerely,

         Alladi Ramakrishnan

         (Alladi Ramakrishnan)

(D)

Letter to Prime Minister Nehru

31 December 1961

TO: Pandit Jawaharlal Nehru
Prime Minister of India
New Delhi

Most Honoured Prime Minister,

At the dawn of the new year, my colleagues and my wife join me in conveying to you our best wishes for your health and happiness which are co-extensive with the well being of the entire nation.

It was perhaps the finest hour of my life when our Finance Minister, a great friend of my father, created an opportunity for me to meet you during your visit to Madras.

The kind words you recently spoke about the members of my group are more an expression of your generosity than an estimate of their achievements. Your wide humanity and deep concern for the prosperity of our country make you see the light of hope even in the feeble efforts of smaller men. As you have on many occasions said, if the cause is great, the shadow of that greatness falls upon all those who strive for it despite their intrinsic smallness.

It has delighted us to distraction that you have consented to be the Patron of the Institute. There is no other way in which we can express our gratitude to you except by effort and achievements in the years to come.

With kindest regards,
Yours sincerely

(sd) Alladi Ramakrishnan

Appendix 21

Speeches at the Inauguration of MATSCIENCE

(A)

PROFESSOR ALLADI RAMAKRISHNAN'S SPEECH ON THE INAUGURATION OF THE MATSCIENCE INSTITUTE

My father Professor Alladi Ramakrishnan was a master of exposition, both in written and spoken form. He was a dynamic speaker, an orator in every sense. Right from my boyhood I had the pleasure to listen to many of his scientific lectures and speeches, and was inspired by his manner of speaking and the power of his oratory. The finest speech he ever gave was perhaps at the inauguration of MATSCIENCE, the Institute of Mathematical Sciences, on January 3, 1962, when his thoughts came pouring out at that very exciting and memorable occasion. As a seven year old boy, I was in the front row of the English Lecture Hall at the Presidency College in Madras when he delivered that speech extempore, as was his custom. The speech was later written up from a tape recording. The speech was printed in the appendix to The Alladi Diary, Vol. I, East-West Books, Madras (2000). It is reprinted here with the kind permission of East-West Books. — Krishnaswami Alladi

The Miracle Has Happened

Alladi Ramakrishnan

So the miracle has happened. By the Grace of God and the will of man, a new situation has been brought into being which augurs to be the starting point of an intellectual renaissance, the nature and magnitude of which cannot be foreseen at the present time. It is incredible that a series of events, each as improbable as the other, should have taken place in such steady and rapid succession. It is as though a chapter of a book of fairy tales has been transmuted into real life and I feel like one who wakes up from a dream to find reality stranger than fantasy.

The dream is so chaste that I have the courage to ask all those present here to share it with me. It originated five years ago in the exotic atmosphere of the quaint old town of Kyoto in Japan where I spent six weeks at the invitation of Professor Yukawa. In the 'domestic' environment of the Yukawa Hall, young Japanese physicists, the hope and pride of their country, just resurrected from the second World War, gathered together in enlightened leisure to discuss the most abstruse problems of modern physics. That strange enchantment drew me into the domain of elementary particle physics and I played with the idea of creating something like the Yukawa Hall in my own home town where my great father made his legendary reputation in another field of intellectual activity.

The enchantment became a passion when a fortuitous circumstance took me to the New World and I had the opportunity to attend the Conference on High Energy Physics at the University of Rochester in the spring of 1956. Within four days I was brought face to face with the rising generation of American physicists. One had only to listen to Gell-Mann and Chew, Feynman and Goldberger, to realise that a new era in American physics had been ushered in. American institutions no longer depended on the guidance of European scientists as they did a decade ago, when due to the chance of war, they were able to offer hospitality to European physicists like Fermi, Segre, and Bethe. American physics leapt from infancy to manhood within this decade and it has now become almost a necessity for European physicists to spend some time in the great American institutions and in the laboratories where things are happening every day and every hour. I felt that such a transformation needs to come in my own country which despite its organised efforts in scientific research has yet to take a place in creative science.

I therefore tried to analyse the causes for our failure. There has always been the conventional argument that there was not enough talent in the country which is not borne out by facts. It is a tragedy too deep for tears that we do not take cognisance of talent or creative work unless it has received recognition outside our frontiers. Sometimes the wait is too long, the response so cold, that it freezes up the all too frail impulses for academic life in our country. What we need is a new generation of scientists, impatient for opportunities, intolerant of mediocrity, full of action, full of manly pride and friendship like their compeers in the new world, who have not only faith in their powers, but in the scientific progress of their country.

I was strengthened in this faith during my stay, at the kind invitation of Professor Oppenheimer, in Princeton, where the most gifted minds in mathematical sciences gather together every year in an atmosphere exhilarating for creative work. It was a momentous year when the work of Yang and Lee marked the greatest advance in physical thought since the birth of quantum mechanics in 1926. I held a watching brief as a representative of our unborn Institute and I returned from Princeton with no other thought dominating my mind except to reproduce, in a small measure at least, the atmosphere for such creative work. Chance and circumstance came to my favour when a small band of students, stricken by the same splendid sickness, gathered round me in goodly friendship. We had no resource at our command except the love of common excitement for doing something new. To this fraternity we gave the name — "Theoretical Physics Seminar". It was located in my family home with the consent of my gracious wife Lalitha. We met in leisured comfort and indulged in the impertinence of attempting to work on the same type of problems as are engaging the attention of theoretical physicists elsewhere. We were encouraged in our efforts by the frequent visits of famous physicists whose friendliness and cooperation were our only sources of strength and sustenance. What a fine hour it was when Bohr and Salam who span the growth of modern physics from atomic physics to gauge theories of elementary particles, evinced an interest which gave us the strength to hope when we were all alone and everything seemed so near despair. We waited and watched for something to happen.

It was one of the fortunate moments of my life when I met the Finance Minister one evening at a gathering of international students. It puzzled me beyond comprehension to find the Minister, who must be more concerned with building dams and bridges, getting interested in the development of mathematical research. I felt a trifle guilty that I had inveigled him into this domain which had intoxicated me and my associates beyond reason. Soon I realised that it became almost a faith with him, a faith which was strengthened by his recent visit to the United States. He returned with the conviction that creative science needed the noble heat of youthful ambition and not the tepid caution of unfeeling mediocrity. Before proceeding to take steps for the creation of an institute for advanced learning, he was anxious to have the blessings and active support of our Prime Minister. It occurred to him in a discussion with my esteemed and genial friend Dr. M. M. Shapiro that all students associated with me should be introduced to the Prime Minister during his visit to Madras. The impression they made on our Prime Minister was more due to his generosity than to their own achievements. It was his wide humanity and deep concern for the prosperity of our country that made him see the light of hope even in the feeble efforts of smaller men. His support by agreeing to be our Patron, gave that final impulse which resulted in the setting up of this Institute.

The final act in this strange dream is even more fantastic than the events that preceded it. I approached Professor S. Chandrasekhar, one of the greatest astrophysicists of our time, who stands so high above the rest of our own common mould, with a request that he should associate himself with the new Institute. It was an insolence on my part to do so when I was assuming the Directorship of the Institute. I suppose you will excuse me for this if I assure you that the spirit in which I did so was animated by that in the greatest of legends when Arjuna approached Lord Krishna for his support. It was accepted with that same legendary grace, and the Institute has honoured itself by his association with it. This band of students, this firstlings of the fold, must consider themselves to be the happy few to have chosen him as their guide.

This then is the genesis of this new Institute which symbolises the hopes and ideals of the entire scientific community in India. The Government of Madras and in particular the Chief and Finance Ministers ably assisted by the Education Secretary, another victim of the splendid sickness, must be congratulated for the most gracious gesture that has ever been made by any administrative authority to the academic community in our country. The best tribute we can pay to our government is to say, "it does not seem to be the red tape — it is the blue riband". Is it not natural that greetings have poured from scientists all over the world, from California in the west to Sydney in the east? To those scientists who visited Madras, whose very presence had introduced the heady atmosphere of Berkeley into the placid environs of my family home, we are deeply grateful, for they kept alive the state of hope till the moment of its realisation. As for myself, it is a period of thanksgiving to my great teachers Professor Bhabha and Professor Bartlett who initiated me into theoretical physics. My only regret is that my parents whose home nursed the happy breed, are not alive today at the crucial moment of my academic life. In recompense I shall pass on to my students their message that the pursuit of science is at its best when it is a part of a way of life. That is the ideal to which this institute is dedicated.

(B)

PRESIDENTIAL ADDRESS BY MR. C. SUBRAMANIAM

I am particularly happy to be associated with this inaugural function. I am happy that I have played some part in bringing this Institute into existence.

I do recall my meeting with Professor Alladi Ramakrishnan at a function where he mentioned that an institute of this sort should be started in Madras. I immediately agreed to work upon it. About eighteen months ago, I wrote a letter to the Prime Minister requesting him to take some interest in the proposal and enable us to start this institution. But unfortunately something or other was happening and it was getting postponed from time to time. I wrote to the chairman of the University Grants Commission and to Dr. Bhabha to give us their blessings. But to my regret I found that encouragement was lacking. Even though I happened to be a humble student of science of this College, still since I am a politician, I wanted to be rather cautious in taking any step in this matter because I might be accused of interfering with the academic activities of the University.

Ultimately, when the Prime Minister visited Madras recently, I suggested that he should meet this band of students — young men and women — who were working under Professor Alladi Ramakrishnan. He immediately agreed and he had a pleasant half an hour with boys and girls who assembled at the Raj Bhavan to meet him. In fact at that meeting he said "I feel humble before you (students) because you are all so intelligent in the field of science". That gave further encouragement to me and I took steps to see that the Institute would be started as early as possible.

It is indeed a novel thing that the Institute should be started apart from the University. But it was the right step; particularly in the field of science, we cannot afford to follow conventional steps. Perhaps unconventional methods have produced better results in science. With his usual common sense, the Chief Minister (of Madras) also approved the idea. The Prime Minister was then approached to become the Patron of the Institute. Even though it was day on which the Goa action was being taken by him, he immediately gave his consent to be the patron of the Institute when we discussed this matter. The starting of the Institute was therefore possible mainly because of the support given by the Prime Minister. He was good enough to mention this to the UGC Chairman, Dr. Bhabha, and others. It was after this that this humble effort of ours got the blessings of everyone who matters in the field of scientific work.

When Professor Chandrasekhar was in the Physics Honours Class, I was studying the pass course in physics. Even then he was a very distinguished student, recognised by the scientific world as a promising young man. He has overfulfilled these expectations. We are happy that he has attained such high eminence. Professor Chandrasekhar has agreed to be an Honorary Professor of this Institute and also to associate himself with the activities of the Institute whenever he visits India. I am sure that by his association, the Institute has been put on a sure footing and has immediately gained world-wide recognition.

Professor Ramakrishnan himself is a well-known figure in the scientific world. Perhaps by his taking up the directorship, The Institute has gained a status in the scientific world; but with Professor Chandrasekhar's association, I think the future is guaranteed.

Professor Ramakrishnan said the miracle has happened. I will say that this is the beginning of the miracle. The justification of the Institute will be in the results to be achieved which is when the miracle will really happen. Having known Professor Ramakrishnan and the band of young men and women working under him, I have no doubt that ere long this Institute will be one of the well recognised scientific institutions in the whole world. In the field of scientific research, we cannot anticipate — anything might happen or might not happen. The chances are there for great discoveries and work of international repute. Out of the young men and women, I hope, a few will earn the Nobel Prize for physics.

I know there were some doubting Thomases with regard to the steps we have taken in starting the Institute. But I have no doubt in my mind, that this is just the correct thing to do, because our young men and women should have full opportunities to do research. During my visit to the United States, I came across many Indians — young men and women — working there. The Professors were all praise for the work they were doing. I thought these young men and women had to come away from our country because opportunities were not made available to them here. Professor Chandrasekhar himself is a classic instance where we have lost — I will not say we have lost — he had to go away to another country to establish himself as a great scientist. If only he had continued his work here and brought fame to the Madras University and not to the University of Chicago! I hope he will come to Madras when this Institute becomes successful. I hope this Institute will justify itself and then it will not require any politician or administrator to give it strength; on the other hand it will give strength to the politician and administrator.

I now request Professor Chandrasekhar to inaugurate this Institute.

(C)

REPORT OF THE INAUGURAL ADDRESS OF
Professor S. Chandrasekhar

*The inauguration of MATSCIENCE was prominently reported in The
Hindu, India's National Newspaper, on January 4, 1962. In doing so, The
Hindu gave a complete report of the inaugural speech of Professor Chan-
drasekhar and that is what is given here.*

Professor Chandrasekhar said that the establishment of the Institute by
the Government of Madras was an event which was unique in many ways.
If one looked at the scientific scene in India, then the greatest impression
was made by the development of national laboratories which are devoted
to various aspects of the applied sciences and in some instance even provide the base for industrial advancement. During his recent visit to these
laboratories, he gathered the impression that in them a determination and
purposefulness in their work prevailed. He said this was a consequence of
the determination and purposefulness with which the Government of India
had been pursuing industrial development of the country since its independence. "On the contrary, if one compares the record of Indian science
during this same decade, I am afraid that the comparison is not only disappointing, it is even discouraging. In saying this, I do not wish to deprecate
in any way the very important work which certain groups and individuals
are doing in their respective fields all over India. But among these groups
and individuals I find very often a sense of frustration and constant irritation derived from it. My feeling is that this frustration and disappointment
among the men and women devoting themselves to pure science is largely
due to the lack of appreciation and understanding of their efforts both at
the Governmental and University levels."

Illustrating this lack of appreciation, he said that in most centres of
learning, even the head of the department could not sign a requisition
exceeding Rs. 50. He pointed out that in America, even a janitor or cleaning woman could make a requisition for amounts exceeding ten times this
without a formal approval. It is quite important that the efforts of pure
scientists should be free from small administrative difficulties of this kind.

Professor Chandrasekhar said that there was no real dichotomy between the pure and applied sciences. On the contrary, they are very closely
related. Many of the spectacular advances of modern living such as the radio, television, air transport, atomic reactors, had all resulted from specific

results derived from the pure sciences — "efforts devoted primarily to our understanding of our physical environment". So also, many of the advances in pure sciences, particularly in the physical and biological sciences, had resulted from advancement in the applied and technical sciences. There should therefore be no misunderstanding that he attached more importance to one or the other. "The important thing is that science is an entire whole, and the different aspects of it are equally essential."

In a country like India, where industrial advancement was essential, it was perhaps easy to appreciate the importance of applied sciences. But in the long run, he said that the lack of corresponding encouragement and support for the efforts of the pure scientist was likely to result in serious harm. It was thus a great thing that Government of Madras should have set up the Institute of Mathematical Sciences, devoted to research in theoretical physics and applied mathematics. The establishment of the Institute was therefore welcome not only in the sense that every new venture in learning was to be welcomed, but also because it represents a new departure from the current scientific scene. He hoped that this "most wholesome development" would be extended to other fields like the biological sciences and also emulated by other states in India.

It is appropriate that the Institute is located in Madras because the record in India in the mathematical sciences is largely due to men who come from this part of the country. "Of course the name of Ramanujan comes to the mind of everyone. He is, in my opinion, without question, the greatest man of science India has produced in recent times. But Ramanujan was nourished during his precious six years in Cambridge. I feel that is a matter of regret that there is still no adequate memorial to Ramanujan in India."

Professor Chandrasekhar said that there were other names to be mentioned in this connection such as R. Vaidhyanathaswamy, T. Vijayaraghavan, and Sivasankaranarayana Pillai. These men did not find adequate support and encouragement which their measure of achievement would have merited.

He learnt that this new Institute would be so organised that it would not suffer from the "hierarchical disease", the presence of which in centres of learning was most harmful. He was happy that the Institute would be organised in such a way that the kind of atmosphere needed for those devoted to research would be present here.

With these words Professor Chandrasekhar declared the Institute open.

(D)

Messages of Greetings and Telegrams

3 January, 1962

TELEGRAM FROM NOBEL LAUREATE NIELS BOHR, COPEN-HAGEN, DENMARK

TO PROFESSOR ALLADI RAMAKRISHNAN, EKAMRA NIVAS, 27 LUZ, MYLAPORE, MADRAS

"At the inauguration of the Institute of Mathematical Sciences, Madras, the whole group of the Copenhagen Institute for Theoretical Physics wants to send its heartiest felicitations. The community of physicists has been impressed by the vigour and zeal with which Prof. Ramakrishnan has been able to educate and inspire his young pupils and collaborators, and the work of the new Institute will be followed with keen expectations. Indeed as an important asset for scientific research in India, the creation of the Madras Institute is eagerly welcomed in that world-wide cooperation in science which offers so great opportunities for promoting the understanding between all peoples."

Similar messages were received from Professors W. Heisenberg (N. L.), T. D. Lee (N. L.) and C. N. Yang (N. L.), Abdus Salam (N. L.), Murray Gell-Mann (N. L.), B. Stromgren, R. Bellman, L. I. Schiff, R. E. Marshak, C. W. Ufford, A. S. Wightman, M. H. Stone, R. H. Dalitz, G. Feldman, O. C. Dahl, B. Rossi, D. Falkoff, G. Gamow, G. Munch, P. Charulli, S. A. Moszkowski, A. T. B. Reid, K. W. Fraser, J. Sucher, A. Copeland, T. Y. Wu, Sir Chales Darwin, M. S. Bartlett, M. J. Lighthill, P. T. Mathews, Longuet-Higgins, D. G. Kendall, L. Rosenfeld, A. Mercier, V. F. Weisskopf, W. Thirring, D'Espagnat and Prentki, W. Heitler, P. Budini, H. Maass, H. Messel, T. G. Room, Sir Mark Oliphant, A. Dessler, L. Janossy, Madame Tonelat, N. F. Mott (N. L.), A. M. Lane, P. A. M. Dirac (N. L.), L. Foldy, W. E. Brittin, K. A. Breuckner, J. M. Burgev, L. Schwartz (Fields Medallist), H. Yukawa (N. L.), S. A. Wothuysen, H. R. Pitt, J. M. Jauch, A. Sandstrom, C. F. Powell (N. L.), H. J. Bhabha, Sir A. L. Mudaliar, Sir C. P. Ramaswami Iyer, and others.

Note

For the scans of the originals of several of these telegrams, see [2], pp. 26–65, or [1]. In the above, N. L. refers to Nobel Laureate already or subsequently.

Appendix 22

Mathematical Institute in Madras — A New Phase in the Current Scientific Scene

The Hindu, Madras, 4 Jan. 1962

Prof. S. Chandrasekhar, the eminent astro-physicist, inaugurated here this morning the Institute of Mathematical Sciences, hailed as "a symbol of the hopes and ideals of the entire scientific community in India" and "a new departure in the current scientific scene". The Institute established under the auspices of the Madras Government, will be devoted to research in theoretical physics and applied mathematics. Fifteen minutes after its inauguration, the Institute began its work when Dr. Chandrasekhar, an Honorary Professor of the Institute gave a lecture on astrophysics.

The function held at Presidency College was presided over by Mr. C. Subramaniam, Finance and Education Minister. Dr. Alladi Ramakrishnan, Director of the Institute described the establishment of the Institute "as a dream come true".

Messages characterising the founding of the Institute as a new encouragement to Indian research and wishing it success were received from eminent professors of physics and mathematics and scientific institutions from various parts of the world. Prime Minister Nehru had also sent his greetings. The Copenhagen Institute for Theoretical Physics in its felicitations mentioned that as an important asset to scientific research in India, the creation of the Madras Institute was eagerly welcomed and added "worldwide co-operation in science offers great opportunities for promoting understanding between all peoples".

Prof. Chandrasekhar attributed the "frustration and disappointment" witnessed in the country now among men and women devoted to research in pure sciences, to the lack of appreciation and understanding of their efforts both at the Governmental and University levels. He said the establishment of the Institute by the Government of Madras was an event unique in many ways. If one looked at the scientific scene in India, then the greatest impression was made by the development of the various national laboratories set up by the Central government. These national laboratories were devoted to various aspects of applied sciences and in some instances even to provide the base for special industrial advancement. During his recent visit to these laboratories he gathered the impression that in them a determination and purposefulness in their work prevailed. He said this was a consequence of the determination and purposefulness with which the Government of India had been pursuing industrial development of the country since independence.

"On the other hand," Prof. Chandrasekhar said, "if one compares the record of Indian science during this same decade, I am afraid the comparison is not only disappointing, it is even discouraging. In saying this, I do not wish to deprecate in any way the very important work which groups and individuals are doing in their respective fields all over India but among these groups and individuals I find very often a sense of frustration and constant irritation derived from it. My feeling is that this frustration and disappointment among the men and women devoting themselves to pure sciences is largely due to the lack of appreciation and understanding of their efforts both at the Governmental and University levels."

Illustrating this lack of appreciation, he said in most centres of learning even the head of the department could not sign a requisition exceeding Rs. 12.50. He pointed out in America even a janitor or cleaning woman could make a requisition for amounts exceeding ten times this without any formal approval. It was quite important that the efforts of the pure scientist should be free from small administrative difficulties of this kind, he said.

Prof. Chandrasekhar said there was no real dichotomy between pure and applied sciences. On the contrary they were very closely related. Many of the spectacular advances of modern living such as radio, television, air transport, atomic reactors, had all resulted from specific results derived from the pursuit of pure sciences, "efforts efforts devoted primarily to our understanding of our physical environment". So also, many of the advances in pure sciences, particularly in physical and biological sciences, had resulted from advances in applied and technical sciences. There should therefore be no misunderstanding that he attached more importance to the one or the other. "The important thing is science is entirely whole and the different aspects of it are equally essential." In a country like India where industrial advancement was essential, it was perhaps easy to appreciate the importance of applied sciences. But in the long run, he said, the lack of corresponding encouragement and support for the efforts of the pure scientist was likely to result in serious harm. From this point of view, it was a great thing that the Government of Madras should have set up the Institute of Mathematical Sciences, devoted to research in theoretical physics, applied mathematics, etc. The establishment of the Institute was therefore welcome not only in the sense that every new venture in learning was to be welcomed, but it was also to be welcomed because it represented a new departure in the current scientific scene. He hoped this most wholesome development would be extended to other fields of science like biological science and also emulated by other States.

Prof. Chandrasekhar said it was appropriate that the Institute was located in Madras "because the record of India in the mathematical science has largely been written by men who come from this area of the country".

GOOD MEMORIAL TO LATE RAMANUJAN URGED

"Of course", he continued, the name of Ramanujan comes to everyone. He is, in my opinion, without question, the greatest man of science India has produced in recent times. But Ramanujan was nourished during his precious six years in Cambridge. I feel it is a matter of regret that there is still no adequate memorial to Ramanujan in India." The professor said there were other names also to be mentioned in this connection such as Mr. R. Vaidyanathaswami, Mr. T. Vijayaraghavan and Mr. S. Sivasankaranarayana Pillai. But all these men did not find that adequate support and encouragement which their measure of achievement would have merited.

He learnt that the Institute would be so organised so that it would not suffer from "hierachical disease" the presence of which in centres of learning was most harmful he said. He was happy that the Institute would be organised in such a way that the kind of atmosphere needed for those devoted to research, would be obtained here.

Mr. Subramaniam said that he was happy that he had played some part on bringing the Institute into existence. He said though the proposal to establish the Institute was mooted 18 months ago it was getting postponed from time to time. He, however, wanted to go about it cautiously, since he might be accused of interference in academic activities. He said the starting of the Institute was mainly possible because of the support given by the Prime Minister. It was a novel thing that an institute should be started apart from the University, he said, "but it was a right step, because, we cannot afford to follow conventional steps, particularly in the field of sciences." The Minister expressed the hope that very soon the Institute would become one of the well-recognised institutions of scientific research in the whole world.

The Constitution of the Institute of Mathematical Sciences (1962)

Annexure I — CONSTITUTION

1. The name of the Institute shall be: THE INSTITUTE OF MATHE-MATICAL SCIENCES, MADRAS.

2. The Institute shall be registered under the Societies Registration Act, 1860 (Central Act XXI of 1860). The Institute shall be located in the City of Madras.

3. The official year of the Institute shall be from the 1st April to the 31st March of the year.

For the purpose of Section 6 of the Registration of Societies Act, (1860), the person in whose name the Society may sue or be sued shall be the Director of the Institute.

4. The aims and objectives of the Institute are:

(i) To create and provide an atmosphere and environment suitable for creative work and pursuit of knowledge and advanced learning in the Mathematical Sciences for its own sake;

(ii) To promote and conduct research and original investigation in fundamental sciences in general with particular emphasis on mathematics, applied mathematics, theoretical physics and astrophysics;

(iii) To foster a rigorous mathematical discipline, to stimulate a zest for creative work and to cultivate a spirit of intellectual collaboration amongst academic workers in pure and applied branches of science;

(iv) To arrange lectures, meetings, seminars and symposia in pursuance of its academic work and for the diffusion of scientific knowledge;

(v) To invite scientists in India and abroad actively engaged in creative work to deliver lectures and participate in its academic activities; and

(vi) To take such other steps as may be necessary and conducive to the advancement of learning in mathematical sciences and for the dissemination of knowledge in these sciences.

5. The following shall be the authorities of the Institute:
(a) The Board of Governors,
(b) The Finance Committee,
(c) The Academic Council,
(d) Any other Standing Committee or Committee which the Board of Governors or Academic Council may set up for discharging any one or more of their functions,
(e) The Director.

6. The income and property of the Institute, however, derived from voluntary donations, subscriptions, grants and endowments shall be applied solely towards the promotion of all or any of the purposes and objects of the Institute as set forth above.

7. The Institute shall have the power to take over and acquire by purchase, gift or otherwise from Government or other public bodies or private individuals willing to transfer, libraries, collections, immovable properties, endowments, laboratories and other funds acceptable to the Board of Governors and not inconsistent with the aims and objects of the Institute.

8. The Government of Madras may appoint one or more persons to review the work and progress of the Institute and to hold enquiries into the affairs thereof and to report thereon, in such manner as the government of Madras may stipulate. Upon receipt of any such report, the Government of Madras may take such action and issue such directions as it may consider necessary in respect of any of the matter dealt with in the report and the Institute shall be bound to comply with such directions.

9. The Institute shall have the power to enter into agreements for co-operation with educational or other institutions in any part of the world having objects wholly or partly similar to those of the Institute, by exchange of teachers and scholars and generally in such manner as may be conducive to their common objects.

10. There shall be a Board of Governors composed as follows:
(i) A Chairman to be nominated by the Government of Madras,
(ii) The Secretary to the Government of Madras in charge of Education (ex-officio),

(iii) A representative of the University of Madras,

(iv) A representative of the University of Madras,

(v) The Director of the Institute (ex-officio),

(vi) One representative of the Academic Council of the Institute other than the Director.

11. The term of office of the Chairman and of the members who are nominated on the Board of Governors in their individual capacity shall be five years. The Chairman and such members shall be eligible for re-nomination.

12. The Institute shall function not withstanding any vacancy in the Board of Governors and not withstanding any defect in the appointment or nomination of any of its members; and no act or proceedings of the Institute shall be invalid or deemed to be invalid merely by reason of existence of any vacancy in the Board or of any defect in the appointment or nomination of any of its members.

13. The Board of Governors shall have the following functions and powers, namely:

(i) To manage the affairs of the Institute and to regulate its expenditure;

(ii) To receive subscriptions and donations for the aims and objectives of the Institute provided that no subscriptions or donations shall be accepted if they are accompanied by conditions which are inconsistent or in conflict with the objects of the Institute;

(iii) To frame its regulations, bye-laws and rules of procedure for the conduct of the affairs of the Institute provided that the rules and regulations relating to appointment, scales of pay and service conditions of staff will be subject to the approval of the Government of Madras;

(iv) To consider the budget estimates submitted by the Finance Committee and to forward it to the Government of Madras for their approval;

(v) To consider the audited accounts and the report thereon and to take decisions relating to them;

(vi) To appoint an auditor (subject to the approval of the Government of Madras) to audit the accounts of the Institute every year.

14. The Board of Governors may make arrangements to carry out special investigations in any scientific matter and such investigations shall not be inconsistent with the objects of the Institute.

15. (a) The quorum for the meeting of the Board of Governors shall be three.

(b) In the absence of the Chairman, a member chosen by the members present shall preside on the occasion.

16. The Board may, by resolution, delegate to a committee or the Chairman or the Director such of its powers of the conduct of its business as it may deem fit, subject to the condition that the action taken by any committee or the Chairman or the Director under the powers delegated to them by this Article shall be reported for confirmation at the next meeting of the Board.

The Board may, by resolution, appoint such committees for such purposes and with such powers as the Board may think fit. The Board may co-opt such persons to these committees as it considers suitable.

17. There shall be a Finance Committee composed as follows:

(a) The Director,

(b) The Secretary to the Government of Madras in charge of education (ex-officio),

(c) A member of the Board of Governors other than (a) and (b), nominated by the Board.

18. The duties of the Finance Committee shall be as follows:

(i) To consider the budget estimates proposed by the Director and to make recommendation to the Board of Governors;

(ii) To consider the budget estimates proposed by the Director and to make recommendation to the Board of Governors before they are considered by the Board;

(iii) To consider the auditor's report and to make recommendations to the Board;

(iv) To review the finances of the Institute from time to time through periodical control statements; and

(v) To give advice and to make recommendations to the Board on any other financial questions affecting the Institute either on its own initiative or on the initiative of the Board or Director.

19. The Board of Governors shall meet at least once in six months for the despatch of business provided that the Chairman may, whenever he thinks fit, call a special meeting. The Finance Committee shall meet as and when necessary.

20. The Director shall be the Chief Administrative and Academic Officer of the Institute. He is responsible for the proper administration of the Institute subject to the supervision and control of the Board of Governors.

He shall make appointments to the posts specified herein and shall exercise also powers of disciplinary control over the staff subject to the regulations and by-laws that may be made by the Board. He shall prepare a budget for the ensuing year for the consideration of the Finance Committee and the Board for being forwarded to Government for approval before the commencement of the next official year. The director shall also be responsible for

(i) the custody of records and of the other property which the Board of Governors may commit to his charge;

(ii) the conduct of official correspondence;

(iii) convening meetings and keeping the minutes of the Board of Governors, Finance Committee Academic Council and other committees;

(iv) the utilisation of all moneys and the maintenance of accounts;

(v) the discharge of such other duties as may be assigned to him by the Board of Governors.

21. Every contract shall be made on behalf of the Board by the Director, provided that:

(a) no contract involving an expenditure exceeding Rs. 5,000 but not exceeding Rs. 10,000 shall be made unless it has been sanctioned by the Finance Committee and the Board of Governors; and

(b) every contract made by the Director involving an expenditure exceeding Rs. 10,000 shall be made with the approval of the Board of Governors and the Government of Madras.

22. The Institute shall consist of the following faculties, namely:

(i) Faculty of Mathematics,

(ii) Faculty of Applied Mathematics,

(iii) Faculty of Theoretical Physics, and

(iv) Faculty of Astrophysics.

(v) Any other Faculty as may be decided by the Board of Governors and approved by the Government of Madras.

23. Each Faculty shall consist of the following members:

(a) Professors,

(b) Members,

(c) Research Fellows.

24. The Director may function as Professor in any Faculty or Faculties.

25. The Director, the Professors, the Members and the Research Fellows of all the Faculties shall together constitute the academic staff of the Institute.

26. Professors shall be of two categories, namely:

(a) Permanent Professors; and

(b) Visiting Professors.

27. Members shall be of four categories, namely:

(a) Permanent Members;

(b) Visiting Members;

(c) Associate Members; and

(d) Temporary Members.

28. Research Fellows shall be of two categories, namely:

(a) Senior Research Fellows; and

(b) Junior Research Fellows.

29. The Academic Council of thee Institute shall consist of the Director, Permanent Professors and Permanent Members. The Academic Council shall be entrusted with the sole and primary responsibility of recommending to the Board of Governors the appointment of Professors, Permanent and Associate Members. It should also in general advise the Board of Governors on all academic matters. It shall meet once every quarter for the purpose of planning the academic programme of the Institute. The Director shall be the ex-officio Chairman of the Academic Council. In the absence of the Director, a member chosen by the members present may preside over the meeting of the Council.

30. (a) The Director shall be appointed by the government in consultation with the Board of Governors.

(b) The appointment of the Permanent Professors and Visiting Professors and Permanent and Associate Members shall be made by the Board of Governors after taking into consideration the recommendations of the Academic Council.

(c) The appointment of the Temporary and Visiting Members shall be made by the Director on the advice of the Academic Council.

(d) In the event of the post of Director remaining vacant for any reason, it shall be open to the Board of Governors to authorise any officer or officers in the service of the Institute to exercise such powers, functions and duties of the Director, as the Board may think fit.

(e) Notwithstanding anything contained in clauses (b) to (d), it shall be open to the Chairman to refer cases of appointments to the Government of Madras for orders where he considers that such orders are necessary.

31. The appointment of Research Fellows shall be made by the Director in consultation with the Professors of the corresponding Faculty.

32. Without prejudice to the generality of the powers conferred on the Board under Article 13, the Board of Governors shall make rules for the purpose of:

(i) conduct of business, convening and conducting the meetings of the Board, the Finance Committee and the Academic Council;

(ii) emoluments and conditions of service of the academic and administrative staff of the Institute subject to the provisions of Article 12 (iii);

(iii) curriculum and course of study and research in the various Faculties;

(iv) acquisition and disposal of the property of the Institute and Investment of its funds;

(v) finances and accounts of the Institute; and

(vi) any other matter relating to the administration of the Institute.

33. Any rule made or decision taken by the Board of Governors or by any authority of the Institute except where the authority acts in accordance with its powers and functions as defined in this Constitution may be amended or set aside by the Board of Governors.

34. The Board of Governors may, by a majority of not less than three-fourths of the members, request the Government of Madras to amend this Constitution in such manner as the Board may decide.

Appendix 24

Chandrasekhar's Letter after MATSCIENCE Inauguration

THE UNIVERSITY OF CHICAGO

YERKES OBSERVATORY
WILLIAMS BAY, WIS.

1962, February 5

Dr. Alladi Ramakrishnan
27 Luz Church Road
Mylapore, Madras 4, India

Dear Ramakrishnan:

I find it hard to believe that it is already a month since we left India. In fact, it is a month ago today that we had the inaugural ceremony of the Institute of Mathematical Sciences. I was very glad that it was possible for me to participate in the occasion; and I want to thank you for treating me with the thoughtful consideration you did.

I should naturally look forward to hearing from you as to how the new institute is faring.

Wentzel returned from India last week after attending the opening of the new premises of the Tata Institute. I suppose you were there too.

I hope that you succeeded in getting Lee to visit Madras to meet with you and your group.

With our best wishes, also to your wife,

Yours sincerely,

S. Chandrasekhar

S. Chandrasekhar

Appendix 25

Registration of MATSCIENCE

(A)

11 June 1962

TO: Professor H. J. Bhabha
Chairman
Atomic Energy Commission
Apollo Pier Road
Bombay -1

Dear Professor Bhabha,

It is my proud privilege to make the following request to you — an event that I have been waiting for all these years. The new Institute of Mathematical Sciences has to be registered under the Societies Registration Act and it is the unanimous desire of the sponsors of the Institute that you should be one of the seven signatories of the document. Professor S. Chandrasekhar has signed the document and I am sending it to you by separate post. With support from you and from Professor S. Chandrasekhar — two of the greatest scientists India has produced — the Institute has had as auspicious a beginning as I could wish for.

Is there a chance of you coming to South India in the near future? If not, we can arrange a meeting of the Board of Governors to suit your convenience.

Dr. Sudarshan is with us now and will be delivering a course of twelve lectures before going to the Geneva Conference.

With kindest regards,
Yours sincerely,

(sd) Alladi Ramakrishnan

PS: I am sending herewith a copy of the Minutes of the Board of Governors of the Institute held on 31 May, 1962.

(B)

सत्यमेव जयते

GOVERNMENT OF INDIA

ATOMIC ENERGY COMMISSION

TELEPHONE: 253724
TELEGRAMS: ATOMERG

APOLLO PIER ROAD
BOMBAY

CHAIRMAN

Ref: 844 -62   June 15, 1962

My dear Alladi,

          Thank you for your letter of June 11, 1962, inviting me to be
one of the seven signatories to the document seeking the registration of
the Institute of Mathematical Sciences, Madras, under the Societies Registration Act. I shall be glad to sign the document when it arrives.

          I am just back from a trip abroad and will shortly be going
to Bangalore in connection with the Summer School in Theoretical Physics
organised by the Tata Institute of Fundamental Research. I do not expect to be able to visit Madras in the near future, but if I do so, I
shall let you know.

                              With kind regards,

                                        Yours sincerely,

Dr Alladi Ramakrishnan,
Director,
Institute of Mathematical Sciences,
Ekamra Nivas, 27 Luz,
MADRAS 4

(C)

Certificate of Registration of Societies

ACT XXI OF 1860.

S. No. 67 of 1962 .

I hereby certify that "THE INSTITUTE OF MATHEMATICAL SCIENCES, MADRAS" has this day been registered under the Societies' Registration Act XXI of 1860.

Given under my hand at MADRAS , this 23rd day of July One thousand nine hundred and SIXTY TWO

Registrar of Assurances.
Madras-Chingleput District.

Old No. Regn. 11-137—2,664—12-8-61.

Appendix 26

Distinguished Scientists who visited Alladi Ramakrishnan's Theoretical Physics Seminar at his home (1954–61)

1) **Professor P. A. M. Dirac**, FRS, Nobel Laureate
   Lucasian Professor, Cambridge University, England (Dec 1954)
2) **Professor Mark Oliphant**, FRS (he was later knighted and became Governor of South Australia)
   Australian National University, Canberra (Jan 1955)
3) **Professor C. F. Powell**, FRS, Nobel Laureate
   Melville Wills Professor, University of Bristol, England (Dec 1955)
4) **Professor T. M. Cherry**, FRS (he was also knighted)
   University of Melbourne, Australia
5) **Professor Harry Messel**
   University of Sydney, Australia (Jan 1957)
6) **Professor W. W. Buechner**
   Massachusetts Institute of Technology, USA
7) **Professor T. G. Room**, FRS
   University of Sydney, Australia
8) **Professor Laurent Schwartz**, Fields Medallist
   University of Paris, France
9) **Professor H. Pitt**, FRS
   University of Leeds, England
10) **Professor C. G. Darwin**, FRS (he was also knighted)
    Former President of The Royal Society, England
11) **Professor L. Janossy**
    Director, Eotvos Institute, Budapest, Hungary (Jan 1959)
12) **Professor Z. Koba**
    Yukawa Hall, Kyoto, Japan (1959)
13) **Dr. T. Kotani**
    University of Tokyo, Japan (Aug 1959)

14) **Professor Andre Mercier**
University of Berne, Switzerland (Sept 1959)

15) **Professor N. Dallaporta**
University of Padova, Italy (Oct 1959)

16) **Professor A. M. Lane**
A. E. Research Establishment, Harwell, England (Dec 1959)

17) **Professor George Gamow**
University of Colorado, Boulder, USA (Dec 1959)

18) **Professor Abdus Salam**, FRS (Salam later won the Nobel Prize)
Imperial College, London, England (Jan 1960)

19) **Professor Niels Bohr**, Nobel Laureate
Bohr Institute of Theoretical Physics, Copenhagen, Denmark (Jan 1960)

20) **Professor Christoff**
University of Sofia, Bulgaria (Feb 1960)

21) **Professor Philip Morrison**
Cornell University, Ithaca, USA (Mar 1960)

22) **Professor A. H. Copeland**
University of Michigan, Ann Arbor, USA (July 1960)

23) **Professor Kampe de Feriet**
University of Lille, France (Jan 1961)

24) **Professor W. Heitler**, FRS
University of Zurich, Switzerland (Feb 1961)

25) **Professor Marshall H. Stone**
Distinguished Service Professor, University of Chicago, USA (Apr 1961)

26) **Professor V. Hlavaty**
Institute of Fluid Dynamics, Indiana, USA

27) **Professor Murray Gell-Mann** (Gell-Mann later won the Nobel Prize)
California Institute of Technology, USA (July 1961)

28) **Professor R. A. Dalitz**
University of Chicago, USA (July 1961)

29) **Professor A. E. Sandstrom**
Uppsala University, Sweden (July 1961)

30) **Professor Donald Glaser**, Nobel Laureate
University of California, Berkeley (Aug 1961)

31) **Dr. Maurice M. Shapiro**
Naval Research Labs., Washington DC, USA (Sept 1961)

32) **Professor S. Chandrasekhar** FRS (he later won the Nobel Prize)
Distinguished Service Professor, University of Chicago, USA (Nov 1961)

33) **Professor McCrea Hazlett**
Vice-President, University of Rochester, USA (Nov 1961)

34) **Professor M. J. Lighthill** (he was later knighted)
Director, Royal Aircraft Establishment, Farnborough, England (Dec 1961)

PART II

Matscience Visitors, Global Academic Travels, L-Matrix Theory and Space-Time Unity

Alladi Ramakrishnan in conversation with C. Subramaniam at Ekamra Nivas. - Editor

DEDICATION

to

my revered teachers

H. J. Bhabha FRS

Sir C. V. Raman (Nobel Laureate)

M. S. Bartlett FRS, **D. G. Kendall** FRS

M. J. Lighthill FRS

S. Chandrasekhar (Nobel Laureate)

my distinguished sponsors

Niels Bohr (Nobel Laureate)

Jawaharlal Nehru (Prime Minister)

C. Subramaniam (Bharata Ratna)

my gracious hosts in the USA

Richard Bellman (Rand)

Maurice Shapiro (Washington)

John Richardson (Rockwell)

Erwin Fenyves (Dallas)

The author with John Richardson (Rockwell).

Erwin Fenyves (Univ. Texas at Dallas).

Author's Preface to Part II

We continue the Alladi Diary from the creation of MATSCIENCE through my period as Director for 21 years, and the equally eventful years until the dawn of the twenty-first century. During this period, opportunities came in succession for my global travels with Lalitha and Krishna, in response to lecture engagements from over two hundred institutions in the USA, Europe, Japan and Australia.

I wish to transmit the excitement through the factual account in chronological sequence, with unconcealed emotion and gratitude, to readers interested in understanding the global enterprise of mathematical research, which knows no frontier in time or in space. The period I describe coincided with Krishna's choice of a mathematical career at an early age, which led him to settle down in the land of opportunities. This in turn resulted in the continuance of our international travel, mainly to spend time with him and his family in Florida, while pursuing my research in active contemplation.

By the Grace of God, such contemplation yielded the incredible revelation that the line, the circle and the ellipse are of direct relevance to special relativity, a departure from the century old 'addiction' to the hyperbola, which should inspire us that the excitement of pursuit, the thrill of discovery, the beauty of logic, and the ecstacy of proof, can be experienced not only by an Euler or a Ramanujan, but by any ambitious aspirant to a scientific career at his own level of creativity.

Alladi Ramakrishnan
Madras, January 2003

Chapter 23

RAND And Round The World (1962)

It was going to be a special overseas trip since we decided to take Krishna with us (1962).
- Author

We left on the 20th of August by the Indian Airlines Viscount for Calcutta enroute to the United States, in response to the kind invitation from Ted Harris[1] and Dick Bellman of the RAND Corporation, California. It was going to be a very special trip since Lalitha and I decided to take Krishna with us. He was just seven years old but a perfect gentleman in table manners and an ideal tourist guide with his faultless memory for faces and places! The entire trip was meticulously arranged by Vaidyanathan of Thomas Cooks.[2]

We stayed overnight at the Grand Hotel in Calcutta. It was still the same symbol of British prestige, even in a free India. We left for Hong Kong by the Air India Boeing 707 and our first stop was Bangkok, when we were met by my brother-in-law, V. M. Subramaniam, a steel expert at ECAFE, and my sister Meenakshi at the Don Muang airport. We arrived at Hong Kong in the evening and stayed at the Ambassador Hotel in Kowloon, with its exotic atmosphere of Oriental luxury and Western efficiency.

We celebrated our sixteenth wedding anniversary watching the dazzling splendour of the Hong Kong harbour at night, from the roof-top restaurant of the Ambassador Hotel. Krishna enjoyed friendly conversation with the liveried waiters, who lisped English in Chinese accent. We took a limousine trip round Kowloon and the new territories, and went to a point from where we watched the boundary of Red China. There was a strange feeling of awe and mystery as we thought of the billion people who stayed on the other side of the frontier, in a world as different from Hong Kong as the Sahara is from the South Pole. The island of Hong Kong is a British jewel set in the silver sea, a demi-paradise for tourists and shoppers. The very comfortable residential areas of the British community clearly demonstrated that this 'blessed plot with its happy breed of smiling shopkeepers' is an important rampart of the British Empire.

From Hong Kong we took the Pan American Boeing 707[3] to Tokyo, where we were received by Dr. Kotani. His father had retired from high judicial service, and so we stayed this time first at the International House and then in a clean and elegant hotel, the San Bancho.

I lectured at the Tokyo University, and came into close contact with Professor Fukuda, an expert on many-body physics. He agreed to come to Madras as a visiting professor at our Institute. We had a wonderful excursion to the Hakone National Park and to the Agami resort on the beach. We went by cable car to the top of a 'mountain lake' and returned to Tokyo after a most memorable trip with the Kotanis.

We took the Pigeon Bus tour round the city, which Krishna enjoyed very much. In the evening we went to the Kokusai theatre, which is a sparkling

imitation of the stage show at Radio City Music Hall in New York. We
dined at the home of Kotani and left by Pan Am 707 for Honolulu.

At Honolulu, we stayed at the comfortable hotel Edgewater and spent
the whole evening on Waikiki beach, where affluent America pays its
homage to the thriftless beauty of Hawaiian sunsets over the seagirt isle. We
saw the Kodak Hula show on Waikiki beach, and took an excursion round
the island watching miles and miles of pineapple plantations, and watched
the picking being performed by harvest machines. After a delightful stay
in Honolulu, we left by a Pan Am 707 for Los Angeles and were received at
the airport by our Indian friends, the Jayaramans,[4] and Dr. Kalaba of the
RAND Corporation. We were taken to the very comfortable Riviera apartments in beautiful Santa Monica. The two months stay at Santa Monica
was an unforgettably delightful experience. It was a pleasure to see my
colleague Vasu at the entrance of the magnificent RAND building, where
I met my gracious hosts Ted Harris and Dick Bellman. During our stay,
Vasu[5] visited us every weekend (from Berkeley), which he stretched by a
day or two consulting for the RAND Corporation. Our trips to Hollywood,
Sunsetstrip, Beverly Hills, Descanso Gardens, Griffith park and other sights
of Los Angeles cast an enchantment, which continues to this day. There
was the inevitable excursion to Disneyland and the seal circus.

At RAND, I worked on the interpretation of the Feynman formalism
from a new point of view. I found that Feynman had not considered the
mathematical consequences of dividing his propagator into positive and
negative energy parts. It therefore became clear that there was still some
interesting work to be done in establishing the correspondence between field
theory and Feynman formalism. Such an attempt looked like 'stammering
where old Chaucer used to sing' but my work proved that new topological
features were revealed by opening the 'jaws of the Feynman propagator.'

In a seminar at the RAND Corporation, I spoke on probability theory
and causation. I drew the attention of mathematicians and statisticians
to the apparently puzzling paradox that to define cause and effect, we
need the concept of probability! Only if there is a choice, can we ascribe
responsibility or guilt for making a particular decision and choice, in turn
implies the concept of probability.

It is impossible to think of a more pleasant and fruitful stay than at
the RAND in all my academic career. My friendship with Bellman and
Harris enriched my academic experience, and provided opportunities for
my colleague Vasu to visit California often, and collaborate with Bellman
on the applications of mathematics to biology and medicine.

One of the most exciting days was October 6th, when we went to Gell-Mann's house in Altadena for lunch beside his swimming pool. Feynman[6] and Zachariasen,[7] and their children were also there. Seven year old Krishna had an animated discussion with Feynman on pre-historic animals! Watching Feynman in such a mood reminded me of Lighthill joining the chorus for nursery rhymes during a Christmas party at Manchester!

I called on Dr. Collbohm,[8] the President of the RAND Corporation, who arranged a visit to the Samuel Goldwyn Studio. It was great excitement to see the filming of 'Irma La Deuce' with Jack Lemmon and Shirley Maclaine. We saw the Bolshoi Ballet, and young Krishna exclaimed that Maya Plisetskaya, the prima donna in Swan Lake, was very beautiful! It only showed that dark eyes and cherry cheeks have universal appeal!

One of the most memorable trips was to the Grand Canyon with Vasu. We left by the comfortable and fast train, the El-Capitan for Flagstaff, and arrived at the Canyon in the morning bus. It is difficult to describe in words our reactions on seeing the Canyon, too ample for human sight — two hundred miles long and five miles wide, a gaping fissure, bizarre in shape, awesome in size, the loveliness of its colours flattered by azure skies and the golden lighting of rising and setting sun. Two years later, when I talked about the Grand Canyon to Marshak, he replied that while America had the Grand Canyon, South India had greater wonders, Belur and Halebid!

I attended Feynman's lectures on gravitation at the Hughes Research Laboratory in Malibu — a magnificent location from where one could watch the Pacific sunsets. There was no doubt that gravitation is an intoxicating subject, made more so by Feynman. He discussed in detail the principle of equivalence between the mass occurring in the Einstein energy relation and the constant in Newton's gravitation law.

I lectured at the Hughes Research Laboratory and gave a Colloquium at UCLA.[9] The customary dinner after the colloquium was at the 'Golden Bull', but it was indeed a friendly gesture on the part of my hosts to provide me with 'Pooris' and spiced vegetables with the help of a Gujarathi student. I did not realise that years later Krishna would work for his PhD in mathematics in that lovely campus of UCLA.

It was during our stay in Santa Monica when everything looked sunny and bright, we heard the startling news on the 22nd of October that China attacked India on a thousand four hundred mile front and advanced thirty-seven miles into Indian territory. Meanwhile, the Cuban crisis also erupted. The Americas wanted to blockade the Russian ships carrying arms to Cuban bases. There was a rumour that Los Angeles may be isolated, and so the supermarkets became empty as customers bought a week's stock in sheer panic. It only showed how comfort and plenty could soften the nerves of the defenders of Bataan and the victors of Iwo Jima! Suddenly, the tension relaxed. U-Thant, the Secretary General of U.N., suggested the dismantling of bases and the Russians agreed not to send ships. Meanwhile, the Chinese continued to advance on the Indian frontier and Nehru appealed for American aid. It was then we left for Stanford, at the kind invitation of Professor Schiff.

We flew by the United Airlines DC-8[10] to San Francisco and Vasu accompanied us. At San Francisco we saw the twin peaks of California, the Golden Gate Bridge and the Bay Area but all the time we thought of war in India. The Nobel Prizeman Robert Hofstadter invited us for dinner at Stanford after my Colloquium lecture. Professor Schiff[11] and his wife took us through Monterey and its environs, the most affluent part of affluent America. A very fruitful experience was my lecture at the weekly colloquium at Berkeley, attended by all the leading lights with a good sprinkling of Nobel Prizemen.[12] I enjoyed every minute of my stay in Berkeley, though it was embarrassing for me to invite scientists to visit India, when China was invading our country!

During that period, the ideas of Chew[13] dominated Berkeley. His 'bootstrap' philosophy, that no particle was more elementary than the others, found votaries who believed that the key to the structure of matter, lay in the study of strong interactions. Years later, it was the possible unification of weak and electromagnetic interactions that claimed considerable attention. I met the Goldhabers, Weinbergs, Rosenfelds and Glassgolds in Berkeley, where I lectured on 'The concept of virtual states in quantum mechanical collisions.'

On Sunday the 18th November came the sad news of the death of Niels Bohr, the father of modern physics. It was God's will that the generosity of this great man, who with Einstein, had set in motion the physics of the twentieth century, should enlighten the lives of eager young physicists in Madras in the last year of his eventful life. And, with what salutary effect — the creation of Matscience!

We left for Washington by American Airlines Boeing 707 and arrived at Baltimore airport to be received by Ugo Fano.[14] We stayed at his comfortable house and it was a kind gesture by his wife to throw open their kitchen for our use. I lectured at Maryland, where I met Professor Toll, who was later to become the President of SUNY at Stonybrook.

We left for New York by a United Airlines Viscount flight and checked in at the Hotel New Yorker. It was a pleasant surprise, when my colleague Ranganathan came with his wife to see us. He was spending a year of post-doctoral study at the Brandeis University. We dined one night at C. V. Narasimhan's apartment in Manhattan, and left for Paris by the Pan American DC-8, and after a comfortable flight we arrived at Orly, to be received by Claude Bloch of Saclay.

Paris in any mood, even in winter, is certainly a tourist's dream. I lectured at Saclay and we had dinner in a restaurant near Champs Elysses, the most famous street in the world, beautiful by day, more so at night. We saw the Palace of Versailles, its gardens covered with a sheet of frost and snow, and The Louvre — too magnificent for description.

We heard that London was under the thickest fog in recent memory, and so we decided to go to India via Geneva. At Geneva, we were received by Bogdan Maglic, who drove us to his house in 'Divonne' on the French side. I met the twin pillars of CERN, van Hove and Weisskopf.[15] While Lalitha and Krishna stayed with the Maglics, I went to Berne at the invitation of Professor Mercier, and we returned to Geneva by car, escaping an oncoming snow storm just in time to catch the plane to Rome.

At Rome, we were received by two young physicists, Zwanziger and Sawicki. I lectured at the University on my recent work on Feynman graphs. Then we went round the city of endless delights, where fiction and history are so interfused, that we could imagine at once Antony's oration on the Roman forum and Mussolini's speeches inflaming the militant crowds! The historic ruins, the piazzas, the museums, the statues and fountains, and above all St. Peters and the Vatican — visible monuments of European history, culture and art, of centuries, all revealed in a few hours of intense excitement. No wonder Keats and Shelley had their resting place in the Eternal City. We arrived in Bombay on the 19th December by Air India and took the Indian Airlines to Madras. The tension of Indo-Chinese hostilities had abated and so we were in the right mood for the music festival at the Madras Music Academy, a fitting finale of melody and rhythm for the year.

Notes

1) Theodore Edward (Ted) Harris (1919–2005) was a noted mathematician who did fundamental work in stochastic processes and Markov chains, some in collaboration with Richard Bellman. Ramakrishnan became aware of connections of his work with that of Bellman-Harris while he was a PhD student in Manchester. Ted Harris was head of the mathematics division at RAND when Ramakrishnan visited there.

Two other noted applied mathematicians at RAND with whom Ramakrishnan interacted were, Robert (Bob) Kalaba and George Adomian.

2) From the 1950s, until the end of his tenure at MATSCIENCE in 1983, Ramakrishnan purchased his air-tickets through Thomas Cooks, a British travel agency of world repute. His travel agent there until the late sixties was Vaidhyanathan, who handled international travel, and so was affectionately called "Vaidhyanathan overseas!"

3) Pan Am operated two round-the-world jet services with their new Boeing 707 and Douglas DC-8 aircraft — one east-bound and one west-bound (Pan Am flights 1 and 2) — with several stops enroute, including one in Delhi. Pan Am was the first airline to launch round-the-world flights. On this trip, Ramakrishnan had two segments of this round-the-world flights from Hong Kong to Tokyo, and Tokyo to Honolulu.

4) In the 1950s, Dr. A. Jayaraman was a Research Assistant of Nobel Laureate Sir C. V. Raman in Bangalore, and that was when Ramakrishnan first met him. Jayaraman left India in the early sixties to settle down in the United States with his family. His first job was in the Department of Geophysics at UCLA, where he worked on metals under high pressure. During the two month visit to RAND in 1962, Ramakrishnan's family and Jayaraman's family got together regularly and thus began a close friendship of Ramakrishnan with Jayaraman, and Lalitha with Jayaraman's wife Kamala.

5) Ramakrishnan's former PhD student R. Vasudevan was visiting the University of California, Berkeley, on leave from MATSCIENCE. Vasu, as he was known to friends, would often come to Los Angeles to work with Ramakrishnan. During such visits, he got to know Bellman and his group, and this resulted in Vasu being invited to the University of Southern California, in the seventies after Bellman moved there.

6) In 1962, Gell-Mann and Feynman were not Nobel Laureates yet, but already were two of the most dominant figures in physics. It was a great gesture on the part of Gell-Mann to host a party at his home, in honour of Ramakrishnan, and to invite Feynman to the party. This was to reciprocate the hospitality he enjoyed at Ekamra Nivas.

7) Fredrik Zachariasen (1931–99) was a brilliant physicist at Caltech, who was noted for his research on theoretical studies of the interaction of elementary particles at high energies. He collaborated with both Gell-Mann and Feynman. In later years, Zachariasen served as Associate Director at the Los Alamos National Lab.

8) The Research and Development (RAND) Corporation was a think-tank formed in 1946, soon after WWII with the initiative of Donald Douglas, founder of the Douglas Aircraft Company. RAND was initially housed within the Douglas Aircraft Company in Santa Monica, and among its founding members was F. R. Collbohm of Douglas. RAND was to offer research and analysis to the US Armed Forces on a variety of scientific problems. In 1948, RAND separated from Douglas and became a corporation, with Collbohm as its Founding President. RAND moved to its own buildings in Santa Monica. but continued to have ties with Douglas Aircraft Company, and so Collbohm arranged a visit for Ramakrishnan, Lalitha and Krishna to see the Douglas assembly line.

9) Ramakrishnan's host at UCLA was Professor Nina Byers (1930–2014) of the Physics and Astronomy Department. Byers was a noted physicist, who joined UCLA as a young assistant professor in 1961, after receiving her PhD from the University of Chicago, where she was Murray Gell-Mann's student during his brief stint there.

10) While in Los Angeles, we saw the Douglas Aircraft factory in Long Beach, and the assembly of the giant Douglas DC-8 jets. So it was an interesting experience to fly in the DC-8, after seeing its assembly. United Airlines was the largest operator of the DC-8, just as Pan Am was of the Boeing 707.

11) Leonard Schiff (1915–71) was one of the very few physicists who made notable contributions to almost every branch of physics. His book on *Quantum Mechanics* became the Bible in the field. Ramakrishnan lectured out of Schiff's Quantum Mechanics in Madras at the Theoretical Physics Seminar. Schiff was Chairman of the physics

department at Stanford, when Ramakrishnan visited in 1962. He was elected to the US National Academy of Sciences in 1957.

12) The Berkeley physics department was teeming with Nobel Laureates — Owen Chamberlain, Emilio Segre, Luis Alvarez, Donald Glaser and others.

13) Geoffrey Chew is an American physicist, whose *bootstrap theory* drew the attention of most of the leading physicists at that time. Chew received the Hughes Prize of the American Physical Society in 1962, for his bootstrap theory.

14) Ugo Fano (1912–2001), in whose home the Ramakrishnan's stayed in Washington DC, was a famous Italian-American theoretical physicist. Fano was working at the National Institute of Standards when Ramakrishnan visited him in 1962. Fano and his wife not only "opened" their home to us, but also graciously hosted a party at their home in Ramakrishnan's honour. One of the guests at the party was Fritz Joachim Weyl, son of the great Hermann Weyl, and himself a mathematician of repute.

15) Victor Weisskopf was not only a great physicist, but also an administrative leader. In 1962, he was the Director General of CERN (Central European Research Nuclear), which has an accelerator used for atomic experiments. Prior to CERN, Weisskopf was at MIT, where Murray Gell-Mann was his PhD student!

L to R: Mrs. Gell-Mann, Richard Feynman, Murray Gell-Mann, Lalitha Ramakrishnan and little Krishna at Gell-Mann's house, in Altadena, California (1962).

Chapter 24

From Infancy To Manhood (1963)

Professor Robert Marshak (Univ. Rochester) delivering the Anniversary Address at
Matscience, January 1963. Seated on the dais from L to R: Alladi Ramakrishnan, Out-
going Education Minister C. Subramaniam, Chief Minister Bhakthavatsalan, the new
Minister for Education R. Venkataraman, Prof. N. Fukuda, and Secretary for Education
K. Srinivasan. - Editor

It was just a year since MATSCIENCE was born, but it leapt from infancy to manhood, thanks to the cooperation of the international scientific
community,[1] the support of the State and Central Governments, and above
all the enthusiasm of my young colleagues.

On the first day of the year, we had a lecture by Professor Fukuda, who
informed us that it was the first time in his life he was working, or made to
work, on a New Year's day! For, in Japanese tradition, it was mandatory
that the holiday is spent at leisure as a day of pleasure! Fukuda attended
the concerts at the Academy, and was impressed not so much by the music
as by the casual festive atmosphere. He used to come in early afternoon,

Ramakrishnan in discussion with the Russian delegation at Matscience in December 1963. Leon Rosenfeld is next to Ramakrishnan. - Editor

and stayed till the end of the final concert at midnight. He gave masterly lectures on many-body problems, in rigorous detail.

On the 5th of January, I performed father's ceremony, which I could not do earlier in October due to my U.S. trip. The next day, I was informed that our distinguished guest, Professor Marshak, arrived in Delhi and I contacted him by phone, welcoming him to Madras.

C. Subramaniam arrived from Delhi on the 13th of January, and we had our meeting of the Board of Governors, when it was decided to institute the Niels Bohr and Ramanujan Visiting Professorships at Matscience. Marshak was named our first Niels Bohr Professor (see Appendix 4).

The Marshaks arrived on the 13th January and were taken straight to the Woodlands Hotel, where we had reserved an entire air-conditioned cottage for them so that they would be impressed with so comfortable an accommodation at such a reasonable price — an almost token rent of Rs. 45, or six dollars a day. Marshak was the architect of the famous Rochester conference and we spared no effort to keep him happy both on and off the lecture hall. We were of course aware we had an obvious advantage, because

Rochester was under a blanket of snow while Madras was basking in the brightest of weather and the sounds of music.

On the 14th January, for the first anniversary of the Institute, we had a public function at the Presidency College with our Chief Minister Bhakthavatsalam presiding, and C. Subramaniam delivering the anniversary address. In the evening, there was a music concert at Matscience by Lalitha's sister Kamalam and a dinner at Ekamra Nivas. The anniversary symposium was held in honour of Professor Marshak, who gave a two-hour talk on R-invariance and weak interactions. We entertained Fukuda and the Marshaks at dinner at Dasaprakash. Marshak made a brief trip to Bombay, while Fukuda continued his lectures at Matscience.

On the 24th January, we received a letter from Vasudevan that he was arriving from California to join the Institute, as its first permanent member. On the 27th, we went to Tirupati with my personal assistant Nambi, to offer prayers to Lord Venkateswara, on the successful completion of the first year of activity at Matscience.

Marshak returned from Bombay and resumed his lectures. Mrs. Marshak found the sunshine irresistible that she browned her skin in the afternoons on the Madras beach, musing how their children would be wrapped-up in winter clothing at Rochester! G. Narasimhan, Editor of The Hindu, was kind enough to take them as his personal guests to the Gymkhana club, where they became temporary members entitled to make full use of the catering and swimming facilities. According to our custom, the Marshaks were taken to Mahabalipuram and the environs of Madras. The most interesting trip was to our village Pudur, where we had a delightful picnic in the mango groves. They took photographs of everything that interested them — little huts, luscious mango trees, stately coconut palms and sprawling rice fields. We showed him the spacious house our father had built, and the older modest one, where our father was born.

Professor Schiff of Stanford arrived on the 26th February, and there was a formal function when he released the Matscience Report I, comprising the proceedings of the anniversary symposium, in honour of Professor Marshak. The Schiffs liked Hotel Dasaprakash and the Indian cuisine there. After Marshak's departure, Schiff gave a series of lectures on gravitation. We took them on an excursion to Mahabalipuram and later to a dance drama 'Valli Parinayam' by Padmini and her troupe, which they enjoyed very much. I had them for dinner at Ekamra Nivas, inviting justice Patanjali Sastri[2] to match the dignity of our distinguished visitors. We later took them to another dance drama 'Sakuntalam' by the Ragini troupe.

On the 12th of March, I went to Delhi to attend a meeting of the Physical Research Committee and met the Prime Minister of India Jawaharlal Nehru, when our patron C. Subramaniam was present. What a pleasant surprise it was when he agreed to make an annual grant of Rs. 70,000 through the CSIR for conducting summer schools with visiting professors!

We decided to conduct the first summer school in Kodaikanal, and I was able to obtain Professors Zemach of Berkeley and Takeda of Tohuku University, as visiting lecturers. Professor Zemach arrived on the 21st May, and we took him directly to the lovely hill station. The summer school was almost a departure from the conventional scene in scientific development in India. For the first time, it was thought necessary and possible — to integrate the lecture programme of the visiting scientists and that of young and active research workers within India. It was also felt that the expository character of the school could be enhanced in value by including in subjects under discussion some of the research work of the participants.

The primary interest of the new group of theoretical physicists in India was high energy interactions of elementary particles. We were fortunate to be able to secure the cooperation of Professors Zemach and G. Takeda to lecture on these twin branches of high energy physics, the strong and weak interactions, respectively. Zemach belonged to the rising generation of American scientists, whose theoretical work is done in close collaboration with the leading experimental centres in the country. As one of the exponents of Chew's 'Bootstrap philosophy' in elementary particle theory, he gave through his series of lectures, a new impetus to the members of our group aspiring for work in such a competitive field. It was a pleasure to hear Takeda talk on his recent work on weak interactions, closely following his well-known contributions to the theory of strong interactions.

My good friend T. S. Santhanam, one of the bosses of the TVS, was kind enough to offer to our distinguished visitors generous hospitality in his lovely guest-house with its exquisitely pretty garden maintained with elegant taste and uncompromising efficiency. It was in every way a most enjoyable session, made twice so, by the bracing air of Kodaikanal, where Heaven's breath smells wooingly amidst fragrant pines and cedars dusk and dim. The weather was at its best in varied excellence, the shimmering lake under the cool transparency of the moon — blanched nights and the flood of yellow gold bursting through cloud-topped hills cast a strange enchantment, as the monsoon held back to the delight of our visitors from the sunny shores on either side of the Pacific.

In July, I had the opportunity to visit the United States at the kind invitation of Professors Marshak and Sudarshan, and participate in the conference on Unified Theory of Fields at Rochester. I left by an Air India 707 and I could not resist staying a few days in Switzerland, visiting Geneva, Interlaken, and Zermatt. From Geneva I went straight to Rochester, and I found that Marshak was in a hospital convalescing after an abdominal operation. The conference was therefore organised by George Sudarshan. I presented my work on splitting of the Feynman propagator into positive and negative energy parts, and how cross-sections for certain processes can be expressed as integrals over those of other processes. My old student V. K. Viswanathan had taken up his first job at Bausch and Lomb in Rochester after his MSc there, and we spent many hours discussing how we could soon form a 'Mylapore-Manhattan' axis in the years to come. McCrea Hazlett, the Provost of the University of Rochester, invited me to dinner at his home, and also took me on a trip to see the Corning Glass Works (Corning Museum of Glass). While returning from Rochester, I stayed with Ram Gnanadesikan[3] of the Bell Telephone Laboratory in New Jersey, and gave a lecture there at his kind invitation.

I flew from New York to London, where I stayed a few days, as the guest of Professor Abdus Salam at the Imperial College, before returning to Madras.

Among the visitors at Matscience was Professor Umezawa, a field theorist of the Heisenberg school who, like Fukuda, enjoyed his stay in Madras. We had at that time two young visitors, Professor Kamefuchi and his Norwegian wife, who found it most enjoyable to spend their honeymoon in Madras, a pleasant surprise to us, who never thought of our city in such romantic light! We had also another distinguished visitor, Professor Paul Roman of the Boston University. He had had a tough time in his home country in Eastern Europe, and was now a happy immigrant in the United States. He was an earnest lecturer and had written an excellent book on elementary particles, which was a favourite among my students. Peter Duerr, a collaborator of Professor Heisenberg, gave a systematic series of lectures on the new Heisenberg theory of elementary particles. He enjoyed cycling through the bazaars of Madras, and it was difficult to find a suitable cycle for him, particularly with his very long legs. He seemed to draw pleasure every minute of his stay in Madras, and travelled everywhere with unconcealed enthusiasm.

On November 21 came the shocking news of the assassination of President Kennedy during his visit to Dallas, Texas, plunging the world into grief, distraction and consternation. America in grateful, tearful, admiration for a President, whose fabled charm had cast over the affluent society, a hypnotic spell which, without wars and victories, rivalled that of Washington and Lincoln, Wilson and Roosevelt, named the gateway to the New World as the John F. Kennedy International Airport (JFK).

I received an invitation to the International Conference on Cosmic Rays[4] at Jaipur, and Professors Jacob and Hagedorn, who visited our Institute, were deputed as delegates along with me. Professor Hagedorn is an almost professional photographer and he was able to capture the charms of the 'Pink City' and its fortresses in his precision camera. Professors Bhabha, Blackett[5] and Powell were the leading lights of the conference.

In December we lost the closest ally to Matscience, K. Srinivasan, the Education Secretary, who died of a sudden heart attack. If C. Subramaniam was the father of our Institution, K. Srinivasan was its god-mother. Under 'her' care, it outgrew its infancy in less than a year and was well set on its high endeavour of creative science.

Later in December we received Marshall Stone as the Ramanujan Professor at our Institute. Among other visitors were Professor Maurice Shapiro, whose spontaneous interest was responsible for the creation of the Institute. All of them were feted to a feast of music at the Academy before they started their lectures. The Russian delegates,[6] from the Cosmic Ray Conference, visited our Institute. We had the pleasure of watching the handshakes and exchange of smiles between the American and the Russian visitors, brought together by the interactions of elementary particles at high energies and the integrating role of mathematical sciences over all space and time! By the end of the year the Institute was so well-established, that we were told with flattering exaggeration by Professor Rosenfeld, who arrived as the Second Niels Bohr Visiting Professor at Matscience, that he did not attend a single conference in Europe without meeting someone or other who had visited or was planning to visit our Institute. It was hard for him to believe that the Institute was functioning in a single room at the Presidency College when he arrived as a Visiting Professor. Fortunately, the next year we were given comfortable accommodation at the top floor of the spacious Central Polytechnic Buildings in Adyar.

We enjoyed the music season with our distinguished visitors, who were convinced that Madras was offering them not only a feast of reason at Matscience but a flow of soul at concerts galore all over the city.

Notes

1) Even when Ramakrishnan was away from September to December 1962, several top scientists visited and lectured at MATSCIENCE. One of the most prominent among them was Emilio Segre from Berkeley, who had won the physics Nobel Prize in 1959.

2) While hosting eminent scientists at home, Ramakrishnan always invited certain members of the Madras elite for dinner such as Justice Patanjali Sastri (1889–1963) who was a junior under Sir Alladi Krishnaswami Iyer. Once, when Patanjali Sastry prostrated before Sir Alladi to get his blessings, Sir Alladi said "May you rise to the level of the Chief Justice of the Supreme Court" — which he did in 1951! He was the second Chief Justice of the Supreme Court. Patanjali Sastri passed away in March 1963, within a month after that dinner in Ekamra Nivas in honour of Professor Schiff!

3) Ramanathan (Ram) Gnanadesikan (1932–2015) was a noted Indian statistician, known for his work in multi-variate analysis. He headed the research group in statistics at Bell Labs (known in the sixties as Bell Telephone Laboratories), and was Ramakrishnan's host throughout the sixties. He visited Matscience in February 1965, and wrote a fine note in the Visitors Book as to how exceptional an institute it is.

4) The International Conference on Cosmic Rays in Jaipur was conducted by the Tata Institute, owing to Bhabha's own strong interest in the subject. Since Ramakrishnan started his work under Bhabha on cosmic rays, he was invited to the conference as one of the main delegates. In December 1963, MATSCIENCE had several visitors from abroad. Some of them, like Rolf Hagedorn and Maurice Jacob, were there in connection with the Second Anniversary Symposium to take place in early January 1964. So Hagedorn and Jacob were invited as delegates of MATSCIENCE to the Cosmic Ray Conference

5) Patrick Maynard Stuart Blackett (1897–1974) was a outstanding British experimental physicist, who did fundamental work on cloud chambers, cosmic rays, and paleomagnetism. During the course of his distinguished career, he had been associated with several universities in England, such as Oxford, Cambridge, the Universities of London and Manchester, and was recognised with many prestigious awards. He received the 1948 Nobel Prize in Physics, for work on cosmic rays while he was at the University of Manchester. Ramakrishnan learnt a lot about cosmic rays from Blackett during the two years he was in Manchester (1949–51), and the Jaipur conference was an opportunity for them to reconnect. In the early seventies, Blackett visited MATSCIENCE briefly for a day and had an interactive session with the faculty and students.

With regard to Professor Powell, referred to by Ramakrishnan as the other leading light of the Jaipur conference, see Part 1, Chapter 13, for description of his visit to Ekamra Nivas in 1954.

6) The Russian delegates who visited MATSCIENCE were E. Sarytsheva and V. S. Murzin of Moscow State University, E. V. Kolomyets of Alma-Ata Stat University, V. J. Yakovlev, M. J. Teotjakova, B. F. Smirnov, Y. Smorodin, N. M. Gerasimova, and E. L. Feinberg of the Lebedev Physical Institute, Moscow, with Feinberg as the Leader of the Delegation. Professor Feinberg wrote the following in the Visitors Book of MATSCIENCE: "The group of physicists from the USSR expresses deepest thanks for the hospitality offered by the Institute. We do not know what makes a greater impression — the paradise of Winter in Madras (they call this Winter!!!) or the variety, the fundamental character and significance of the science developed here. Best wishes to every member of this remarkable Institute, and especially to its enthusiastic, optimistic, and talented leader — Professor Ramakrishnan."

Chapter 25

My First Visit To Russia (1964)

Ramakrishnan handing over to C. Subramaniam a copy of his book "Elementary Particles and Cosmic Rays". Professor Rosenfeld looks on. Matscience (1964). - Editor

The year 1964 opened with the anniversary symposium on the third of Januar, inaugurated by the Patron, C. Subramaniam, with Professor Rosenfeld,[1] the Second Niels Bohr Professor of Matscience, delivering the opening lecture on Bohr's contribution to twentieth century physics. There was participation by scientists from various centres in Europe and America, Hagedorn[2] from CERN, Maurice Jacob[2] from Paris, Hugh Dewitt, Barucha Reid,[3] George Sudarshan, Marshall Stone[4] and Bruno Zumino from the U.S., O'Raifeartaigh from Dublin, and Yamada from Japan. Hagedorn lectured on large angle electron scattering at high amplitudes, while Jacob spoke about crossing relations and spin states. Marshall Stone, one of the most influential mathematicians of the 20th century, gave a general lecture on "Current trends in mathematical research." Sudarshan gave an in-depth analysis of the origin of internal symmetries.

Prof. Marshall Stone in conversation with Chief Minister Bhakthavatsalam while Mrs. Stone is in conversation with Lalitha, 1964. - Editor

The Hazletts arrived from Rochester on the 12th of January, and Professor Hazlett delivered the anniversary address in the presence of the Chief Minister of Madras, Bhakthavatsalam. Later in January we had the visit of Professor Jensen,[5] which coincided with that of General Cariappa, the outspoken ex-Commander-in-Chief of the Indian army.[5] That was the year following the triumphant march of the SU(3) symmetry since it was first formulated by Gell-Mann in 1961, and so the lectures of the visiting scientists centred round the subject.

The scientific excitement was so great that I did not care to attend the cricket match between the MCC and India — a sacrilege on my part, an incredible departure from the Madras tradition of abandoning home and office to rave and rant over the classic elegance of a cover drive or the bizarre beauty of a shattered wicket. This was only a temporary lapse, for my interest in cricket was to be soon restored, as our Institute emerged from infancy to manhood within a year with the co-operation of the scientific community throughout the world.

Professor Rosenfeld had heard so much of our Institute and our visiting scientists programme, that he was expecting an impressive building and it was a surprise to him to see that we were working in a single room behind the Presidency College! Fortunately, a month later, the then Director of Technical Education, Muthian, offered us accommodation on the first floor of the spacious Central Polytechnic buildings.[6] About the middle of February, we started moving to this new premises.

During a visit to Delhi for the Boson Symposium, I met Hussain Zaheer, the then Director-General of the C.S.I.R. and suggested the conduct of annual summer schools. Though this scheme was not adopted by the C.S.I.R. directly, it was incorporated into the regular academic schedule of our Institute, through the generous support of our Prime Minister.

In March, Henri Stapp from Berkeley visited us. His lectures dealt with the spinor structure of Dirac matrices, on which I was able to do some work from a quite different angle, only five years later. Stapp created interest in me for the CPT theorem, and three years later, I obtained the representation of CPT operation in Feynman formalism, without recourse to sophisticated arguments.

The 18th of March was a memorable day for the Institute, when Professor Bhabha attended the meeting of the Board of Governors and had dinner with the members of the Institute and visiting scientists. Our Chairman, R. Venkataraman, asked his frank opinion about the visiting scientists programme. Bhabha readily complimented us on its success and the discerning choice of the visiting professors.[7]

We had also the visit of Professor Novoshilov from Russia, while Lukierski from Poland joined us as a visiting member. During this period I tried many strange and impossible methods of interpreting the hyper-charge quantum number and understanding the Gell-Mann-Nishijima relation in a deeper way. It seems as if I was making all the mistakes then so that they were eliminated in later effort! For, after three years came the answer in a totally different way.

In mid-summer (May), we made a grand trip by car to the south of India, since we planned to attend wedding in Trivandrum of Lalitha's niece Girija to S. Srinivasan, who worked as a partner in his father's well-known firm Subrahmanyam & Co. in Madras. Lalitha's sister Annapoorni and brother-in-law Vaidhyanathan had arranged the wedding in the spacious grounds of Lalitha's maternal grandmother. On the way to Trivandrum, we went via the Vaigai dam and Kumili, to the game sanctuary of Thekkady. We drove through Peermade and Munnar, with their lush tea gardens, a heritage from the British to the enterprising Indian businessmen. We then went on to Trivandrum via Varkala, where we offered prayers at the sacred Janardhana temple with its magnificent location, where sun and sky, ocean and land compete to offer their charms before the Lord of Lords. We visited Kovalam beach, still unspoiled before the onset of overzealous tourist development. The grand finale to this magnificent trip was watching the dazzling sunsets over the triple oceans at Cape Comorin.

Returning to Madras, I found in May I had to go to Ooty on personal business. During my brief stay, we heard the tragic news that our beloved prime minister, Pandit Nehru passed away. Due to sudden stoppage of all traffic, I was about to be stranded in Ooty but somehow I managed to take the plane from Coimbatore and reach Madras without delay.

At Matscience, we passed a condolence resolution at the irreparable loss of Jawaharlal Nehru, our great sponsor and patron. I wrote on that occasion:

"Few lives in the history of mankind have shone with such many-splendoured hues. Born to wealth and prosperity, he became the idol of peasant India, its hope and its redeemer. Educated in the most exclusive of English institutions, he stood against the might of the British Empire with dauntless courage and in open defiance. He emerged a victor after decades of travail and remained the staunchest friend of England, its people and its gracious Queen. As the architect of a new republic, he designed it within the stable structure of the great Commonwealth. Rational in his beliefs to the limit of secularism, he wore the mantle of Gandhi, the man of God, with grace and dignity. He had irrepressible faith in a socialistic economy but his very name is synonymous with the freedom of the human spirit. His patriotism was a flaming passion, which burnt out the vestiges of foreign rule from our sacred land, yet he was a true citizen of the world, that his counsels were sought in the United Nations. An individualist in thought, he loved people to an extent that every child looked to him as a father and every Indian felt his benign influence. His princely graces set him apart from our too common mould, but he was the darling of modern India and there is no single home from the southern Cape to the Himalayan heights, where the very mention of his name did not inspire love and affection.

The best years of his youth were spent in isolation behind prison bars, but that was the period he wrote his dearest letters, transmitted the warmest feelings and nourished the most sanguine hopes. Those years left no wrinkle on that handsome brow, which age could not wither nor anxieties strain.

His life was dedicated to improving the standard of life of the common man, but he remained a restless intellectual, a votary for the advancement of science in our country. Amidst the tumult of politics and the anxieties of administration, he found time to exhort scientists to greater achievement and the aspirant youth to the pursuit of knowledge."

In early July, Professor Caianiello from Italy, and Fuji from Japan arrived as visiting scientists. I had to leave them to the care of my colleagues at Matscience since I had to go to Russia for the High Energy Physics Conference at Dubna.

On the eve of my departure to Moscow, I won the doubles tennis tournament at our local club with Veekay Venkatesan as my partner — the only prize I ever won in tennis, which I therefore remember with unconcealed warmth and unbecoming pride. My acquaintance with Venkatesan grew into a deep friendship, which had the salutary consequence of his turning out to be the 'architect' of Krishna's marriage years later!

On the 26th July, I left for Paris via Geneva enroute to Moscow. After a brief halt at CERN visiting Hagedorn,[1] I arrived in Paris to be received by Maurice Jacob. I gave a talk at Saclay at the kind invitation of Professors Dominicis and Bloch. Jacob took me on a pleasant excursion by car to Chartres, where we visited an exquisitely beautiful cathedral before leaving for Moscow. On arrival, I was received by a representative of our Ambassador to Russia, T.N. Kaul,[8] and I checked in at the hotel Ukraine. The first visit to the communist country was a very strange and thrilling experience, particularly walking at night through the famous Red Square before the Kremlin. My mind travelled back to the early thirties, when we heard of the economic successes after the Soviet revolution, the emergence of Stalin, mightier than his master, the massive build-up of military power behind the Iron curtain comparable with the German Wehrmacht, and then the confrontation through the bloodiest, bitterest, boldest war ever waged on earth, when the blood of Nordic hordes and the numberless millions of Russia "incarnadined the multidinuous snows making the white one red," from the flaming inferno of Stalingrad to the beleaguered fortress of Leningrad.

A shiver passed through me as I stood on the Red Square. Was this the city that was saved from doom by an incredible miracle as the German panzers came to a grinding halt with the fatal onset of the Russian winter? Was this the nation that survived the blasting fury of the German conquest and even worse, the nameless horrors of Nazi occupation, which made hell seem mercy before the genocidal rapacity of Hitler's fiends? Yet, that

night all seemed peaceful and the tranquil silence seemed incredible as I
watched the silhouette of the Kremlin walls, wondering at the resurrection
of Russia and the redemption of the human race. I just couldn't believe I
was in Lenin's country, which looked so distant and inaccessible when we
were discussing the meaning of the revolution in the placid environment of
Mylapore in my student days.

We visited the Pushkin museum before leaving for Dubna[9] by motor
coach. Among our companions were Schiff and Hofstadter from Stanford,
who were talking about CP or T violation, to be announced at the conference by the Princeton group. The excitement over this discovery was
perhaps disproportionate to its importance, unlike the earlier discovery of
parity violation. I presided over a session on pion-nucleon scattering and
participated in a press conference led by Goldberger of Princeton. It was
gratifying that my hosts at Dubna took very good care of my vegetarian
palate by providing cheese blintzes and sour cream as the main course for
lunch and dinner every day! Of course they were amused to find that
Caviar and Vodka held no attraction for me, while the American visitors
were scrambling for these Russian delicacies!

What I heard outside the conference was even more exciting than in its
actual sessions. I found a few physicists, like Salam and Glashow, talking
of the necessity of introducing additional quantum numbers into the Gell-
Mann–Nishijima relation like charm or triality, the problems which engaged
my attention a few years later.

Salam with characteristic grace and friendship, suggested I should visit
the new International Centre for Theoretical Physics at Trieste, of which he
was the first director. Nothing pleased me more, for I not only admired the
contributions of Salam to modern physics but also his role as a generous
host to aspiring scientists from developing countries.

I left Moscow by the Air India and returned to Madras via Delhi, to be
received by Lalitha and Krishna, Vasu and Nambi. I was just in time to
celebrate the eighteenth anniversary of our wedding in good style since it
coincided with the Varalakshmi Vratham. The visiting scientists, including
Calogero who had arrived from Rome, came for dinner to Ekamra Nivas.

The Matscience Summer School was held in Bangalore, inaugurated by
the Director General of the C.S.I.R. Since the summer school was immediately after the Dubna Conference, there was a good discussion among
the participants about the results that were presented in Dubna. The
foreign delegates to the Summer School were E. Symanzik (New York
University), E. R. Caianiello (Naples, Italy), F. Calogero (University of

Rome), W. Brenig (Max Planck Institute, Germany), A. Fujii (Tokyo) and J. Lukierski (Warsaw, Poland). The papers presented at this Summer school were published as Volume 3 of the Plenum series (see Notes item 10 below). Among the visitors was B. D. Laroia, a very enthusiastic official of the University Grants Commission with a passion for science and appreciative of the ideals of our Institute. How we wished we had such a person as the Chairman of the University Grants Commission or the Head of the C.S.I.R.! However, he gave us a good insight into the working of these organisations, which was very useful to me during the period of growth of our Institute.

In September, I had a pleasant surprise when Mr. Coleman, the President of the Plenum Press,[10] arrived in Madras and called on me at the Institute. He did not take much time to agree to publish the proceedings of our symposia as a Plenum series and the formal contract was signed[11] after I took him to a Bharatha Natyam performance. I could not help feeling that he signed the agreement under the serpentine spell of a dance to the tune of 'Punnagavarali' and may perhaps refute it when restored to pragmatic mood in the Board Room of a New York publishing house!

In October there was a cricket match between Australia and India, and this time I could not resist the spell of the Madras fever. Pataudi's sparkling innings sent the crowds raving with delight and we certainly had our full share of it.

Navarathri was celebrated in the usual style, followed by Deepavali when we had Lukierski as one of our guests. During this period, we had the opportunity to receive Mr. and Mrs. Munshi at our Institute. It gave me great satisfaction to show them round, since they had offered me their spontaneous hospitality during the period of my study under Bhabha in Bombay. As the founders of the world-wide Bharathiya Vidya Bhavan, they could appreciate the emergence of the Tata Institute and Matscience from the womb of Kenilworth and Ekamra Nivas.

Equally satisfying were the visits of my old teachers at P. S. High School, G. Srinivasachari and R. Narasimhachari, to whom I owe my continuing passion for the English language and mathematical sciences.

December was quite eventful with the visits of Victor Weisskopf, the Director-General of CERN, Alf Sjolander of Sweden, and the French scientists, Gourdin, and Dominicis, who introduced a breath of Paris into the sedate atmosphere of Mylapore and Adyar. We expressed our appreciation by regaling them with the sound and sight festival of music and dance at the Academy during the Christmas season.

Notes

1) Leon Rosenfeld (1904–1974) was a famous Belgian physicist who was a close collaborator of Niels Bohr. He was Niels Bohr Visiting Professor at Matscience in 1964. In his long career, he held positions at several academic institutions — in Liege, Manchester, Copenhagen, and Utrecht. Ramakrishnan had heard the lectures of Rosenfeld at Manchester. Rosenfeld was much impressed that so much was being done at Matscience with such modest accommodations and with an even more modest budget. He wrote the following in the Visitors Book of Matscience on 25 January, 1964: "I am most grateful to an audience composed of eager young researchers, not afraid of asking questions to the point of cornering the lecturer!"

2) Rolf Hagedorn (1919–2003) was a German theoretical physicist, who from 1954 onwards was at CERN. He actually fought for the Germans in WWII, and was an officer in Rommel's Afrika Corps until 1943, when he was captured. After the war ended, he returned to Germany in 1946, and obtained his PhD in 1952 at the Max Planck Institute in Gottingen. Hagedorn did important work in theoretical physics, which were confirmed by experiments at CERN. He was so impressed with his visit to Madras that he wrote in the Visitors Book of MATSCIENCE that if everything in India would be like MATSCIENCE (and the Madras Music Academy), then India would have no problems!

Maurice Jacob (1933–2007) was a French theoretical physicist, who was at Saclay near Paris in the early sixties, and from 1967 onwards at CERN. He was a leader of the European particle physics community, and as per the Obituary in the CERN Courier of July 2007, "he was one of the scientific pillars of CERN, working closely with experimental colleagues in predicting and interpreting results from successive CERN colliders." Jacob and Ramakrishnan were close friends and so they travelled together from Paris to Russia for the Dubna conference after Ramakrishnan visited him in Saclay.

3) A. T. Barucha-Reid (1927–85) was a well-known American probabilist, who wrote influential books on the subject. In his 1960 book "Elements of Markov processes and their applications," he made significant references to the work of Ramakrishnan and his students (P. M. Mathews, S. K. Srinivasan, R. Vasudevan and N. R. Ranganathan) on product densities in stochastic processes. Barucha-Reid also discusses the Bhabha-Ramakrishnan equations.

4) As the First Ramanujan Visiting Professor at MATSCIENCE, Professor Marshall Stone gave a general lecture entitled "Some current trends in mathematical research," which appeared in Volume 2 of the Symposia in Theoretical Physics and Mathematics (Plenum) edited by Ramakrishnan (see Notes item 10 below).

5) Johannes Hans Daniel Jensen (1907–1973) was a German nuclear physicist, who shared 1/2 of 1963 Nobel Prize in physics along with Maria Groppert Mayer for work on the shell model; the other half of the 1963 Physics Nobel Prize was awarded to Eugene P. Wigner of Princeton University for work unrelated to Jensen's. Jensen was at the University of Heidelberg at that time, and visited MATSCIENCE a few weeks after receiving the Nobel Prize.

From time to time, Ramakrishnan invited leaders in other professions to visit the Institute so that the message of the mission of MATSCIENCE would spread to the public at large. General Cariappa visited MATSCIENCE the same day as Jensen. Ramakrishnan and Lalitha met General Cariappa in Australia in 1954 (see Part 1, Chapter 13), and renewed contact with him after the creation of MATSCIENCE.

Both Professor Jensen and General Cariappa were at Ekamra Nivas that evening for dinner.

6) MATSCIENCE occupied the rooms offered by the Central Polytechnic from 1964 until 1969. In 1969, MATSCIENCE moved to its (now) permanent home — its own

buildings, but still located in the Central Polytechnic campus area, known now as Taramani.

7) It was very important for Ramakrishnan that Homi Bhabha should attend the Board of Governors Meeting and be impressed with the progress of the Institute. The early years of MATSCIENCE were the finest in its history. Ironically, that was the only meeting of the Board of Governors of MATSCIENCE that Bhabha attended, because he died in the Air India Boeing 707 crash in January 1966.

8) T. N. Kaul was one of India's foremost diplomats. He was Indian Ambassador to Russia during 1962–66 and 1986–89. Ramakrishnan had good association with leaders of the Indian political and diplomatic cadres, and so he was received personally at Moscow airport by representatives from the Ambassador's office.

9) Dubna, known as the science city, is located about 120 miles to the north of Moscow. Dubna as a city/town was officially opened in 1956, with the launch of the Joint Institute for Nuclear Research (JINR). Several elementary particles have been discovered there, and one of them was named *dubnium*. Dubna was proud to host the prestigious International Conference on High Energy Physics (the Rochester Conference) in 1964.

10) Earl M. Coleman (1916–2009), a poet, ventured into the world of publishing in the late 1940s by starting Consultants Bureau, which specialised in doing translations. He then developed it into one of the largest translators and publishers of scientific work. He formally renamed his company in 1965 as "Plenum" meaning full, rich or plentiful, but in 1964 when he met Ramakrishnan in Madras, the name Plenum was already in use. Mr. Coleman retired from Plenum in 1977 and formed Earl M. Coleman Enterprises, a new publishing venture. Plenum was bought by the Dutch publishing company Kluwer in 1998, and after Kluwer's merger with Springer in 2004, it has become part of Springer.

11) A total of ten volumes under the title "Symposia in Theoretical Physics" were published during 1965–70 by Plenum Press, with Ramakrishnan as the Editor. The contents of these volumes were primarily lectures delivered at the MATSCIENCE Anniversary Symposia from 1963 to 1968, but also included were some important lectures given during each of the years at MATSCIENCE outside of the anniversary symposia, such as in the MATSCIENCE Summer Schools.

Alladi Ramakrishnan with Col. K.M. Cariappa and Nobel Laureate Jensen at Matscience (1964).

Chapter 26

Visitors Galore And An European Serenade (1965)

1965 was a year of unqualified success of the Visiting Scientists Programme at Matscience and an exhilarating European experience for me, Lalitha and Krishna. As usual, the year opened with the Matscience anniversary symposium, which was inaugurated by Professor Victor Weisskopf,[1] the Director-General of CERN, Geneva, who delivered an introductory lecture on the status of elementary particle theory, similar to that given by Oppenheimer on the 'sub-nuclear zoo,' I had heard years before. It was clear to me that the Gell-Mann–Nishijima relation and unitary symmetry were taken as sacrosanct and no real attempt was made to understand the deeper meaning of internal quantum numbers. Among the participants to the symposium were R. Blankenbecler from Princeton, Dominicis, Gourdin and Meyer from Paris,[2] and Marshall Stone, a great friend of the Indian academic community, who gave stimulating lectures on Axioms and Models. Blanckenbecler gave an inspiring talk on a new approach to scattering theory. We had the privilege of receiving Harish-Chandra, the world renowned Indian mathematician, and hearing two of his lectures on characters of semi-simple Lie groups, which were incorporated in the proceedings of the Matscience symposia Vol. 4, where all the lectures of the Third Anniversary Symposium were published. I had heard so much of Harish-Chandra during my student days, that to me, he was a legend come to life.

We took all the visiting scientists to a government house party of Republic Day to have a glimpse of the leisured atmosphere of the privileged classes, reminiscent of the spacious days of the British Raj.

In late February, I used an opportunity to go to Delhi to meet Mr. T. T. Krishnamachari, the Finance Minister, Mr. Chagla, the Education Minister, and the new Prime Minister Lal Bahadur Shastri, to augment the support to our Institute.

I received a handsome invitation from Professor Salam to spend a month at the new International Centre for Theoretical Physics,[3] whose conception I witnessed five years ago at the Miramare gardens! As was customary, we offered prayers at Tirumala to the Lord of the Seven Hills before undertaking the journey to Europe.[4]

On the 19th of March, we left for Rome by the direct Air India flight from Madras. We were received by Calogero in Rome. We feasted our eyes with the sights of the Eternal City — from the Pantheon and Coliseum to the Vatican and St. Peters. After my lecture in Rome, we left for Naples by a rapid express. All the praises of the beauty of Naples and Capri seem grossly inadequate — Naples, an Elysian city and Capri, the mountain island, set like sparkling jewels against the azure sky and the crystalline ocean. We stayed at the comfortable hotel Bretagne with its kingsize rooms and enjoyed the gracious hospitality of Professor Caianiello, a Neapolitan by birth, choice and temperament. We saw the famous Blue Grotto and made an excursion to Pompeii with Professor Umezawa, which brought to my mind that beautiful novel of Lord Bulwer Lytton, which I read at school — the ageless majesty of Vesuvius bridging the gaps of centuries.

During the two months in Trieste, I was intrigued by the Gell-Mann–Nishijima relation. I asked myself the question: Is there a symmetric method of writing the Gell-Mann–Nishijima relation such that the relationship between the various quantum numbers follows in a natural way?

Years later I was able to do some really interesting mathematical work relating to the Gell-Mann–Nishijima relation from an entirely different approach, which can however be traced to this initial effort at Trieste.

I met my friend M. G. K. Menon of the Tata Institute, who was visiting Trieste, and discussed with him at length about the state and status of Indian science. We agreed that it was a curable but ignored malady that a scientist in India would not recognise the merit in another till external recognition compelled him to do so!

There was a glittering Mayoral reception at the Hotel Excelsior, when Salam complimented ten year old Krishna for his gentleman look, dressed in black tie and dark suit. Krishna justified the compliment by reading a book quietly in a corner as the guests warmed up the conversation swayed to the vineal flow of Italian hospitality.

We visited the Postagna caves in Yugoslavia, one of the natural wonders of the world, revealing the spectacular symmetry of pendent stalactites and projecting stalagmites. The drive in a sleek Mercedes tourist bus gave us a glimpse of Toto's country, which stands as a buffer between the communist and the free worlds. We also made an excursion to the Dolomites and compared their scenic grandeur with that of Swiss and French Alps.

I gave lectures on the Gell-Mann–Nishijima relation at the University of Padua at the kind invitation of Professor Dallaporta, a gracious host whose human qualities match his excellence as a physicist.

It was during our stay in Trieste that hostilities broke out between India and Pakistan, but like true scientists, Salam and I made no reference to it while discussing his work on the possible unification of internal and dynamic quantum numbers!

From the Trieste, we went to Venice to get a brief glimpse of the beauty of the 'nursling of the Adriatic and then its queen.' We later flew to Milan, Zurich and Geneva. It was a pleasant surprise to see Gell-Mann as one of the passengers carrying his little son on his back! I gave a seminar at CERN, at the kind invitation of Professor Hagedorn, on fundamental multiplets. That night, Professor Weisskopf invited us for dinner, when we met Mrs. Sulamith Goldhaber[5] of Berkeley. During our conversation, we found Weisskopf and Sulamith nostalgic about Viennese opera in just the same manner as a Mylaporean feels about Carnatic music.

The Hagedorns took us on an excursion to the Chamonix Valley and Mont Blanc — sights of snow-clad splendour of still and solemn power. Weather this time was not good, and in fact, there was almost a snowstorm when we went to Pic Du Midi, which still revealed its serene and tranquil beauty — remote, and inaccessible to Shelley when he wrote his immortal verse, but which could now be reached by electric cable cars by impatient tourists! We returned to see a slide film show at Hagedorn's house and enjoyed some long playing records of Carnatic music, particularly, 'Sarojadalanethri' by M.S. in the raga Sankarabharanam. I explained to my eager friends the peculiar charm of the 'Neraval' or sonorous repetition with variation in carnatic music.

We left for Paris and were received at the Orly airport by the charming French trio, Dominicis, Gourdin and Meyer. At Saclay, I talked on the symmetries of the Gell-Mann–Nishijima relation. The dinner at Professor Gourdin's apartment was really a banquet, at the end of which we heard some delightful Carnatic music Sambho Mahadeva of M.S. from the collection of records so zealously bought by him during his visit to Madras. He had decorated his apartment with brass vessels and souvenirs he bought in Mylapore, things we ignored by their surfeit in South India! I gave a seminar at the Institute of Henri Poincare at the invitation of Professor Kichennassamy, an Indian by birth and Parisian by spirit and domicile.

We were taken by Dominicis to the Chateau Rambulliet, the magnificence of which was matched only by the thriftless loveliness of its terraced gardens, arousing in the visitor fantasies of Napoleonic glory, which dazzled the eyes and drained the blood of a distracted Europe from the crimson dawn of Austerlitz to the closing sunset at Waterloo.

We had a delectable lunch in the little garden of the house of Dominicis, tasting the widest variety of cheeses for dessert. Later we had dinner with Claude Bloch, and drove straight to the Paris Opera with the Meyers. The next day was so bright and sunny that we went up the Eiffel tower with Kichennassamy to inhale the splendour of the most beautiful city in the world. Vasu arrived in Paris from India and joined us at the hotel Louvre. We had a 'rice and vegetable' dinner in the small restaurant opposite the hotel, using all the spices we had brought from home. We took a boat trip on the Sienne, and spent the whole night talking about various events from Matscience to Berkeley. The next day we flew out of Le Bourget airport to London by a BEA Trident, in which the Instrument Landing System (ILS) was being used for the first time![6]

We went on an excursion to Windsor Castle to feel the enchantment of its regal splendour and had a glimpse of the lovely gardens of Eton and the playgrounds of the victors of Waterloo and El Alamein! We had tea with Professor Lighthill before going to the popular musical 'The sound of music,' as much a feast to the eye as to the ear.

We left for Athens via Rome by the British European Airways Comet IV. The weather was superb and the atmosphere was so clear that we saw the historic island of Elba, the legendary isle of Monte Cristo, the sparkling profile of the Grecian coast, an enthralling spectacle which was not repeated again in all my travels. Athens is the timeless link with the origins of human civilisation, a city of such breath-taking beauty "as vision builds from clouds in derision of kingliest masonry." The Parthenon on the Acropolis has a situation which no other edifice has in the entire world. We saw the Acropolis at moonlight and also witnessed the 'Sonet Lumiere' show which unfolded in a few magic minutes the glory and grandeur of Hellenic centuries. The excursion to the Grecian islands of Aegina, Porus and Hydra was an unforgettably exhilarating experience with the bloom of heaven reflected on the azure waters of Aegean sea. The best tribute I could pay was to build a terrace in my new house in Madras in imitation of the luxury hotel which overlooked the Aegean sea! We took the Olympic Airways to Beirut, and returned to Madras by the Air India through Bombay.

The next few months at Matscience were devoted to seminars, with emphasis on unitary symmetry, inspired by the excitement over Salam's work at Trieste. We had quite a few visitors at the Institute, Grossman from Paris, Capps, Jastrow, Symon, Teplitz, Oakes and Nielsen from the U.S., and Ruegg from Geneva. The contact with Keith Symon was to grow into a deep friendship in the years to come. I took all our visiting scientists to meet the Governor of Madras — the Maharajah of Mysore.[7] It was a great excitement for the American visitors to see a Maharajah in flesh and blood since they had heard so much about Indian Maharajahs, and legends both real and imaginary that had grown about them!

Our summer school in Bangalore was inaugurated by the Chief Minister of Madras, Mr. Bhaktavathsalam. It was such a great success that an informal suggestion was made by Balachandra Nair, Chief Secretary of the Government of Mysore, that Matscience should function at Bangalore simultaneously. It was a flattering idea too good to be executed by the tardy administrative machinery of Indian Science. R. J. Oakes of Stanford University gave a series of lectures on weak interactions, and V. L. Teplitz of MIT gave his lecture series on applications of algebraic topology to Feynman integrals. All papers presented at this Summer school appeared in Volume 5 of the Plenum series. The seminar wound up with a most pleasant excursion to Nandi Hills, Jog Falls and Sharavathi projects, the drive through the valley at sunset being as spectacular as the roaring cataract itself! We returned visiting Belur, Halebid and Shravanabelgola. I was reminded of the observation of Marshak — 'while America has the Grand Canyon and the Old Faithful, South India has the irreproducible grandeur of Belur and Halebid.' We returned to Madras after pleasant excursions to Bandipur, Mysore and Ootacamund.

Back at Matscience, we had seminars by Teplitz and lectures on gravitation by Kichennassamy of Paris. On October 4, I met Sir C.P. who assured us that our Institute will be recognised by the Inter-University Board as a centre for Ph.D.[8] He was to fulfil his promise by announcing the resolution of the Board a few months later at the Anniversary symposium in January 1966.

I spent the next few months trying to understand the meaning of $SU(4)$, Quarks and the Gell-Mann–Nishijima relation governing them. Little did I realise at that time, that the generalisation of the Gell-Mann–Nishijima relation was to come from an entirely different approach based on the generalisation of the Clifford Algebra.

On Sunday the 17th of October, I complained of excruciating pain and it was found that an emergency operation was necessary to remove a strangulated hernia. The operation was performed by my friend and family surgeon Dr. Mohan Rao, at his nursing home and the operation theatre was specially got ready on a holiday — a gesture which could not be executed in private nursing homes in countries like America. On returning home from the nursing home, I conducted seminars at my house for about a month trying to resurrect the memories of the exciting days before the birth of our Institute! The Goldhabers from Berkeley arrived in early December, but their visit ended tragically when Sulamith Goldhaber died suddenly in Madras due to a brain tumor.[5] Among the visitors for December were Bruno Gruber and Takahashi from Dublin. My association with Bruno grew into a deep friendship, strengthened by our frequent visits to his home in the United States, where he migrated with his gracious wife Burgyl. The visiting scientists enjoyed as usual the annual music festival during Christmas, which included the inevitables and incomparables — the concerts of M.S., Semmangudi and Madurai Mani Iyer.

It had been a year of 'grand alliance' transcending national boundaries, of collaboration between Matscience and the great centres of learning in the world, from Stanford and Berkeley to Dubna and Moscow.

Notes

1) Professor Weisskopf was so much impressed with his visit to Madras that he wrote in the Visitors Book of MATSCIENCE, the following:

"We are overwhelmed by the beauty of Madras, the kindness of its people, the depth of its culture, and last but not least, the great work of Alladi Ramakrishnan, who has created a great institute out of nothing! Alladi saw the opportunity and made it work."

2) Ramakrishnan had close contact with several French theoretical physicists, in and around Paris. Besides Claude Bloch and Maurice Jacob, his hosts included Michael Gourdin, Philip Meyer and Dominicis. They all visited MATSCIENCE — especially Gourdin, who visited several times with his wife.

3) The International Centre for Theoretical Physics (ICTP) was founded by Abdus Salam in 1964, with support from UNESCO. Salam, like Ramakrishnan, tried initially to launch an institute in his native country (Pakistan), but since that did not bear fruit, he successfully founded the ICTP in Trieste, Italy, and served as its Director until his death. The Institute is now named as the Abdus Salam ICTP.

When the ICTP started, it was housed for a few years on the upper floors of a building in Piazza Oberdan, in the heart of Trieste. A few years later, the ICTP moved to its permanent home — a magnificent campus in Grignano, outside Trieste, near the Castle Miramare. Ramakrishnan attended a meeting at the Castle Miramare in 1960, when the idea of the ICTP was first promoted by Salam, and Ramakrishnan enthusiastically endorsed Salam's idea. As soon as the ICTP was founded, Salam graciously invited Ramakrishnan and hosted him magnificently.

4) Lord Venkateswara of Thirupathi is the Alladi family deity. Until the seventies, Ramakrishnan would make a pilgrimage trip to Thirupathi before undertaking any major assignment abroad. To reach the Venkateswara temple, one has to cross seven hills; so Lord Venkateswara is known as the Lord of the Seven Hills.

5) Gerson Goldhaber and his wife Sulamith were highly reputed physicists, both of them on the faculty of the University of California Berkeley. When Ramakrishnan met Sulamith at Weisskopf's home, he invited her and her husband to visit MATSCIENCE, which they did later in the year. The Goldhabers were on sabbatical leave from Berkeley that year, and so after visiting CERN, they spent some time at the Hebrew University in Jerusalem, before coming to MATSCIENCE. During their stay in Jerusalem, they prepared their notes for their lectures in Madras, which unfortunately they could not deliver due to a tragic development. Within 24 hours of arrival in Madras, Sulamith complained of uneasiness; she was diagnosed as having a brain tumor, and died a few days later in Madras. It was gracious of Professor Gerson Goldhaber to have given MATSCIENCE the text of the lectures that he and Sulamith were scheduled to deliver, and their joint paper was published in Vol. 6 of the Plenum series.

6) The Trident was a three engined short to medium distance jetliner designed by de Havilland, but by the time the Trident came into operation in the early sixties, de Havilland was taken over by Hawker Siddeley. BEA was the primary operator of the Trident. The ILS was termed as Autoland by the British, and the Trident was the aircraft used to test the ILS.

7) When India secured its independence from Britain, all princely states were abolished and so the Maharajahs were reduced to figureheads. There was a growing feeling that the Maharajahs had enjoyed great wealth and luxury at the expense of the poor. It was only a matter of time before corruption in political circles in free India became so rampant, that politicians amassed wealth several times higher than what the Maharajahs accumulated! A few Maharajahs were given ceremonial positions, like the Maharajah of Mysore being appointed the Governor of Madras.

8) Sir C. P. Ramaswami Iyer at that time was Chairman of the Inter University Board. He arranged for it to be possible for a student of any state in India to do his/her doctoral work at MATSCIENCE, but be granted a PhD degree by any university in the home state of that student. This was to stimulate the enrollment of PhD students from all over India, which it did.

9) One important visitor to Matscience, that Ramakrishnan forgot to mention in his narrative, was Jayant Vishnu Narlikar, a famous cosmologist from Kings College, Cambridge University, during Feb 23–25, 1965. There is a very interesting incident related to his visit, and so I have included it in this Note.

Narlikar had shot to fame due to his joint work with Fred Hoyle on their newly proposed *Steady State Theory of the Universe*. The day before Narlikar's first lecture, Ramakrishnan had sent an announcement to The Hindu newspaper that Narlikar would be talking about a new theory of the creation of the universe. Narlikar's lecture was scheduled at 2:30 p.m. on February 23. About an hour before the lecture, a vast crowd of eager listeners began entering the Matscience premises, obviously having seen the announcement in The Hindu. I quote from Ramakrishnan's annual diary of 1966: "1500 persons surge into Matscience!" A decision was immediately taken to shift the lecture to the nearby Nehru Auditorium of the Central Polytechnic campus, and so Narlikar could give his talk to the audience of 1500 who had come to hear him.

Narlikar gave two more lectures on cosmology at Matscience, on February 24 and 25, in the regular seminar room of the Institute. He was much impressed with his visit and wrote a fine note in the Visitors Book. Narlikar visited Matscience again in 1966.

Prof. Victor Weisskopf, Director General CERN, in conversation with Education Minister R. Venkataraman at Ekamra Nivas, January 1965. The ladies in the picture are (from L to R) Lalitha, Mrs. Weisskopf, Lalitha's sister Kamalam, unknown, and Mrs. Marshall Stone.

Alladi Ramakrishnan introducing Swedish physicist Gunnar Kallen to Sir C. P. Ramaswami Iyer at MATSCIENCE (Jan 1966). Minister for Education R. Venkataraman is looking on.

Chapter 27

Syracuse And Round The World (1966)

The Anniversary symposium was inaugurated on the 2nd January, 1966 by Sir C. P. Ramaswamy Iyer. I realised my ambition of receiving at our Institute, that versatile statesman of my father's generation, whom I admired since boyhood as one endowed with the remarkable gift of holding the attention of any audience, any time, on any subject.[1] With his characteristic penchant for surprise, he announced that the Inter-University Board under his Chairmanship has approved of Matscience as a national centre for PhD studies in mathematical sciences. Professor Gunnar Kallen,[2] the stormy petrel of European physics, gave the first lecture on the state of modern physical theories. Among the participants in the symposium were Rzewuski from Poland, Takahashi from Dublin, Kotani from Tokyo, Gruber from the U.S.A., Ruegg and Mercier from Switzerland and Nozieres from France. Professor Kallen gave a review full of sizzling humour and exuberant confidence and left the audience convinced about the unsatisfactory state of high energy physics! When I invited him as a visiting professor, many warned me that he was hard to please and unsparing in academic criticism. We found him friendly in conversation and enthusiastic about his visiting assignment. But he frankly told me that in my desire to play a solicitous host, I should not impose Carnatic music on his irresponsive ears or offer spiced South Indian dishes to his uncompromising European palate!

On January 10, we heard the Tashkent declaration of peace by Prime Minister Lal Bahadur Shastri and Ayub Khan, a triumph for Soviet diplomacy. But alas our Prime Minister did not live to watch its success, for he died in Tashkent within hours of signing the treaty. I wrote on that occasion:

"It was only sixteen months ago that the mantle of Nehru, the apostle of a resurgent and re-awakened Asia, fell on him. He bore it with courage and confidence and under his leadership, the country emerged with hope and success through one of the greatest ordeals it had ever faced in its almost ageless history. Conscious of his responsibilities to a nation threatened by the invasion of its time-honoured frontiers, he displayed great magnanimity in agreeing to discuss his country's problems at the instance of the Soviet Government. He has left us the manner of a 'Krithakarya' with the message that it is now time to 'Unarm, for the long day's work is done.' Thereby he has earned the gratitude of thousands of millions of our country, who can now look forward to the satisfactions of safety assured, of peace restored, honour preserved, of the comforts of fruitful industry, of the home-coming of the soldiers, of the smiles of their wives and children. With all these will be mingled the ache for him who, in his hour of greatest triumph, could not come home."

Wednesday, the 19th of January was a momentous day in Indian history. Mrs. Indira Gandhi was elected the leader of the Congress Party and the Prime Minister of India.

On the 24th January came the news that Dr. Bhabha was a victim of an air-crash of Air India over Mont Blanc.[3] His death ended an era in the scientific life of our country. I wrote about my great teacher, the following:

"He was born to be a theoretical physicist of world-wide reputation. He sought and grasped opportunities with undiminished vigour and uninterrupted success, characteristic of a man destined to fame and fortune. A true product of the triumphant era of modern science, his early association was with the great masters — Pauli in Zurich, Dirac in Cambridge, Bohr at Copenhagen and Kramers in Holland. He entered the Indian scientific scene in his early thirties, fresh from his laurels after the formulation of the cascade theory and the award of the Fellowship of the Royal Society. Every scientist knew that something miraculous was going to happen from the small beginnings of the cosmic ray research unit under him in the Indian Institute of Science. They had not to wait too long to see the gigantic edifice of the Tata Institute of Fundamental Research rise, overlooking the lapping waters of the Arabian sea. What is more, this ancient land where methods of agriculture have not changed since the dawn of civilisation, under his leadership joined the worldwide effort for harnessing atomic energy for peaceful purposes. This revolution in thought and in deed, is bound to alter the course of the economic development of this country of teeming millions. This movement has the specific consequence that the pursuit of science, instead of being confined to the ivory tower, has now become a profession, the achievements of which will affect the daily lives of our people.

The legacy of Bhabha is the desire for excellence in the Mathematical and Physical Sciences, and more generally in the fascinating endeavour of understanding nature."

In February, Professor Walter Hayman, the eminent mathematician from Imperial College, London, visited us and initiated research in pure mathematics at Matscience.[4] He gave a series of lectures on "An introduction to Nevanlinna theory," and we included the text of his talks in Vol. 6 of the Plenum series.

On February 4th, the Soviets announced the first unmanned soft-landing on the moon by Lunar 9, an achievement followed by a similar one on June 2 by U.S. Surveyor 1. The race for triumphs in space was now on, leading to 'the giant leap for mankind' on Apollo mission 11, three years later.

There was a conference on Cosmology and Gravitation at the National Physical Laboratory in Delhi. After participating in it, I met the Prime Minister Mrs. Indira Gandhi, the Director-General of CSIR, the Secretary of the Planning Commission, and the Education Minister, hoping to augment Central aid for our Institute.

In March, I lectured at the Indian Institute of Science in Bangalore, on my new work on Feynman graphs. Burton Moyer of Berkeley visited our Institute, and in long conversations we compared the place of pure science in India and the United States.

On April 6, we had a pleasant surprise when Mr. Wermer, a photographer for the CBS TV Network of America, came to the Institute to take pictures! The next day, he and his crew took pictures of Ekamra Nivas.

I completed my work on energy denominators and established the detailed correspondence between Feynman and field theoretical formalisms. I completed my paper on new topological features of Feynman graphs and the CPT in Feynman formalism. It was like sticking a plume on the Feynman crest, for it was hitherto believed that the Feynman formalism was so complete that nothing worth mentioning could be added. But I was convinced that any work, which was new and correct, will sooner or later enforce attention like the harmony of chaste music.

On the 5th of June, the rupee was devalued and the dollar was set equal to Rs. 7.5. The possible consequences of devaluation have been studied all over the world by Harvard economists, Indian politicians, British administrators, Swiss bankers and Japanese businessmen. The actual outcome depends on the time and extent of the devaluation, and particular conditions prevailing in the country. Devaluation results in a sudden increase in debt to foreign countries, but also enhances the possibility of export since goods are available at cheaper rates to foreign buyers. The beneficial effects depend essentially on how much advantage the country can take from exporting its goods in spite of the increased cost of imports. Anyway, the psychological effect in 1966 was disastrous. Desire for foreign goods became a craze, smuggling increased to rival normal channels of trade, black marketing and black currency created a parallel economy, with transactions outside banking operations and beyond the reach of law and taxation.

The brain drain from the country increased. The lure for the dollar and the lack of faith in Indian money, compelled talented scientists to migrate to happier climes. Since foreign travel became tighter and official delegations made more demands on the exchequer, this was just the beginning of the decline, which in three decades brought down the rupee to as low a value of two cents, that is fifty rupees per dollar!

Krishna's upanayanam[5] was performed on the 24th June. It was one of the happiest days in my life since the ceremony was done in true traditional style, just as my own, when my father 'initiated' me in 1938. I recalled to Krishna how my father explained to me that the sacred thread represented the timeless traditions and wisdom of our ancient land of hallowed saints and rivers. It had always to be worn closest to the heart for it carries the message, the 'Upadesam,'[2] the most valuable inheritance from father to son, is the 'Gayathri manthra' invokes the blessings of God for the true wealth and happiness of enlightenment through knowledge and awareness.

In response to an invitation from Sudarshan to visit the University of Syracuse, I planned a trip to the United States.

We left for the U.S. by the Air India on the night of Friday, the 8th of July. We arrived at New York seven hours late, and were received by my former student Viswanathan, now a prosperous New Yorker. It was a most gracious gesture, in true Hindu tradition, for my 'chela' Viswanathan to vacate his apartment for his 'Guru' and stay in a neighbouring hotel. From New York, we moved to Syracuse by the Mohawk Airlines and stayed five weeks there when the Sudarshans played gracious hosts. We made an interesting trip to the Cornell University, Ithaca and had a picnic on the shore of Lake Cayuga. I gave my lectures at Syracuse on perturbation theory and also on stochastic processes.

Marshak invited me to Rochester and took us on a glorious trip to the Niagara. We lunched on the Sheraton Brock and went up the Skylon tower. We left Syracuse by car on the 4th of August for Washington, where I lectured at the Naval Research Laboratory. I enjoyed being guided by my 'co-pilot' Krishna as I drove along the Beltway to reach Maurice Shapiro's house for a gracious party arranged by him for us to meet his friends.

From Washington we drove to Princeton to visit Robert Goheen,[6] the President of the Princeton University, who invited us for dinner at the Nassau Inn. We returned to Syracuse after seeing the lovely Catskill game farm enroute.

After a memorable stay in Syracuse, we left for San Francisco by American Airlines and were received by Professor Schiff, who arranged our stay at the comfortable Tan Plaza apartments. Schiff invited me to spend three weeks at Stanford University, where I gave several lectures on my research. We celebrated our wedding anniversary by going to San Francisco along the sky-line route and having a dinner in an Indian restaurant with the Schiffs.

During our stay in Stanford, we attended the 'Rochester conference' at Berkeley and the conference banquet at the Claremont hotel. There was also a dinner at the home of Nobel Laureate Emilio Segre, to which we were invited. It was during the banquet that I realised how the dashing ambitious youth of America are annoyed at leisurely speeches. When Marshak was recalling in nostalgic spirit the birth of the Rochester conferences, the audience expressed their impatience by clapping hands before the completion of sentences. Marshak with sparkling wit took the sting out of their disapproval by just remarking with a smile 'I get it,' sending the audience into peals of laughter!

We went on an excursion by car with Sudarshan to the Yosemite Valley, the pride of California. The drive through the valley was exhilarating — the balmy winds laden with the scent of fragrant pines and cedars, verdurous valleys and stately mountains. At night, there was the awesome spectacle of the 'firefall,' when hundreds of fiery logs were thrown over a cliff and plunged more than a thousand feet into the valley below. Dawn broke, with splashes of light over the mountain tops spraying the valley with a roseate hue, wakening the wild buds and flowers wet with dew.

We left for Los Angeles by United DC-8 and were received by my good friend Sridhar (an associate professor in electrical engineering) of Caltech, who entertained us frequently at Mexican restaurants. We visited Dick Bellman, who had moved from RAND to the University of Southern California, where he invited me to give a lecture. We left for Honolulu by Pan American 707 on the 13th of September and spent a most delightful holiday there. From Honolulu, we took the the JAL DC-8 to Hong Kong via Tokyo. The airline graciously accommodated us for a night each at the stately Prince Hotel in Tokyo and the spacious Hotel Miramare in Hong Kong. From Hong Kong to Singapore we flew by Cathay Pacific and spent a day there. We were back in Madras by Air India on September 21.

Within a week of my return from the US, I left for Bangalore for the Matscience Summer School in Mathematical Sciences (September 27–October 2). The conference featured a series of lectures on multiple scattering processes by F. Pham of Saclay and CERN, and on superconductivity by G. Rickayzen of the University of Kent in Canterbury. The lectures of this summer school appeared in Volume 7 of the Plenum series.

In early October, M. S. Subbulakshmi and her husband T. Sadasivam came to Ekamra Nivas to get some addresses and contacts in the United States before their departure to America in connection with concert of MS in the United Nations on October 23.[7] In particular, I asked to them contact my former student V. K. Viswanathan (VKV as he was known in music circles) in New York, and he was of immense help to them.

The next few weeks in Madras I spent in 'enlightened leisure' thinking of basic problems in elementary particle physics, since all my students had completed their doctorate degrees and the new group for the PhD had not yet arrived. This period of idleness turned out to be very significant. At first, it looked like a waste of time to think of the old and unsolved problem of the mathematical transition, from Pauli to Dirac matrices. These questions, though unanswered, had become 'stale as a garment out of fashion' but I was intrigued by the peculiar placement of the Pauli matrices in the Dirac structure. I made a hundred attempts to solve the mystery and mused as I heard the enchanting concerts at the Madras Music Academy in December, arriving at 'false' solutions which teased my mind and confused my thought. I could not anticipate that the dawn of the new year was to bring in a sudden revelation, which dispelled the mystery in a single flash!

Notes

1) Appropriate to the occasion, Sir C.P. gave a magnificent talk on the role of the Hindus in the discovery of zero. The decimal system of notation invented by the Hindus made crucial use of the concept of zero (*poojya* in Sanskrit) that is due to the Hindus.

2) Gunnar Kallen (1926–1968) was a world famous Swedish physicist, who did fundamental, work in quantum field theory and elementary particle physics. He worked for several years at the Bohr Institute in Copenhagen. He tragically died in a plane crash, while piloting his own plane to CERN near Geneva.

3) The Air India aircraft that crashed was a Boeing 707 named *Kanchenjunga*; all Air India Boeing 707s were named after Himalayan peaks. Ironically, this aircraft crashed just a few hundred metres away from the 1950 crash of the Air India Lockheed Constellation called *The Malabar Princess*!

4) Walter K. Hayman, FRS (born in 1926) is a world renowned mathematician, who has done fundamental work in the area of analysis. He was appointed as the first Professor of Pure Mathematics at Imperial College in 1956, where he ran an outstanding program in complex analysis for 30 years. At MATSCIENCE, he gave a series of lectures focussing on Nevalinna theory in complex analysis, a report of which appeared in Vol. 6

of the Symposia on Theoretical Physics and Mathematics (Plenum), which was edited by Ramakrishnan. It was at his suggestion that MATSCIENCE began a program in pure mathematics, with the appointment of Professor K. R. Unni.

5) The *Upanayanam*, or the sacred thread ceremony, is for Brahmin boys for their initiation into the world of learning. During the Upanayanam, the father (the first Guru), whispers the *Gayathri* mantra into the ears of his young son. The boy is given a sacred thread that he wears for the rest of his life. In ancient times, after the Upanayanam, the boy would go to the house of a Guru and spend several years there learning the vedas.

6) Robert Goheen (1919–2008) was an assistant professor at Princeton University from 1950 to 1957, when he was appointed as the University President! At the age of 37, he was one of the youngest ever to achieve the Presidentship of a US University. Goheen was born in South India, and attended one of the best high schools in Kodaikanal, in the state of Madras. Later, under the Jimmy Carter Presidency, he served as US Ambassador to India. Goheen had a deep understanding of India, both culturally and economically, and Ramakrishnan and he were able to discuss several aspects of the development of science in India.

7) M. S. Subbulakshmi, one of the greatest Carnatic vocalists, was honoured with the invitation to sing at the UN General Assembly. In connection with the UN concert of MS, Ramakrishnan and Lalitha hosted a dinner at Ekamra Nivas.

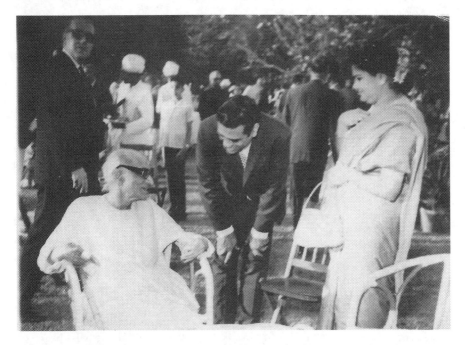

Alladi and Lalitha Ramakrishnan talking to C. Rajagopalachari (Rajaji) at Raj Bhavan Madras (Jan, 1966). The Ramakrishnans use to attend the Republic Day party at Raj Bhavan every year.

Chapter 28

Round The World
On *L*-Matrix Theory (1967)

Alladi Ramakrishnan with Prime Minister Indira Gandhi and Madras Chief Minister Bhakthavatsalam at MATSCIENCE (Jan 1967).

Eleven year old Krishna presenting a bouquet to Prime Minister Indira Gandhi (MATSCINCE, Jan 1967).

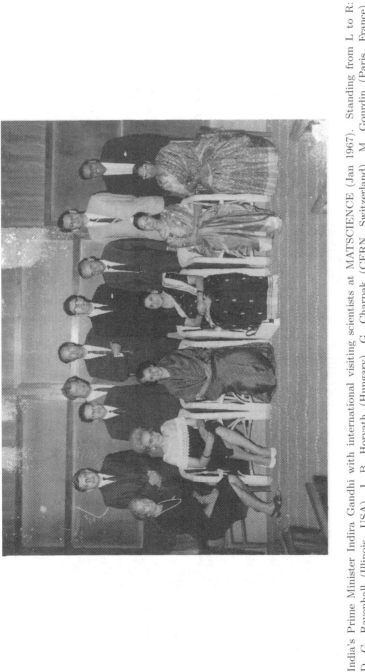

India's Prime Minister Indira Gandhi with international visiting scientists at MATSCIENCE (Jan 1967). Standing from L to R: D. G. Ravenhall (Illinois, USA), J. B. Horvath (Hungary), G. Charpak (CERN, Switzerland), M. Gourdin (Paris, France), J. H. Williamson (York, UK), N. Dallaporta (Padua, Italy), M. Moshinsky (UNAM, Mexico), Alladi Ramakrishnan. Seated L to R: Mrs. Dallaporta, Mrs. Moshinsky, Indira Gandhi, Mrs. Ramakrishnan, Mrs. Gourdin, and Sarojini Varadappan.

The year 1967 turned out to be very fruitful and eventful in my scientific life. But on New Year's day, there was nothing to indicate that it was going to be so. The year started in the usual festive mood. The anniversary symposium was inaugurated by our Chairman R. Venkataraman, with an array of visiting scientists from abroad — Michel Gourdin from France, Ravenhall from the USA, Charpak from CERN, Dallaporta from Italy, Horvath from Hungary, Williamson from England, Mercier from Switzerland, Joachim from Belgium, Eliezer from Malaya and Moshinsky from Mexico. Dallaporta gave the inaugural address on the theme "Fundamental problems on quasars." Andre Mercier, well-known for his study of the philosophical foundations of modern physics, critically examined various attempts, such as of Einstein, to formulate unified field theory. (The talks of this anniversary symposium were published in Volume 8 of the Plenum series.)

For the past few months, I was making frantic and futile attempts at understanding the transition from Pauli to Dirac matrices, which had eluded Pauli himself! What intrigued me was the position of the Pauli matrices within the Dirac structure, which made me ask whether there was a natural way of understanding the 'architecture' of such anti-commuting matrices?

It is hard for anyone, particularly a scientist, to believe that a new idea could occur in a dream. But this did happen on the morning of the 4th of January, when I dreamt that the Dirac matrices can be obtained from the two dimensional Pauli matrices by a particular operation, which I named the 'sigma' operation in the dream itself! This occurred at about four in the morning, and I jumped out of bed to write out on a piece of paper the results of the revelation lest I should forget the details![1] Within a few hours, I found that the sigma operation was perfectly valid and indeed was not explicitly mentioned in the literature. With great enthusiasm, I talked about this in a lecture at the symposium the next day (Jan 5).

What a pleasant surprise, when on the following day I had a phone call that the Prime Minister, who had just arrived in Madras, was willing to visit our Institute at 9:30 p.m.! The meeting was arranged by C. Subramaniam, when he was travelling with the Prime Minister, with his characteristic solicitude for, and abiding interest in, our Institute. It was indeed a privilege to receive the Chief Minister Bhakthavatsalam, along with the Prime Minister and C. Subramaniam. Our visiting scientists[2] were excited to see a woman Prime Minister, powerful beyond ambition in a land where it was generally believed by the Western world that a woman had a subordinate, almost servile place in society and had to seek redemption only through Anglo-Saxon laws.

On the 8th, we had a music performance for the symposium by the violinist T. N. Krishnan, which was enjoyed with enthusiasm by the distinguished visitors. On the 11th, there was a dinner at Ekamra Nivas in honour of Professor J. H. Williamson and the Nobel prizeman Linus Pauling, who was in Madras as Sir C. V. Raman Professor at the University.

It was a pleasant circumstance that the cricket Test match between the West Indies and India was played in Madras during the Pongal holidays. The match ended in a draw, but we saw some splendid cricket and had the opportunity to watch the West Indian legend Gary Sobers in action.

The lectures of some of the visiting scientists continued and I took them to the Republic Day reception at the Raj Bhavan on the 26th January. We had an interesting experience with the mathematician Harold S. Shapiro from Ann Arbor.[3] He was a vegetarian with the rational conviction that raising cattle is an expensive and cruel way of converting inedible grass into nutrient protein and human ingenuity should really be capable of devising more effective and humane means of doing so. He preferred to stay at a Western style hotel but had his lunch and dinner at the Woodlands or even in the smaller restaurants at Luz Corner. I took him to the Thyagaraja Aradhana festival, when he followed the *unchavrithi* (begging bowl) procession[4] and enjoyed listening to the Pancharathna krithis.

In February, I received a kind invitation from Salam to visit Trieste, and I started planning a trip round the world since I had many invitations to lecture on my new work, which I called the L-matrix theory.

The third general election in India was held during March, and there was great excitement and tension in our country. As expected, the Congress won at the centre but it was a bolt from the blue when it suffered total defeat in Madras, after its uninterrupted success since the dawn of freedom. The D.M.K., a new party, swept the polls, while Indira Gandhi continued as the Prime Minister of India.

At Matscience, we had an excitement in our work when I saw a paper by Steven Weinberg, in which he expressed that an adequate comparison had not been made between Feynman formalism and old fashioned field theory. He was unaware of our papers on this subject, though they were published as a series in the well-known Journal of Mathematical Analysis and Applications of the Academic Press (see Appendices, Part II). It only shows how in this age when communication is fast and complete, the proliferation of journals makes it impossible for a person to be informed of all the developments in his own field! I also wrote a paper on a new form of the Feynman propagator and published it later in the new IIT Journal.

We had lectures by Professor Gaier, a mathematician from Geissen, a perfect example of German efficiency and thoroughness. In March we had as a visitor, a most pleasant mannered Russian, Professor Novoshilov, a perfect gentleman in the most English sense of term. I discussed with him the algebra of L-matrices, when I discovered that while it is conventionally known that spin is imbedded in the relativistic Dirac equation, it is even more natural to think of helicity as being accommodated along with energy. What is more, helicity and energy are similar dichotomous quantum numbers. In the pleasure of the discovery, I wanted to express my admiration for my great teacher Bhabha, a votary of matrix theory, by talking about his contributions on Rama Navami day, the anniversary of the Theoretical Physics seminar.

I decided to leave for the USA with Lalitha and Krishna in early June, and before our departure I finalised my plans for the 'Alladi House' to be built next to Ekamra Nivas, and asked my architect Narayana Rao and the contractors Subramaniam & Co. to go ahead with the construction.

We left for Rome by Air India, and on arrival we were received by Calogero, who arranged my seminar at the University. After a day of sight seeing in the Eternal City, we flew to Trieste, where we were to spend a month at the International Centre of Theoretical Physics (ICTP). We enjoyed our stay there, spending most of our leisure on boat trips to Sistiania, walks in the Miramare gardens and the petite streets of Trieste. I gave two lectures at the ICTP, one on the hierarchy of matrices and the other on new topological features of Feynman graphs. We made a weekend trip to Padua, where Dallaporta played a gracious host. We stayed at the luxurious Hotel Plaza and I lectured at the ancient University before a select audience, which included Professor Fry of Wisconsin.

On the 10th of July, we received a letter from C. Subramaniam, who was in Europe, that he would be visiting Trieste in response to my request. Professor Salam, with characteristic generosity, informed me that he would receive C. Subramaniam as a distinguished guest of the Institute. We received C. Subramaniam at the Ronchi airport and took him straight to the comfortable Jolly Hotel. Subramaniam visited the ICTP and had a meeting with Salam, who placed at our disposal his own car, a lovely Peugeot, and his chauffeur, the handsome Joe, to take CS around on excursions. The next day we took CS to the fabulous Postagna caves and had a picnic in the Yugoslav countryside. We made short trips to Sistiania by boat, watching the sunsets on the Adriatic and shopping for souvenirs in Grado. We drove to Venice, where we watched Gondolas gliding through the Bridge of Sighs, looking for a breaded Shylock and a love lorn Bassanio on the famed Rialto, feeding the pigeons on St. Mark's square beside the palace of Doges, before seeing CS off at the Marco Polo airport in Venice!

After a fruitful month in Trieste, we left for Geneva, where we were received by my friends, Maglic and Jacob. I gave a seminar at the University of Geneva at the kind invitation of Professors Jauch and Misra. We made a six hour boat trip on the lake of Geneva, admiring the tranquil beauty of the liquid crystal of Europe, the envy of less endowed lands. We made a trip to the countryside of Geneva, of such exquisite beauty, which only the happy land of Switzerland could offer. We had lunch with Professor Weisskopf, the Director General of CERN, and tea with Hagedorn before leaving for Paris.

I gave a seminar at Saclay, which was followed by a lavish lunch with Bloch and Dominicis and their colleagues. We had dinner with Malcolm Adiseshiah, who was then the Deputy Director General of UNESCO, before we left for London. After a few hours in London, when Krishna got his first glimpse of the Buckingham Palace and Nelson's Column at Trafalgar Square, we took the BOAC VC-10 for Montreal, where we were received by Dr. Lee of Air Canada at the Dorval airport. In Montreal, I gave a seminar at the Air Canada Research Centre at his kind invitation and enjoyed the hospitality of my Indian friend K. Srinivasacharyulu of the University of Montreal. We stayed at the stately Sheraton Mt. Royal Hotel and saw the Expo 67, a magnificent spectacle of the achievements of human ingenuity and a demonstration of international cooperation and goodwill. One wonders why mankind should be smitten so often by the scourge of war and torn by racial hatreds and self-imposed conflicts, when there was so much to share from the fruits of science and technology.

We left Montreal for New York, where I spent some time with Mr. Earl Coleman, the publisher of the Plenum Press, discussing the contents of the volumes of the Matscience symposia. I gave a seminar at the Bell Telephone Laboratories at the instance of Ram Gnandesikan. We left for Washington, where we stayed with V. K. Balasubramaniam, a cosmic ray physicist at NASA, and the elder brother of my former student V. K. Viswanathan. I gave a seminar at the Naval Research Laboratory, and our gracious host Maurice Shapiro regaled us at lunch and dinner.

From Washington we flew to Denver by the United Airlines flight, and drove to Boulder to participate in a summer seminar, where to our pleasant surprise we met our good friends the Grubers. They took us on a lovely excursion to Bear Like, and I admired how Gruber drove in an expert manner, though he had lost the use of one hand since childhood! The Indian meteorologist Kotiswaran was in Boulder, and he took us on a two hundred mile motor trip through the Rocky Mountain National Park.

From Denver we flew to Chicago by TWA to be greeted by Keith Symon, who took us to his lovely home at Madison. August 21 was the day of our wedding anniversary, and one of my old students Devanathan, who was in Madison, invited us for dinner to his apartment. I lectured at Madison, and one afternoon Symon took us to his farm, where Krishna could not control his excitement riding on a tractor! Professor Symon drove us from Madison to Chicago, where we visited a group of scientists working on the plans for the national accelerator to be built at Batavia under the directorship of Robert Wilson. The project was later completed ahead of schedule — a supreme example of the American spirit of resolute pragmatism and restless initiative. 'The job is done, let us get on with the next.' Professor Ravenhall took us from Chicago to Urbana, where I gave a seminar. We left for Lafayette, where I lectured at Purdue University at the kind invitation of Professor Abhyankar, the eminent Indian mathematician. The Abhyankars drove us to Chicago from Lafayette from where we went to Rochester by the to attend the international conference on particles and fields.

The conference opened with the magnificent lectures of Feynman and Schwinger, the architects of quantum electro dynamics, now on the track of new ideas to meet new challenges in modem physics. But the greatest of men cannot reproduce or repeat their greatest performances. Perhaps the explanation lies in that all creative work has to be spontaneous, but famous men so consciously attempt to match their previous work, that the spontaneity is inevitably lost. Except for Newton or Einstein, there are few examples where famous men have put forward two equally brilliant ideas. The conference was well-organised in the Marshak tradition, and the weather in Rochester was heavenly in late summer.

Kotra Krishnamurthy met us at Rochester and drove us to Buffalo. We had a wonderful day at the Niagara, watching the roaring waters in their restless task of rolling down the frontier between United States and Canada. We drove to Toronto along the highway skirting the Niagara, and I lectured at the University there at the invitation of Professor Pugh. We left Tomoto for Chicago by an Air Canada flight, and then by North Central to Milwaukee, where I was the guest of Professor Umezawa who treated us to lavish hospitality.

After my lecture at Milwaukee, we left for Los Angeles by a direct United Airlines flight and Sridhar received us at the airport and took us to his comfortable apartment in Pasadena. We spent a whole day at Disney-land, as guests of the Douglas Aircraft Corporation, arranged by Dr. Dave Pandres, the Director of the Research Laboratory at Douglas. He is dis-armingly frank in speech, touchingly sincere in action and embarrassingly generous in hospitality. I gave a lecture at the Douglas Research Labora-tory the next day, and in the evening we were regaled at the 'Islander' on Restaurant Row by our generous host Sridhar.

We left for San Francisco by United Airlines and drove to the Hotel Durant. I visited the 'RADLAB' up the hill[5] and had dinner with Gerson Goldhaber. I lectured at the Stanford University and drove back to Berkeley in time for dinner at Professor Moyer's.[6] It was indeed a privilege for me to lecture at Berkeley, at the invitation of Professor K. M. Watson,[7] before the entire staff of the 'RADLAB' and it was flattering to see Nobel prize winners, Owen Chamberlain and Emilio Segre[8] in the audience. We dined with (Nobel Laureate) Professor Glaser in his magnificent house, when he recalled his visit to Madras and Ekamra Nivas.

We left by United Airlines for Vancouver, the most English of the Cana-dian cities though farthest from the mother country, and stayed at the luxurious St. Georges hotel. I lectured to the Simon Fraser University at the instance of an Indian physicist K. S. Viswanathan. We left for Seattle by United Airlines flight and stayed at the magnificent Olympic hotel. I gave a lecture at the Boeing Research Laboratory at the invitation of Dr. Drum-mond, a plasma physicist. We dined at an Italian restaurant, when to my pleasant surprise I met George Marsaglia, who was my colleague at Manchester twenty years ago! I lectured at the Washington University at the invitation of Professor Geballe and we made an afternoon trip to the Boeing plant, where there was 'mock up' of the Boeing supersonic aircraft of the future. Later, it turned out that America abandoned the supersonic programme while Europe pursued the Concorde.

From Seattle we had to get back quickly to Madras because I had to perform my father's annual ceremony.[9] So we took Pan American on a diagonal flight across the Pacific, from Seattle to Singapore. We left for Honolulu by a Pan Am Boeing 707 and stayed at the exotic Hotel Wakikian, and the next day flew through Guam, Manila, and Saigon to Singapore. Landing in Guam reminded me of the fateful months of 1942, when we heard with bated breath the news of Midway, which turned the tide of war in the Pacific. At Manila, I was reminded of Bataan and Corrigedor and Douglas MacArthur's heroic defence of the Phillipines. Of course things had changed, and the Americans had left the Phillipines but freedom had not brought democratic ideals to the nation of a thousand isles. Landing in Saigon, we felt as if we were at the mouth of a volcano, for the Vietnam war was on, though America was rapidly phasing out its troops.

We arrived in Singapore and slept continuously for twenty hours to remove the jet lag and set the biological clock in unison with the solar time! After a whole day in Singapore, in the company of my good friend Ramanathan, we left by Air India for Madras.

On return from America, I completed a paper on a comparison between *L*-matrix theory and the Cartan spinors. I was always intrigued by the transformation properties of spinors and it was hard to believe that there was a close correspondence between the eigenvectors of *L*-matrices and the Cartan spinors. I lectured on my work at various institutions in Madras, like the IIT and the A.C. College (of the Madras University). It was then that I succeeded in collaboration with my colleagues, in extending Clifford conditions in higher powers and formulating the Generalised Clifford Algebra (G.C.A).

On the 16th November, we had a strange visitor, a captain of a Japanese ship, friend of the physicist Kamefuchi, who had told him about the Margosa tree[10] at Ekamra Nivas! He wished to take photographs of the tropical trees in our garden. Like a prophet, who is not honoured in his own country, beautiful things in our backyard seem to escape our attention.

On the 7th of December, I was able to complete a paper entitled 'The higher dimensional Dirac Hamiltonian,' which yielded in addition to helicity and energy, another quantum number identified as chirality. Later in December, the Matscience seminar on mathematical analysis was inaugurated by our Chairman Nedunchezhiyan (Education Minister), with Professors Fuchs from Cornell and Rubel from Urbana as invited lecturers.

Notes

1) The Indian mathematical genius Srinivasa Ramanujan (1887–1920) often used to get up in the middle of the night to record mathematical formulae he would get in his dreams. The story is that the Hindu Goddess Namagiri would come in Ramanujan's dreams and give him incredible formulae. When he got a formula in his dreams, he would get up and write it on the slate using a chalk, lest he would forget it in the morning! He would later record his results in notebooks that he maintained.

2) The visiting scientists who met Prime Minister Indira Gandhi at MATSCIENCE were J. H. Williamson (York, England), Michel Gourdin (Orsay, France), N. Dallaporta (Padua, Italy), G. Charpak (CERN, Switzerland), M. Moshinsky (UNAM, Mexico), D. G. Ravenhall (Illinois, USA), and J. B. Horvath (Hungary). Charpak later won the 1992 Nobel Prize for his invention and development of particle detectors, in particular the multivariate proportional chamber. Gourdin, Dallaporta, and Moshinsky were accompanied by their spouses.

3) Harold S. Shapiro (born 1928) is an analyst of world repute, best known for inventing the Rudin-Shapiro polynomials. He was at the University of Michigan in Ann Arbor, from 1963 to 72, which is when he visited MATSCIENCE.

4) Saint Thyagaraja, who lived in the 18th century, was the greatest of the Carnatic music composers. Among his more than one thousand beautiful compositions, there are five called the *Pancharatnas*, which in Sanskrit means five gems, that are sung at the music festivals, which are held annually in his memory (The Thyagaraja Aradhanas). Like many other saints, he went from door to door asking for food *unchavrithi*), and so a unchavrithi procession also takes place on his birthday each year.

5) The Lawrence Radiation Laboratory (popularly known as Radlab) is located in the hills behind the campus of the University of California, Berkeley. It was founded in 1931 by Ernest Lawrence, who a few years later won the Nobel Prize in 1939 for the creation of the cyclotron, a type of particle accelerator. Radlab was renamed the Lawrence Berkeley Laboratory — the LBL — as it is known today.

6) Burton J. Moyer (1912–73) was an internationally acclaimed high energy physicist, and was also an authority on radiation safety. He was Chair of the Berkeley physics department from 1962 to 1968, and was Ramakrishnan's host at the LBL and at the physics department. He had visited MATSCIENCE earlier in 1962.

7) Kenneth M. Watson (born 1921) was a highly regarded physicist, who was on the faculty of the University of California, Berkeley from 1951 to 1981. He was Ramakrishnan's host at Berkeley in 1967 and earlier in 1962. Watson had arranged for Ramakrishnan's student R. Vasudevan to spend an extended period in 1962 as a post-doc under him in Berkeley.

8) Emilio Segre (1905–1989) was an Italian-American physicist, who was awarded the 1959 Nobel Prize for the discovery of the anti-proton, and the elements technetium and astatine. He had visited MATSCIENCE and contributed an article on his talk to the Symposia in Theoretical Physics, Vol. 2, that Ramakrishnan had edited.

9) Even though Ramakrishnan travelled extensively and frequently, unless he was on a long assignment abroad, he never missed performing the annual ceremonies (*sraddhams* in memory of his parents at Ekamra Nivas. So in 1967, he shot back to Madras via Singapore by the Pan Am flight diagonally across the Pacific; he skipped his lecture assignment in Japan, to be in time for his father's annual ceremony in Madras.

10) The Margosa or Neem tree is known for medicinal properties. The leaves, the stalks, the flowers and fruits of the tree are bitter and the juices have an antiseptic property. Neem toothpaste is considered very healthy, and neem leaves are considered good for the digestive system.

Chapter 29

USA, Trieste, Russia And
A Conference In Vienna (1968)

Alladi Ramakrishnan (in front row center) attending a conference lecture at the International Center for Theoretical Physics, Trieste (1968). - Editor

We had a stream of visitors at the dawn of the new year. The eminent statistician and my close friend Professor C. R. Rao[1] arrived from Calcutta on January 11. Professors Lee Rubel (Illinois) and Wolfgang Fuchs (Cornell) were already there for the Seminar in Analysis, while Gordon Shaw from Irvine, Roland Good from Iowa, Shreeram Abhyankar from Purdue, Okubo from Rochester and Alec Lee from Montreal, joined us to participate in the anniversary symposium, which was inaugurated on the 17th of January. Abhyankar stayed with us at Ekamra Nivas and gave a series of

357

Alladi Ramakrishnan with Nobel Laureate Hans Bethe and other physicists at Cornell University (1968). - Editor

lectures at the Institute and a talk on Ramanujan at the C. P. Foundation, when I presided over the function. In honour of the visiting scientists we had two concerts, a vocal recital by K. V. Narayanaswamy on January 13 and a violin concert by T. N. Krishnan on January 16. Alec Lee gave a series of lectures on Operational Research and particularly impressive was the one in which he described how the BEA terminal in London was designed to ensure a smooth flow of traffic.

On the 20th of January, I entertained all the visiting scientists at dinner at Ekamra Nivas. As was customary, I took them to the Republic Day reception at Raj Bhavan, where they saw the widest variety among the social elite of Madras — westernised Indians, smiling diplomats, orthodox conservatives, scientists and engineers, high officials, successful businessmen, established and aspiring artists and politicians of all shades from the extreme left to the extreme right. In early February, I went to Delhi for a meeting of the Executive Council of the National Physical Laboratory and lectured on L-matrices in the revered presence of Professor S. N. Bose. I remembered his favourite assertion that he would prefer brash originality to scholarship, even bad new work to good old stuff! I returned via Hyderabad, where I spent a day with my elder brother Kuppuswami, who had become a Judge of the Andhra High Court, and also attended the function at the Advocate's Association, when we presented my father's books to the library. I lectured at the Regional Research Laboratories in Hyderabad at the instance of my friend and classmate Thammu Achaya.

At Matscience, we had lectures by Gordon Shaw during February and he enjoyed attending Carnatic music concerts. At the end of February when Gordon left, Sudarshan arrived and we had a one day symposium on 'faster than light particles.'[2] The lectures attracted interest and were attended by a good sprinkling of the elite of Madras, anxious to understand the 'imaginary' intrusion into the hallowed domain of Einstein's work. At that time I also felt that propagators can be defined with such a mass. A few years later, I realised that a space-like interval need not be associated with the motion of a single particle with imaginary mass but can be interpreted as separating two different particles of real mass.

I visited Kalpakkam at the invitation of Mr. Siva, the local chief of the Madras Atomic Power Project (MAPP). I was wondering how such projects, which claimed highest priority on the national exchequer, had their origin in the mind of a single man, the most pragmatic visionary of India since Tata — Homi J. Bhabha, who as a creative scientist fresh from Cambridge, had the glorious madness or imaginative boldness to conceive of nuclear energy in a bullock cart country.

In March I received a handsome invitation from David Pandres of the Douglas Aircraft Company. We started planning for our trip round the world since I had received many invitations to lecture on L-matrix theory.

At the Christian College, I gave a lecture on 'The creative mind' emphasing that original ideas spring essentially in a sudden flash but against the background of prolonged contemplation.

On the 27th of March, we took delivery of the new Ambassador car MSQ 4110 and went to Tirupati, as was our general custom, before leaving for U.S. We got the P-forms sanctioned by the Reserve Bank in the last hour of the working day on 25th of April, and the next morning left for Singapore by an Air India 707. On arrival at Paya Lebar airport, we were taken to the comfortable and elegant Lady Hill Hotel at night. After spending a pleasant forenoon in Singapore, we left for Hong Kong by Malaysian Airways. At Hong Kong, we stayed at the luxurious President Hotel, where Krishna enjoyed ordering vegetarian Chinese dishes, with adult assurance. We bought a Minolta cine camera, which has captured all the beautiful sights during our travels since then. From Hong Kong we flew by JAL DC-8 to Tokyo, staying at the luxurious Kokusai Kaikan. The city had become thrice as expensive since my first visit on 1956, though the Japanese currency had become harder and stabler. We flew by JAL DC-8 to Honolulu, where we stayed with Professor T. K. Menon, an astrophysicist. At his invitation, I lectured at the University of Hawaii, which has obviously one of the most beautiful locations in the world, perhaps excelled only by Santa Barbara.

We arrived at Seattle and were received by Professor Lawrene Wilets, a soft spoken physicist at the University of Washington, who reminded me so much of Professor Schiff. Dr. Drummond of Boeing arranged a visit to the Jumbo Jet plant at Everett. We saw a mockup of the 'mansion of the air,' which was scheduled to be ready for flight early 1970. We wondered how such a large aircraft could be designed to get off the ground but indeed it did on schedule with Pan Am as the first airline to fly the 747. From Seattle, we flew by United Airlines to San Francisco, where I lectured at Stanford at the invitation of Professor Schiff, and rambled round the environs of the Golden Gate Bridge at sunset musing how man's achievement was flattered by nature's thriftless bounty.

From San Francisco, we flew to Los Angeles by United Airlines DC-8. We stayed with our friend Sridhar, who took us out for dinner arranged in honour of G. Kasthuri,[3] the Editor of the Hindu, who was on a tour of the United States. I lectured at the Douglas DC-8 factory at Long Beach. While Krishna was excited at seeing the gigantic factory, he wondered at the small number of workers for such a great enterprise.

Gordon Shaw took us to Laguna Beach, and I lectured at Irvine at his invitation. He drove to San Diego and we enjoyed the fast ride on the freeway and the visit to the famous zoo. We then left for St. Louis by TWA 727. Bruno Gruber received us and I lectured at his University. I remember his classifying the streets of St. Louis into dangerous, more dangerous, and not so dangerous for walking even during day time! He took us on a delightful excursion to the Mirarnec caverns, which reminded us of our visit to the Postagna caves in Yugoslavia. We left St. Louis by an Ozark Airlines DC-9, arriving at Milwaukee, where we stayed as usual at the colourful Milwaukee Inn. I lectured there and was treated to generous hospitality by my host Professor Umezawa. We then left for Buffalo via Chicago, where we were received by the flourishing group of Indians — P. L. Jain, Gopinath Kartha and Kotra V. Krishnamurthy. My seminar at Buffalo was well-arranged by Jain and well-attended by students and faculty. Jain drove us via Niagara to Toronto, where we took the Air Canada to Montreal. My old student Radhakrishnan was there to receive us at the Dorval airport. I lectured at the McGill University, and we spent a whole day at the fair, 'Man and his world' the surviving pageant of the famous Expo 67. We left for Boston by the Northeast Airlines and stayed with Teplitz in his lovely home. I lectured at the M.I.T. Physics Department, arranged by Teplitz and also at the Boston University at the invitation of Paul Roman.

We left for New York to be received by Viswanathan and were taken to his comfortable apartment in Manhattan. I lectured at the New York University with its active theoretical physics group under Professors Zumino and Symanzik, and at Yeshiva at the instance of Professor Lebowitz. We were taken to a play 'Cyrano de Bergerac' at the Lincoln Center, as the guests of Mr. Coleman, the President of the Plenum Press. We left for Washington where, I lectured as usual at the Naval Research Laboratory, after which our gracious host Shapiro took us to a theatre. While Lalitha and Krishna stayed on in Washington, I left for Ithaca to visit Professor Wolfgang Fuchs at Cornell. I lectured at Cornell University, and Nobel Laureate Professor Bethe graced the occasion with his presence. It was an irresistible impertinence to talk on 'New features of Feynman graphs' because Feynman and Bethe did their famous work on electro-dynamics at Cornell! I returned to Washington to join Lalitha and Krishna.

We were taken to Dulles Airport to board the Pan Am DC-8 for London. There was a slight excitement when we were told that our reservations were not recorded on the computer and therefore we were wait-listed and should take our chance after the plane arrived from Atlanta. Meanwhile on checking, it was found that we were listed under K as Krishnans since, while making the phone reservations I had over-explained my name by pronouncing Rama and Krishnan separately! Fortunately, we had our vegetarian meals reserved under the name Krishnan and we enjoyed our travel to London. After a few hours at the London airport, where Krishna eagerly watched the air traffic, we left for Milan by Alitalia, and then continued on to Trieste.

The International Center for Theoretical Physics (ICTP) had just moved into its magnificent new (and permanent) premises in Grignano outside of Trieste. So we stayed at the lovely Adriatico Palace Hotel, close to the ICTP, from where we could watch the splendid sunsets over the blue Adriatic from our balcony. I attended the International Conference on Contemporary Physics, that was held to commemorate the new buildings of the ICTP. The conference featured a generous splash of lectures by Nobel Prizemen — Crick, Dirac, Schwinger, Lee and Wigner. After that inspiring conference, we left for Milan, where we saw the exquisitely beautiful

Cathedral (The Duomo) and the famous painting 'The Last Supper' of Leonardo Da Vinci. Krishna teased me that I had to be induced by Dirac to admire the work of Leonardo Da Vinci, though he had been talking about the painting even during our earlier visits to Italy!

We flew by Alitalia DC-8 to Moscow, where we were received by a pleasant young Russian, Dr. Roysen. Our stay was very well-organised by Professor Khalatnikov[4] of the Academy of Sciences, who had visited MATSCIENCE earlier in the year. We spent a whole day seeing the exhibition of the technology of the USSR, and visiting the Kremlin. Lalitha's pink silk saree earned us precedence over all the visitors in the queue for Lenin's tomb!

At the circus in Moscow, Lalitha was treated as a specially distinguished guest and Krishna enjoyed every minute of the spectacular show, which included several Siberian tigers. We spent another day at the Moscow University, the largest single building in the world. We lunched at the Hotel Russia and waited for a hour to get our vegetarian meal but it was worth doing so for the delicious cheese blintzes with sour cream.

We left for Leningrad by a night train and visited the Hermitage, which contains some of the greatest art treasures in the world. Like Naples, it has to be seen and admired at least once within a life-time. I met Professor Gribov at the University of Leningrad, for I had heard Gell-Mann quoting him so often in his lectures.

We rambled on the coast of the Baltic sea, watching the ships in the harbour, wondering how we were right inside the Iron Curtain, which in Churchill's view has descended across the continent from Stettin in the Baltic to Trieste in the Adriatic, after the triumphant entry of Russian forces into Berlin. We returned to Moscow, and that evening watched the Bolshoi Ballet in a mammoth theatre. At the end of the show, we went directly from the theatre to the airport to catch the Air India Boeing 707 to Delhi.

Krishna had an opportunity to go round the Indian capital, and I recalled memories of my stay with my father during the great days of the Constituent Assembly and showed him the home where we had stayed — 13C, Pheroz Shah Road. We returned to Madras after a short stay with my brothers in Hyderabad.

In early July, there was a meeting of the Aeronautical Study Group and we left for Bangalore by car and met C.S. at the lovely Birla Guest house, a fantasy in a dream location outside Bangalore. We made the inevitable trip to Mysore and returned to Madras after attending the meeting in Bangalore.

It was then I worked on the meaning of parity transformation in L-matrix theory and wrote a paper entitled 'Should we revise our notions on Spin and Parity?'

I decided to go to Vienna for the High Energy Physics conference and attended a music concert by M.S. in connection with Semmangudi's sixtieth birthday celebration, later followed by a performance by the maestro himself. There was an inauguration of a local Sabha called Nadopasana, of which I became a Life Member. These Sabhas are a special feature of Madras life, for they provide first rate concerts at moderate prices. This is possible because these Sabhas are able to get front rank musicians in their 'idle' intervals between major engagements.

On August 2 occurred to me a new idea, the generalisation of Helicity matrices, the higher roots of the unit matrix, leading to the Omega Commutation as the generalisation of anti-commutation. At first it looked like a mad idea but what a thrilling surprise it was then I found that this commutation relation was meaningful and there were just three Pauli-like matrices in the lowest representation and that the Sigma operation was applicable to these matrices also, leading to higher dimensional matrices with the same commutation relation — perhaps and excellent example of luck visiting those who seek it. I lectured on my work at my Alma Mater the Presidency College, where it was always a pleasure to announce my new results.

I left Madras by Air India on the 25th July for Geneva and stayed with an old student of mine, Chidambaranathan, who had become the Secretary General of the World University Service. I flew by Austrian Airlines DC-9, and was received by Gruber at the Vienna airport and taken to the very comfortable Vienna Central Hotel. It was a very interesting conference with Gell-Mann and Weinberg as its leading lights. I had the opportunity to discuss my new work on the Generalisation Clifford Algebra with Professor Wightman, who was rather sceptical about the role of the roots of unity in the description of elementary particles.

Who would visit Vienna without attending an Opera or visiting the Schonbrunn Palace?! The city was the Royal cradle of European history, and though its power and glory have now vanished, the beauty and elegance of its palaces and gardens remain, perhaps enhanced by the absence of such treasures in more modern and affluent cities of this too commercial age.

I left Vienna by Scandinavian Airlines DC-9 for Athens, and took the QANTAS Boeing 707 at Athens for Tehran. Earth has nothing to show more glittering than Athens by night. Los Angeles is more dazzling and glamorous, but too contemporary and cannot borrow tradition and history and the treasured memory of a vanished civilisation.

I arrived at Teheran to be greeted by Professor Riahi, who took me to the Arya Mehr University. It was the time when the prestige of the Shah of Persia and his charming queen was at its highest. I lectured at the Arya Mehr University on L-matrices and left for Delhi by Air India, reaching Madras after visiting Hyderabad.

We conducted the Matscience summer school in Madras itself, when we had Professor Prigogine, the Nobel Laureate, as our visiting scientist. In his honour, we had a flute performance by N. Ramani. Prigogine gave a brilliant series of lectures, which involved advanced concepts in probability theory in statistical mechanics. It was also the time of Navarathri but we could not enjoy it as much as we normally did since Krishna was confined to bed with jaundice fever. I arranged a meeting of Professor Prigogine with our patron C.S. before he left for Bombay. By the middle of October, I completed some new work on L-matrix theory connecting the Sigma operation with Rashevsky's theorem.

In September, we had the visit of Professor Robert Rankin of the University of Glasgow, an expert on the work of the Indian mathematical genius Srinivasa Ramanujan. He gave a beautiful lecture of the Ramanujan τ function.[5]

On the 20th October, I went to Hyderabad to witness the unveiling of the portrait of my father by Hidayatullah, the Chief Justice of India. It was then that Krishna Reddy, a young friend of my elder brother, made an excellent speech summarising the life of our father.

28th of October was a significant day when we discovered the connection between the Clifford and Lie Algebra, demonstrating that a suitable linear combination of Clifford algebra matrices obeyed the Lie Algebraic relation.

In November we received a telegram from the Prime Minister Mrs. Gandhi, inviting us to attend a dinner in New Delhi in honour of Professor Chandrasekhar, who was in India as a Nehru lecturer. I took Lalitha with me to Delhi since such an occasion was unprecedented and would not occur again. Watching Indira Gandhi play the gracious hostess at dinner, I could understand why she had the confidence, the ability and the charisma to hold the destiny of millions in the palm of her hand, direct it with the stroke of her pen.

We stayed with our friend V. R. Reddy at Delhi, and after a dinner at the Asoka Hotel we called on the Chief Justice Hidayatullah, a great friend of our father. I returned to Madras after a day's stay at Hyderabad.

In November, I worked on the Generalised Gell-Mann–Nishijima relation with my colleagues and completed the paper, which was sent for publication to the JMAA. On the 25th of November, we attended a lecture by Chandrasekhar at the Ramanujan Institute, and later a pleasing Russian cultural show, easy on the eye and the ear.

There was a one day seminar at Matscience on 'Mathematics as a stimulus to physical thought,' when I presented a summary of our new work connecting Clifford and Lie algebras.

We greeted the birth of the new year, which was to become momentous in human history, ushering the era of space travel, as 1945 which initiated the atomic age.

Notes

1) Professor C. Radhakrishna Rao (born 1920) is arguably the most eminent statistician in the world! He was the Director of the Indian Statistical Institute in Calcutta for many years, and his period of directorship there intersected with Ramakrishnan's period as Director of MATSCIENCE. Ramakrishnan and Rao were very close friends and conversed in Telugu, the language of the state of Andhra Pradesh.

2) George Sudharshan was the proponent of the theory of *tachyons* — faster than light particles. He was very pleased that MATSCIENCE conducted a one day symposium on tachyons, with him as the lead speaker. Sudarshan wrote the following in the Visitors Book of MATSCIENCE:

"Appreciation of one's creativity is most cherished when it is critical; I find my visits to Madras and this Institute inspiring in the sense that it makes me strive to do my best. May I have many more opportunities to visit and let not the speed of light put a barrier on our thinking."

3) Mr. G. Kasthuri, Editor of *The Hindu* newspaper, was a classmate of Ramakrishnan at the P. S. High School. The Hindu has a great reputation for accuracy in reporting, and for not being tempted to provide sensationalism in news reporting. In the summer of 1968, Mr. Kasthuri was visiting the USA to receive an award for The Hindu.

4) Professor I. M. Khalatnikov, Ramakrishnan's host in Moscow, is one of the foremost Russian physicists. He is especially known for the BKL Conjecture, one of the greatest open problems in the theory of gravitation. Khalatnikov was very much impressed with MATSCIENCE, when he visited Madras early in the year, and so he invited Ramakrishnan to Russia under the auspices of the Russian Academy of Sciences.

5) Rankin's paper on the Ramanujan τ function and Abhyankar's talk earlier in the year on Ramanujan, both appeared in Volume 10 of the Symposia in Theoretical Physics and Mathematics (Plenum) edited by Ramakrishnan.

Shreeram Abhyankar (1930–2012), one of the most eminent Indian mathematicians of the 20th century, was a world authority in the deep and difficult field of algebraic geometry. He was known for very significant contributions to the resolution of singularities problem.

Robert Rankin (1915–2001), was a Scottish mathematician who was a leading expert in the theory of modular forms. He made significant contributions to the study of the τ function of Ramanujan, and it was on this topic that he lectured in Madras.

Chapter 30

Round The World For Me
As Man Lands On The Moon (1969)

The year 1969 was the most momentous in human history, when man set his footprints on the moon and gazed at his own planet in distant wonder; it was also an eventful year for me — a trip across the Pacific and round the world with Lalitha and Krishna!

The year started in an auspicious manner with C.S. visiting, on New Year's Day, my new house which was nearing completion next to Ekamra Nivas. When I told him that the house will be rented while I continue to stay in Ekamra Nivas, with lambent humour he remarked "Alladi, fools build houses, wise men live in them." The same day I also completed my paper on 'Kemmer Algebra and Clifford matrices.'

A few days later we had a family gathering in our village Pudur, sixty-eight miles from Madras. It is a village unchanged in temper and tempo of life for over a century, except for the sound of a few tractors and the hum of electric pumps. No ambition mocks the useful toil of its peasants, each bound to his paternal acres and pursuing his family occupation with rustic faith and genial contentment. They still reckon time by the movement of the sun in the azure sky, respond to the cock's shrill clarion to greet the dawn on their furrowed lands. Their untutored minds do not admit any thought of astronauts and lunar modules, but are more concerned with the day's labour, of scattering seeds on sun-drenched fields. As I watched the dust raised by the hoofs of the lowing herds winding home in the lingering light of the setting sun, I mused on the ceaseless shine of the automobiles on the Los Angeles freeways and the roar of the jets taking off for far off destinations. It is all one world but the contrast is so real and seems stranger than the wildest fantasy.

We returned to Madras in time for the seventh anniversary symposium on mathematical analysis. Among the visitors to Matscience were the mathematicians, Nagy from Hungary, Krysz from Poland, Lehto from Finland[1] and Riahi from Tehran.[1] The new buildings of Matscience were also completed, and on the 20th of January we had the inaugural function and the dedication ceremony. Marshal Stone gave the first lecture in the seminar hall of the new building, at a symposium which emphasised the impact of science on technology, in which representatives from industry participated. Each day of the symposium was devoted to a certain area, such as aerospace engineering, biological sciences, elementary particles, and mathematical physics. Each day, we had a leader in that area preside over the symposium. We also had in January a series of lectures by the eminent cosmologist J. V. Narlikar.

On the 7th of February, we had the Grihapravesam or 'housewarming' of the 'Alladi House.' Among those present, besides our relatives and some close family friends, were Mr. C. Subramaniam, the Ex-Chief Minister Bhakthavatsalam, and my friends in the American Consulate. Besides the religious ceremonies on such an occasion, there was also Veda Parayanam (recitation of Vedas) in the library hall of the new house by a group of Ghanapaatis (venerable Vedic scholars) led by our family priests, Nataraja and Pattabhirama Sastrigals. My father was so proud of the Vedic heritage of our family, that I was particular that the Alladi House should be sanctified by Vedic incantations.

On the 22nd of February, we left for Ooty and proceeded directly to Krishnaraja Sagar dam near Mysore, where we stayed overnight. We watched the golden hues of the sunset over the sprawling reservoir, before enjoying the paradise of illumined fountains over the well-laid gardens. My mind travelled back to my boyhood days when I accompanied my father, who was invited to an observation spot overlooking the river Cauvery by Sir M. Visweswarayya who, with vision and foresight, conceived of such a dam across the river. There was a story going round at that time that his brain was insured and was dedicated to posterity for medical study as to the source of its excellence! After our visit to Ooty — always a pleasant experience of nostalgic memories, I lectured at Mysore and later at the Indian Institute of Science, Bangalore before returning to Madras.

I decided to go to America during early summer, in response to some kind invitations. Accompanied by Krishna and Lalitha, I left by the Air India for Singapore on the 20th April. We stayed at the Lion City Hotel and romped about that fascinating city. We left for Bangkok by the Malaysian Airlines and after a four hour stay at the Asia Hotel, we went to Hong Kong by a Cathay Pacific Convair 880. After the fleeting fantasy of an overnight stay in Hong Kong, we flew to Tokyo by JAL DC-8.

We stayed at the spacious Palace Hotel in Tokyo, and Professors Takeda and Kamefuchi took us to the Meiji Shrine and gardens. I lectured at the Tokyo University on the generalisation of the Gell-Mann–Nishijima relation, and we left for Honolulu by Pan American 707. We were received in Honolulu by my friend Professor T. K. Menon, with whom we stayed. I lectured at the University of Hawaii also on the Gell-Mann–Nishijima relation, and saw the usual sights at the American haven on the Pacific. We left for Los Angeles, where we were received by Dave Pandres of the Douglas Aircraft Company and Sridhar of Caltech. I gave lectures at the Douglas Research Laboratories, at UCLA, and at the Long Beach State College. I gave three lectures at UCLA, on a new generalisation of Pauli matrices, on unitary groups and Clifford algebras, and on the Kemmer algebra and Clifford matrices. I enjoyed discussions with my UCLA host, the eminent control-theorist Professor A. V. Balakrishnan. While in Los Angeles, I met the new President of the Rand, Harry Rowen, who invited me to resume my consultancy at the Rand Corporation. I gave a seminar lecture on unitary groups and Clifford matrices at the University of California, Irvine at the invitation of Gordon Shaw, and we stayed at his lovely home in Laguna Beach. Later, I had an opportunity to visit and lecture at the Systems Development Corporation, an off-shoot of the Rand in Santa Monica.

We moved to Stanford, where I gave a seminar on the concept of integration and differentiation of random functions at the Statistics Department at the University, at the invitation of Professor Emanuel Parzen. I lectured on matrix theory at the Lockheed Research Centre in Stanford and at the University of California at Berkeley. We had a dinner with Professor Roland Good at Stanford, while Professor Schiff invited us for an evening party, in which the Nobel Laureate Hofstadter joined, having had his lunch earlier in the day in Chicago! Then we moved to Seattle, where I lectured at the Boeing Research Laboratory. The next day we saw the mock-up of the Jumbo Jet 747 at the Everett plant, wondering whether such a large aircraft could really get off the ground despite the ingenuity of designers and the skill of the technologists. Indeed it did the following year, ushering in the era of mass transportation in the air on a global scale. we had a delightful excursion to Mt. Rainier and returned just in time to catch the United Airlines DC-8 to Vancouver, from where we left by Air Canada for Edmonton.

Takahashi and Bhatia were my hosts at the University of Alberta at Edmonton, where I gave a course of three lectures on the Grammar of Dirac Matrices. It was during our stay there that we saw on television the successful trip round the moon of Apollo 10 and the safe landing of the astronauts in the Pacific. This practice flight, before the landing of Apollo 11 on the moon in July, was a triumph of American genius and organisation in science and technology, and above all the faith in competitive effort stimulated by the launching of the Russian Sputnik in 1957.

We left for Ottawa by a Canadian Pacific DC-8. As we approached Toronto, we had a frightening experience as the plane rocked and rolled in a turbulent domain and was almost struck by lightning. I lectured at the Carleton University at the invitation of Professor Puttaswamy. We had dinner with my good friend Shanmugadasan and enjoyed a picnic at the Gatineau Park with Puttaswamys. We left for Buffalo, where I lectured at the invitation of the amiable Professor P. L. Jain, who took us the next day to the Niagara falls, where we feasted our eyes watching the roaring waters.

We then moved to Washington, where I lectured at the Naval Research Laboratory and enjoyed the gracious hospitality of Maurice Shapiro. At Dayton, I was the guest of the statistician Dr. P. Krishnaiah at the Wright-Patterson Aerospace Centre. We went to New York, where I gave a seminar at the Yeshiva University at the instance of Professor Lebowitz and stayed with my former student V. K. Viswanathan, our untiring host, before going to Syracuse, where we were invited by Sudarshan. We made a trip to Cornell at the kind invitation of Professor Hans Bethe, the Nobel Laureate. I was able to induce Bethe to accept the Niels Bohr Visiting Professorship at our Institute and deliver the anniversary address in January 1970.

We left for Detroit, where we were received by Butterworth of the General Motors Research Division and Professor Garrett Birkhoff of Harvard University, a consultant there. It was a memorable day at General Motors, when my lecture on "New facets in matrix theory" was well-attended and received. They were good enough to arrange a visit to the Cadillac plant, and that made such an impression on Krishna, that after return to Madras, he wrote an article for his school magazine entitled 'The mark of excellence,' the motto of Cadillac. While in Detroit, I also participated in an International Conference on Elementary Particles by presenting my work on the generalised Gell-Mann–Nishijima relation. The conference was held at the Wayne State University, Detroit.

A visit to Montreal has been a great attraction in my travels. It has a cosmopolitan atmosphere, an attractive mixture of French, British and American influences, with native Canadian pride. At Montreal, two of my friends, Dr. Radhakrishnanan, an old student of mine, and

Dr. Srinivasacharyulu, a mathematician at the University of Montreal, where I gave a seminar, vied with each other to offer warm and friendly hospitality. From Montreal, we flew to London by a BOAC VC-IO and were met by Professor Lighthill, who had invited me as a guest of the Imperial College. His gracious hospitality and friendliness added a charming halo to his genius as a mathematician. Krishna heard with admiration the British accent of Lighthill's impeccable English, a language more universal and lasting than the Empire itself. Lighthill arranged my talk on Clifford algebras at the University of London.

From there we moved to Paris, where we were received by Dominicis and Bloch, my gracious hosts, endowed with the savoir vivre of true Parsians, 'native there and to the manner — born.' There was a magnificent lunch at Saclay after my lecture. We were invited by Dr. Malcolm Adiseshiah, then acting Director-General of the UNESCO, and his gracious wife for a delectable dinner, which was the starting point of an intimate friendship which has had a salutary effect on the growth of our Institute.

On our way to the Orly airport, we saw a mock-up of the Concord, which soon was to usher in the supersonic age in air travel. It is a tragic irony that as airplanes are becoming faster and more comfortable, the haunting fear of violence in the skies is robbing us of the pleasures of travel. Security checks have destroyed the charm of friendly farewells as passengers are searched, hustled, huddled and isolated for their own security.

We spent a week at Bern, the sweet home of Tobler, one of the pleasantest sojourns in our travels. The aroma of freshly brewed coffee with warm rolls and Swiss butter for breakfast at the impeccably clean and 'sauber' Hotel Regina, still haunts my senses. Professor Mercier, our gracious host, took us on a delightful motor trip and picnic to the countryside. After my lectures at the University, we went to Interlaken and made a boat trip on the Lake of Thun. We returned via Geneva, to the welcome heat of Madras with Swiss chocolates melting in our suitcases under the watchful gaze of the customs officials, who leave no chocolate unturned or a gift-wrap unscathed in the zealous performance of their duties.

Then occurred the most significant event in all human history — Man's landing on the moon! At Cape Kennedy, Apollo 11 was fired into space at 7 a.m. on the 16th of July, and it raced towards the moon for two days before entering the lunar orbit. The first men, Armstrong and Aldrin landed on the moon on Sunday the 22nd of July, when Astronaut Neil Armstrong announced "The Eagle has landed" with typical American humour recalling in triumphant contrast the title of the best-seller war story of an attempt on Churchill's message.

"That's one small step for a man, one giant leap for mankind."

It is amusing that in the message the article 'a' was actually either omitted by Armstrong or missed in transmission! That aphoristic statement marking mankind's most glorious moment will enlighten human history like the immortal tribute of Winston Churchill to the valiant 'few' in England's finest hour. The Astronauts safely arrived on earth on the 24th of July. We were fortunate to see the newsreel about the Lunar Odyssey at the U.S. Consulate at Madras.

August 16 was a significant day in my academic life when, as has often happened with me, suddenly before the break of dawn, the idea of the composition of helicity matrices flashed across my mind. I quickly completed a paper and sent it to the Journal of Mathematical Analysis and Applications.

On the 10th of September, the U.S. Consul General Recknagel released a volume of the Matscience symposia published by the Plenum Press. That month I met the President of India, His Excellency Mr. V. V. Giri, in Madras and he consented to inaugurate the eighth anniversary symposium.

At the end of September, we had the first Mastech (acronym for Mathematics in Science and Technology) conference in Bangalore, a unique experiment which I started, to induce a spirit of competition among the younger generation of students in our country. It was open to those who wished to present research papers, and an award of Rs. 1,000 was to be presented at the end of the conference for the best paper, as judged by the participants themselves! We had also a visitor Professor Klepikov from Russia, a most pleasant mannered and soft spoken scientist, who evinced considerable interest in *L*-matrix theory. The prize was awarded to Zargamme of Iran at a public function by Dr. Atma Ram, the Director-General of the CSIR, which sponsored the conference.

The Apollo 12 was launched on the 14th November and the lunar module Intrepid carrying the Astronauts Charles Conrad and Alan Bean landed on the moon at the 'Ocean of Storms' on Wednesday the 19th November. They stayed there for about thirty hours, while their colleague Richard Gordon orbited forty-five times round the moon in the command module Yankee Clipper waiting for 'the kids to get back from school.' Little did I realise that we would have an opportunity to receive the astronaut Alan Bean seven years later, as a distinguished guest at Matscience. On the 15th of November, I gave a lecture at the C. P. Foundation on "Planck to Gell-Mann" and noticed that the audience was electrified by the man's second lunar adventure, which had begun the previous day. Later in December, I had an opportunity to lecture again on the topic "Planck to Gell-Mann" at the National College at Trichy before an audience of over a thousand students, whose interest in science was roused more by the Apollo flights than by the systematic teaching in their colleges.

On the 24th of November, I expressed my own admiration at the achievement of the lunar landing at the U.S.I.S., where I was invited to speak. In November, I was also invited to speak on the role of mathematics in life at the conference on 'Science for the Citizen' at Delhi. There I met the President of India, to fix January 3rd for the Eighth Anniversary Symposium of MATSCIENCE that he had graciously agreed to inaugurate.

In early December there was a luncheon at our Institute in honour of Paul Morris, the Director of USIS, who was leaving India after a long record of meritorious service. He is a perfect gentleman, blessed with a smiling wife, who shares with him that delightful American quality — the desire for fresh human contacts, while strengthening old friendships.

Professor Hans Bethe arrived as the Niels Bohr Professor at Matscience, just in time for Christmas and the New Year in Madras.[2] The feast of music at the Academy and Bethe's lectures at Matscience provided a fitting finale to a year, which witnessed the giant leap of mankind into space and time, and our own small steps forward in understanding the mathematical basis of a four-dimensional world.

Notes

1) Professor Olli Lehto (born 1925) is a distinguished mathematician from Finland, who specialises in function theory. He received his PhD under the direction of Rolf Nevanlinna, a giant in the area of complex analysis. After serving as Professor at the University of Helsinki from 1961 to 1988, he rose to the position as Chancellor of the University.

Physicist F. Riahi was Ramakrishnan's host at the Arya Mehr University in Tehran in 1968. During that visit, Ramakrishnan invited him to MATSCIENCE and Riahi arrived in January 1969 for the Seventh Anniversary Symposium.

2) Hans Bethe (1906–2005) was one of the most influential physicists of the twentieth century. He received the 1967 Nobel Prize for his theory of stellar nucleosynthesis that explains how stars change due to nuclear reactions in their core. When Ramakrishnan visited Cornell University in 1968 at the invitation of mathematician Wolfgang Fuchs, Bethe attended Ramakrishnan's lecture. In 1969, Bethe himself was Ramakrishnan's host in Cornell and during that visit, Ramakrishnan invited Bethe to visit MATSCIENCE as Niels Bohr Visiting Professor.

Alladi Ramakrishnan speaking at the inauguration of the 1970 Anniversary Symposium, the first at the new permanent buildings of Matscience. On the dais from L to R: Governor Sardar Ujjal Singh, President of India V. V. Giri, Minister for Education V. R. Nedunchezhiyan, and Nobel Laureate Hans Bethe. - Editor

Alladi Ramakrishnan handing his book "L-Matrix Theory or the Grammar of Dirac Matrices" to President of India V. V. Giri for it to be released by him at Matscience on 23 December 1972. - Editor

Chapter 31

A Divine Lease Of Life (1970)

In the year 1970, I was granted a new lease of life by the grace of God after a near fatal motor accident and so I was able to make a round the world academic trip lecturing on further developments in L-matrix theory. The year started with the magnificent series of lectures on nuclear matter by Hans Bethe as the Niels Bohr visiting professor at Matscience.[1] Since the word went round that his work was closely connected with fusion and the hydrogen bomb, there was a wave of excitement and the first lecture was attended even by some judges of the Madras High Court and the elite of the city! By the time he came to the core of nuclear matter in his fourth lecture, the audience was reduced to the hard core of Matscience physicists!

The anniversary symposium was inaugurated on January 6 by his Excellency V. V. Giri, President of India, when Professor Bethe lectured on his 'Life in Physics.' Lalitha sang the invocation song Thripurasundari in Amrithavarshini, which she had learnt from the great Chembai Vaidhyanatha Bhagavatar. There was a dinner at the Raj Bhavan in honour of the Nobel Laureate, to which other visitors to Matscience were also invited. I had the privilege of lecturing on relativity in quantum mechanics in the presence of Hans Bethe and other visitors — James E. Drummond from Boeing, Dave Pandres from Douglas, Michel Gourdin from Paris and A. O. Morris from Wales. When I asked Professor Bethe whether he would like to attend a Carnatic music concert, he said he would listen to just one if it were the best. My good friend T. Sadasivam was gracious enough to arrange a private concert by his wife M. S. Subbulakshmi[2] at his house on January 12 in honour of the Nobel Laureate. In thanking the Sadasivams, I said that if Bethe's life lies in physics, M.S. and music are one and the same.

In mid-January, there was a music performance by Chembai Vaidhyanatha Bhagavatarar, a doyen of Carnatic music, at Matscience in honour of the visitors, and a public lecture by Bethe at the Ranade Hall on the rise of American physics. There was dinner at Ekamra Nivas, in honour of Bethe and other visitors, that was attended by Malcolm Adiseshah, C. Subramaniam, and H. V. R. Iyengar. Shortly before leaving for Bombay, Bethe unveiled the portrait of my teacher Professor Bhabha at the Presidency College.

The Mastech Conference was held for one week (Jan 13–20) at MATSCIENCE and Professor Tom Kailath[3] of Stanford University, who was Madras won the award for the best paper, which he received from the Governor of Madras.

On the 21st of January, we left for Bangalore by car for the Matscience conference on 'Elementary particles' held at the auditorium at the National Aeronautical Laboratory, with Drummond from Boeing, Pandres from Douglas, and Morris as the principal lectures. After the conference, we left for Mysore on January 25 and it was then that a terrible accident occurred, when our car swayed off the road and hit a tree a few miles before Ram Nagar and I was thrown out unconscious (story told by others). Lalitha and Krishna were unhurt but were dazed to see me unconscious, their only comforts being that our loyal staff, Govindan and Bakkayya, were there. Fortunately, they were able to stop a passing car and it was a fortuitous circumstance that Dr. Verma, an eminent neuro-surgeon was travelling in it. He suggested that I should immediately be admitted at the Kalappa Nursing Home in Bangalore. He assured Lalitha that he would return from Mysore to attend on me later in the evening. Govindan arranged a taxi from Ram Nagar, and I was told that all through the journey to Bangalore, I was asserting in a semi-conscious state that I was well and there was nothing to worry about! I was admitted to the Kalappa Nursing Home and for forty-eight hours I was conscious but had no memory. Lalitha and Krishna had a nerve-wrecking suspense since they could not know whether I would recover and even if I did, whether I would get back my memory unimpaired. I remember clearly how on opening my eyes I found myself in a strange Nursing Home, with Lalitha and Krishna in front of me. Somehow, I could not place the time of my life but there was a drowsy numbness and stupor to save me from the shock at the sight of 'strange surroundings.' Within three hours, after a great struggle, my memory returned and I tested myself by deriving some of the results of L-matrix theory in my mind, without a sense of sequence and slowly they ordered themselves, and I could understand what the present was! We summoned our family doctor Vinayaka Rao from Madras, whose healing presence hastened my progress.

For Krishna and Lalitha, the agony of suspense was dispelled by the ecstasy at my recovery. In two days, I was well enough to travel by air with all of them to Madras.

Soon after I returned to Madras, I found there was painful abscess in my back, which could not be controlled by anti-biotics. On consulting Dr. Mohan Rao, our family surgeon, I was advised to undergo an operation, which I did at his nursing home on the 20th of February. The convalescence was slow but sure, and I felt encouraged to plan my trip round the world in response to many invitations to lecture in American Universities and the Rand Corporation of California. During this period of recovery, for relaxation, I attended music concerts, particularly of M.S., M.L.V. and Chembai. On the 22nd March, I went to Kancheepuram to pay my respects to the older and younger Sankaracharyas[4] and receive their blessings. Returning from Kancheepuram, I was well enough to give a talk that same evening at the Hostel Day at the I.I.T., when Lalitha distributed the prizes.

On the 31st of March, the IMPACT Conference was inaugurated with Mr. A. Sivasailam, a leading industrialist, and Tamarkin,[5] a Russian Mathematician, as guest speakers since the object was to bring industry and science together. Tamarkin gave a series of three lectures at Matscience. At the technical session of the conference, I spoke on 'Cayley, Hamilton and Clifford in the light of modern physics.'

On the 10th April, I gave a talk at the USIS at the opening of the exhibition of Apollo 13. That Apollo mission was unsuccessful due to an accident in space, but the Astronauts returned safely — as much a miracle as the successful landing on the moon!

We planned to travel via the Pacific, attending the Expo 70 at Osaka and reaching Los Angeles via Honolulu, and returning to India via Europe. The 19th of April turned out to be a day of shock and anxiety, when Lalitha casually consulted a lady doctor and was told that she should undergo an operation immediately, though for a benign condition. We consulted Sir A. L. Mudaliar[6] and he confirmed the diagnosis and advised the surgery to be done in the United States to avoid complications of blood transfusion. He also advised us to go to New York directly, and so we now reversed our route and made reservations by the Air India flight via Europe.[7]

We left for New York from Bombay by an Air India 707 on the night of Saturday, the 25th of April. Lalitha would have normally enjoyed such travel, but this time she was oblivious of the journey, all the time thinking of the forthcoming operation. The flight took us through Kuwait, Rome, Zurich, London and at JFK we were received by Viswanathan, who drove us to his comfortable home in New City. Our friends the Jayaramans fixed an appointment with Dr. Morton Beer[8] at the Morristown Memorial Hospital in Morristown, New Jersey. We shifted to Jayaraman's house in Murray Hill, New Jersey, in the evening and consulted Dr. Morton Beer the next day. He confirmed the diagnosis of Sir A. L. Mudaliar and agreed to perform the operation on the 6th of May. He was a doctor of such charm, confidence and ability, that we considered ourselves fortunate to have consulted him. Before admitting Lalitha at the hospital, I lectured at Rutgers University at the invitation of Bogdan Maglic, who had now shifted to the U.S. from Europe as a professor. Lalitha was admitted to the hospital on the 5th May, and the next day the operation was performed. I prayed and read a few verses from the Ramayana, outside the operation theatre and my prayers were well-answered, when the doctor informed me that the operation was successful and there were nothing to worry about. Jayaraman and Kamala, like true friends, were spontaneous and generous in rendering us help in this period of trial by offering to have Lalitha (and Krishna) stay with them during the period of Lalitha's recovery; without them, we would not have liked to have the operation done in New Jersey.

Since Lalitha was recovering rapidly, I decided to leave on my lecture tour on the 14th of May[7] and flew by American Airlines 707 to Los Angeles from Newark airport. I was taken directly from the airport to Long Beach State College, where I gave a lecture. Gordon Shaw met me there to take me to his comfortable home at Laguna Beach. The Shaws are such friendly hosts that I enjoyed every minute of my stay with them. I visited Pasadena, where I missed my wonderful host Sridhar, who was stricken by a kidney ailment and was resting in Santa Monica, where I visited him later. I called on Dr. Quade at the Rand Corporation, and he arranged for my visit a month later as a consultant. From Los Angeles, I flew to Phoenix and gave a lecture at the Arizona State University in Tempe. From there, I left for Dallas and was received by Dr. Kobe, who had invited me to the North Texas University at Denton. I left for St. Louis immediately after my lecture at Denton, and was received at the airport by my good friend Bruno Gruber. After my lecture at St. Louis, I moved to Lafayette by Alleghany Airlines and spent a week with my old student Radhakrishnan, who had shifted from Montreal to Purdue University. It was music all day at his apartment through recorded tapes! I gave two lectures at Lafayette, one on stochastic processes at the department of electrical engineering and the other on Clifford algebras in the physics department. I made a short trip to Madison via Chicago by the Allegheny and North Central Airlines, and was taken by Keith Symon to his lovely home. There was a dinner at a Mexican restaurant and a coffee party at his house. I returned to Lafayette the next day.

A friend, Panchapakesan took me to Urbana by car and I lectured there at the invitation of Professor Ravenhall. We took the opportunity to call on the Nobel Prize winner John Bardeen who, as expected, was at work in his office perusing a current journal, but easily accessible to visitors even without appointment. It is this informal atmosphere, where high eminence enjoys confrontation with ambitious youth that has made the American Universities the breeding ground of Nobel Laureates, Field medallists and Fermi Prizemen.

While returning, we had a nerve-shattering experience when on the freeway while driving about sixty miles per hour; the automatic transmission of Panchapakesan's car broke down and the accelerator failed completely, and the car went on free-wheeling. It was dark at about 9 p.m. and it would have been a fatal disaster if the car stopped on the highway amidst the torrent of traffic. It was a fortunate chance that there was an exit just when we hoped for one and with inexplicable presence of mind I urged Panchapakesan to swerve the car out of the freeway. Actually, to stop in an unknown place in the darkness of night is as dangerous as on a highway, for we have heard of gruesome incidents of stranded persons being attacked or even killed. It was another chance in a million that the free-wheeling car came to a stop, just in front of the only motel in that area! The restaurant was just being closed and the maids there refused to open the door imagining that we may be dangerous intruders! One look at our pathetic vegetarian Mylapore faces convinced them that we were harmless and they admitted us into the motel and allowed us to make a call to our friend Ranga Rao, a mathematician at Urbana, who came in his new Dodge car and 'rescued us' promptly from our unfortunate plight.

After returning to Panchapakesan's apartment, we relaxed in a spirit of thanksgiving listening to the music of M.S. all day. I then left for Dayton with a friend of Radhakrishnan by car and reached the house of P. Krishnaiah by evening. At Dayton, I lectured on stochastic processes and Clifford algebra at the University of Dayton and left for Washington by TWA. From the National Airport, I drove straight to the Naval Research Laboratory, where I gave a lecture and later stayed with Maurice Shapiro. One evening, I had dinner with my friend Paul Morris, who was on an assignment at the capital after retirement from foreign service. From Washington, I flew to Knoxville by American Airlines and lectured at Oakridge at the invitation of Dr. V. R. R. Uppuluri. I returned to New York the same evening by an American Airlines 727. It was an overcast sky breathing thunder and lightning, and both Lalitha and Krishna, who received me with Viswanathan, spent an anxious hour before the plane landed. I took Krishna and Lalitha to Manhattan to visit the Air India office and arrange our itinerary via the Pacific.

I left for Dallas by American Airlines and then by the colourful Braniff Airways for Austin, where I stayed in the comfortable motel 'Forty Acres.' I gave two lectures at the University of Texas — one on stochastic processes and another at the University at the invitation of Professor Sudarshan, who had moved to Texas from Syracuse. I returned to Dallas, where I lectured at the South West Centre (which later become the University of Texas at Dallas) at the invitation of its President, Professor Francis Johnson and returned to New York by an American Airlines 727. This time the weather was good and I watched with wonder and delight the magnificent sight of Manhattan at sunset, so touching in its majesty as would have inspired Wordsworth to greater eloquence than the vision of London. We moved back to Jayaraman's home and met Dr. Morton Beer the next day. He gave Lalitha a clean bill of health that we could travel by air via the Pacific and even visit the Expo 70 at Osaka enroute!

I went to Buffalo by American Airlines and after making a trip to the Niagara, I lectured the biophysics department of the SUNY Buffalo at the invitation of Prof. V. Vaidyanathan. There was an inquiry whether I would consider a consultancy in cancer research, but the idea of taking up a statistical study of terminal cases, although very important, did not appeal to me. I flew to Toronto by Mohawk Airlines, and by an Air Canada DC-9 to Ottawa, where I stayed at the Skyline Hotel and lectured at the invitation of Professor Sundaresan. From there, I flew to Montreal by Air Canada DC-9, and met Alec Lee of Air Canada before returning to New York.

We left New York for Los Angeles by a TWA Boeing 747 and it was our first flight on a jumbo jet. One can understand our anxiety, especially when we had just heard that the 747 from Los Angeles arrived only after an emergency landing in Washington because of fuel shortage! But we summoned enough confidence to take the 747 from New York, and it was quite amusing when Lalitha asked the smiling air hostess whether the flight on such a large aircraft was safe, she replied 'Certainly' though we later found that she was also making her first journey by a 747!

The two weeks at Santa Monica were one of the pleasantest sojourns in our travels. We stayed in a lovely apartment motel on the marine drive and I visited RAND as a consultant for a week and gave three lectures on Clifford algebras. Vasu had just arrived from Madras for his two year assignment at the University of Southern California and took us to his apartment for dinner. My grand student Sankaranarayanan took us round the sights of Los Angeles. During my stay at RAND, I completed a paper on 'The weak interaction Hamiltonian from L-matrix theory' projecting the role of the 'Gamma five.' It was then I understood the mystery surrounding that matrix, which worried me ever since Oppenheimer handed to me Feynman and Gell-Mann's paper in 1957 at Princeton.

We left for Honolulu by the United Air Lines Super DC8 and after a very comfortable flight we were received by Professor T. K. Menon at the airport and taken to his lovely home in Aina Haina in Honolulu. We enjoyed the honeyed breath of the azurous air, lolling in the limpid waters of the Hanauma Bay, which lay imparadised in the arms of the winding hills under the flushing light of a Hawaiian sunset. We watched the Kodak Hula Show and a delightful musical 'Portrait of America' at the Waikiki Shell. We left for Tokyo by a BOAC 707 (flight 911) which made a technical stop at Wake Island, which reminded me of my first visit in 1956. We stayed overnight at the luxurious Palace Hotel in Tokyo, and left for Osaka the next day by a JAL 727, to be received by the Kotanis. We spent a whole day at the Expo 70, which rivalled in magnificence the earlier Expo 67 at Montreal.

We left Osaka for Bangkok by Thai International DC-8 and were received at the Don Meuang Airport by my sister Meenakshi and her husband V. M. Subramaninam, a U.N. official at ECAFE. Going round Bangkok, we understood why the Americans have an obvious preference for the city, which is a bizarre blend of Oriental magnificence and Western comfort. We saw the zoo and the marble Buddha in the morning, and left for Singapore by JAL Super DC-8. After shopping in Singapore with the Ramanathans, we left for Madras by Air India, to be received by friends and relatives at Meenambakkam airport.

We reported to Sir A. L. Mudaliar, who was very happy that Lalitha was restored to normal health.

I made a trip to Delhi to attend a meeting of the Board of Scientific and Industrial Research, of which I was a member. I met the President of India, the Prime Minister, and Vikram Sarabhai, the Chairman of the Atomic Energy Commission, all in pursuance of obtaining more support for our Institute.

On the 21st of August, on our wedding anniversary there was a Veena concert organised by our academic group, when the USIS Director,

Mr. Henry released the Volume 10 (the final volume) of the Matscience Symposia by the Plenum Press.

On the 31st of August, I attended a lunch at the U.S. Consul General's house in honour of the popular American Ambassador, Chester Bowles. A week later, there was a dinner at Ekamra Nivas in honour of the U.S. Consul General, Mr. Recknagel, when I invited many friends, like Mr. T. V. Viswanathan, P. A. Menon and the Sadasivams.

We left for Ooty by car on the 12th of September, and after an overnight stay at Bangalore, we reached Ooty for the inauguration of the Matscience conference on "Frontiers of Physics" at the Aranmore Palace. I lectured on Bhabha's contribution to stochastic theory and also on the Clifford algebra. We had tea with the old Maharajah of Porbandar, a dignified representative of a bygone age. Watching him and his old but well-preserved palace, my mind travelled back to the glorious era of pomp and splendour of the Native States, when Ooty was the summer resort of the royal families.

On the 20th of September, we had an IMPACT conference at MATSCIENCE, when we had the privilege of receiving T. S. Krishna of the TVS, Unni of Air India and Nagaraja from Dayton, Ohio, of the Wright Patterson Air Force Base. In welcoming Mr. T. S. Krishna to attend a conference where mathematical equations were discussed, I said that it was like offering cucumber salad to a Bengal tiger! To which he graciously replied that he was not a tiger and what was being offered was certainly not cucumber!

We listened to a splendid music concert of Semmangudi on 30th of September before celebrating the Navarathri for nine days, when as usual we invited our friends and relatives for the religious and social functions. In October, Lalitha sang for the Dikshitar Day at the Thyagaraja Vidwat Samajam, and at my request rended "Minakshi Mae Mudam" in Poorvikalyani, which carried the full spirit of Alexander Pope's dictum that the sound must seem an echo to the sense.

On the 14th of October, I lectured at the Christian College 'On the challenge of the Seventies' speaking of the unsolved problems of modern physics. I was not prophetic enough to anticipate the startling discoveries of 1974 — the psi particle and the charm quantum number.

On the 2nd of November, I devised a method of recovering helicity matrices from any set of anti-commuting matrices. So my work on the shell structure of L-matrix was now complete, and I communicated my results to the Journal of Mathematical Analysis and Applications. I sent my work on the shell structure of L-matrices to my teachers, M. S. Bartlett and D. G. Kendall, in England for their critical assessment.

In November, Lalitha decided that we should make a trip to Guruvayur, which we had been postponing for some reason or other during the past few years. It occurred to me to consult Chembai, whose name was synonymous with devotion to 'Guruvayurappan' or Lord Krishna of Guruvayur. When we went to his house, he was performing his evening Sandhyavandanam and he asked me whether I had come to summon him to Guruvayur. I was taken aback, wondering how he knew about our plans, when he told me that he had a dream the previous night that somebody with a car would come to take him to Guruvayur!

We started the next day but after travelling a hundred miles, our car suddenly stopped and our chauffeur Govindan had to go to nearby Vaniyambadi to fetch a mechanic. During this period, Chembai asked Lalitha to join him in singing a few songs, sitting on the roadside pial oblivious of the interruption to our travel and the presence of gazing bystanders! After the car was set right, we reached Mettur in the evening, where we stayed as the guests of my old classmate Ramani, the Managing Director of the Mettur Chemicals. Next day we drove to Guruvayur via Palghat and had a thrilling darshan of Lord Krishna. Lalitha joined Chembai in singing before the Lord the song 'Rakshamam Saranagatham' and went round the temple doing 'bhajan' in praise of Lord Krishna. After a 'Thulabaram seva' we left for Alwaye, where we stayed at the house of Lalitha's sister Janu and her husband Jayaram. We left Palghat for Chembai's village, where we had a darshan of his personal deity, Lord Krishna. We returned via Pollachi and Palani, where we offered prayers to Lord Subrahmanya.

On the 6th of December, Professor Mercier arrived from Bern and gave a lecture at Matscience on 'Concepts of space.'

We decided to conduct the Anniversary symposium in honour of Professor Bhabha during early January. I wanted to express my gratitude to my great teacher for having introduced me to the enchantment of theoretical physics.

On the 15th December, the Supreme Court struck down the Princely Order erasing a chapter out of India's agreements as null and void. There was an unmanly triumph throughout the nation over fallen greatness and impaired wealth, at the cancelling of privy purses of the vanishing race even before they vanished.

The Music Academy season started with the sonorous concert of Chembai on the 20th and ran through its magnificent course. We entered into the spirit of the New Year, which was to take me twice round the world to major international conferences to propagate L-matrix theory.

Notes

1) Hans Bethe was so impressed with MATSCIENCE that he wrote in the Visitors Book: "This is an important center for the advancement of research free of bureaucratic interference."

2) M. S. Subbulakshmi (1916–2004) — popularly known as MS — was one of the greatest vocalists in Carnatic Music during the 20th century. She had an enchanting voice that captivated audiences the world over. She was the only Carnatic musician to be awarded the title of *Bharata Ratna* (Jewel of India), the highest civilian honour in India. Her husband T. Sadasivam was her friend, philosopher, guide, and manager. It was rare opportunity to listen to an MS concert at her home. Imagine being invited to the home of Luciano Pavrotti to hear him perform!

3) Thomas Kailath (born 1935) of Stanford University is a world authority in the area of control theory in electrical engineering. In the sixties, Kailath was a rising star when he presented his work at the Mastech Conference. He was awarded the National Medal of Science in 2014.

4) The Sankaracharya is to the Smartha's of the Hindu religion what the Pope is to the Catholics. While is difficult to get an appointment with the Sankaracharya to see him in person for his blessings, it is not nearly as difficult as for a Catholic to have a similar meeting with the Pope! Each senior Sankaracharya selects his successor (the Junior Sankaracharya), whom he mentors with care.

5) Tamarkin wrote the following in the Visitors book, first in Russian and then its English translation: "I would like to express my gratitude to Professors Alladi Ramakrishnan and K. R. Unni for inviting me to the Institute. For the period of one week I spent at the Institute, I had the opportunity to convince myself that MATSCIENCE is a beautiful mathematics centre and it would become one of the great mathematical centres in the world."

6) Besides being known for his role as Vice-Chancellor of the University of Madras for 27 years, Sir A. L. Mudaliar was a physician of world repute, specialising in gynaecology. He served as Chairman of the World Health Organisation (WHO) Executive Board in 1949 and 1950.

7) Those were days before mass travel on jumbo jets, when tickets were in full fare (and reasonably priced), and changes in itinerary, both in terms of airlines and routes, were easy, and without much extra charge. In just a few days, Ramakrishnan completely changed his route from round-the-world in an easterly direction to westbound. Similarly, after Lalitha's surgery in New Jersey, he made changes for his flights within the USA on his academic trip. Such changes these days would be more expensive than the hospital costs for surgical care!

8) Dr. Morton A. Beer (1920–2011) was a distinguished surgeon. After receiving his MD from Cornell in 1944, he served during WWII as Captain of the US Army Medical Corps. He began his practice at Morristown Memorial Hospital in 1955, where for many years he was the Chairman of the Ob-Gyn Department.

Chapter 32

USA, Christchurch, And IBM (1971)

1971 was the year of extensive travel to propagate my research work at Matscience. The year started with preparations for the anniversary symposium on the 13th of January. I realised the long nursed ambition of receiving Sir A. L. Mudaliar[1] at our Institute, when he unveiled the portrait of my great teacher Professor Bhabha. The eminent Indian statistician, Professor C. R. Rao, released my Matscience Report on Dirac Matrices. The symposium was devoted to Bhabha's contribution to modern physics. To commemorate such a happy occasion, we arranged a concert by the grand old maestro of Carnatic music, Chembai Vaidyanatha Bhagavatar.

The third Mastech conference took place in Bangalore and after attending it we left for Ooty for the Matscience symposium on Clifford Algebra. We had invited the Japanese mathematicians — Takayuki Nono from Fukuoka and K. Yamazaki from Tokyo — who had independently initiated mathematical work on the generalised Clifford algebra. We realised that our work was based on solid mathematical foundation, though we were unaware of it till we listened to them.

On the 18th February, at a seminar in the C. P. Ramaswami Iyer Foundation, the Japanese scientists and I discussed the concept of beauty in mathematics, coming to the old conclusion that truth is beauty and truth implies consistency!

On the 22nd February, when I was lecturing at the Institute, I got a telephone call that our dear doctor Vinayaka Rao[2] was admitted to hospital after a sudden heart attack. My wife and I rushed to see him and prayed for his speedy recovery. I have never seen one like him, a living symbol of gentleness and affection, with a deep concern to relieve human suffering and unconscious unconcern for worldly gains, an ornament to the healing profession, with a charmed hand that brought a thousand pulses back to life. God answered the prayers of his family and friends, and he was restored to normal health. The anxious community of grateful patients expressed their relief at his recovery by consulting him for their ailments, as soon as he returned home for rest and convalescence!

On the 11th March, came the news of the Indian elections. The D.M.K. won in Madras, while the new Congress under Indira Gandhi swept the polls throughout India. Within a month, C. Subramaniam was appointed a cabinet minister, ending his period of 'exile.'

In March, I had the opportunity to visit Chidambaram at the invitation of the Annamalai University. We had darshan at the famous Nataraja temple, which I wanted to visit ever since I heard in my boyhood days that enthralling song 'Natanamadinar' in the melting strains of the raga 'Vasantha.' Watching the architectural beauty of the pillars in the Mandapam, I found that symmetry could be appreciated better when viewed in a diagonal manner! This soon led me to a new decomposition theorem in matrix theory, where matrices are viewed as diagonals, shifted and unshifted.

After my return from Chidambaram, I received invitations from over a dozen centres in the U.S. to lecture on my recent work. Krishna was appearing for the Matriculation examination and since his entrance to the University and the choice of the college will have to be made in June, he wished to stay behind in Madras. Lalitha preferred staying with Krishna instead of accompanying me to the United States. Krishna did his examination very well, standing first in his school, winning the gold medal and securing cent percent in Mathematics.

On the 28th March, I gave the commemoration address at the P. S. High School, recalling that the origin of my interest in mathematics could be traced to the geometry classes in the sixth form (see [1]).

In early April, I obtained the precise statement of the new decomposition theorem of a matrix, which could be expanded in terms of all possible products of all possible powers of two base matrices with a special commutation relation.

On the 3rd of April, I left for New York by an Air India 707 from Madras, and 747 from Bombay and arrived in Frankfurt the next morning. I proceeded to Bonn by the Lufthansa 737, and I was received at the airport by Atkinson, a colleague of my host Professor Dietz. The week at Bonn was enjoyable and fruitful, and my lectures were well-received. It was indeed a gracious gesture that the Chancellor of the University, Mr. Wahlers, and Professor Dietz, personally escorted me round the main university building, an old palace noted for its architectural elegance and splendour.

From Bonn, I went to Liege by train in response to the kind invitation from Professor Garnir. The University campus was most impressive, spreading over a few thousand wooded acres. Garnir and Serpe played gracious hosts. I talked on matrix theory to the mathematicians, and the generalisation of Gell-Mann-Nishijima relation to the physicists.

After returning to Bonn, I left for New York by Lufthansa 747. Before going to the airport, I visited the famous Cathedral at Cologne with the fantastic stained glass windows, which I was told, were preserved in vaults from the ravages of war.

At New York, I lectured at the Yeshiva University at the instance of Joel Lebowitz, and then proceeded to Washington, where as usual I was the guest of Maurice Shapiro of the Naval Research Laboratory. I got the opportunity to talk at the Howard University, summarising my research work. Then I proceeded to Raliegh by an Eastern Airlines DC-9 from Washington. I had been desirous of going to the University of North Carolina, once the stronghold of Indian statisticians, since it was Professor Bartlett's North Carolina lectures that created my interest in stochastic processes twenty years earlier. I lectured there on Feynman graphs and kernel functions. Professor Leadbetter took me round the environs of the twin cities of Raliegh and Durham. Then I went to St. Louis, as a guest of Professor Gruber. This time I saw the Memorial Arch, the gateway to the West, a towering triumph of American architecture and technology. The flight by a DC-9 to Milwaukee by Ozark Airlines was delightful and Professor Umezawa and his friends made my stay quite comfortable. They were interested in my work on matrix theory. Next, I went to Iowa University at the kind invitation of Professor Roland Good. Our interest in Dirac matrices started with a close and critical study of Roland Good's article in the Reviews of Modern Physics in 1958.

Then I proceeded to Logan, Utah via Denver and Salt Lake City. As the plane flew in clear weather, I had a magnificent view of the Rocky mountains and the sprawling Salt Lake. Professor Elich and the members of the department were hospitable beyond measure, and so I made up my mind to bring Lalitha and Krishna at the next opportunity to Utah. Then I moved to San Francisco and gave a seminar at Berkeley, which was attended by Henry Stapp and Mandelstam. I lectured at the Stanford engineering department, on Stochastic processes at the instance of Tom Kailath, a young Indian professor and one of the leading men in 'control theory.' At Los Angeles, I spent a fortnight at Dick Bellman's invitation at the University of Southern California and had a very enjoyable time with Vasu, who was on a visiting assignment there. It was then that I decided to accept the invitation to participate in the Rutherford Conference in New Zealand. I decided to go by the longer route through New York and India, almost thrice the direct distance to Christchurch via the Pacific, because I had to stop in Madras enroute. During this period, Gordon Shaw invited me to Irvine to give a seminar and I stayed the weekend at his lovely house in Laguna Beach. I enjoyed equally my visit to the University of California at Riverside, where Bipin Desai played a charming and friendly host. I then left for Dallas to lecture at the North Texas University in Denton at the instance of Donald Kobe, a soft-spoken young American, and at the University of Texas at Dallas in response to the gracious invitation of its acting President, Francis Johnson, an eminent space scientist, and Erwin Fenyves, a cosmic ray physicist. It was a curious feeling to be televised while lecturing, more so to see the televised picture before me as I lectured! My association with Fenyves at Dallas ripened into a deep friendship, and a visit to Dallas and his beautiful house became a pleasant annual routine. He is a perfect gentleman, unconscious that he is one, with a natural winsome smile and sparkling sense of humour.

Then I moved to Dayton, this time lecturing at the Wright State University at the instance of my friend Nagaraja, a flourishing scientist at the Wright Patterson Air Force Laboratory. I came back to New York and stayed at the luxurious Americana Hotel in the heart of Manhattan, which can enliven the spirits of even a man tired of London and of life. I presented my results at Rutgers University at a well-attended lecture, organised by good friend, Bogdan Maglic. From there I proceeded to Montreal, and at the Sir George Williams University I talked on stochastic processes. Montreal has always attracted me with its cosmopolitan atmosphere, a triple flavour of American affluence, the Commonwealth spirit and the milieu of France. My hosts, Sankar and Srinivasacharyulu, added a fourth flavour — the spice of South Indian life! From Montreal, I flew to London by a BOAC VC-10, and then by Air India via Moscow to Delhi and Bombay. It was cyclonic weather at Bombay, and the airport was a total mess and confusion. After a few hours of indescribable inconvenience, I was back again in Madras amidst the leisured comfort of E.N. on the 24th of June.

I prepared the paper on new internal quantum numbers for the Rutherford Conference. I left for Singapore by the Air India on the 3rd July. After staying overnight at Hotel Singapura, I arrived in Sydney on the 4th July morning and stayed at the fabulous Wentworth Hotel.

I flew from Sydney to Christchurch by QANTAS, watching the sunset over the snow-bound mountains. There was a strange feeling of tremulous triumph, looking down on the congregation of mountain peaks inaccessible to the boldest climbers. I watched the awesome grandeur from varying angles as the QANTAS Boeing swept past the mountain range under the assuring whine of the whirling engines.

The Rutherford Centennial Conference was one of the best organised conferences I had attended and it was gratifying that my lecture "On a new approach to internal quantum numbers" was well received.[3] It was my tribute to the 'grandfather' of modern physics, whose experimental insight was as deep as Einstein's genius. I spoke of the triple justification for my lecture — firstly, Rutherford's discovery that matter was almost empty except for point nucleii and point electrons was made in Manchester, where I had the privilege of obtaining my doctorate. Secondly, his great protege Niels Bohr, the father of modern Physics, whose daring ideas initiated the present Quantum theory of matter, had been the principal sponsor of our Institute. Thirdly, Rutherford was an innovator, the only thing conservative about him was his love of tea, his attachment to the walls of the Cavendish, the gardens of Cambridge, and of course the language of Shakespeare and Milton. I had something new to present at the conference, perhaps not too profound, yet a departure from convention. In my paper, I suggested that the real internal quantum numbers of elementary particles, which puzzled the physicist, may be linear combinations of mathematically meaningful complex eigenvalues of a class of matrices.

Chirstchurch is a very pretty city with an English atmosphere and even the American style Avon Motor Hotel, where I stayed, provided leisurely breakfast and afternoon tea. Professor Wybourne, our friendly host, took us on an excursion to the mountains, which was a thrilling experience since I was conscious that it was the southernmost part of the globe I had visited.

After the conference, I flew to Melbourne, which is an unbelievably sprawling city, looking spectacularly beautiful at night as the plane landed at Tullamarine Airport. I lectured at the La Trobe University and stayed with my hospitable host Professor Eliezer. At Canberra, my stay at the University House reminded me of my visit with Lalitha in 1954 to the lovely capital. Of course Canberra had grown into a modern city with an American flavour of prosperity. My lecture at the Australian National University was arranged by Professor Le Couter, whose early work I had discussed twenty years ago from a stochastic point of view in my doctorate thesis.

Back in Sydney, I lectured at Messel's School of Physics at the Sydney University. S. T. Butler, the eminent nuclear physicist, entertained me before I left for Singapore. The most enchanting aspect of the flight from Sydney to Singapore by Air New Zealand DC-8 was the sight of the Australian coast as we flew over the northern part of that continent.

In Singapore, I stayed at the Hotel Equatorial, jostling with Japanese tourists. Singapore looks cleaner and greener each time I visit it and holds a great attraction for me for its triple feature — British discipline, Indian informality and Chinese capacity for hard work.

I took the Air India to Madras to be received by Lalitha, Krishna and my colleagues at Matscience. For two weeks, I relaxed in Madras enjoying the sweet intoxication of South Indian music available almost everyday at the local sabhas.

I left for New York by Air India on the 31st July to attend the IBM conference on 'Stochastic point processes' at Yorktown Heights. I used the overnight stop at London to make a three hour excursion to Windsor Castle, to feel the spell of England's regal charm and splendour before gazing at the towering affluence of New York.

The flight by Pan American 747 from London to New York was a sheer beauty and the landing at JFK was a 'sitter,' a tribute to Boeing technology and the million-miler's skill. On my way from JFK to the White Plain's hotel, a thunderstorm of incredible intensity broke out and I was lucky enough to reach my hotel without inconvenience. What a pleasant surprise it was to receive a phone call from Vasu, that he arrived in New York and was a participant in the IBM conference from the University of Southern California! It was an equal pleasure meeting my teacher Professor Bartlett, serene, composed and friendly as ever. In my lecture, I summarised the methods developed during the period 1937–1971, of course emphasising our contributions from 1949–1971.[4] It was gratifying to find that my work on product densities, like old wine, was in good demand with the increasing identification of many physical and biological phenomena as random point processes. After a pleasant weekend with my old student and flourishing New Yorker, Viswanathan, I left for Providence, in response to an invitation from the University of Rhode Island. From there, I flew by American Airlines 707 to Stanford, for a seminar arranged by Walecka. Then I went to Los Angeles to stay with Vasu, before returning to India via the Pacific, visiting Honolulu and Tokyo.

The flight by Pan American 747 was just superb, and I enjoyed the visit to Honolulu, where I stayed at the Reef Tower hotel to feel the honeyed breath of Hawaii and the unwearying charms of Waikiki. After an overnight stay at the magnificent new hotel Keio Plaza Intercontinental in Tokyo, I flew to Hong Kong. We had an exciting (frightening) experience before landing in Hong Kong. The city was being lashed by a vicious typhoon and the skies above Kaitak Airport were overcast, breathing thunder and lightning. The Boeing weathered the storm magnificently and with the Pan American pilot's skill and God's will, we landed safely at Kaitak after a two-hour suspense above a sinister blanket of swirling clouds. The journey from Hong Kong to Bangkok and Singapore by JAL DC-8, and by Air India to Madras, was made in pleasant weather. At Singapore, I stayed at Hotel King but spent most of the time with my friend Ramanathan, embarrassingly hospitable as ever.

Since 1971 turned out to be a year of global travel par excellence, I felt like changing the tenor and spirit of Kipling's rhyme, whistling the tune,

East to West or West to East,

The twain shall always be a treat.

Notes

1) It was very important for Ramakrishnan, that Sir A. L. Mudaliar be brought to MATSCIENCE and shown what was accomplished. Sir Mudaliar graciously accepted Ramakrishnan's invitation to visit the Institute and unveil the portrait of Homi Bhabha. Mudaliar said the following in the Visitors Book in his calligraphic handwriting:

"I was privileged to be present at the Ninth Anniversary Celebrations of the Institute. During this period, the Institute has made great progress under the dynamic personality of its Director, Dr. Alladi Ramakrishnan. It is a pleasure to note that within the last few years, research of a high calibre has been carried on ... and through highly qualified scientists from all over the world visiting the Institute and influencing the work in the different fields of Matscience."

2) Ganti Vinayaka Rao, our family doctor (physician), was a saint. He wore the traditional panchagajan dhothi and angavastram, like the priests. He never charged any fees for his patients. They just left whatever money they felt appropriate, or could afford, in a small tin box near the door. In his presence, like being in the presence of a saint, patients forgot their ailments, however serious they might be. Those of us who have benefited from his care, feel that he was unique among the doctors in this world. There was, and never will be, a doctor like him.

3) Lord Ernest Rutherford (1871–1937) is considered to be the greatest experimentalist since Michael Faraday. He won the Nobel Prize in Chemistry in 1908, and was appointed the Director of the Cavendish Laboratory in 1919. To mark his birth centenary, the Rutherford conference was held in 1971 Christchurch, New Zealand, his birthplace. Ramakrishnan's paper presented at the conference is: "A new approach to internal quantum numbers", *Proc. Rutherford Centennial Conf.*, Christchurch (1971), 150–156.

4) Ramakrishnan's paper at the IBM Conference is: "Stochastic theory of evolutionary processes, 1937–1971", in *Stochastic Point Processes*, Proc. IBM Conf. on Point Processes (J. Lewis, Ed.), Interscience, John Wiley (1971), 533–548.

Chapter 33

Eventful Sequel To World Travel (1971)

The five months after my world travel were crowded with events of significance to my domestic and academic life and one of them tragic, the loss of
my personal assistant, Nambi Iyengar.

On arrival in Madras on the 17th August, I was told that our dear
Nambi Iyengar was admitted to the General Hospital and was receiving
the best medical attention. There was no indication that the illness was so
serious as to take his life a few months later.

On the 21st of August, Lalitha and I celebrated our Silver Wedding
Anniversary. It was really gracious on the part of Chembai, Lalgudi Jayaraman, Maharajapuram Santhanam, Vellore Ramabadran and Lalitha's
sister Kamalam to have made it a music festival,[1] to the delight of our
guests, which included several distinguished invitees. In such a mood, I
completed an essay on music entitled 'Bridge of Sound and Light' and
handed it over to Lalgudi Jayaraman, who was leaving on a concert tour
to U.S. arranged by my former student V. K. Viswanathan.

I gave a talk at the Rotary Club on the 'Economics of creative activity,'
emphasising that creative thought is as much a source of wealth, like oil and
precious metal, and should be harnessed and not ignored or suppressed.

On the 11th October, we had a visitor to Matscience — Dr. Rasche
from the University of Zurich, who spoke on the history of Isotopic Spin.
Introducing him, I recalled how enthusiastically I taught the concept to
young honours students in Madras.

Meanwhile, Nambi Iyengar's illness took a serious turn and he passed
away on the 27th of October. He was part of Ekamra Nivas and Matscience,
our hopes, and our ambitions. Ten hours a day for more than ten years
he was my companion, friend, ministering to my needs, great and small,
listening to my conversation, speeches and exhortations.[2]

He lost his life in the service of the Institute and his devotion to advanced research. He participated in the conception, creation, evolution, growth of Matscience.

He protected the interests of Matscience, as if he shared its possessions. Thirty thousand letters to my dictation, notes and jottings ranging from research papers to paint shop bills, communications with friends, scientists and Nobel Laureates, arranging the itineraries of legendary scientists, like Hans Bethe and Paul Dirac, and settling accounts with the plumber — in short, everything needed his attention provided it was the work of Matscience! He was part of the stuff and texture of our academic and scientific life. He inspired love and affection, and above all, trust in all those who came in contact with him. His flamelike honesty in financial matters made all of us feel secure that our matters were in responsible hands.

It is hard to imagine a Nambi-less Matscience, and even the healing force of memory cannot compensate for such a loss. Nambi was part of the great tradition, which makes Matscience a haven of the finest ideals of the academic way of life.

In early November, I wrote an article on 'C.R. and Alladi — Unison in contrast' in response to a request from my friend Sadasivam of 'Kalki'.[3] I wrote the article recalling my school days, when I watched C.R. and my father discuss at leisure in the office room at Ekamra Nivas. (For the full article, see [1].)

On the 4th of November, we went to Bangalore for the Matscience conference on Gravitation. This was followed by the Matscience Conference on Optics that started on the 11th of November.

The 14th November, which was Nehru's birthday, was a proud day for our family when at the prize distribution at Vidya Mandir, Krishna received a gold medal for standing first in his school in the Matriculation (school final) Examination. It was in November that it occurred to me that the velocity transformation formula was the entrance to the theory of relativity and the Lorentz formula should be derived therefrom. This, of course, meant that the transformation applied in the first instance only to time-like intervals, which by an analytic continuation can be extended to space-like intervals. It yields the most important result that time-like intervals are always associated with a single particle, while space-like intervals have to be associated with two particles, thus rendering the concept of the Tachyon (faster than light particle) meaningless.

On the 26th of November, I despatched the paper on "Einstein is a natural completion of Newton" to Dick Bellman for the Solomon Bochner memorial volume of the Journal of Mathematical Analysis and Applications (see Appendix 9). It was the successful end of my thirty year quest to understand the meaning of special relativity.

Meanwhile in India, Indo-Pakistan relations worsened over Bangladesh. Pakistan declared war on the 4th of November, but in two weeks the Indian troops advanced to Dacca under General Maneckshah and ceasefire was declared on the 17th of December. Thus, we could enjoy the music season at the Academy with our visiting scientists. The year ended with a one day Matscience conference on Mathematical Education in honour of V. Ramaswamy Aiyer,[4] the founder of the Indian Mathematical Society.

Notes

1) The musicians who performed at Ekamra Nivas — Chembai Vaidhyanatha Bhagavatar (vocal), Maharajapuram Santhanam, Lalgudi Jayaraman (violin), Vellore Ramabhadran and Trichy Sankaran (Mridangam), the volinist trio, L. Subramaniam, L. Shankar, and L. Vaidhyanathan, were all giants in their respective domains of Carnatic music. Ramakrishnan and Lalitha were highly regarded in the Carnatic music circles in Madras, and so these stalwarts enthusiastically performed for their 25th wedding anniversary in Ekamra Nivas. Imagine getting Luciano Pavrotti, Placido Dimingo, and Yehudi Menuhin, to all perform for an event like this at a home in the United States!

2) To run any operation successfully, it is necessary to have loyal and efficient staff. Nambi Iyengar and his family lived in Ekamra Nivas — they had their separate lodgings within the Ekamra Nivas property. Thus, Nambi Iyengar was not only like a family member, but he was with Ramakrishnan right from the fifties, when the Theoretical Physics Seminar was being conducted, and he was ready to happily assist Ramakrishnan ANY TIME of the day (or night).

3) T. Sadasivam, the husband of M. S. Subbulakshmi, and an Editor of the magazine *Kalki*, was an admirer and follower of the great C. Rajagopalachari. Sadasivam wanted to bring out a volume in honour of C. R., and he invited Ramakrishnan to contribute to that volume (see [1] for Ramakrishnan's article on C. R. and (Sir) Alladi). C. R. died in 1972, but this volume appeared the previous year and so C. R. could read the many fine articles written in his honour.

4) V. Ramaswamy Aiyer (1871–1936) was a civil servant, but it was he who founded The Indian Mathematical Society (IMS) in 1907, acted as its first Secretary until 1910, and served as its President during 1926–1930. The mathematical genius Srinivasa Ramanujan sought his patronage. When Ramaswamy Iyer went through the Notebooks of Ramanujan, he was struck by the spectacular results contained there, and drew the attention of several of his friends to the phenomenal work of Ramanujan. He helped publish Ramanujan's early work in the Journal of the Indian Mathematical Society.

Chapter 34

Freewill And Destiny (1972)

In early 1972, Krishna expressed his uncompromising desire for a career in Aerodynamics. But before the year was out, by a series of unforseen and irresistible circumstances, he had set his heart on Number Theory.

The New Year started in a festive mood with our attending a magnificent music concert by M. S. Subbulakshmi at the Indian Fine Arts Society, followed by that of Lalitha's sister Kamalam at the same venue. I took my good friend Gordon Shaw from the University of California, Irvine, to the performance and he not only enjoyed the music but the casual informal atmosphere, so different from that of a Western concert.

On January 8, Governor K. K. Shah, inaugurated the Einstein Symposium at Matscience to commemorate the tenth anniversary of the Institute. The most important question raised therein was: If the special theory of relativity is based on the Lorentz transformation, why is it associated so much with Einstein? The symposium concluded with Lalgudi Jayaraman's splendid violin concert and a dinner on the lawns of the Institute.

In early February, Lalitha and I went to Trichy in response to kind invitations from various colleges to lecture on my recent work. At the Seethalakshmi Ramaswamy Women's College, I gave a talk on 'Einstein is a natural completion of Newton,' and at the Jamal Mohammad College on 'Pauli to Dirac,' emphasising that the tradition from two to four dimensional matrices in the wave equation for the electron was not anticipated even by Pauli! At the National College, I talked on 'Dirac to Gell-Mann' and we returned to Madras after a visit to the Sri Ranganatha Temple, where we had a magnificent darshan of the Lord.

In late February, we had a pleasant surprise when we heard that my distinguished friend Dr. McCrea Hazlett of Rocheter was in Madras.[1] I met him at the U.S. Consulate and invited him to inaugurate the Twin

Matscience Symposia in Bangalore. I also met the President of India, who agreed to visit our Institute for the second time in late December.

There was a pleasant hostel function at the Madras Institute of Technology (M.I.T.) in Chrompet when I talked to an audience of a thousand students to justify my role as a chief guest at a delightful dinner.

It was during this period that Krishna made his first attempts at research by 'playing with progressions.' I encouraged him to publish a paper in the local journal 'The Mathematics Teacher' with a sympathetic editor who encouraged young talent. He had also completed by that time an essay on the geometry of flight, which clearly indicated that he had an analytic insight into motion in three dimensions. In his first year in College, Krishna won a mathematical talent prize awarded by the Association of Mathematics Teachers of India (AMTI) just before his PUC examination, which stimulating in him a desire to take to research.

On 23rd March, Rama Navami, at the anniversary of the Theoretical Physics Seminar, I unveiled Nambi Iyengar's portrait at Matscience and spoke about his untiring assistance to me in the creation of the Institute.

In early April, after Krishna completed his PUC examination with confidence, we departed for the United States, where I had received several lecture invitations.

We left for Bombay and spent a few pleasant hours at the Sun and Sand Hotel before taking the TWA 707 flight to Rome, where we boarded the TWA Boeing 747 for New York. We flew to Los Angeles directly from New York JFK airport and were received by Vasudevan, who took us to the comfortable apartment he had arranged for us next to his own. Vasu was on an extended visit to the University of Southern California to work with Dick Bellman, who graciously arranged my visit to USC for a month.

I had busy schedule at the USC where I talked about my work on Einstein's special relativity. During our stay there, I made a day trip to Stanford and gave a lecture at the kind invitation of Tom Kailath, and in true 'commuter spirit' returned to Los Angeles the same evening. While in Los Angeles, at the invitation of Professor Bipin Desai, I lectured at the University of California, Riverside.

We left for Salt Lake City by Western Airlines and were received by Professor Joe Elich, who took us to Logan, Utah, for my lectures. The university has a lovely location from where one could watch with tireless wonder the magnificent sunsets enlightening the multi-coloured hills. From Salt Lake City we flew to Denver to be received by a group of young Indian mathematicians. My lecture at Laramie at the University of Wyoming was very well-received, and Dr. Bhowmik was kind enough to take us on a day trip to the Rocky Mountain National Park, where we enjoyed varied weather — snow and rain, cloud-spangled skies amidst flashes of golden sunshine. We flew back to Denver and left for New Jersey and were received by our good friend Dr. Jayaraman of Bell Labs at the Newark Airport. We called on Dr. Morton Beer at the Morristown Memorial Hospital, and he was happy to see Lalitha in normal health.

Since Krishna had evinced interest in a career on aerodynamics, I took him to Professor David Hazen at Princeton University, who was quite impressed with Krishna's enthusiasm in aviation, but frankly told him that he should first take a B.Sc. in mathematics as a good preparation if he were to study later at the United States.

Krishna and Lalitha left for India on the 17th of May, and I saw them off at the JFK airport. I left for Buffalo by an American Airlines 727 and lectured at SUNY at the invitation of Professor Soong. Following this, I gave a lecture at Syracuse at the invitation of Professor Rohrlich. From there, I went to Montreal by Eastern Airlines DC-9 to be greeted by my old student K. Ananthanarayanan and my friend Marsaglia of my Manchester days. I gave a lecture on Matrix theory at the Computer Centre and also at the University of Montreal at the invitation of K. Srinivasacharyulu. I then left for Boston by North East Airlines DC-9 and was taken by Professor Lakshmikantam to the University of Rhode Island, where my lecture was very well-attended. I proceeded to Washington from Providence, Rhode Island. After my lecture at the Naval Research Laboratory at the instance of Professor Maurice Shapiro, I dined with the Caldwells, who had served in the American Consulate in Madras.

From Washington National Airport, I flew to State College by a small aircraft of the Pennsylvania Commuter Service. Professors Patil and Vedam were very gracious hosts at the Pennsylvania State University, which looked to me one of the greenest campuses in the United States.

From there, I flew to New York to stay with my former student V. K. Viswanathan. I lectured at the Yeshiva University at the kind invitation of Professor Lebowitz, and later checked in at the New York Hilton, as the guest of the Rutgers University. At Rutgers, I gave a lecture organised by my great friend Bogdan Maglich, whose 'bon vivre' had increased in intensity with his arrival in U.S. On my way back to Manhattan, I called on Bela Julesz at the Bell Telephone Laboratories to discuss my nephew Ramu's work on binocular vision.

I left New York by Air India on the 8th of June for Madras via Bombay. Krishna did not lose time in describing to me some new work he had done on Fibonacci numbers. I sent his notes to three 'referees' — Professors M. J. Lighthill at Cambridge, Helmut Hasse in Germany and Morris Newman at Washington, all of whom expressed warm appreciation that the contributions were unusually original, particularly from a lad of sixteen, who should be encouraged to further creative effort. Krishna felt so elated that he shifted his affections from Aerodynamics to Number Theory. As a father, I first felt that he was choosing his career too early in life abjuring other possible avenues for his talents. But, it is given only to a chosen few to woo with success the Queen of Sciences and why deny Krishna his claim to the incomparable charms of creative work?

We made a trip to Tirumalai, where as usual we had a good darshan of Lord Venkateswara. But on this trip to Tirupati, our car broke down on three occasions and finally came to a halt at the foot of the hills, and we had to go up by a taxi provided by the Executive Officer. It was then I told Lalitha that in Los Angeles I had dreamt that I was travelling in a Boeing 747, which developed engine trouble and prayed that if it landed safely I will perform Sahasrakalasa Abhishekam at Tirumalai. It was indeed strange that I should have thought of this in the dream, though I was actually unaware that such a 'seva' could be performed. With customary Hindu faith, we interpreted the unexpected trouble in our car as a reminder that we should keep our vow. So we came again to Tirumala to perform the Sahasrakalasa Abishekam on the 24th October and had a magnificent darshan of the Lord.

At Matscience, we had an eminent visitor, Professor B. Volkmann, a number theorist from the University of Stuttgart, who evinced keen interest in Krishna's career and encouraged him in his efforts at research.

I visited Chidambaram and gave two lectures at the Annamalai University. Krishna and Lalitha accompanied me and it turned out to be a memorable trip for Krishna. With incredible generosity, unusual among the scientific community in India, Lakshmanan Chettiar, the professor of Chemistry, gave an opportunity to Krishna to give a lecture before an audience of mathematicians and scientists, when he presented his undergraduate research work on the Farey-Fibonacci sequence.

On the 3rd of October, I attended the opening of the buildings of the Ramanujan Institute.[2] We enjoyed listening to the profuse eulogies about Ramanujan, though everyone was aware that the plight of mathematicians in India was such that any young undiscovered Ramanujan today would face the same difficulties!

During late October, we had a distinguished visitor, Professor Littauer, a friend of Hans Bethe, from Cornell. He was so tall that it was quite amusing to watch Krishna looking up to explain to him his work on the black board in the seminar room. I was reminded of my father standing beside Sir Maurice Gwyer, when showing him the library at Ekamra Nivas.

I was invited by the Calicut University to give some lectures. The trip turned out to be one of the most memorable in our travelling career. It was on the Calicut road from Gundlupet that we had one of the most exciting experiences in our car journeys. We were stopped at the Kerala-Mysore border due to a total 'bandh' (strike) by the workers in the Wynad tea estates, who expressed their discontent by placing large boulders and logs of wood across the road making transport impossible. There seems to be some moral code even in such strikes for after sunset, as agreed, the 'bandh' was voluntarily lifted. Cars and buses started moving in a convoy as the passengers themselves periodically got down to remove the obstacles. While waiting for the bandh to end, we got into conversation with an Assistant Engineer from Calicut by name Ganapathi Iyer, an acquaintance which soon proved to be very useful. After a few miles, there were no obstacles and the traffic started speeding up. We followed closely a fast moving bus, which we thought would act as our pilot vehicle! Suddenly, we heard a deafening sound and found that a large stone, which did not affect the bus with its high chassis, struck the crank-case of our car, which was completely shattered. The car came to a grinding halt, and we found ourselves stranded at mid-night, miles away from civilisation, wondering how we are going to get out of such an unfortunate mess. What a fortuitous circumstance when a speeding car stopped at our request, and it was Ganapathi Iyer who emerged from it! He was ahead of us earlier but had halted for his coffee on the way. He was so friendly that he invited us to get into his car. He took us to the guest house at the University of Calicut. It was indeed a most generous gesture by K. A. Jaleel, the Principal of the Farook College, to have invited sixteen year old Krishna to give a formal lecture on the Farey Fibonacci sequences, and even more generous on the part of the Pro-Vice Chancellor Mr. Madhava Menon to have attended it, and the Vice-Chancellor Mr. Ghani to have invited him for dinner.

After our car was repaired, we left for Mangalore, where we enjoyed the exuberant hospitality of an enterprising businessman Mr. Aroor, whose acquaintance we had made earlier in Madras. From Mangalore, we moved to Kollur to offer prayers at the temple of Mukambika, with its magnificent location chosen by Adi Sankara himself, where the gleam of dawn and the blaze of sunsets enhance the pristine beauty of the Western hills.

On our way back, we offered prayers to Lord Krishna at Udipi and what a fortuitous circumstance that the Head of the Adhmar Mutt should be there to welcome us! We had the privilege of attending his sacred prayers on 'Kaisika Ekadasi' day. He invited Lalitha to sing in the Mantapam, where two hundred devotees had gathered to listen to his discourse. He gave his benediction to Krishna for a fruitful mathematical career.

We stayed at the Valley View Hotel at Manipal near Udipi and from there we went to Mercara, perhaps one of the most beautiful areas of our country, comparable in scenic grandeur with the Blue Mountains of New South Wales. We had a little excitement with three successive flat tyres and had to take a jeep ride of such hazardous nature that we considered ourselves fortunate to have survived it without injury. From Mercara, we drove back to Madras via Mysore and Bangalore.

On the 16th December, Professor J. H. Williamson arrived in Madras from York, England, well ahead of time for the International Conference on Mathematical Analysis at Matscience. On the 23rd December, my book on 'L-matrix Theory' or the Grammar of Dirac Matrices was released by His Excellency Sir V. V. Giri, the President of India, with the Governor of Madras K. K. Shah presiding. It was a satisfactory culmination of my effort to bring various papers on L-matrix theory together. I thanked the Tata-McGraw Hill company for compiling a book of original contributions instead of insisting on an expository survey by the author.

It turned out to be a momentous year when destiny guided human will — Krishna, who wanted to design airplanes, had now set his heart on a mathematical career, while I, who resolved to write an expository book on the symmetry principles in elementary particle physics for the Academic Press, ended up with a monograph of original papers on L-matrix theory.

Notes

1) McCrea Hazzlett, who previously visited Ekamra Nivas in 1961 and 1964 as Provost at the University of Rochester, was this time in India as the Cultural Affairs Officer at the US Embassy in New Delhi, a position he accepted in 1971.

2) The Ramanujan Institute, named after the mathematical genius Srinivasa Ramanujan, is currently the Department of Mathematics of the University of Madras. The Mathematics Department was formed when the University was created in 1857. In the 1950s, the Ramanujan Institute for Advanced Study in Mathematics was created in the University, but it functioned independently of the Mathematics Department from 1957 to 1966. The two were merged in 1967. The Ramanujan Institute moved into its own separate building in 1972.

Chapter 35

A Year Of Opportunities (1973)

A flood of friendship burst upon Matscience on New Year's day, as over sixty mathematicians from various parts of the world arrived for the inauguration of the International Conference on Functional Analysis, conducted by Professor Unni. It was an event unprecedented in the academic life of Madras, which greeted them with the finest of weather. It was bright, sunny and cool, and the function was organised in open air on the lawns of the Institute, at a time when Harvard and Princeton were blanketed by snow. Mr. C. Subramaniam inaugurated the conference and Dr. M. Adiseshiah delivered the Anniversary Address. Lalitha rendered the invocation song, and the scientific sessions began the next day, when in the evening we had a delectable performance by the eminent flautist N. Ramani.

On the second day of the conference, Lalitha[1] and her teacher Mannargudi Sambasiva Bhagavathar gave a concert attended by all the delegates, and that grandmaster of Carnatic music, Semmangudi Srinivasa Iyer. Each song was 'dedicated' to a country, the delegates of which were introduced to the gathering amidst cheers. This was followed by a dinner, attended by about two hundred invitees, on the lawns of the Institute. We arranged an excursion to Kancheepuram and Mahabalipuram — the 'Musts' for tourists to Madras, like the Vatican or the Coliseum in Rome. On the last day, we had the 'lighting of the lamp' ceremony, symbolising the spirit that knowledge, unlike wealth, can be imparted without loss to the donor.

After the conference, Krishna gave an informal lecture on his work before a few distinguished delegates, like Professor I. J. Schoenberg and Louis Nirenberg[2] who expressed appreciation of his efforts at research.

The Matscience symposium on Nuclear Physics was held at Mysore in early March. We called on the Maharaja of Mysore, who was ailing from a serious illness which claimed his life a few months later. The gifted Prince looked serene and composed, true to the quality of his Royal blood, imbued with the culture and tradition of his ancient forbears. The Matscience seminar on Numerical Analysis was held in Bangalore and inaugurated by C. Subramaniam.

On March 14 Krishna received an invitation from Professor D. J. Lewis to participate in the three months Summer Institute on Number Theory at Ann Arbor, Michigan. It was a gracious gesture, characteristic of the spirit and vitality of the land of opportunities.

Next, on March 26, Krishna received a handsome letter from Professor B. H. Neumann[3] inviting him as a visiting scholar to the Australian National University in Canberra to work under the eminent mathematician Kurt Mahler.[3] I also received letters of invitation from various universities in Australia, in particular from Eliezer at La Trobe University, Melbourne. Lalitha and I were naturally very happy that we could accompany Krishna to Australia.

On the 8th of April, there was a function at the Jain College when I unveiled the portrait of Professor Lakshminarasimhan, my old teacher who taught us chemistry when he was at the Loyola College. The dramatic manner in which he used to describe the properties of the hierarchy of Halogen elements in a sing-song tone still rings in my ears.

We left for the United States by an Air India 747 from Bombay and were received by Jayaraman and Viswanathan at the JFK airport. While Lalitha went to New Jersey with the Jayaramans, Krishna and I accompanied Viswanathan to his New City home. I lectured at the IBM at Yorktown Heights at the invitation of Professor T. J. Rivlin, and at the Courant Institute of New York University as the guest of Professor Professor Nirenberg.[2] Both had visited Matscience in January for the International Conference on Functional Analysis. I made a day trip to State College and lectured at Pennsylvania State University, where I visited Professor Roland Good. We had a pleasant time at New Jersey with the Jayaramans, and flew to Montreal by the Alleghany DC-9 from New York. I lectured at Sir George Williams University as a guest of Dr. Sankar, the son-in-law of our family friend Sanskrit Professor Raghavan.

We flew from Montreal via Chicago to Carbondale, where I lectured at the Southern Illinois University at the invitation of my friend Bruno Gruber. From there, we went to Dayton, where we enjoyed the hospitality of Professor K. S. Nagaraja, who arranged my lectures at the Wright Patterson Air Force Base and the University of Dayton on 'Stochastic Processes' and on 'Einstein is a natural completion of Newton.'

From Dayton we flew to Dallas by an American Airlines 727, where on arrival I gave a seminar before lunch at the University of Texas at the invitation of Professor Fenyves.[4] We had tea with Professor Ivor Robinson,[4] who had migrated from Cambridge to the sunnier clime of Dallas. Krishna watched with obvious delight the impeccable accent and English humour of the famous scientist, while Lalitha and I admired his taste for comfortable living in a Texan home with an indoor swimming pool. After visiting Professor Robinson, we saw the place where Kennedy was assassinated, wondering how frail and vulnerable is the life of a President, who could sway the lives of billions by a stroke of his pen.

We took the American Airlines DC-10 flight to Los Angeles and we spent two wonderful weeks with Vasudevan, repeating the schedule of lectures at the University of Southern California. While in Los Angeles, I made a short visit to the University of California, Irvine, where Gordon Shaw was a solicitous host. We stayed at his lovely home in Laguna beach. I lectured at Irvine and returned to Los Angeles, where Krishna and I took the opportunity to visit Professor Leveque, a number theorist at the Claremont College, who evinced considerable interest in Krishna's work. Dick Bellman, the generous host he always is, took us out to dinner at the 'Aware Inn' on Sunset Strip, a glamour spot of the glamorous city. That was the last time I saw him in good health, for after a few months, we heard the sad news that he was stricken with partial paralysis and had lost his power of speech.

Krishna and I flew to San Jose by PSA 727 to meet Professor Hoggatt, the Editor of the Fibonacci Quarterly. Krishna and Vern formed an instant bond — a relationship which was strengthened a year later over rallies on the tennis court between a seasoned old-timer, who derived his toughness from active service during the second World War, and a buoyant teenage volleyer, who was yet to learn the facts of life!

We left for Detroit by the American Airlines DC-10. On arrival, we were received by Professor M. S. Ramanujan, who had settled down as a senior mathematician at Ann Arbor. We stayed in a comfortable apartment at the University Towers, near campus of the University of Michigan. Krishna participated in the summer school, which started on the 18th of June, and Lalitha kept house for him while I moved around many places in response to invitations for lectures.

I left for Washington by the Northwest Airlines from Detroit and was received by my friend Kotra Krishnamoorthy and my old student Devanathan, who was visiting the Catholic University. I lectured at the Naval

Research laboratory and my gracious host Shapiro took us for dinner to an Indian restaurant. The next day I lectured at the Catholic University, arranged by Devanathan, and also lectured on Feynman graphs at a Conference on Graph Theory, at George Washington University.

Washington was in the grip of Watergate fever. For the first time in American history, the President, the first citizen of the most powerful nation in the world and the commander-in-chief of its armed forces, was in dire trouble, in danger of impeachment for 'covering up' guilty men who broke into the office of the Democratic convention a year before.

I left by the Eastern Airlines DC-9 'shuttle' for New York, where I lectured at the Yeshiva University at the instance of Professor Lebowitz. I returned to Detroit by an American Airlines 727, to be back with Krishna and Lalitha for a few days in Ann Arbor. We had a delectable picnic with the Ramanujans, when we met quite a few flourishing Indians employed in Ford and General Motors. I recalled my boyhood days, when in Madras we used to argue with our chauffeur Sounderraj over the relative merits of Ford and Chevrolet. It was hard to believe that we were right in Detroit, where these cars are manufactured by the millions per year.

I left for Madison by the Northwest Airlines 727 and enjoyed the kind hospitality of Professor I. J. Schoenberg.[2] My lecture at the Mathematics Research Centre of the University of Wisconsin turned out to be in honour of Schoenberg, who was retiring the same day to turn 'Emeritus' at the University! At a dinner at the Cuban Club, I met many of his colleagues, who had heard about Matscience from him after his visit to Madras earlier in the year.

I returned to Ann Arbor by the Northwest Orient for a brief break before leaving on the 4th of July for Winnipeg through Minneapolis, in response to a kind invitation from Professor B. K. Kale. There I met other young Indian mathematicians, Venkataraman, Padmanabhan, Parameswaran and Shivakumar, who migrated to Canada carrying with them the spiced flavour of South Indian life. On my way back, the Northwest plane stopped at Milwaukee, when suddenly the aircraft was filled with very fat persons and as I was wondering at this statistical miracle, the person next to me whispered that there was a Bariatrics Conference of over-weight persons, where medals were awarded to those who succeeded in losing the maximum of pounds! Some were given tickets to Europe as special prizes! Weight-watching is a typical American occupation, just as discussion on weather is an old English tradition.

On July 9 Krishna gave a seminar at the Summer Institute, a great opportunity for him as an undergraduate to lecture before a gathering of full blooded mathematicians. A few days later, I left for Bufffalo to lecture at the invitation of Professor Parzen at the Statistics Department of SUNY. That evening, there was a delectable dinner at the home of Professor Bernard Gelbaum, who had visited Madras Matscience for the Conference on Functional Analysis, and a delightful party later at Parzen's. I did not miss the opportunity to visit the Niagara, and returned to Ann Arbor by Allegheny Airlines DC-9.

We left Ann Arbor for Detroit from where we flew by a Northwest 747 to arrive at JFK, in time to catch the Air India to Bombay and to Madras. Upon return, Krishna lectured at Matscience on what he had learnt at Ann Arbor and also on his own work, which he had presented there.

In late July, we received the sad news that our dear Kailas, the husband of Lalitha's sister Kamalam, passed away. We left for Trivandrum by car so that Lalitha could be with her sister in her period of great grief. Kailas was a gentleman in the true sense of the term that he never spoke ill of another.

We left for Australia by Air India from Madras on the 16th of October and our first stop was Singapore where we stayed at the Hotel Imperial. From Singapore we flew by BOAC VC-10 to Brisbane, where we stayed at the comfortable Gateway Inn for one night before taking Ansett Airlines to Sydney, where we stayed at the comfortable home of my sister Rajeswari in the suburb of Killara which had the pleasant atmosphere of an English countryside.

20th October was one of the greatest days in Australia, when Queen Elizabeth of England arrived to open the fabulous Opera House, a symbol of new architecture and Australian affluence. Sydney went mad with delight, offering its charms and hospitality, presenting its gayest plumage. We went on a motor tip to the lovely parks around Sydney and to the Koala reserve. We visited the Commodore Point, where sea and sky contended to display their azure charms under the radiance of the noonday sun.

We left for Canberra to be greeted at the airport by Professor B. H. Neumann. While Lalitha and I stayed at the comfortable University Guest House, Krishna took residence in the dormitory since he had to stay longer as a research scholar. I lectured on stochastic processes and it was a pleasant surprise to meet professor Bartlett, my old teacher, who was on a visiting assignment at Canberra. I was thrilled when he also attended my lecture on 'Einstein is a natural completion of Newton.'

Canberra had grown immensely and was so different from the elegant little town we had seen twenty years ago, combining the splendour of Delhi and the affluence of Washington. Lalitha and I left for Sydney by a Trans Australian Airlines (TAA) DC-9, while Krishna stayed back in Canberra.

Krishna had the privilege of being introduced to transcendental numbers by the famous number theorist Professor Kurt Mahler, who on retirement from Manchester had settled down in Canberra in consonance with the general trend of retired eminence seeking milder climes and being sought by younger universities in such favoured locations. Like all truly great men, he was generous (Udara) and easily accessible (Sowlabhya) to young aspirants and Krishna was particularly fortunate to deserve his attention.

On Deepavali day, we exchanged greetings with Krishna in Canberra and my brother Prabhu in Melbourne. I visited Sydney University as the guest of my good friend Harry Messel, who invited me to lecture on Einstein. The talk was well-attended by Messel, Hanbury Brown and McCusker.

I met Professor George Szekeres, the famous number theorist and friend of Erdös, at the University of New South Wales to discuss Krishna's work and naturally felt happy when he had everything nice to say about it and arranged for Krishna to give a seminar. Messel invited me for dinner in his fabulous apartment overlooking the Sydney Harbour bridge.

We made an excursion to the Lion Country Safari and the Waragal dam. It was then our pleasant surprise we heard that a world tennis championship was being played at the Hordern indoor stadium. By a fortunate circumstance, we met one of the sponsors of the matches — Mr. Tony Hodges who, as a tribute to the colourful elegance of a Conjeevaram saree which Lalitha wore, gave us complimentary tickets for a VIP box, where we were also provided with lunch and refreshments. He must have regretted his generosity when he found us insisting on vegetable sandwiches since he was a rancher promoting the export of prime Australian beef!

It was a feast of tennis at its best, and we could not have a surfeit of it through watching seven hours a day for seven days! Twenty years before, we had watched such tennis at Wimbledon in the early era of 'serve and volley,' and a few years later at the White City Stadium when Ken Rosewall and Lew Hoad entered Australian tennis as youthful 'debutantes.' Now, Rosewall of forty summers was still delighting the crowds with his roseate smile and immaculate return of serve. Rod Laver, true to his golden mane, was now the king of tennis with four Wimbledon crowns and two Grand Slams under his belt, his lissome limbs swinging and swaying in perfect rhythm for a midcourt forehand swipe or a lashing backhand volley, or while galloping backwards, as only he could do, for an overhead smash.

On the 12th November we left for Melbourne, where I lectured at the Monash University and at the University of Melbourne. We watched the

Davis Cup match between Czechoslavakia and Australia, when on the first
day, Jiri Hrebec sprang a surprise by beating John Newcombe! Laver proved
the grand master and conquering hero by winning all his matches, two
singles and the doubles, with Newcombe as his partner.

One of my distinguished physicist friends, Professor Mark Oliphant,[5]
was the Governor of South Australia and he invited me to spend a week
as his personal guest at the Government House in Adelaide. Meanwhile,
Professor Hurst invited me for lectures and I left for Adelaide to be received
by the A.D.C. to the Governor and taken to the Government House by a
Rolls Royce with a liveried chauffeur. It was perhaps one of the most
thrilling experience in my life when I stayed in a suite of rooms, which was
once used by the Duke of York. As the Governor's guest, my daily activities
were announced in a printed circular!

One could not but admire the perfect manner in which British tradition is preserved fifteen thousand miles away from Buckingham Palace and
Windsor Castle, in the floral beauty of an 'English garden,' the velvet sheen
of its untrodden lawns, the serene dignity of the impassive butler and the
hallowed precedence of guests at the dinner table. The emergence of the
commonwealth is essentially an English phenomenon, for it is the gradual
evolution from the world wide empire to suit the democratic ideas of the
twentieth century. Even Nehru and Gandhi, who swore for Puma Swaraj,
found it consistent to imbed the Indian republic into the larger framework
of the Commonwealth.

Sir Mark Oliphant took me to a tennis match when I watched with
a tinge of sadness, mixed with wonder, the grandmasters Sedgman and
Gonzales, who had lost the power but retained the skill of 'serve and volley,'
made weak by time and fate, yet rich in art and grace.

At the University of Adelaide, my lecture on Einstein stressing that
the postulate of faster than light particles was unnecessary was given to a
group which had just received a special grant to search for tachyons! It was
like preaching the benefits of abstinence to purveyors of wine and virtues
of celibacy to newly-wed couples. That's the nature of the scientific game.
The charm lies in controversy and the greater charm in its resolution. Hurst
and Green saw me off at the airport and I returned to Melbourne with tales
to tell of gubernatorial hospitality to Lalitha and Krishna.

I left for Perth by an Ansett 727 and stayed at the lovely Mount-
bay Lodge, as the guest of University of Western Australia. I lectured
on stochastic processes at the kind invitation of Professor Buckingham.
Lalitha and Krishna arrived from Melbourne by an Ansett flight and joined

me on a motor trip to the Freemantle harbour, reviving nostalgic memories of our trip to Australia in 1954. We left Perth for Singapore by Singapore Airlines and stayed again at the Imperial Hotel before arriving in Madras by Air India on December 1.

On the 28th of December, I gave a talk on 'the theory of evolution of physical theories' at the All India Science Teacher's Association in the Hindu High School, emphasising despite changes in a theory, knowledge always advances monotonically.

Notes

1) Ramakrishnan's wife Lalitha hailed from a musically talented family. She and her four sisters were initiated into Carnatic music in Trivandrum by the disciples of Semmangudi Srinivasa Iyer. After marriage to Ramakrishnan and her move to Madras, she continued her music training under Mayavaram Krishnan (a student of the great Maharajapuram Viswanatha Iyer), who taught all the ladies of the Alladi family in the late forties. After a gap of nearly two decades, Lalitha resumed her music practice under the tutelage of Mannargudi Sambasiva Bhagavatar, also a disciple of Viswanatha Iyer, who groomed Lalitha into an accomplished musician, who became a graded artiste of the All India Radio. During the days of the Theoretical Physics Seminar, Lalitha supported Ramakrishnan by hosting all the visitors at Ekamra Nivas. After MATSCIENCE was created, Lalitha contributed by giving musical concerts at the Institute on special occasions.

2) Professors I. J. Schoenberg and Louis Nirenberg were two of the eminent mathematicians, who were at the International Conference on Functional Analysis. One of Schoenberg's fundamental contributions is the idea of approximation of continuous function by *splines*, which are piecewise combinations of polynomials, and he spoke about Cardinal Spline Interpolation at the conference.

Nirenberg is a leader in the important area of partial differential equations, and has been recognised with numerous major awards, including the Abel Prize in 2015.

3) B. H. Neumann (FRS) was a group theorist of world repute. After working in England for many years, he migrated to Australia in 1962 to occupy the Foundation Chair in Mathematics at the Australian National University in Canberra. He was Chairman of the Mathematics Department at ANU until his retirement in 1974.

Kurt Mahler is known for seminal contributions to Transcendental Number Theory. He was at the University of Manchester from 1937 to 1963, and although the statistics and mathematics groups were in the same department, there was little interaction between the two groups; in particular, Ramakrishnan and Mahler did not interact in Manchester. Mahler moved to the Australian National University in 1963.

4) Erwin Fenyves was Ramakrishnan's host at the University of Texas at Dallas annually from 1971. A regular attendee of Ramakrishnan's lectures in Dallas was Ivor Robinson, the distinguished relativistic astrophysicist, who built the group in general relativity and cosmology at UT Dallas. Thus, Ramakrishnan spoke on Special Relativity several times in Dallas.

5) Sir Mark Oliphant, a very distinguished physicist, was Director of the School of Physical Sciences at the Australian National University, and after retirement was Emeritus Professor there from the mid-sixties. He was appointed as Governor of South Australia in 1971 and served a five year term. Ramakrishnan first met Oliphant in 1954 during his visit to Australia.

Chapter 36

The Spring Of Hope (1974)

The year 1974 was the spring of hope for Krishna, who aspired for a mathematical career. He was blessed with opportunities to meet the 'masters' and visit many centres of mathematical learning in the United States. He found his 'Guru' in Paul Erdös, who visited Madras at the end of the year and led him to Professor Straus at UCLA.

The year opened with the Anniversary Symposium on Computational Methods and Numerical Analysis, with Professor George Marsaglia from Montreal as the chief participant. This was followed by the Matscience seminar in February at Bangalore on 'Mathematics and Medicine.'

On February 15, I gave a radio talk on Satyendranath Bose,[1] a 'native' Indian scientist of first magnitude with an international reputation. I recalled my association with him at Mussoorie, where he directed the summer school with his uncompromising preference for original ideas.

I received an invitation from Professor Miller to be one of the principal lecturers at the International Conference on Numerical Analysis in Dublin from July 29 to August 6. I accepted it with pleasure since it would give me an opportunity to present my work on matrix theory to a group of numerical analysts.

On the 5th of May, I gave a talk on Swathi Thirunal the prince-composer of Kerala at a local musical association making a comparison of his compositions with those of the Trinity, Thyagaraja, Dikshitar and Shyama Sastri.

It was a pleasant surprise when Krishna received an invitation from Professor Hoggatt[1] to spend a month with him as a Fibonacci Scholar in San Jose, California. Since I received many invitations in the U.S., I decided to accompany him, while Lalitha chose to stay back in Ekamra Nivas.

Krishna and I left for the USA on the 10th of May via Bombay, where we stayed for a few hours at the Sun and Sand Hotel before taking the Air

India 747 to New York. There was a suburban railway strike in Bombay and hundreds of thousands of people were transported by trucks and lorries, one of the strangest sights I have seen in all my travels. The roads were jammed with overloaded vehicles that it was an incredible feat for the taxi driver to get us to the airport in time.

On arrival in New York, we were received by an old student of mine, Thunga and her husband, an official in the U.N., who took us to their comfortable home in Jamaica. The next day we left by American Airlines 707 for Los Angeles, where we were received by Gordon Shaw, who took us to his lovely home in Laguna Beach, where we relaxed for a few days. We then went to San Francisco by the United Airlines Commuter service to be received by Professor Hoggatt, who took us to San Jose, where Krishna was to stay with him for a month as a Fibonacci Scholar. He arranged a comfortable room for Krishna in his neighbour Speasock's house and Mrs. Hoggatt provided delicious vegetarian food, particularly cheese dishes adapted to satisfy the spiced palate of a Mylapore youth.

I left for Winnipeg via Vancouver on an Air Canada DC-8 flight. I gave a series of lectures on matrix theory at the invitation of Professor Kale of the Statistics department. It was during my stay in Winnipeg we heard the exciting news that India entered the Nuclear Club by the successful implosion at Pokharan. Hitherto the paradox of India was that it could reconcile a Rolls Royce civilisation with a bullock cart economy. Now, we felt proud of an atomic blast in a country where domestic gas supply was a luxury, and firewood was still the common kitchen fuel!

From Winnipeg I flew to Columbus via Minneapolis and Chicago, and stayed at the Holiday Inn, which was supposed to be owned by one of the astronauts who had settled down to comfortable terrestrial living after the conquest of space! I gave a colloquium at Ohio State University at the invitation of Professor Edwards, and I met many of the professors at leisure at a delightful party in Professor Heer's house. And then I moved to Tallahassee, where I lectured in the Statistics department of Florida State University at the invitation of Professor Sethuraman.[2] The University there had gained new prestige by the presence of Professor Dirac who, after his retirement, preferred the relaxed atmosphere of sunny Florida with its eternal summers to the sombre dignity of Cambridge with its long drawn winters. I met Dirac and reported to him my work on the Dirac matrices and the generalisation of the Clifford Algebra.

From Tallahassee I went to Oak Ridge, where I lectured at the Statistics Division a the instance of Professor Uppuluri. He took me on a delightful excursion to the Smoky Mountains and Gatlinbeg.

I returned to Washington to lecture at the Naval Research Laboratory in response to Maurice Shapiro's invitation. The capital was agog with 'Watergate fever,' which demonstrated that even the President was not above the law, that the vitality of American democracy could survive the shock of his guilt.

There was the inevitable visit to UT Dallas to lecture at the invitation of my gracious host, Professor Fenyves. From Dallas I flew to San Francisco and stayed at Hotel Durant in Berkeley. Krishna came to Berkeley with Hoggatt, and we met Professor Max Rosenlicht, who explained to Krishna the possibilities of graduate study in his department.

During the month's stay in San Jose, Krishna had advanced not only in his research but in his tennis since, playing with Hoggatt, he had to stretch his limbs and his patience to keep the score even. It is impossible to meet more generous hosts than the Hoggatts, and I particularly enjoyed hearing Mrs. Hoggatt calling Krishna in the correct Sanskritic intonation!

From San Jose, we moved to Los Angeles to be received at the airport by Professor Straus of UCLA. Krishna's interview with Straus[3] turned out to be one of the most fortuitous events in his career. He was to join UCLA a year later for his graduate studies on the advice of Professor Erdös, a great friend of Professor Straus. After my lecture at the University of Southern California, we flew from Los Angeles to Washington by American Airlines, where Krishna met Professors Larry Goldstein and Bill Adams at the University of Maryland, who had evinced great interest in his work.

Our next destination was Boston to meet Professors John Tate and Garrett Birkoff at Harvard, and Professor Stark at MIT. We stayed with the Viswanathans, who had now become 'Boston Brahmins' shedding their ten year old addiction to Manhattan. We left for New York by American Airlines 727 and returned to Madras by Air India 747.

Within a week of returning to Madras, I left on the 25th July by Air India 747 and arrived in London the next morning. I spent a day in London, reviving memories of previous visits as I walked along Trafalgar Square and around the Buckingham Palace. Arriving in Dublin the next day, I had a 'Mercedes' reception and was taken directly to Trinity College, a lovely old fashioned building set amidst beautiful lawns and gardens.

At the entrance to the Trinity College (Dublin), there is a statue of Oliver Goldsmith, who had enlightened the world and enriched its literature with 'The Vicar of Wakefield' as durable as the English language itself. The week-long conference was exciting to me in every way, making new contacts with numerical analysts from all over the world at the academic sessions and also during the pleasant excursions to the Irish countryside. There is a strange paradox in Ireland in preserving the monuments and relics of historical significance. On the one hand, they represented the origins of Irish traditions so well-preserved in daily life. On the other, they were reminders of Irish subservience to British rule and military occupation! A similar situation was created in India after the dawn of freedom in preserving the historic monuments of British rule.

In my lecture, I presented the new approach to matrix theory to a new group of mathematicians. There was a dinner at the house of the parents of Professor Miller, the convener of the conference, when I really understood the amount of effort and passion that goes into the maintenance of an 'English garden,' a delightful piece of manicured nature in which every leaf and blade of grass has its appointed place in a setting of exquisite floral beauty accentuated by man's nourishing care.

Then I flew by BEA Trident to Manchester and travelled by train to York, musing how my good old Manchester had acquired an American flavour of prosperity. At York, I was the guest of Professor Williamson, a most gracious host, a gentleman in the most English sense of the term, a steadfast friend ready to help but firm in convictions, warm hearted without being exuberant, shy of praise but responsive to goodness, confident of his powers but circumspect in general. The University is set amidst lovely surroundings with the open air smell of the English countryside. He took me on an excursion to the lovely Lake district, which looks even more beautiful to persons familiar with Wordsworth's luminous verse. During my stay in York, I had the pleasant surprise of seeing a paper by the eminent mathematician Norman Levinson[4] of MIT in the Journal of Mathematical Analysis and its Applications entitled 'Ramakrishnan's approach to relativity' (see Appendix 10), certainly a title flattering to my vanity!

From York I went to London by train and then by Swissair DC-9 to Geneva. The weather was so good that I saw the white cliffs of Dover, the coastline of France and the exquisite loveliness of the Swiss countryside. At Geneva I spent a few days at CERN, where one can hear more physics being discussed over lunch than in year-long courses of conventional lectures. This time there was another excitement to compete with high energy physics, the Watergate scandal in American politics! Would Nixon resign? A leading American physicist predicted that he would, and he did within forty-eight hours of asserting he would not!

There was an inevitable trip on the Lake of Geneva, with haunting memories of our stay in Switzerland in 1950 and 1951. Then I moved to Trieste, where I spent a week with Professor Salam at the International Centre for Theoretical physics. I arrived in Madras on the 14th August, just in time to attend the reception at Raj Bhavan with Lalitha on Independence Day, the 15th August.

We had the Matscience conference on 'Green's functions' in Mysore. From there we went on one of the most delightful motor trips via Mercara and Mangalore to offer prayers at the Mukkambika temple in Kollur, where there is a legend that the Goddess Mukkambika as the name implies, answers the prayers of the dumb and restores to them their power of speech. Lalitha sang in the temple following the advice of Chembai, who ascribed the indestructible timbre of his resonant voice to Divine blessing. On our way back to Mangalore, we offered prayers to Lord Krishna at Udipi, and returned to Madras via Bangalore.

It occurred to Krishna that we should contact Professor Erdös, the world famous Hungarian number theorist, who had a reputation for discovering and encouraging youthful talent. But he was always on the move, an untiring jet traveller, and we were wondering how to reach him. So I just wrote to his 'nominal' Hungarian address, and to my pleasant surprise we received a reply from Canada that he would be in India in December to participate in Mahalanobis symposium at the Indian Statistical Institute in Calcutta, and would like Krishna to meet him there.

I visited Calcutta to participate in that International Symposium, but Krishna could not accompany me since he was busy preparing for his final B.Sc. examination. I presented his paper on number theory. Professor Erdös,[5] who was present at the lecture, did not conceal his disappointment in meeting me instead of Krishna, and said that he could reroute his trip from Calcutta to Sydney by stopping in Madras enroute to see Krishna. I felt it would be appropriate to invite him to Madras and give a talk at Matscience. After the conference, at the invitation of Professor Dutta, a pleasant and friendly host, I gave the Cullis memorial lecture at the Calcutta Mathematical Society on my new approach to matrix theory.

During Christmas and the New Year season, we had a stream of visitors at Matscience diverted from the Calcutta conference — Krishnaiah from Dayton, Kallianpur from Minnesota, Hammersley from Oxford and above all, Paul Erdös from Hungary.

I took Professor Erdös to the Governor, who was deeply interested in fostering creative mathematics and supporting young talent. Erdös took out a sheaf of currency notes from his coat pocket — the entire honorarium he had received from Matscience and the University of Madras — and handed them to Governor Shah for his special educational fund: when the Governor solicitously inquired how Professor Erdös, with his frail physique, could stand the strain of international travel, he replied that he was now in Madras on his way from Canada to Australia, before getting back to Los Angeles in February! He was completing a paper every week, collaborating with 'local talent' at every stop! Erdös spent a whole day with Krishna listening to the exposition of his work. With percipient concern, he told me that Krishna should start work without delay under Straus at UCLA. He planted the seeds of ambition on an impressionable mind, the greatest gift a man of achievement can bestow on one who aspires to achieve.

Notes

1) Verner Hoggatt Jr. (1921–80) of San Hose State University was the Founding Editor of the journal *The Fibonacci Quarterly*. He was a Fibonacci enthusiast and devoted all his research to the study of Fibonacci numbers.

2) J. Sethuraman was Distinguished Professor (now Emeritus) of Statistics at Florida State University in Tallahassee. He did his BSc Honours in statistics at the Presidency College, Madras during 1954–57. It was announced then that Ramakrishnan would be giving lectures in the Inter-Collegiate Lecture Series for the Honours students. Word went around that Ramakrishnan had recently returned to India and would be lecturing on new topics. This kindled Sethuraman's interest and he attended Ramakrishnan's lectures on Stochastic Processes.

3) Ernst Gabor Straus (1922–83) was a versatile mathematician, who started with research in the mathematical aspects of relativity in New York and Princeton, but after he moved to UCLA, shifted his focus of research to combinatorics, number theory, and analytic functions. He was the last student and assistant of Albert Einstein. In combinatorics and number theory, he collaborated with the great Paul Erdös, and the two were very close friends.

4) Norman Levinson (1912–75) was a brilliant mathematician with a wide ranging expertise. He started with both his BS and MS degrees in Electrical Engineering at MIT, but in that process took a lot of mathematics courses. He switched to mathematics and received his PhD at MIT under the direction of Norbert Wiener. His fundamental contributions spanned complex analysis, differential equations, number theory, Fourier analysis and signal processing. He and Ramakrishnan were Associate Editors of the Journal of Mathematical Analysis and Applications (JMAA), where Ramakrishnan's paper "Einstein — a natural completion of Newton" appeared. Even though Levinson did not work in Relativity, this purely mathematical formulation of Ramakrishnan attracted his attention and motivated him to write his paper "Ramakrishnan's approach to Relativity," which appeared in the JMAA in 1974 (see Appendix 10). Shortly after this, Levinson died suddenly of a brain tumour in 1975.

5) Paul Erdös (1913–97) was one the most influential mathematicians of the 20th century, who made seminal contributions to number theory, combinatorics, set theory, and geometry. Author of over 1500 papers (most of which were in collaboration), Erdös was constantly on the move, rarely spending more than a week at one place. His life mission was to spot students with mathematical talent and encourage them.

Chapter 37

Tidings From Trieste And California (1975)

1975 was a significant year for Krishna when he was provided with opportunities to pursue a career of research through the triple sponsorship of Professors Erdös, Salam and Straus.

The year started with our Anniversary Symposium with Professors Hammersley from Oxford and Vinze from Hungary as our distinguished visitors. It was incredible that we had the opportunity to receive Mrs. Vijayalakshmi Pandit as the chief guest on the evening when Lalitha gave a concert at Matscience in honour of the participants. We were naturally delighted to receive so gracious a lady on whom God has bestowed all the talents and the rewards worthy of one who was elected once as the President of the General Assembly of the United Nations. In introducing her, I said that age had not withered nor the responsibilities affected her charm and poise.

Professor Hammersley gave an excellent lecture in Oxford accent and style. When I introduced him as the Head of the Statistics department, he politely corrected me with characteristic British humour that he was neither the head nor the tail since he was the only one in his Department!

On the 25th of January, Krishna received a letter informing him of his admission to UCLA. On the same day, I received an invitation to participate in the International Symposium on Statistics at Dayton, Ohio and felt happy that Lalitha and I could accompany him to the US.

On the 31st of January, the eminent statistician C. R. Rao visited our Institute and this was followed by a visit of Phatarfod of Australia.

The 7th February was a happy day for Krishna when he received a kind letter from the great mathematician Professor Hirzebruch, offering him a research scholarship in at the University of Bonn, Germany, where he could work with eminent number theorist Professor Don Zagier.

In February I had the opportunity to receive Professor Kale, my host at Winnipeg, at our Institute when he gave a series of lectures. It was a pleasant surprise when on the 5th of March Krishna received an offer from ICTP to participate in an International Summer Institute on Complex Analysis for a period of three months — May to August — offering both airfare and a liberal sustenance allowance. This was a gracious gesture to an undergraduate student. Next, on the 17th of March, which turned out to be the most significant day of his career, Krishna received the award of Chancellor's fellowship from UCLA, offering complete support for four years. In view of these developments, he could not visit Bonn to work with Professor Zagier.

The 30th of March was a memorable day at Ekamra Nivas when the Governor K. K. Shah visited our family house and stayed for three hours to listen to a concert by Ramani and had dinner along with about hundred guests.

Krishna passed his B.Sc. Examination in flying colours. After the exams, he gave a lecture at Max Mueller Bhavan at the invitation of its Director Lennertz, on the Riemann hypothesis as a tribute to one of the greatest mathematicians of all time. Before our departure to the US, we made a trip to Guruvayoor to offer prayers to Lord Krishna, and to Tirupathi to invoke the blessings of our family deity, Lord Venkateswara.

Chapter 38

Trieste And U.C.L.A. (1975)

On the eve of our departure to Italy and the United States, in a most gracious gesture, the Governor of Madras, His Excellency K. K. Shah, a mathematics enthusiast, invited Krishna for dinner at the Government house.

We left Madras on the 15th May by the Air India Boeing 707 and arrived in Bombay to take the Air India Boeing 747 to Europe. On arrival in Rome, we were told that the flight to Trieste was possible only the next day because of a strike. We stayed at the pretty Golden Beach Hotel near the Rome Fumicino airport, and the next day we took an Alitalia DC-9 to Trieste. On arrival, we were taken to our apartment in lovely Sistiana overlooking the azure waters of the Adriatic. The two and a half months stay in Trieste, at the kind invitation of Salam, turned out to be a one of the most enjoyable and fruitful sojourns abroad in all our travels. Krishna, at the age of nineteen, was the youngest participant at the International Conference on Complex Analysis, which opened with a magnificent address by Professor James Eells of Warwick.[1] He not only attended over two hundred lectures by over twenty professors, but presented his own research in number theory. On my part, I gave four seminars on my work at the ICTP. In between our active academic programme, we had plenty of tennis and frequent walks in the gardens of the Castle Miramare nearby.

During this period, I made a short trip to Germany, Belgium and England on lecture assignments, while Lalitha and Krishna stayed on in Trieste. I went to Wurzburg in Germany by train, and the journey took me through lovely Austria.

Wurzburg is a lovely German town, set amidst wooded hills and vineyards, the ancient castle reminding us of the turbulent years when dukes and abbots held immense power and influence in feudal Germany. I made a delightful excursion to Rothenburg, the city of castles, which was spared by the Americans during the great bombing, a reminder that there is ethics even in war! The lecture at Wurzburg was well-attended and it was a pride to note that it was the University where Roentgen first discovered X-rays.

I left for Frankfurt and then went to Liege at the kind invitation of Professor Garnir. At the University, I lectured on my new work in the special theory of relativity, which was received so well that we continued the discussions over a pizza dinner in a pretty tavern-like restaurant with a typical European atmosphere. I flew to London by Sabena Boeing 727 and proceeded directly to Cambridge by train.

The weekend in Cambridge was spent with my nephew Ramu[2] at Trinity College, in relaxed leisure and I enjoyed the casual walks with him through the lawns and gardens, colleges and corridors. Every nook and corner was sanctified — here was the window from which Newton gazed at the heavens, there was the urn which brought the gold from the Royal Treasury to build the College Chapel, that was the blackboard where Bohr explained to Rutherford the stability of the atomic orbits and this was the table on which Dirac worked out his relativistic wave equation.

I thought of my own country, where the intellectual and cultural traditions can be traced to a period where mankind elsewhere was in its swaddling clothes. Yet, in a fitful fever for modernisation, we were abandoning, sometimes erasing with a vandal's fervour, such a rich and precious heritage. While of course tradition should not stifle progress, it should be kept alive to sustain present achievements and inspire future effort. England was as proud of its Cavendish as of its monarchy — and what a line of 'Kings,' Maxwell, Lord Raleigh, J.J. and Rutherford!

Ramu and I called on Sir James Lighthill, who was occupying Newton's Chair as the Lucasian Professor of Mathematics. I had the opportunity to hear a hundred lectures of his at Manchester twenty-five years before. He inquired about Krishna, and felt happy that he had secured a place for work for his PhD in Mathematics at the University of California. With his characteristic goodwill he remarked 'Los Angeles, it is warm and lovely there' to which I replied 'not as tranquil as Cambridge,' though, like Krishna, I was as much in love with the 'City of Angels' in the Golden State as with the tranquil charm of Isis and Cam. I also made it a point to call on Professor Alan Baker because Krishna was interested in his new book on Transcendental Number Theory had just appeared.[3]

Later I left for York, where I stayed at the Vanbrugh College enjoying the hospitality of my gracious host Professor Williamson. He took me on a delightful motor trip through lovely country, where we saw the ruins of an old abbey, reminders of the historic confrontations of Papal supremacy and the sovereignty of Henry VIII. It was during my journey by train back to London from York on the 26th of June that I heard of the fateful happenings in India and the declaration of emergency. In the idle comfort of the restaurant car I watched the moving panorama of the pleasant English countryside. I mused with envy and admiration at England's great democratic traditions. I thought of my father, who revelled and regaled in comparing the merits of British and American democracies, of the decisive role of the Supreme Court in the interpretation of the written constitution of that happy land of Jefferson and Lincoln. Had Gandhi and Nehru sought and wrought in vain to bring us freedom, which we could not preserve? God alone could save our country and He would answer our prayers only if we learnt the twin lessons of democracy — freedom without discipline is chaos, discipline without freedom is tyranny. And so, with unmanly discretion, I resolved not to think of these 'difficult' problems I could not solve. I took a cab from Victoria station to Heathrow to take the Alitalia to Milan enroute to Trieste.

At Trieste, we watched the Wimbledon tennis finals on television when Arthur Ashe beat Jimmy Connors, a triumph of patient effort over power and speed.

We left Trieste on the 9th August by train for Milan since the Alitalia flights were cancelled due to a strike. At Milan, we took the British European Airways Trident for London, where we stayed at the luxurious Posthouse Hotel near the airport. Even British hotels were being run on American style under the pressure of inflation — gone were the days of the leisurely breakfast with table service. The only thing that remained British in the hotel was the name 'Buttery' for a restaurant where breakfast was served in buffet style.

We took a three hour tour of London by omnibus to see the famous sights of the historic city. The flight by Air India from London to New York was a delightful experience and the fifteen days we spent in that fabulous city were most enjoyable. I gave a lecture at the Yeshiva University at the instance of Professor Joel Lebowitz, gracious and hospitable as ever. Lalitha gave a concert at the Ganesha Temple, arranged by V. K. Viswanathan, my New Yorker student. Gomathi Sundaram, Lalitha's childhood friend from Trivandrum, now settled in New York, provided violin accompaniment. From New York, we flew to Boston by American Airlines to be received by my nephew Sivaprasad, who was comfortably settled as a professor in the University of New Hampshire in Dover. He had a lovely house in Greenland in the neighbourhood of Dover, and he took us on an excursion to the White Mountains. I lectured at his University and then we moved to Montreal by Eastern Airlines DC-9. On the day we were to leave for Calgary by Air Canada, there was total electricity failure for a few hours and we wondered at the 'frailty' of modern civilisation, which would 'crumble' at the loss of electric power. Fortunately power was restored, the airport was opened and we made a comfortable flight to Calgary, where we hired a car to make an excursion to the National Parks at Banff and Jasper. While the Himalayas could be admired only in distant wonder from Dalhousie or Darjeeling, we could travel through the gorgeous valleys in the Canadian Rockies in the supreme comfort of an automobile.

From Calgary we flew to West Yellowstone and spent two days driving through Yellowstone and the Grand Teton National Parks. They fulfilled our expectations and Lalitha was particularly excited about the hot springs, wonders of nature reminding us of the explosive power of volcanoes. The national parks of the United States display the prodigality of Nature. Yellowstone, the oldest and largest, has all the variety and profusion of natural beauty — towering mountains and plunging canyons, lush forests and brush of bizarre colours, spectacular waterfalls and bubbling hot springs and above all that wonder of wonders — the world's greatest geyser, the Old Faithful, gushing at inexplicably regular intervals.

We then flew by Western Airlines to San Francisco via Salt Lake City. Professor Hoggatt received us and took us to the comfortable Edgewater motel. At San Jose, we had plenty of tennis and excellent food cooked by Mrs. Hoggatt, who displayed her culinary skill as a Rumanian born. I lectured at San Jose and we flew to Los Angeles by United Airlines DC-8. We were received at the airport by Gordon Shaw, who took us in his minibus to his lovely house at Laguna beach. On the way we stopped at Sproul Hall, where Krishna was to stay during the first year of his college life. At Laguna Beach, we enjoyed the lavish hospitality of Gordon and his family.

Krishna joined UCLA on the 24th of September and we stayed with our friends the Sankas for a few days at Canoga Park. It was a most touching moment when we took the plane at Los Angeles for Honolulu. The approach to Honolulu was breathtakingly beautiful — this cluster of islands is God's gift to the affluent nation, which has made a heaven of these little pieces of earth amidst the awesome expanse of the mighty Pacific. We stayed at the glamorous Princess Kauilani Hotel, enjoying the magic atmosphere of Hawaii.

From Honolulu we flew to Tokyo by Pan Am 747 and stayed at the luxurious Pacific Hotel, where we had dinner in the glass-walled hall overlooking the Japanese garden with artificial waterfalls. Then to Hongkong, the pearl of the Orient, a honeymooner's dream of an island paradise. The Hotel Miramare was a lavish combination of oriental opulence and Western comfort. It was a particular pleasure to lecture at the University to a group of Chinese post-graduate students. Later, they took us to a Chinese vegetarian restaurant, the only one of its kind in that city.

From Hong Kong we flew by JAL DC-8 to Singapore, where we stayed at the magnificent Marco Polo Hotel. Singapore is a shopper's paradise and one of the cleanest cities in the world, getting cleaner every year. We arrived in Madras on the 4th of October 1975, just in time for Navarathri.

In late November, I had an opportunity to visit Hyderabad to lecture at the Osmania University on Stochastic Processes and Special Relativity. It was a handsome gesture on the part of the Vice Chancellor, Jaganmohan Reddi, to have attended my lecture on relativity, when I mentioned that the concept of simultaneity was slightly modified without affecting the plea of 'alibi' available to an accused in criminal law!

In early December, we arranged a music concert by flute Ramani and a dinner at Ekamra Nivas in honour of Professor Emery, a guest of the British Council. At the end of December, Gordon Shaw from California and Gourdin from Paris arrived to take part in the Anniversary Symposium at the dawn of the next year.

Notes

1) The International Center for Theoretical Physics had made great strides in its first decade, and so Salam was interested in adding a Mathematics Department to it. With this in mind, the ICTP conducted a few Summer Institutes, the first of which was on Complex Analysis in 1975, organised by Professor James Eells of Warwick. In 1986, he was appointed as the first Director of the Mathematics Section of the ICTP.

2) V. S. Ramachandran (alias Ramu), now at UC San Diego, Ramakrishnan's nephew, is a neuroscientist of world repute. In 1975, he was doing his PhD at Cambridge University when Ramakrishnan visited there. Even as a medical student in Madras, Ramachandran was conducting research on the binocular theory of vision at his home, and at the suggestion of Ramakrishnan, sent his work to *Nature*, where it was published. Ramakrishnan supported Ramachandran in his decision to take to a research career, when his father wanted him to become a medical doctor. There was always a special bond between Ramakrishnan and his brilliant nephew Ramachandran.

3) Professor Alan Baker (FRS), did revolutionary work in Transcendental Number Theory that earned him the Fields Medal in 1970. Baker wrote a definitive book on the subject, which was published by Cambridge University Press in 1974. While visiting Cambridge University in 1975, Ramakrishnan purchased this book there and on returning to Trieste, presented it to Krishna with the inscription: "To Krishna, in the hope that he will attain transcendence in number theory."

Lalitha's concert in honour of Madame Vijayalakshmi Pandit, Jan. 1975. - Author

Alladi and Lalitha Ramakrishnan with Astronaut Alan Bean at Matscience, 7 March, 1976. - Editor

Chapter 39

A Year Of Strange And Significant Events And America's Bicentennial (1976)

1976 was the year of America's Bicentenary. It turned out to be a year of strange events at Madras and significant events in my personal life.

The Fourteenth Anniversary Symposium was inaugurated on the January 6 by our Chairman Nedunchezhiyan, with lectures by Gourdin and Beauregard of France, Bleuler and Sandhas from Germany, and Pisent from Italy. It was a privilege to have come into contact with Professor Bleuler, who is not only a great scientist but a gentleman imbued with the finest of graces of European culture. My friendship with him was to be reinforced by my visit to Bonn a year later. Beauregard[1] was keenly interested in my work on special relativity, for he had written a book on the subject.

The Prime Minister Indira Gandhi arrived in the city for the Pugwash meeting, and it was a pleasant surprise to see N. R. Puthran, the registrar of the Tata Institute, who arrived earlier to make arrangements for an exhibition in honour of the visitors of the Pughwash meeting at the Taj Coromandel Hotel. I knew him when he was a handsome young accountant to Professor Bhabha, and so I invited him to visit our Institute. He must have been amused to see us happy and contented with the modest buildings and more modest budget of Matscience, while the Tata Institute had grown to a mammoth edifice with a matching budget in the last twenty-five years.

During Pongal Holidays, I got some new results on a matrix decomposition theorem. Just before Republic Day, I visited Hyderabad at the invitation of my brother Prabhu to lecture to the computer society on matrix theory, expressing my hope that the work on the decomposition theorem was relevant to Numerical Analysis. Returning to Madras, I enjoyed a magnificent concert by Semmangudi before attending the Republic Day Reception at the Raj Bhavan.

On the 31st of January, without the slightest warning, there occurred one of the strangest events in Indian history — the sudden dismissal of the local government by the Centre and the immediate installation of President's Rule through the appointment of two advisers. Actually, the news came when I was attending a dinner at the Sudarshan International Hotel arranged by my friend Viswanathan of Enfield, while Lalitha was participating in a music festival at the Tyagaraja Vidwat Samajam singing the Krithanas of Purandaradasa!

Against these strange happenings, under the auspices of Government of Madras, I conducted a seminar at Matscience on 'Careers in mathematical sciences.' I drew attention to two circumstances that scared most students away from mathematical careers — firstly, for the same emoluments, mathematicians have to meet international standards in creative research, while those in other professions need satisfy only local criteria for efficiency. Secondly, mathematicians in India are severely critical of one another, to the point of jealousy, while experts in other professions 'stick together' in defence of their common rights and privileges.

It was then that a tragedy occurred in our family when my sister's husband V. M. Subramaniam, who had just settled down in Madras after a long tenure as a U.N. Official at Bangkok, died after a heart attack. He was a supreme example of a self-made man who, after an outstanding academic career, had to work hard and wait patiently before moving into a well-deserved position as a steel expert in the ECAFE. He was not only very efficient, but intolerant of inefficiency, and unsparing in judgment.

On the 4th of March occurred an incredible event when we had the opportunity to receive 'a man from the moon' — Astronaut Alan Bean — at our Institute. This fortuitous event occurred after my visit to the U.S. Consulate for dinner, when my friend La Salle had invited me to meet the astronaut. When I asked him how the astronauts were selected out of a quarter billion Americans, he promptly replied, with the apparent casualness of a pragmatic American, that it was just a principle of 'logical selection' of the toughest and ablest person who could do the job! Presenting him with the Matscience Annual Report for 1969, the year of man's journey to moon, I wrote the inscription:

"What can I offer to a man who has walked the surface of the moon, an achievement which even Shakespearean diction is inadequate to describe and Einstein's imagination finds impossible to comprehend. He seems to smile and talk just like any of us, but the very thought that he has looked at the earth from the Moon's surface gives me a thrill too deep for contemplation. I wish to pay my tribute through the simple thought that mathematical reasoning convinces me that it is impossible to think of space and time

without including the concept of matter, and this is the integral nature of God's creation."

An impressive incident occurred at the U.S. Consulate after his lecture, when someone asked him whether profound philosophical thoughts passed through his mind while watching the distant earth from the Lunar surface. He replied that he had very little time to contemplate on philosophical questions for he had a job on hand, to collect the moon rocks and rush to the Lunar Module hoping that the instruments would work without fail and he would be back safe on earth! It was a question of survival and there was no time for idle thoughts, which he could engage in at leisure after reaching earth! Even this he did not do, for he was put in-charge of the SKYLAB mission, another job to do in the true American sense of the term.

On the 19th May, I left for New York by Air India Boeing 707 from Madras and the Boeing 747 'Emperor Vikramaditya' from Bombay. Lalitha decided to remain in Madras since she felt she would be 'imprisoned' in the hotel room when Krishna was staying in the dormitory, and I would be moving round on a hectic lecture schedule. After an all day journey through Kuwait, Frankfurt and London, I arrived in New York and stayed with my former student, Thunga and her husband Satyapal, in their comfortable home in Jamaica, New York near JFK Airport.

I lectured at SUNY, Albany on 'Einstein is a natural completion of Newton' and enjoyed walking round the fabulous multi-million dollar plaza built by Rockfeller when he was Governor of New York. I spent a few hours at the Rensselaer Polytechnic in Troy before leaving for Rochester, where Professor Okubo arranged my lecture on internal quantum numbers. From Rochester I flew to Toronto by Allegheny Airlines and on to Winnipeg by Air Canada DC-8. My lecture on new concepts in matrix theory was well-received and I enjoyed the hospitality of the Indian of mathematicians in Winnipeg at the University of Manitoba.

From Winnipeg I went by Air Canada DC-8 to Calgary, where I lectured at the University. It was a pleasure to meet Professor Erdös, with whom I had lunch and dinner. Professor Aggarwala took me on a delightful excursion to Banff National Park and Lake Louise, the pride of Canada.

From Calgary I flew by Western Airlines 737 to Denver. Professor All-gower received me and took me to Fort Collins, a lovely town at the foot of the magnificent Rockies where I lectured at the Colorado State University on Matrix theory.

From Denver I flew to Los Angeles by United Airlines DC-10 and was received by Krishna, Venki and Vasu! I was taken by Venki to the Hotel

Miramare in lovely Santa Monica, overlooking the Pacific ocean. Krishna shifted from the dormitory to stay with me. I visited UCLA and met Krishna's teachers, Ernst Straus, Harry Niederreiter and Richard Arens.

From Los Angeles I flew to Dayton to participate in an International Conference on Statistics organised by Paruchuri Krishnaiah, when I met my good friends Professors C. R. Rao and Uppuluri. It was during one of the sessions I heard an interesting lecture on 'pattern recognition' by my old friend Richardson, whom I knew through Dick Bellman in 1962. He was now working at Rockwell International at Malibu and he invited me to lecture there during my next visit to California.

I left for Washington by TWA 727, where I lectured the kind invitation of Professor Uberall at the Catholic University after a delightful lunch at an Indian restaurant, and left for New York by American Airlines. I was received by V. K. Viswanathan, who took me to his new home at Yorktown Heights. Krishna arrived the next day from Los Angeles, and I joined him at Jayaraman's home in New Jersey before leaving for Princeton as the guest of my great friend Bogdan Maglic. I lectured at the Fusion Research Laboratory, where Maglic had an active set-up investigating his new MIGMA method for clean fusion. After the lecture, Krishna and I left Princeton by car to reach JFK Airport, just in time to catch the Air India 747 for Bombay, where we connected to Indian Airlines to Madras.

Krishna plunged into tennis at the Mylapore Club and a busy lecture schedule, but within a week of his arrival in Madras, he complained of pain in the back and on examination by Dr. Sundararaman, the eminent surgeon, he had to undergo surgery for an abscess called pilonidal sinus. He was confined to bed for two weeks, which took a large slice of his vacation. Within two weeks, he resumed his lecture engagements and his tennis with renewed vigour.

On the 4th of July, America celebrated its Bicentennial with festivals and parades galore all over the world. To mark America's entry into its third century, the Viking I made a successful landing on Mars on July 20, followed by Viking II in September on the Utopian plains. Reports from the unmanned space crafts revealed rocky reddish surfaces and blue skies and some evidence of possible biological activity.

On the 29th of July, at the kind invitation of my friend Dr. Santappa, I lectured at the Central Leather Research Institute (CLRI) on Einstein — a strange subject to a group engaged in the smoothening of hides and the toughening of skins! On the 4th of August, I gave a talk at the All India Radio on 'Mathematical Sciences in India' emphasising that half-hearted support to advanced institutions resulted only in a waste of public funds, continuous brain-drain to more hospitable countries and frustration among those who remained in India. After convalescing, Krishna gave his first lecture at the C. P. Foundation 'On Analogues to the Hardy Ramanujan Theorems' and a radio talk on 'My association with Paul Erdös.' He said: "Paul Erdös is so friendly that people feel it is always a pleasure to be his host. It would be improper to call him a nomad, for a nomad by definition is a homeless man. On the other hand, to Professor Erdös every University is his home. So we may justly describe him as a mathematician eternally on the move from one home to another."

During August, Professor Sudarshan and I were invited by the University to inspect the Department of Physics and submit a report. I naturally wanted to further the interests of my old department, which I had the privilege to initiate in June 1952.

It was then I heard a lecture by Hyman Bass at the Ramanujan Institute and invited him to visit Matscience. Krishna had the opportunity to speak to him about his work, and Professor Bass reacted favourably by inviting him to lecture at the Columbia University on his way back to California.

While Krishna was in India, we went on a southern tour in response to kind invitations to lecture at Madurai and Pollachi. We then left for Coimbatore and made a trip to Guruvayoor to have a darshan of Lord Krishna. It was a wonderful trip through the verdant forests and valleys of the Palghat district. While Krishna lectured at the P.S.G. College of Technology in Coimbatore, I addressed a 'mass' meeting of graduates and under-graduates of the P.S.G. Arts College on special relativity. From Coimbatore, we went to Mysore via Ooty for the Matscience conference on Matrix Analysis and Number Theory. Krishna gave a talk at the conference on his joint work with Paul Erdös, work on number theory which was to appear in the Pacific Journal of Mathematics. Following that conference, in Bangalore on our way back to Madras, he gave a talk at the Indian Institute of Science at the kind invitation of Professor Mukunda.

Notes

1) Olivier Costa de Beauregard (1911–2007) was a famous French physicist, who did fundamental work in relativistic and quantum physics. His visit to Matscience sparked his interest in Ramakrishnan's work in Relativity. Ten years after meeting Ramakrishnan, he wrote a paper on the Lorentz transformation and the velocity composition approach to Relativity, in which he emphasised Ramakrishnan's approach to Relativity. The paper appeared in *Foundations of Physics*, Vol. 16, (1986), pp. 1153–1157 (see also narrative in Chapter 54 and Appendix 11).

Alladi Ramakrishnan giving a lecture on the theory of probability. His handwriting was beautiful. - Editor

Professor John Bardeen, two times Nobel Prize Winner, with the author at Matscience (Jan 1977). - Editor

Chapter 40

Rhapsody In California (1977)

On the 15th of September 1976, Krishna left for the United States and he lectured at Columbia University in New York, and at the University of Maryland before reaching Los Angeles. He informed us that he had enough of dormitory life and would be moving to an apartment.

In October, I formulated a multiplicative conservation law for velocities equivalent to the velocity transformation of Einstein. I was able to complete a paper entitled 'Unnoticed symmetries in the theory of relativity,' which was communicated to the Journal of Mathematical Analysis and its Applications. I lectured at the Jain College, at a meeting of young students and teachers, to whom I talked on new symmetries in relativity.

On the 11th of November, I received a kind letter from Helmut Hasse, the famous German mathematician, paying a fine tribute to Krishna's work.

In early December, we had a visitor to Matscience — Professor Eliezer from LaTrobe University, Melbourne, who gave two lectures with the quality and timbre of the Cambridge tradition.

On New Year's day 1977, we attended Semmangudi's magnificent concert at the Music Academy. The next day we heard Ravi Shankar, the sitar maestro, and though I did enjoy the uncloying sweetness of his concert, I must confess I love Keerthanas of Carnatic music more.

In mid-January, in Madras, we witnessed the Test Match between the M.C.C. and India. After dull dour batting, England was all out for 262, while India made a poor reply of 162. The match ended in a victory for England. With India all out for an ignominious 80. The Madras crowd was so sporting that it carried Tony Grieg, the English captain, on its shoulders, an honour he enjoyed with obvious delight. I was wondering how the imperturbable Jardine would have reacted on such an occasion!

The same day came the sensational announcement by the Prime Minister that the national elections will be held in March. Why she did it when she was in absolute control and command of the situation puzzled everyone. Clearly she wanted the people's mandate for a further lease of power.

The 24th January, the fifteenth anniversary symposium at Matscience was inaugurated. I gave lecture on 'Lorentz to Gell-Mann' the next day at the anniversary function, Professor John Bardeen, the Nobel Prize winner from Urbana, Illinois, visited the Institute; he attended Lalitha's concert in the evening and stayed for dinner. Malcolm Adiseshiah, the Vice-Chancellor of Madras University, delivered the anniversary address, while Santappa unveiled the portraits of Lorentz, Gell-Mann and Feynman.

On the 5th of February, we left for Pollachi via by car in response to the kind invitation by my good friend N. Mahalingam, a leading businessman of South India, a classmate of mine, who was the enlightened sponsor of well-endowed institutions of higher learning in Pollachi. I gave my lecture on 'Lorentz to Gell-Mann,' with Mahalingam presiding over the function. We had a dinner at the magnificent house of Mahalingam, where he has so well-preserved the revered memory of his illustrious father. At his request, Lalitha gave an open air concert well-appreciated by a large audience of students, professors and citizens of Pollachi.

We left Pollachi for Madurai, where I gave two lectures at Madurai University, one on 'Lorentz to Gell-Mann' and on 'Unnoticed symmetries in Einstein's special relativity theory.'

We had a pleasant surprise in February when Mrs. Sharon Marcus, the mathematics Librarian at UCLA and her husband (good friends of Krishna), who were on a tour to India, visited Ekamra Nivas. We took them to a flute concert at a local sabha and later to a Western concert on the spaciously lighted terrace of the Alladi House, sponsored by Lennertz, the Director of Max Muller Bhavan, who was our tenant.

The nation rocked with excitement as the national elections were held on the 16th and 17th March, and we had the greatest surprise in Indian political life when we heard over the radio the new Janata Party winning seat after seat against Indira Gandhi's congress. It was 'Janata wave' and we could not believe our ears when at midnight it was announced that Indira Gandhi had lost the election! The Janata party obtained absolute majority and lifted the emergency rule that was enforced by Indira Gandhi. Morarji Desai was elected the leader of the Janata Party and was sworn in as the Prime Minister.

It was in such an atmosphere of incredible excitement we left for New York by Air India on the 26th of March. I gave a lecture at City College of New York, of which Robert Marshak was the President.

It was then we heard of the worst disaster in aviation history claiming 531 lives — the incredible collision of a K.L.M. Boeing 747 as it took off, with a Pan Am 747 on the runway at Tenerife on Canary Islands. The tragedy was all the more poignant since the disaster need not have happened at all. It was just avoidable human error for the K.L.M. pilot just heard the word 'OK,' while the actual message from the control tower was 'OK, wait five minutes' to allow the Pan American Boeing to cross the runway! Incredible, impossible in this era of the most sophisticated means of communication making conversation possible even with an astronaut on the moon! Yet it happened, and the world received the stunning news through television and press, with disbelief.

We left for Boston by an American Airlines 727 and were greeted by my nephew Prasad, who took us to his comfortable home in Greenlands after driving in heavy rain. We had a pleasant time with Prasad and Indira, and I lectured at M.I.T. at the kind invitation of Professor Gian Carlo Rota,[1] one of the most eminent mathematicians.

We left Boston for Washington by an American Airlines 727 and we called on Shapiro, who was in hospital and he greeted us with his usual smile, invariant in sickness and in health! In Washington, I gave a lecture at the Catholic University at the invitation of Professor Uberall, who treated us to a delightful lunch in an Indian restaurant, where several of his colleagues joined. We then left for St. Louis to be received by Bruno Gruber, who took us by car to Carbondale. I gave three lectures at the Southern Illinois University in Carbondale — one on the 'Generalised Clifford Algebra,' one on 'Evolutionary stochastic process' and another on 'Unnoticed symmetries in special relativity.' We drove to St. Louis, and after a stay at the luxurious Executive International Inn, left by on Ozark DC-9 for Dallas, where we were received by Professor Fenyves. After my lecture at the University of Texas at Dallas, we went with Lakshmikantam to the University of Texas at Arlington, where I gave a lecture on the same day. We returned to Dallas and stayed at the spacious house of Fenyves, which had a design similar to that of the Alladi House at Madras. From Dallas we flew by American Airlines DC-10 to Los Angeles and were happy to be received by Krishna and Venki at the Los Angeles airport.

It was a most enjoyable stay for about a month and a half at Los Angeles in Krishna's elegant apartment, a Californian rhapsody as enjoyable as our first stay in Santa Monica in 1962. Lalitha and I went to the Santa Monica beach area everyday, while Krishna was busy with his research work. He really worked very hard, both at his courses and his research.

Leaving Lalitha in Los Angeles, I made a short trip to Albuquerque and gave a lecture at the University of New Mexico. Later, I went to Santa Fe to stay with Viswanathan for a few days. He had shifted from New York to Los Alamos, where he accepted a tenured position in the famous laboratory. It was thrilling to visit the laboratory, wondering how the history of the world was altered by what happened there under the guidance of 'a few egg heads and long hairs.' The spirit of Oppenheimer reigned everywhere, and I could not conceal my wonder, excitement and exaltation at visiting the birth place of atomic energy.

We used to accompany Krishna to UCLA and spend most of the time in the library in the morning, and relaxing in Santa Monica during the evenings. Venki entertained us at the faculty Club for lunch and took us on frequent drives along the Pacific coast. There was a seminar at a California State University on the Indian elections of 1977, to which Sharon Marcus took me, and I participated in the discussion talking on my father's role in the Indian Constitution. Krishna took us to all the restaurants he used to visit when he was a resident in the 'dorm.' We called on Dick Bellman during our stay at Santa Monica. I made a visit to Santa Barbara at the invitation of Dr. Uppuluri, who was visiting there and gave a talk on Clifford Algebras at the University, which has the most beautiful location in the entire world.

Krishna informed us on the 29th May that he had passed all his qualifying examinations. Following that, we had Professor Straus and his wife, and Krishna's classmate Charles Grinstead for dinner in Krishna's apartment. On another occasion, we invited Professor Richard Arens and his wife, and Professor Bruce Rothchild of UCLA for dinner, so deliciously and solicitously prepared by Lalitha.

Of course there was the inevitable visit to Laguna Beach at the irresistible invitation of Gordon Shaw and his wife Lorna, who regaled us with dinners in Mexican restaurants, tennis on the lovely courts and a visit to the Lion Country Safari. What really touched me was the spontaneous manner in which they also 'went vegetarian' during our entire stay, when Lorna solicitously prepared our favourite Italian dishes. We were also invited for dinner and tennis by Sharon and Marcus at the Country Club with its plush lounges and lavish facilities, lush environment and rolling golf downs.

On the 3rd of June, Krishna passed his oral qualifying examination when he was asked to lecture before his examiners, and I had the pleasure of attending his talk.

I left for San Francisco for a week and visited the Stanford Linear Accelerator Center (SLAC) at the kind invitation of Professor Drell. SLAC is a unique place where theory and experiment flourish side by side. I had dinner at Hoggatt's when Mrs. Hoggatt pampered me with rich cheese dishes. I returned to Los Angeles to join Krishna and Lalitha.

On the 10th June, at the invitation of Richardson, I gave a lecture at the Rockwell Science Centre, an excellent example of business enterprise fostering scientific research, which in turn made business more efficient and profitable.

When Krishna's classes for the academic year were over in mid-June, he joined us on our return to India, but we made a few stops in USA and Canada enroute due to my lecture engagements. From Los Angeles we flew to Denver on the 15th of June, and at Fort Collins I lectured on stochastic processes at the invitation of Professor Srivastava. He took us on an afternoon excursion to the magnificent Estes and the Rocky Mountain National Parks. We left by Frontier Airlines 737 for Winnipeg, and I gave a lecture on matrix theory at the invitation of Professors Shivakumar and Venkataraman at the University of Manitoba. Lalitha gave an informal concert before a few friends at Venkataraman's house.

We left Winnipeg by a Northwest Orient DC-10 to Minneapolis and 747 to JFK, where we were received by Lalitha's cousin Narayanamurthi of Air India. After a day's stay at his apartment, we flew by Air India to Bombay from JFK. After a brief stay at the Centaur Hotel in Bombay, we left by the Indian Airlines Airbus for Madras.

I started my jogging schedule on the 27th of June, a habit which had taken America by storm — from the college students at the campuses, to the tycoons at the country clubs, from the sprinters on the race track, to the President at the White House.

On the 9th of July, Professor Marshal Stone, the distinguished mathematician, visited Madras and was happy to note that Krishna was progressing well in graduate school at UCLA. For me, July was a productive month since I got a new idea, which I called External Relative Velocity, in the special theory of Relativity.

Kichenassamy of Paris visited our Institute on the 19th of July and addressed the monthly meeting of the Tamil Nadu Science Academy. We had a lively discussion with him on Hindu marriages, none of us realising that Krishna's marriage was soon to be fixed within the next few weeks!

On the 26th of July, while playing tennis in the morning, my friend Venkatesan casually mentioned to me that a close friend of his,

N. C. Krishnan,[2] a senior auditor in Madras, had a daughter who he was thinking of proposing to Krishna. The next day, Venkatesan called at Ekamra Nivas and mentioned this to Lalitha, who told him that Krishna was very young, but we would still like to meet the girl and her parents informally at his house.

August 4th was a significant day for Krishna when he met Madhu (Mathura), his future partner in life, at Venkatesan's house. We had many get togethers and Krishna reacted favourably to the proposal for marriage and agreed that it should take place after the completion of his PhD next year. We consulted our family friend 'Chitra' Narayanaswamy, who heartily endorsed the 'alliance'.[3]

During this period, Krishna gave lectures at local institutions — at the A.C. College at the invitation of Professor Laddha, at Devanathan's Nuclear Physics department in the University, at the Ramanujan Institute and at the CLRI as the guest of Santappa. He also delivered a series of lectures at Matscience.

On the 17th of August, we left on our Southern tour on which Madurai was our first stop. We stayed at the Madurai University Guest House and I enjoyed jogging in the morning, inhaling the fresh 'mountain air.' When I complimented the professors for their choice of such a fine environment, they informed me it was a snake infested locality, and if we ignore such insidious dangers, walking could certainly be a pleasure! My lecture at the Madurai University on new concepts in relativity was well-received. We made a trip to Guruvayoor, and the rains held back to make our trip enjoyable and memorable. We had a wonderful darshan of Lord Krishna at the Guruvayoor Temple and invoked His blessings for Krishna's marriage. On the 21st August, the day our wedding anniversary, we enjoyed the motor trip from Coimbatore to Trichy, where we offered prayers at the Akhilandeswari temple.

On Varalakshmi puja day on the 26th August, Madhu visited Ekamra Nivas for the first time since it was considered auspicious for a prospective bride to invoke the blessings of the Goddess on such an occasion.

A few days later, we left for Bangalore and stayed that evening at Sankaranarayanan's house. It was he who had suggested in 1972 that Krishna should meet Prof. Hazen of Princeton University. We left for Mysore and made the delightful excursion to the Bandipur and Mudumalai wildlife sanctuaries, where we had a pleasant surprise in seeing the American Ambassador Robert Goheen being given a traditional style 'elephantine reception!' Back in Madras, I was at my friend Fry's house to meet

Robert Goheen, the American Ambassador to India. I felt flattered when
he recalled with pleasure our earlier meeting at Princeton, when he invited
me for dinner at the Nassau Inn as the President of Princeton University.

On the 17th of September, Krishna left for the United States by Air
India from Madras. He phoned to us from California that his paper with
Erdös appeared in the August issue of the Pacific Journal of Mathematics. This was the first of his five papers with Erdös, and his first major
publication.

I had received invitations to participate in a conference in Oberwolfach
and to visit Universities in Germany under the auspices of DAAD — the
German Exchange Service. On the 17th of November, I left by Indian
Airlines for Delhi and after an overnight stay there, I took the Air India
747 and arrived at Frankfurt. From there I flew to Paris by Air France and
arrived at the new Charles DeGaulle Airport, a spectacular show place. I
lectured at the Institute of Henri Poincare at the invitation of Professor
Kichenassamy. Paris bared its beauty in glorious weather and I made the
most of it by walking near the Eiffel tower and shopping for perfumes for
Lalitha near the Opera House. I left for Brussels by Sabena 727 and went
to Liege, where I gave a lecture at the kind invitation of Professor Garnir. I
returned to London and went directly to Cambridge to spend couple of days
with Ramu, who had made arrangements for my stay at Trinity College. I
took the opportunity to meet Professors Lighthill, Cassels and also Eliezer,
who was staying at Madingley Hall, which was a residence of Edward VII,
when as the Prince of Wales, he studied in Cambridge. Eliezer took me on a
delightful excursion to the Ely Cathedral. We met Alan Baker, the famous
number theorist in Cambridge, who occupied the same room where Hardy
stayed at Trinity College. I left London for Frankfurt by British Airways
and made a short trip to Wurzburg to lecture at the invitation of Professor
Hink and my gracious friend Gruber, who had now shifted to Germany.

I enjoyed the journey by 'Schnell-Zug' (fast train) to Oberwolfach,
watching with wonder and admiration the beauty and prosperity of the
'Vaterland,' a country which had lost the war but gained the peace and
achieved prosperity, rivalling that of the 'Affluent Society' in less than two
decades. I got down at the little town of Hausach and went by a Mercedes
cab to the Institute at Oberwolfach, an exquisitely beautiful village amidst
the thriftless loveliness of the Black forest, now looking absolutely white
under a blanket of snow. The Institute is devoted to the organisation of
conferences throughout the year and the elegant building, built and donated
by Volkswagen, is a haven of comfort and cleanliness, with well-furnished

rooms and elegant balconies, from which one could watch with delight the dawn illumined hills, star-spangled skies and moon-blanched valleys — just the environment for which Einstein yearned, of silence inspiring thought and tranquillity that fosters imagination.

The Oberwolfach conference was one of the best organised meetings I have ever attended. My lecture on relativity was well-appreciated and I very much enjoyed conversing with various scientists at informal leisure over breakfast, lunch and dinner and during walking tours.

At the conclusion of the conference, I returned to Bonn, made a short trip to Clausthal in the Harz mountains and stayed at a hotel opposite the station. Professor Doebner played a wonderful host, and my lectures at Clausthal were well-received. I returned to Bonn and took the flight by Lufthansa 747 from Frankfurt to Delhi. Soon after arriving in Madras, I despatched my contribution to the Proceedings of the Oberwolfach conference before the New Year.

Notes

1) Gian-Carlo Rota (1932–99) was both a mathematician and a philosopher. He had the rare distinction of being both a Professor of Applied Mathematics, and Professor of Philosophy, at the Massachusetts Institute of Technology (MIT). IIe was one of the most eminent and influential combinatorialists, and together with Richard Stanley, created a school of combinatorics that was hailed as the greatest in the world.

2) Mr. N. C. Krishnan, Mathura's father, was one the leading Chartered Accountants (CPAs) of India. He was one of the senior partners of Viswanathan & Co., one of the premier accounting firms in Madras. He had the great honour of being elected as the President of the Institute of Chartered Accountants of India.

3) S. Narayanaswamy, popularly known as "Chitra" Narayanaswamy, was one of the most well-known and lovable citizens of Madras. He had association with the Alladi family since the days of Sir Alladi. As the Chairman of the Board of several corporations in India, he became a close friend of Mathura's father, Mr. N. C. Krishnan, who was himself the CPA for several major companies in India. When two families in India get together to discuss the marriage of their sons and daughters, it is customary to consult a well-known family friend before confirming the alliance. Chitra Narayanaswamy was a close friend of both the Alladi and the Krishnan families, and so he was consulted, just as Semmangudi Srinivasa Iyer was consulted by Sir Alladi when the alliance of Ramakrishnan with Lalitha was considered.

Chapter 41

Wedding Bells —
La Trobe To Los Angeles,
Madras To Ann Arbor (1978)

Wedding photograph of Krishna and Mathura, 29 June, 1978. - Author

Mathura Alladi — dance for the Max Mueller Bhavan, Madras (1978). - Author

1978 — What a glorious year of domestic felicity and academic assignments — Krishna's marriage, following by the completion of his PhD at U.C.L.A. and followed by the Hildebrandt Research Assistant Professorship at Ann Arbor, my own travels to Australia and later to Europe and America with Lalitha, all these against the panorama of momentous world events ranging from the space endurance tests of Soviet Cosmonauts, to the reversal of history by America's shift from Taipei to Peking!

The year started with pleasant surprises — the visit to Madras of our friends the Morrises for the first time after their retirement from the American foreign service. We also enjoyed the performance of the music season before starting our preparations for the sixteenth anniversary symposium, the theme of which was 'New Particles, do they imply new Physics?' The anniversary function was held on the January 19, with the unveiling of the portraits of Bartlett and Bellman. There was a concert by Lalitha, which was graced by the presence of Governor Prabhudas Patwari.[1]

It was then that even a world used to space conquests was astounded by the endurance tests of the Soviet cosmonauts on the Soyuz and Salyut missions. Meanwhile, the American president completed a nine day eighteen thousand mile diplomatic odyssey, which took him through Poland, Iran, India, Saudi Arabia, Egypt, France and Belgium. There were new prospects of peace in the Middle East following Sadat's spectacular efforts after his incredible journey to Jerusalam, which stunned even the protean diplomats out of their wits.

On the January 24, Krishna called me from Los Angeles to inform me that he received an offer from Berkeley of an Assistant Professorship. This was followed by another call a few hours later that he was also offered the Hildebrandt Research Assistant Professorship at Ann Arbor, Michigan. He decided in favour of the Hildebrandt position, well aware that the weather at Ann Arbor certainly could not compete with the incomparable attractions of the Golden State.

At that time we had a most pleasant surprise in the visit of my distinguished friend Harry Messel from Sydney, who came to Madras on a most unexpected assignment to participate in a 'Wild life preservation conference' near Mahabalipuram. It was impossible for us to emulate his generosity as a host and offer hospitality consistent with his style of life, but we invited him for an open air dinner on the lawns of the Institute, which he enjoyed with customary friendliness.

On the 7th of February, in consultation with our family priest Nataraja Sastrigal, Mathura's parents, and a few well-wishers, we decided that the 29th of June would be the date of Krishna's marriage to Mathura, after referring to four *Panchangams* (Hindu religious calendars), which differed slightly in the timings of the auspicious hour for the marriage!

The Matscience symposium was held in Mysore from the 11th to the 13th of February. Soon after our return, we had a phone call from Krishna that he would arrive in Madras as early as April since his thesis was completed ahead of time and his professors allowed him to leave for India. The immediate effect was Lalitha's prompt and unhesitating withdrawal from the Australian trip so that she could be in Madras with Krishna! So I decided to 'go it alone' as regards my assignment at La Trobe University in Melbourne.

I left for Singapore by a Malaysian Airlines 707 on March 6. After a beautiful flight via Kuala Lumpur, I arrived at Singapore and checked in at the Oberoi Imperial. I offered prayers at the Murugan temple in Singapore and left for Sydney by a Singapore Airlines 747. After a delectable breakfast on board watching the sunrise above the clouds, I arrived at Kingsford-Smith Airport to be greeted by my sister Rajeswari and her husband Hari, who took me to their lovely home in Turramurra, a charming suburb of Sydney.

It was glorious weather during the weekend, and we feasted our eyes watching the azure ocean under cloudless skies from the sun-drenched beaches of Sydney, picnicking rambling and driving along the winding highways skirting the golden sands. Two days later, I left by T.A.A. for Melbourne, where I was received by my gracious host, Professor Eliezer. I checked in at the comfortable apartment at the Glenn College, nestling sensuously amidst fragrant groves of stately Eucalyptus trees. To my pleasant surprise, I found the refrigerator stocked with a week's provisions by my thoughtful host Mrs. Eliezer. On Sundays, the La Trobe Campus is deserted and I felt like Robinson Crusoe walking along the lush green grounds in the company of my own thoughts, in speechless communion with surrounding Nature. I started my lectures the next day on Quantum Mechanics, when Eliezer introduced me to the students. The two months stay at La Trobe turned out to be one of the most enjoyable and significant assignments in my academic career. I prepared assiduously for the lectures on Quantum Mechanics, feeling once again like a college student, recalling memories of my youth at the Presidency College. It was an equally great pleasure to watch the happy faces of young and healthy students, relaxing on the lawns between lectures or conversing in the restaurants at the Agora. I spent long hours in the magnificent library, where a wide spectrum of literature was available, from ancient Indian history and the American heritage, to the conquest of space, and the most recent advances in mathematical sciences.

What strikes the attention most of an Indian in Australia is the relatively small number of people employed for service in any organisation due to the high cost of labour. The La Trobe library was being run by about a dozen persons, who issued and received books, arranged and reshelved over a thousand volumes a day. I enjoyed my evenings browsing through a variety of books, from the nineteenth century diary of the Earl of Moira, to Stanislav Ulam's recently published autobiography of a mathematician.

It was a fortunate circumstance that I met quite a few Indians, teachers and doctors, who provided me relief from salad lunches through spiced South Indian food. Though Melbourne weather is considered variable, I was fortunate in enjoying a very sunny autumn. Every weekend I went out on excursions with Professor Basawa, a statistician at La Trobe, to the Kangaroo parks and Eucalyptus forests.

For the Easter Holiday, I flew to Sydney to spend the weekend with my sister. It was heavenly weather, and just right time to see the cattle and horse show at the Royal Easter Fair, and I could understand why the land was so prosperous, blessed by God and preserved by man. The excursion to the Blue Mountains and the lovely picnics at Katoomba convinced me that my sister had made the right decision in choosing Sydney as her home away from home. Back at Melbourne, after a delightful flight on so clear a day that I could see the glistening profile of the golden beaches, I found a flood of letters waiting for me, a natural consequence of the termination of a two week old postal strike!

On a bright sunny day, Basava and I made an excursion to the Sovereign Hills, an extinct mine near Ballarat, which brought to my mind a famous story of Conan Doyle, in which Sherlock Homes discovered that the three letters 'rat' were part of the fuller word Ballarat, where the criminal had made his pile as a miner.

I took an opportunity to visit Adelaide at the kind invitation from Professor H. S. Green. To me, Adelaide always meant the home of Don Bradman, who had captured the hearts and minds of my generation in the early thirties. This time I could not enjoy the gubernatorial hospitality of Professor Oliphant, as I had done when he was a Governor five years ago, but had to be content with the standard comforts of the Grosvenor Hotel. It was then I met a friendly young German physicist Professor Kreuzer from Edmonton, with whom I made a date for a visit to the University of Alberta later that year. After a delectable lunch at Professor Jagannath Mazumdar's home, I left for Melbourne to return to the lonely comforts of my rooms at Glenn College.

Among the lectures I heard at La Trobe, the most interesting was the video-tape seminar of Professor Thorne of Caltech, on the experimental tests of gravitation. I recalled Feynman's lectures on gravitation fifteen years ago at Hughes in California, and Chandrasekhar's oft repeated remark that it is an intoxicating subject, in which only the boldest and the most gifted could hope for success in research.

One weekend, I went to Geelong to visit my friends Mr. and Mrs. Walls. Peggy (Mrs. Walls) was Bartlett's secretary at Manchester, when I was a student there and had typed my thesis as a friendly gesture. We therefore talked many hours over the eventful days at Manchester, when Lighthill and Mahler, Bartlett and Rosenfeld, were its leading lights. Her husband was a senior scientist at the CSIRO and had made a very significant invention 'the double twist,' improving woollen textile industry.

During the second week of April, Krishna phoned to me that he was offered a visiting membership of the Institute for Advanced Study in Princeton and an assistant professorship at Urbana. Just before my final lecture on Quantum Mechanics, I received his thesis by airmail, which contained a touching dedication to his parents and a grateful acknowledgement to his teachers, Straus and Erdös. The pleasure was enhanced by the gracious gesture of all my students at La Trobe signing a brochure gifted to me at the conclusion of my lectures.

Eliezer hosted me at a magnificent lunch at the Country Club in Melbourne. Observing the serene atmosphere of affluence and elegance, I understood how a private club is considered England's gift to civilisation so well-described by Sir C. P. Ramaswami Iyer as "a haven where the people get together, while respecting the right to be left alone!" My assignment at La Trobe was over on the 12th of May, and I left Glenn College with pleasant memories of my stay in the lovely campus, the busy schedule of lectures, regular jogging over the lush green grounds, inhaling the scented air at sunset from the clusters of tall Eucalyptus trees. At the Canberra airport, I was greeted by Professor Lecouter, and my student and colleague Santhanam, my colleague who was spending a year at the Australian National University. It was a pleasant surprise when I was told that C. Subramaniam was there on an academic assignment staying at the University Guest House. He joined us at dinner at Santhanam's house, where other Indian friends also gathered for a pleasant evening.

It was heavenly weather and Canberra looked like a smiling bride on a honeymoon trip. I gave a seminar in Le Couteur's department, after which I was taken on a leisurely drive round the capital by a friend, who gave me the precious gift of a recorded cassette containing Chembai's rendering of Thyagaraja's immortal song — Raghuvara in Panturvarali — which sends me into trance of bliss every time I hear it or even think of its theme!

Flying to Sydney, I seem to have carried the glorious weather with me, for the city had an unusual spell of rain for a week earlier. The few days in Sydney before I left for India were spent on excursions to the southern beaches down to Woolangong and Kiama. We watched the blowhole bursting into life every few minutes with a spectacular spray, a fantastic sight against the background of the magnificent ocean and the blue unclouded sky. It was such splendid weather that we equally enjoyed the trips by car in and out of metropolitan Sydney, watching the titanic grandeur of the famous harbour bridge against the soaring skyline of Sydney. I gave a talk at the University at the kind invitation by my great friend Harry Messel. My lecture was attended by the famous mathematician George Szekeres, who had evinced interest in Krishna's work in 1973.

We had the pleasure of receiving C.S. in Sydney at my sister's house. I left for Singapore by a Singapore Airlines 747. It was a beautiful flight, the best part of which was the splendid sight of the argent profile of the northern coastline of Australia against the blue expanse of the Indian ocean as we left the continent a few hours before reaching Singapore. As usual, I checked in at Oberoi Imperial. Who would waste his time without buying the good things of life at this shoppers' paradise, and I certainly did not on the eve of Krishna's marriage! I arrived in Madras to be received by Krishna and Lalitha, Madhu and her parents, everyone stricken with wedding fever.

A South Indian marriage is a wholesome combination of old-world tradition, time-honoured religious observances and receptions with modern paraphernalia of entertainment. Invitations are sent to any acquaintance, for a Hindu wedding ceremony is an announcement in the absence of official registration. There is an air of festivity, almost like that at a country fair and all formalities are confined only to the religious rites by the couple, directed by the priest in the presence of their parents.

A few weeks before the wedding, Mathura[2] gave a delightful dance performance at the elegant auditorium of the House of Soviet Culture in Mylapore, attended by all our friends and relatives, a fitting prelude to the 'Hochzeit' or the high moment of marriage ceremony.

The 'Nischithartham' (betrothal ceremony) took place in Ekamra Nivas, in the same *koodam* (hall) adjacent to the *puja* (worship) room, where Krishna had his Aksharabhyasam and Upanayanam. It was attended by over three hundred guests — near relatives and close friends — who were entertained in the well-lit gardens of Ekamra Nivas.

The marriage function started on the morning of the 28th June when we shifted from Ekamra Nivas to the Rajeswari Kalyanamantapam (marriage hall). The first ceremony was the *Vratham* or the purification of bride and bridegroom and their parents with the blessings of elders and Vedic scholars. In the evening there was the *Janavasam* (procession of the bridegroom), a combination of secular functions and prayers at the temple. The procession was from a temple to the Mantapam to the accompaniment of a police band and Nadaswaram by Namagiripettai Krishnan and his team. The Nadaswaram vidwan elaborated the famous song 'Mohana Rama,' the theme of which defies effective description except by Thyagaraja in the enchanting strains of Mohana raga. Dignitaries, like C. Subramaniam and R. Venkataraman, walked with the procession.

After a ten course dinner, which included a king-size Mandey, a delicacy from Mangalore at the suggestion of our friend, the famous architect Narayana Rao, the entire group of guests feasted their ears with Ramani's flute recital. The mellifluous strains of Sankarabharanam cast an enchantment on the eve of the marriage ceremony.

On June 29, the wedding day, the religious ceremonies started at dawn under the direction of Nataraja and Pattabhirama Sastrigals, our family priests. The *Muhurtham* (the time to tie the marriage knot) was fixed at 12:52 p.m. — the most auspicious moment of the auspicious day. In the morning, there was the colourful semi-secular ceremony with the bride and bridegroom seated on the *Unjal* (swing) and smothered with flowers and blessings by singing mangala keerthanam (songs). All the rituals — Samithadanam, Madhuparkam, Padapuja and Kannikadanam before the sacred fire — were over by 12 noon. The *Mangalyam* or the sacred thread tied by the husband, and worn around the neck by the wife, is valued as dearer than life. It represents the indissoluble bond, decreed by God, uniting the couple with the blessings of parents and relatives. Three thousand guests assembled in the mantapam sharing the excitement of the hour. At the appointed moment, Krishna and Mathura were married in the presence of the parents, before the sacred fire invoking the blessings of God through Vedic incantations. All the guests streamed in to congratulate the couple, while we expressed our thanks to the Venkatesans, who brought the families together. Marriages are determined in Heaven but are realised through human agencies with the cooperation and goodwill of friends and relatives.

In the evening, we held a reception when we had M. L. Vasanthakumari's concert. While music addicts listened to the concert with undivided attention, the concert provided a pleasant background for conversation on such a festive occasion, as guests arrived and departed in casual informality. The presence of our beloved Governor, who had a paternal interest in Krishna's career, completed the picture. The Muhurtham lunch and the reception dinner were attended by a thousand guests, who were served in traditional plantain-leaf style by a crack-squad of super-swift commandos under the direction of the master cook, conscious of his indispensable role in a South Indian marriage.

Later in the evening we had the *Grihapravesam*, the formal entry of Madhu into Ekamra Nivas. The next day, we had a vidwat sadas when a group of Vedic scholars explained the meaning of marriage and blessed Krishna and Madhu. We had also a Harikatha — 'Rukmini Kalyanam' by Lalitha's music teacher Mannargudi Sambasiva Bhagavathar, a fitting finale to a wedding, which was celebrated in the true Madras tradition of sonorous religious ritual and exuberant informality.

We offered our prayers to Lord Venkateswara at Tirupati in reverential gratitude for the successful completion of the wedding function. We had a triple darshan at Ekantha seva, Thomala seva, and Archana.

On the 2nd of July, we had the wedding dinner at Ekamra Nivas attended by over two hundred guests, who were accommodated in our Koodam (hall) and the terrace, where years before we had the dinner on the occasion of my father's Shastiabhapoorthi. Among the guests were C. Subramaniam and R. Venkataraman, who was soon to rise to the level of President of India.

Krishna and Madhu spent their honeymoon in Kodaikanal — the lovely hill resort with its beautiful lake and plethora of picnic spots.

On the 20th of July, we left on a South Indian tour with the dual purpose of offering prayers at various temples and fulfilling lecture engagements. Our first stop was at Pennadam in the comfortable Aruna Sugars guest house. The next morning, we offered prayers at Siruvachur shrine and later at the temples of Rajagopalaswami at Mannargudi and Kothanda Rama at Vadugur. By evening, we reached the TVS guest house at Madura, a visible symbol of the efficiency of the TVS organisation, excelling in comfort and cleanliness and in the floral elegance of the gardens, even five star intercontinental hotels. We offered prayers at the Meenakshi temple. The drive from Madura to Pollachi was a pleasureable experience, giving a glimpse of the beauty of the South Indian rural scene. We were received with warmth by the enterprising Principal Namasivayam and the amiable Professor Gundu Rao. Krishna gave an inspiring lecture on Number Theory and we were treated to thriftless hospitality at Mahalingam's house. My earlier description of Mahalingam's generosity seemed inadequate to Krishna, who actually experienced it in lavish intensity. The next day we went to the Ayilar dam and on a picnic to the waterfalls, a cosy spot of exquisite loveliness. I lectured at the Polytechnic on 'Planning a new campus,' emphasising that the physical environment of a university was as important as the intellectual. Lalitha gave a concert at the Gandhi Mantapam on the final day of our stay at Pollachi.

From Pollachi, we went to Guruvayoor to invoke the blessings of Lord Krishna. It was the monsoon season but the clouds held back their showers and we had a wonderful drive through the verdant tree studded valleys of the Palghat gap in the Western ghats. We performed the *Tulabaram* of Krishna and Madhu at Guruvayoor and offered prayers to Lord Krishna. We returned to Pollachi to enjoy again the hospitality of the Mahalingams. The next day we left for Ooty via Coimbatore and reached by evening the HPF Guest House, from where we could watch the glowing sunset over the hills, a sight which captured the mind of the Nobel Laureate Hans Bethe, who considered the golf links area of Ootacamund one of the loveliest attractions of South India.

From Ooty we moved to Madumalai, Krishna's favourite excursion spot and went out on jeep and elephant rides into the forest. We were fortunate enough to see bison and sambur, peacocks and wild dogs, a lone Tusker and a herd of wild elephants. Through Bandipur we reached Mysore, where we offered prayers at the Chamundeswari temple atop a hill. We returned to Madras after a brief halt in Bangalore.

The two weeks in Madras before Krishna's departure to Ann Arbor were spent in preparations for his journey. On the day of Varalakshmi Vratham, Madhu was inducted into the traditional pooja in our family.

On the fourteenth of August, the eve of Independence day, Madhu gave a delightful dance performance at the Ballroom of the Connemara Hotel, under the auspices of the Max Muellar Bhavan at the instance of its gracious director Gunther Lennertz, my distinguished friend and neighbour. Professor Uberall arrived from Washington, just in time to watch the performance, his best introduction to the cultural attractions of South India.

Krishna and Madhu left for the US via the Pacific, halting in Singapore, Hong Kong, Tokyo and Hawaii. Krishna's fascination for Honolulu was understandable, for he had enjoyed on his world tours with me the strolls along the Waikiki, the sensuous warmth of the crystal waters of the Hanauma Bay and the resplendent sunsets bathing the skyline of Honolulu with a gorgeous crimson glow. But Hilo on the Big Island of Hawaii stole his heart and Mathura's — its lush vegetation, the volcanic earth and the foaming waves rolling down to the palm fringed coast with its myriad species of fragrant ferns, winning over the five star comforts of Waikiki! We were happy to hear that they reached Ann Arbor after visiting Professor Straus at Los Angeles, who delivered to Krishna personally the Alumni Medal for one of five best PhD theses at UCLA in 1978 (all subjects).

On August 21, we celebrated our wedding anniversary this time at a dinner in Devanathan's house, just as we did ten years earlier at Wisconsin in his apartment! Uberall, the joint guest of the University and Matscience, was with us that evening.

We left for Mysore on the 27th August for the Matscience symposium, a delightful journey as usual through Bangalore. This time, Lalitha had an opportunity to sing at the local sabha in Mysore, attended by some representatives of the older generation who had heard the great masters and remembered them with nostalgic zeal. It was gratifying when they complimented her that she had rendered the kirthanas (songs) according to 'Sampradaya' (tradition).

On 4th September was a significant day for the city of Madras — the opening of the Anna University[3] and the Centenary of *The Hindu* in the presence of the President of India. The latter was a superb example of efficiency and affluence, worthy of its unique position in international journalism, its significant role in India's evolution into a sovereign republic and in the effective propagation of the cultural traditions of our country.

On the 22nd September, there was a function at Kesari High School, where the Governor of Madras unveiled a portrait of my father. The Chief Justice of Madras, Ramaprasad Rao, who presided made a touching reference to his association with him. While my elder brother spoke about my father's legal eminence and human qualities, I emphasised that he was a natural genius without the slightest trace of pretence or the need for it, and how wealth and success came to him as incidental consequences of his primary passion, the pursuit of excellence in law and learning.

During this period came the news of the incredible admission of an East German, Jahn, into the Russian 'cosmonaut club' following the earlier inclusion of Polish and Czech partners in the Intercosmos programme. It was similar sense an effective answer to the collaboration between the NASA and ESA (European Space Agency), initiated at the beginning of the year.

Meanwhile 'on earth's human shores' even more incredible events were happening, bringing 'irreconcilables' together. In the lush secluded luxury of the plush presidential retreat at Camp David, Carter brought Anwar Sadat and Menachim Begin together to work out a formula for a stable peace in the Middle East. Time magazine described so aptly as a 'virginal experience' in diplomatic history, a meeting without an agenda, a vow of goodwill between longstanding foes in the presence of a common friend.

I received an invitation to participate in a conference at Oberwolfach — an irresistible temptation to any one interested in mathematics or the natural attractions of the Schwarzwald. Professor Bleuler also invited me to visit his department at Bonn, and I found it possible to respond to these invitations due to the generous support from the DAAD, the German Academic Exchange service.

Lalitha and I left Madras on the 12th of October and took the Lufthansa 747 from Bombay the next morning. Since there was a heavy fog over Frankfurt, the plane was diverted to Bonn, which was indeed our destination! So things were made easier and Professor Petri, who received us at the airport, took us to the comfortable private hotel Schwann near the University. The same evening we attended a reception to Robert Wilson, who gave a lecture on the future of High Energy Accelerators. As the erstwhile director of the Fermi Laboratory in Batavia, he naturally propagated the view that the expense on high energy accelerators was justifiable. I felt that an optimisation between theory and experiment was necessary, for the secrets of nature could be probed both by mathematical analysis and by experimentation. My friends, Professor Dietz and Sandhas, took me out on an excursion to a lovely lake and a monastery, in which Adenauer was supposed to have taken refuge during the war.

We left for Oberwolfach on Sunday by the Dortmund Konstanz express and reached the place by evening, after a most enjoyable journey. It was glorious weather in golden autumn, with a riot of colours on the lush tree-studded valleys and mountains. Oberwolfach looked like a fragment of heaven under a golden sunset. Our stay there was one of the most enjoyable in my travelling career, and I was in the right mood to present my new work on stochastic processes at the conference on Operational Research. I talked on a new concept in evolutionary stochastic processes by defining a quantity called 'activity' as a measure of the change in a probability distribution and conjectured a theorem that activity decreases with time — certainly true in human life! Lalitha particularly enjoyed the walks through the woods. It was just like Ootacamund and the comforts at the lovely guest house left nothing to be desired. It was a wholesome combination of German thoroughness, Swiss cleanliness, American informality, British discipline and French elegance combined into one!

We were back in Bonn, where I gave a lecture at the kind invitation of Professor Bleuler. I took the opportunity to visit Professors Hirzebruch and Zagier in the Mathematics Department. Later in the evening, Bleuler took me and Lalitha for dinner to a lovely restaurant on a hill-top with a spectacular view of the city of Bonn. Professors Sandhas, Dietz and Gehlen joined us, and it was flattering when they recalled their visits to

Matscience with pleasure. During a dinner in a friend's apartment, we watched on the television the impressive installation of the first non-Italian Pope from Poland. One wonders, how great a hold religion has on the minds of the people even in this rationalised, commercial world of scientific and technological triumphs. It also demonstrated the vital role of ritual in preserving Catholic traditions, as in the Hindu way of life, which unfortunately is being eroded under the desiccating influence of misconceived rationalism.

We left Frankfurt for New York by a Lufthansa DC-10. The transAtlantic flight was a sheer beauty and on arrival at JFK, we checked in at the Hilton Inn near the airport. After a one day in New York, we left by Northwest Airlines for Detroit to be received by Krishna and Madhu, who took us in their new Pontiac LeMans to their lovely apartment in Woodbury Gardens in charming Ann Arbor. It was autumn weather at its best, just before the onset of winter and we made most of it by visits to the botanical gardens. We celebrated Deepavali on the 30th of October and had our Gangasnanam under hot showers and wore new clothes brought from Madras. At the University of Michigan, I met Professors Don Lewis (Krishna's mentor) and Fred Gehring (Department Chairman), before leaving on my academic tour round U.S. and Canada, leaving Lalitha with Krishna and Madhu.

My first stop was the Rennsaeler Polytechnic Institute, where my colloquium on Relativity was well-attended, and it was gratifying that my new work claimed attention. I left for Washington by Alleghany Airlines to be received at the National airport by Uberall, at whose instance I gave a lecture at the Catholic University, this time on stochastic processes. A visit to Washington, the capital of the most powerful nation in the world, is always an exhilarating experience.

After an enjoyable flight by an Eastern Airlines DC-9, I arrived in Durham to be received by my gracious host Paul Morris. The weekend with the Morrises at Chapel Hill was one of the happiest in all my travels. The weather was glorious and Paul and Ruth treated me to such flattering hospitality that I felt Krishna and Lalitha should have shared it in equal measure. I left for Dallas by Eastern Airlines to be received by Fenyves, an incomparable host and a delightful friend.

At Dallas, I lectured at the University of Texas, whose campus seems to grow every day in size and in prestige. I had the pleasure of meeting Professor Ivor Robinson, a delectable conversationalist besides being an eminent gravitationist. I also lectured at the Southern Methodist University at the kind invitation of Professor Milton Wing, who entertained me later at the exotic Trader Vic restaurant. From Dallas I flew to Oklahoma and was taken by car to Stillwater by my good friend Professor N. V. V. J. Swamy. After a seminar there I was treated to a lavish dinner by Swamy, who had invited his distinguished friends.

I left Oklahoma City for Albuquerque, where I took the Ross Aviation Commuter service to Los Alamos to be received at midnight by Viswanathan. The three days at Los Alamos were unforgettably pleasant, enjoying the generous hospitality of Viswanathan and Selvi, and watching the gleam of dawn and the glow of sunset on the hills from the deck of their lovely home. My lecture at Los Alamos was held in the main auditorium, where I spoke on 'Mathematical Sciences in India' to an eager and attentive audience. I also visited the Los Alamos Fusion Laboratory, one of three major centres where research on the possibility of controlled fusion was going on, the other two being Livermore and Princeton. Viswanathan and I heard a talk by Professor Teller on the uncertainty principle, and I felt happy how in my own lectures to graduate students at La Trobe, I had emphasised similar concepts. Later, we had a dash of tennis, enjoying the bracing coolness of the mountain air and basking in the sunshine under a blue sky.

I flew by the small commuter aircraft to Albuquerque on a sunny afternoon, when I saw the valley of Rio Grande, its bizarre beauty bared by the mid-day sun under a cloudless sky. My mind travelled back to the momentous month of August 1945, when we heard the stunning news of the atomic bomb and the story of Oppenheimer and his team engaged in secret work on the fateful project. While the world was unaware of these efforts, its destiny was determined by the scientific efforts of a small band of scientists under Oppenheimer and what a band(!) — Bethe, Fermi, Wilson, von-Neumann and Teller! From Albuquerque to Los Angeles I flew by a TWA 727. It was thanks-giving day and Los Angeles airport was almost deserted. Krishna's friend, Srivatsan and I relaxed there for a few hours before I went to the Santa Ana airport to be received by Gordon Shaw. The three days at Laguna Beach were enjoyable as usual. I lectured at UC Irvine on 'New symmetries in special relativity.' Gordon Shaw drove me to Los Angeles, where Richardson was waiting for me at Diamond Jim's restaurant in Wilshire Boulevard, to take me to the Rockwell International, where I lectured this time on 'Time reversal in evolutionary stochastic processes.' After my lecture, I was taken by a chauffeured car to the Los Angeles airport, from where I flew to San Francisco by a United Airlines

737. I spent three fruitful days at SLAC at the kind invitation of Professor Sidney Drell, inhaling the mystery of the atmosphere where 'charm and beauty' are being discovered in unobservable quarks! The visit to Stanford and the lunch with Professor Hofstadter at the Faculty Club brought back memories of the pleasant time I had with the late Professor Schiff years before.

I left for Edmonton by Air Canada on a Sunday, to be received by Professor Kreuzer, a splendid host whose smile warmed up my spirits when the temperature outside was 15° below zero! I gave a lecture at Edmonton, after which Professor Takahashi drove me to Kreuzer's house. After an anxious drive through heavy snow, it was an enhanced pleasure to enjoy a delectable dinner in the cosy comfort of Kreuzer's well-designed house. The next day, I left for Winnipeg and was met at the airport by friends, Professors Shivakumar and Kale. My lecture at Winnipeg was well-received and was followed by dinner in Shivakumar's house.

The flight by Northwest DC-10 to Minnesota and 747 to Detroit was a perfect example of comfort, speed and efficiency. Back in Ann Arbor, I gave a talk at the Mathematics Club, attended by Professors Lewis and Gehring and others.

Our visit to U.S. and stay with Madhu and Krishna turned out to be the most fruitful and exhilarating experience in a quarter century of international travel. It was also the period of great changes in the world-wide diplomatic scene, the most astounding of which was the *volte face* in American policy towards China. It was indeed incredible that the billion people of Communist China should turn with friendly eyes towards the U.S.A., whom they considered till then as the mighty colossus of the capitalist world. Nobel prizes were awarded to Begin and Sadat, in the hope that it will further the cause of international peace.

There was equal excitement in the world of science and mathematics. Kapitza of Russia, and Penzias and Wilson of the Bell Labs, received Nobel awards in Physics. Krishna was naturally more enthusiastic about the Field medallists at the recent International Congress of Mathematicians — Deligne of Paris, Fefferman of Princeton, Margulis of Moscow and Quillen of M.I.T. Even more exciting to him was the proof reported at that conference of the irrationality of $\zeta(3)$ by a French mathematician Apery. It was then that I received a letter from Don Zagier and Gunther Harder of Bonn that they were willing to visit Matscience in early January before a conference at the Tata Institute. Influenced by Krishna, I requested Zagier to give a lecture critically estimating Apery's spectacular proof.

During my stay in Ann Arbor, I read a Arthur Ashe's book on his tennis circuit, which included his visit to South Africa. It made a great impression on me and I felt he had done a useful service to the hapless, voiceless millions in South Africa by a factual account of his visit to Soweto.

Krishna was able to slake his passion for tennis by joining an indoor club, the facilities of which were ample compensation for his missing the golden sunshine of Los Angeles, where tennis could be played throughout the year. We enjoyed a long drawn knock on the lovely courts, agreeing to play without scoring, a tribute which youth and power paid to an ageing veteran. Leaving Krishna and Madhu behind in Ann Arbor, we felt for the first time in our lives that we were leaving home, when actually returning to Ekamra Nivas! The flights by Northwest 747 to New York and by Lufthansa to Bombay via Frankfurt were sheer delight. On arrival in Bombay, we had an unprecedented experience when the Indian Airlines refused us seats despite confirmed reservation on the flight to Madras. We were stranded at the Centaur hotel as the self-appointed guests of the Lufthansa! We arrived in Madras in time for the festival at the Music Academy. It was a grand finale to an eventful year because of Krishna's doctorate, his marriage, his first job, and for our trans-Oceanic travel on an academic circuit through four continents, with wedding bells ringing from La Trobe to Los Angeles and Madras to Ann Arbor!

Notes

1) Prabhudas Patwari (1909–85) was a lawyer and freedom fighter who was a follower of Mahatma Gandhi. He was Governor of Madras during 1977–80. He admired Sir Alladi's brilliance as a lawyer, and wrote an article about Sir Alladi on the occasion of his Centenary (see [1]).

2) Mathura is an accomplished Bharathanatyam dancer. She started her training, at the age of five, under Srimathi Kausalya, a disciple of Vazhuvoor Ramiah Pillai, one of the greatest teachers of Bharathanatyam. Mathura completed her *Arangetram* (full debut performance) at the age of eight, and has continued to be active in Bharathanatyam as a performer and a teacher.

3) Anna University of Technology, named after the late DMK leader C. N. Annadurai, is a technical university formed in 1978 by combining the Madras Institute of Technology, the Guindy Engineering College, the AC College of Technology, and the School of Architecture and Planning. The name Anna University was adopted in 1982.

Chapter 42

The Einstein Centenary, And Strange Events In India (1979)

It was the year of the Einstein Centenary when the world paid homage to the architect of space-time unity and of the conversion of mass into energy which initiated the atomic age. I got an opportunity, as his ardent votary, to participate in an international conference in the U.S. devoted to his memory.

At Madras we greeted two eminent mathematicians, Don Zagier and Gunther Harder from the University of Bonn. They arrived on the 3rd of January, the official birthday of Matscience. Professor Zagier gave a brilliant lecture on the recent sensational result of Apery on the irrationality of $\zeta(3)$. In the evening, we had a dinner on the lawns of Matscience in honour of Professors Zagier and Harder.

On the 9th of January we had a lecture by a visiting Czech scientist on Queuing Theory, a subject which had demanded the attention of many statisticians but had not yielded commensurate practical results.

The Matscience Anniversary Symposium was inaugurated on the 22nd January by the Governor of Madras, Sri Prabhudas Patwari, who unveiled the portraits of Oppenheimer, Fermi and Bethe, the 'complete physicists.'

On the 6th of February I had a pleasant surprise in receiving a letter from Dr. Barbour, an associate of Professor D. G. Kendall of Cambridge, that my conjecture on the activity of stochastic processes was proved by him to be correct!

On the eve of my departure to the United States, I gave a talk on 'Campuses round the world' at JIPMER (Jawaharlal Institute of Post-Graduate Medical Research) in Pondichery, well attended and received by large group of enthusiastic young doctors.

On the 22nd of February I left for the U.S. for the Einstein Symposium. The flight by Lufthansa from Bombay to Frankfurt was a sheer beauty and I experienced the enchantment of watching the lights of Baghdad and Ankara in the darkness of the night hearing the assuring hum of the Rolls Royce engines of the Boeing 747. I flew from New York to Detroit by a TWA 727 and was to be received by Krishna and Madhu at the Metropolitan airport. Krishna took me the next day to the Ford Museum which had an impressive collection of 'vintage cars.' We were not able to see the model of our 1933 Buick though we were able to identify it from Souvenir books as model 90L, manufactured specially for export during the recession years.

I flew to Carbondale via St. Louis and checked in at the comfortable Holiday Inn. It was a pleasure to meet Professor Doebner at the St. Louis Airport when he renewed his invitation to visit Clausthal.

The Einstein Symposium opened on the morning of the 28th of February. At the Conference I presented my paper on L-Matrix theory. On the last day Gell-Mann gave a spectacular talk emphasising that the unobservable nature of the quark was a logical necessity and in fact if observed it will throw physics into a quandary!

I flew back to Ann Arbor and spent a week there at the invitation of Professor Gehring, the chairman of the Mathematics Department. Winter had lost its sting in the ides of March and we visited the Ford Mustang plant in nearby Dearborn. I left for India by Lufthansa and after a pleasant refresher stop at the Centaur hotel in Bombay I returned to Madras.

On the 24th of March I gave a talk presiding over the Einstein Symposium at the Loyola College. At the end of March we went to Mysore for the Matscience Conference on Stochastic Processes when I spoke on the 'Approach to Stationarity.'

In early April we had the privilege of receiving the eminent number theorist Professor Richert from Ulm, Germany at Matscience. He was friendly and easily accessible, at the same time very frank, thorough and critical in his estimate of mathematical contributions.

On the 21st of May I inaugurated a summer school in at MIT (Madras Institute of Technology) when I exhorted the participants to imbibe the seminar spirit, characteristic of American institutions though we could not provide the physical amenities available at such campuses.

On the 24th of May we performed the Vasantha Uthsavam[1] at the Mylapore temple to respect the desire of our revered mother.

Krishna informed us that he was participating in the Oberwolfach conference in Germany on his way to India for the summer vacation. It gave him an opportunity not only to contact the leading number theorists of Germany, but also to present his recent work on irrationality inspired by

Apery's spectacular proof of the irrationality of Zeta three. After his arrival in Madras in early June, he gave a series of lectures at Matscience on his recent work in irrationality. We celebrated the first wedding anniversary of Krishna and Madhu on the 29th of June with a flute concert by Ramani followed by a dinner on the gardens of Ekamra Nivas.

We left on our Southern tour on the 18th of July reaching Pollachi and enjoying the hospitality of the Mahalingams. The best part was the excursion to the Anamalai Hills which stimulated the insatiable enthusiasm of Krishna for wildlife. We left Pollachi for Ooty and stayed at the comfortable HPF guest house and returned to Madras after visiting the Mudumalai Wildlife Sanctuary.

We watched the Wimbledon finals on the television (videotaped) when Borg won the Crown for the fourth successive year, out-topping the achievements of Rod Laver 'the greatest of them all.'

We left for Mysore on the 17th of August and visited the Mudumalai wildlife sanctuary before the Conference on Number Theory. It was the first time senior mathematicians of the Tata Institute participated in a Matscience conference. Professor K. Ramachandra[2] and Krishna both gave a series of lectures on Number Theory. After the conference Krishna and Madhu left for the US on August 29. Enroute they spent a day at the Tata Institute visiting Professor Ramachandra.

Meanwhile strange things were happening in the Indian political scene. The Prime Minister Morarji Desai had to step down and there was a caretaker Government under his colleague Charan Singh. The ruling Janata Party was riven by internal dissensions which inexorably led to its doom.

We read also detailed descriptions of the Mountbatten's funeral at Westminister Abbey. The test of civilisation is the manner in which a country honours its dead and Britain, true to time honoured tradition, paid its grateful tribute to a person who held the South East Asia Command during the World War and gracefully presided over the dissolution of the Indian empire and initiated India's freedom.

As usual Navarathri was celebrated traditionally at home and it was strange irony that we should complete the surrender of our excess lands under the Rural Land Ceiling Act on the eve of Vijayadasami, the day of victory! I drew comfort from Mountbatten's example as the imperial representative who handed over the reins of power with a royal smile to Jawaharlal Nehru.

On October 10, I gave a lecture at the Jain College on 'Einstein — Magic or Logic' as part of my new mission to teach relativity to college students. I then went to Hyderabad to deliver the Silver Jubilee Lecture on the same topic. When I referred to the recent developments in physics by mentioning casually that Salam and Weinberg would be awarded the Nobel Prize for that year, a member of the audience shouted that he had just heard on the radio that my 'prediction' was correct!

On the 22nd of November I gave an after dinner speech at the Cosmopolitan Club on 'Free will or destiny' trying to convince the enlightened citizens of Madras of my favourite theme that the doctrine of Karma implied free will and not determinism, paradoxical as it may seem.

We left for Mysore on November 24 for the Matscience Conference on 'Elementary Particles.' This time Lalitha had on opportunity to give a concert at the traditional Bidari Krishnappa Sabha in Rama Mandir.

In December we had the pleasure of receiving an old Matscience visitor, Professor P. T. Landsberg who recalled with pleasure and pride his association with our Institute.[3]

Notes

1) The main temple in Madras where the Alladi family worshipped regularly is the *Kapaleeswara Temple* in Mylapore. In all temples, various festivals are held in honour of its deities. One of the festivals of the Kapaleeswara Temple is *Vasantha Utsavam*, in honour of Lord Subramanya. The Vasantha Utsavam runs for several days, and Ramakrishnan's mother, Mrs. Alladi Venkalakshmi, began sponsoring the event for one full day, a practice that the Alladi family has continued to this day.

2) Professor Kanakanahalli Ramachandra (1933–2011) of the Tata Institute of Fundamental Research, may be considered as the father of analytic number theory in India. He was especially pleased that MATSCIENCE began organising conferences in 1979 emphasising classical aspects of number theory, and enthusiastically took part in them. Ramachandra's fundamental contributions extended beyond analytic number theory to algebraic number theory and transcendental number theory.

3) Peter Landsberg (1922–2010) was a English physicist who had done significant work in a variety of fields including black hole thermodynamics and entropy. He visited Matscience in 1965, and his lecture is published in Volume 4 of the Symposia on Theoretical Physics, Plenum (1967), that Ramakrishnan edited.

Chapter 43

The Tryst At Ann Arbor (1980)

January 1980 turned out to be one of the most momentous periods in Indian political history since the dawn of freedom. Indira Gandhi was voted back to power and the Janata Party was defeated in the polls within two years of its total triumph, which itself had astounded the world.

We had the pleasure of receiving at Matscience, Sir Charles Frank, the eminent physicist from Bristol. I introduced all the members of our academic staff who explained their work to him.

On the 21st of January we celebrated the Eighteenth Anniversary of the Institute when the Governor unveiled the portraits of Newton, Salam and Hilbert. It was appropriate that Newton's portrait should adorn our library since he was the father of modern science. Hilbert influenced the mathematics of the twentieth century with his famous problems which have challenged the ablest minds to achievement or distraction. Salam had recently won the Nobel Prize and was in every sense a sponsor of the Institute along with Niels Bohr.

The Republic Day reception (January 26) at the Raj Bhavan was of significance against the background of a new government at the Centre.

In early February we went to Tirupathi to offer our prayers on the eve of our departure to the U.S. and returned to Madras after visiting our village and enjoying an open air bath in the mango gardens.

We left for the U.S. on February 17 by Lufthansa from Delhi and after a beautiful flight arrived in Frankfurt where to our surprise the German papers announced that nine State Governments in India were dismissed, President's rule was established, and that elections were to take place in June.

After a delightful flight by Lufthansa DC-10 across the Atlantic, we arrived at JFK and flew to Detroit by a TWA 707. Ann Arbor was in the grip of a lingering winter and spring seemed to be far behind. We relaxed for one full month before my tour of the U.S. and Canada on lecture engagements.

We played plenty of tennis but one day due to sheer bad luck Krishna slipped and suffered a hairline fracture on his right wrist which disabled him for a few months from writing and playing tennis but he managed to drive even on the freeways using only his left hand!

On the 4th of March occurred a momentous event in the world political scene, when the Blacks won the elections in Rhodesia and Mugabe, hitherto a hunted guerilla, was voted to power. It was a reversal of history, marking the end of White rule which only a few years ago looked like continuing till the end of time. What was more astounding, he abjured the policy of revenge and invited White cooperation with the new government.

On the 17th of March Krishna received a letter from Don Zagier inviting him to spend a few days at Bonn and lecture on his work. This restored his spirits which were a little affected due to his fractured wrist.

I left on the first part of my lecture tour of the U.S. on the 18th of March. My first talk was at the Case Institute in Cleveland where I was received with the most gracious hospitality by Professor Takacs and the Chairman, Professor Young. Professor Takacs took me round Shaker Heights, the wealthiest locality in Cleveland, where the Fords and Rockfellers had their mansions. From Cleveland I flew by a Northwest 727 to Washington where I enjoyed the hospitality of Professor Uberall and his colleague Nagi.

From Washington I flew to Raleigh where the Morrises played gracious hosts in their delightful home, I watched the birth of spring in the gardens of University of North Carolina as the sweet magnolias, like lids of Junos eyes, opened at the tender breath of scented vernal air. I lectured at the University at the invitation of Professors Merzbacher and Biedenharn.

From Raleigh I flew by Piedmont Airlines to Knoxville where I was received by Professor Balram Rajput who took me to his spacious new house. My lecture at the University of Tennessee was well attended and received. From there I flew by American Airlines to Dallas and as usual I was received by Erwin Fenyves. I gave lectures both at University of Texas at Dallas arranged by Fenyves and at the South Methodist University at the invitation of Professors Bhatt and Kapadia. I returned to Detroit by American Airlines to spend a week with Krishna, Madhu and Lalitha before starting on my major lecture tour.

We invited Krishna's friends for dinner and among them was Professor Bruce Berndt[1] who was making a special study of the life of Ramanujan and his notebooks! Berndt was in Ann Arbor to deliver a colloquium at the invitation of Krishna.

On the 3rd of April I left by Eastern Airlines for Florida and was received at the very elegant solar airconditioned Gainesville airport by Professor P. V. Rao.[2] I was treated to the most generous hospitality in the statistics department by the Chairman, Professor Schaeffer, Professor Rao, and their colleagues. My lecture on the Approach to Stationarity was very well-organised. The next day in delightful Florida weather, Professor Charudattan,[3] a young Indian botanist took me to a State park containing the geological phenomenon of a 'sunken forest.'

From Gainesville I flew to Denver by Eastern Airlines via Atlanta to be received by Professor Jacques Marchand on behalf of Stan Gudder, the Chairman of the Department of Physics at the University of Denver. Marchand is a perfect gentleman in the true European tradition and I was flattered when he showed considerable interest in reading the chapters of the Alladi Diary on Ekamra Nivas that I had begun writing in 1978.

I stayed for a week at the Holiday Inn and my room was in the tenth floor from which I had a magnificent view of the Rockies. I gave two lectures at the University of Denver, one on stochastic processes and the other on the Applications of Probability to Quantum Mechanics.

I left by Hughes Air West for Los Angeles and during the wonderful flight I had a beautiful view of the snowcapped Rockies and the tawny Canyon landscape of Arizona. I was received at the Burbank Airport by Professor Venkateswaran of UCLA. The week with Venki in his lovely house in Northridge in the San Fernando Valley was one of the happiest interludes in my travelling career. Venki is a gracious host, lavish to the point of embarrassing even a demanding guest. He had acted as a 'Godfather' to Krishna during his graduate days at UCLA, providing him friendship, philosophy and guidance in a city of distracting delights and attractions. I met Krishna's teachers Professors Straus and Arens. Krishna's friends Stan and Sharon Marcus entertained me at lunch in the exotic Scandia restaurant. My good friend Richardson invited me to deliver a lecture at Rockwell, always a pleasant and fruitful experience.

On the 17th April I shifted to Laguna beach to enjoy the friendly hospitality of Gordon Shaw and his wife Lorna. I lectured at Irvine following which Gordon hosted a lunch at a vegetarian restaurant called the 'Good Earth'. I shifted back to Venki's residence and before departing the next day for lectures in Canada.

I took the Air Canada for Edmonton and was received by Professor Kreuzer who took me to the comfortable guest rooms of the Mckenzie Hall. I was treated to gracious hospitality by the Chairman, Professor A. N. Kamal and his colleagues. I enjoyed my conversations with him at lunch and dinner at the Faculty Club after my lecture in his department. Professor Subbarao, a Number Theorist at Alberta, invited me to give a talk the next day in the mathematics department and immediately after the lecture I left for Winnipeg by Air Canada. At the University of Manitoba I gave a talk at the invitation of Professor Sivakumar.

From Winnipeg I flew by a Northwest DC 10 to Minneapolis and a 747 to Detroit. The flights were a sheer beauty, with nothing left to be desired, a far cry from the rough and tumble of the Bombay airport and the traumatic experience of checking in at the Indian Airlines counter with excess baggage or without confirmed reservation!

On the 30th of April, I flew to Madison where I gave a lecture the University of Wisconsin at the invitation of Professor Keith Symon. It was a pleasant surprise when I found that an old colleague of mine at Princeton, Dr. Loyal Durand, was in charge of the seminars at Madison!

It was an enjoyable stay with Krishna and Madhu, a tryst at Ann Arbor. Lalitha and I left for India by a Lufthansa 747 on the 12th of May and arrived at Frankfurt after a delightful flight. We checked in at the luxurious Frankfurt Sheraton hotel where we had a pleasant rest before taking the flight to Bombay. After the customary refresher stop at the Centaur Hotel in Bombay Airport, we arrived at Meenambakkam Airport, Madras.

Notes

1) Professor Bruce Berndt of the University of Illinois, Urbana, is one of the greatest experts on the work of Srinivasa Ramanujan. He is especially well-known for having edited Ramanujan's Notebooks in five volumes published by Springer, a task he started working on in 1977 and completed in 2005. He received the Steele Prize of the American Mathematical Society for this monumental contribution. The Colloquium at the University of Michigan that Berndt gave in 1980 at Krishna's invitation, was one of his first on Ramanujan's Notebooks.

2) Pejaver V. Rao (1935–2015) — known as PV — was a noted statistician at the University of Florida. He completed his BA at the University of Madras in 1954 and there he attended some lectures of Ramakrishnan who had joined as a Reader in 1952. PV received his PhD from Georgia in 1963 and joined the newly formed statistics department at Florida which he helped develop. PV was Ramakrishnan's host in Florida in 1980. PV contributed an article in 2010 to the volume in Ramakrishnan's memory (see [5]).

3) R. Charudattan (alias Charu), an authority in the field of plant pathology, is Professor Emeritus at the University of Florida. Charu's father, the late Sanskrit Professor Raghavan at the University of Madras, was a close fiend of Ramakrishnan and Lalitha, and they interacted regularly at the Madras Music Academy where Raghavan was Secretary. When Ramakrishnan met Charu in 1980, he was a young post-doc at Florida.

Chapter 44

Promises Kept And Miles Traversed (1980)

During Krishna's three month vacation in India we had many promises to keep (Prarthanas) and miles to traverse. We wished to perform a *Kalyana Utsavam*[1] at Tirumala, a *Thulabharam*[1] at Guruvayur and offer prayers at the Chamundi temple in Mysore. We wanted to spend a few days in the wild life sanctuaries of Parambikulam and Mudumalai and take part in the second Matscience Number Theory Conference at Ooty.

Krishna informed us by phone from Germany that he was having a pleasant and fruitful time with Don Zagier and the members of Hirzebruch's group at Bonn. Later he lectured at Frankfurt at the invitation of Professor Wolfgang Schwartz before taking the plane to Madras. A day after their arrival in Madras, Krishna and Madhu went to Kodaikanal, the hill resort which rivals in charm the Queen of hill stations, Ootacamund. On his return from Kodaikanal, Krishna gave a series of lectures at Matscience.

On the 20th of June a new concept in probability theory occurred to me and I named it 'Disparity in stochastic processes' which included the concept of 'activity' which I had formulated before. It brought to a successful conclusion my thirty year effort to understand the concept of direction of evolution of a stochastic process.

On the 24th of June we heard the sad news that our beloved ex-President Mr. V. V. Giri[2] passed away. Besides being the official head of the State he represented the historic bygone era of the moral giants, Gandhiji and Nehru, the nonviolent struggle of the few for the redemption of the millions from the omnipotent British Government. He had a fond corner for me and was particularly favourable to Matscience and responded to our invitation twice during his tenure as the President of India.

Since Krishna's fracture had completely healed we enjoyed regular tennis at the Mylapore Club and were in the right mood for celebrating his wedding anniversary on the 29th of June at Ekamra Nivas, with Ramani's flute concert followed by dinner attended by our friends and relatives.

In early July we offered prayers at Tirumala temple and performed Kalyana Utsavam with the customary Hindu faith that watching the marriage ceremony of Lord Srinivasa would imbue our human bonds with a touch of sanctity and Divine Grace.

Madras entered the international network of television on the 4th of July when for the first time the Wimbledon matches were projected 'live' on the television. What a magnificent stroke of luck that we should watch the greatest match ever played at Wimbledon, a five set duel when, in a cliff-hanger, heart-throbber, nerve thriller tiebreaker in the fourth set, McEnroe stemmed seven match points against him but ultimately lost to the charmed hand and magic wand of Bjorn Borg in the fifth set.

Madhu gave an elegant dance performance on the 17th of July at the Mylapore Fine Arts Club which was well attended and well reviewed later by the press. Krishna gave a series of lectures at the Mathematics Department of the Anna University on 'Transcendence and irrationality' explaining how in recent years the theory of transcendental numbers attained pre-eminence in mathematics through the work of Alan Baker followed by a resurgence of interest in irrationality with Apery's proof of the irrationality of $\zeta(3)$, namely the value of the Riemann zeta function at 3.

The 18th of July was a significant day in Indian scientific history when a Satellite was launched for the first time from the Indian rocket station at Sriharikota. Later we had an opportunity to visit the Madras Atomic Power Plant at Kalpakkam at the kind invitation of a young scientist, Dr. Gopinath. What impressed me most was the high morale of the Indian scientists there and their justifiable faith in the success of their venture.

We left on our southern tour by car on the 28th of July. Madhu did not accompany us since she had her *prarthana* (promise) to fulfil to visit Vaideeswaran temple near Tanjore along with her parents. Krishna's friend, G. P. Krishnamurthy took her place! We had a delightful stay at Parambikulan in the Anamalai Sanctuary. Krishna felt that the view at Parambikulam would compare favourably with that of Lake Louise at Banff National Park in Canada. We took an early morning jeep ride into the forest and were fortunate to see herds of Indian Gaur. We later spent two days at Pollachi in the palatial house of my good friend Mahalingam whose hospitality is as lavish even in his absence! We made a trip to Guruvayur and as usual we had good luck that in the midst of monsoon it was a sunny day. After offering prayers to Lord Krishna and performing Thulabharam[1] for Krishna we returned to Pollachi admiring the scenic beauty of Kerala. From Pollachi we drove to Ootacamund via Mettupalayam and Coonoor and reached the HPF guest house, the venue of our Matscience Conference. We were happy to find that Professor K. Ramachandra of the Tata Institute just arrived to participate in the Conference as a principal lecturer. After the conference we descended to Mudumalai by the steep Masanagudi route and stayed in the lovely and secluded Kargudi guest house. a place Krishna dreamed of in distant Ann Arbor. True to our expectations, from the secure comfort of the travellers bungalow we watched a lone tusker (male elephant with prominent tusks) move around the guest house at night and rub its tusks against the trunk of the mango tree in the front lawn of the guest house!

We moved to Mysore where we stayed at the elegant CFTRI guest house. We offered prayers at the Chamundi temple, where Lalitha sang the immortal composition of Mysore Vasudevachariar in honeyed strains of Bilahari. We visited Krishnarajasagar with its fountains outspraying the charms of Versailles, and returned to Madras after a most pleasant journey.

On August 16, Krishna and Madhu left for Ann Arbor by Air India to be in time for the summer meeting of American Mathematical Society at which Krishna was presenting a paper in a Number Theory session organised by Bruce Berndt.

Notes

1) A *Kalyana Utsavam* is a ceremony that concerns the marriage of one of the Hindu gods. In Hindu temples devotees perform and witness the re-enactment of the marriage of the main deity, and this is considered auspicious for married couples. Among all Kalyana Utsavams, the one in Thirupathi for Lord Venkateswara, is most famous.

A *Thulabharam* is a Hindu ceremony in which the devotee sits on a large (common) balance, and fruits, vegetables, or grain is set on the other side equal in weight to the devotee, and the cost of this is paid to the temple. Devotees who can afford much more, can place more expensive items on the balance! Among all thulabharams, the one in the Lord Krishna temple in Guruvayur is most famous.

2) His Excellency V. V. Giri (1894–1980) was the fourth President of India (1969–74). He played an active part in organisations like the Indian National Congress in an effort to gain independence for India from the British rule. He had a very high regard for Ramakrishnan's father Sir Alladi, and an almost parental affection for Ramakrishnan. He visited MATSCIENCE twice during his term as President of India.

Chapter 45

A Strange Melange Of Stirring Events (1980)

The next few months the world witnessed a strange melange of stirring events — Presidential elections and the Voyager's entry into Saturnian domains, Middle East wars and space sojourns of Soviet cosmonauts, visits of an enchanting Prince and a puissant President. At home in Ekamra Nivas there was the tranquil calm of Mylapore life and at Matscience the quiet efficiency in research work.

On the 18th of August I opened the buildings of the State Bank of India, Mylapore, when I took the occasion to draw the attention of banks in India that loans for housing should be given on the basis of salary and not only against property as security, so that the benefit will be available to a larger number of people. This is the basis of the housing schemes in the U.S.

It was a memorable day when I visited the Space Research Centre (SHAR) at Sriharikota along with Lalitha. What an incredible achievement that within ten years, India was able to enter the Space Age and obtain the capability of launching rockets for atmospheric studies!

On the 5th of September I sent my paper on 'Disparity' for publication in the Journal of Mathematical Analysis and Applications. Since I had already defined the 'difference' between two distributions, I wanted to understand how the 'difference' in initial conditions is 'propagated' in time. I found that the earlier concept of Activity was embedded in Disparity and the proof that it decreases with time was quite easy if we adopt this viewpoint.

On the eve of Navarathri, we attended a splendid concert of Semmangudi when we particularly enjoyed his rendering of Thyagaraja's Manamuleda in Hamir Kalyani. The spirit of the song is characteristic of South Indian tradition and finds no place in Western culture. It is an example of Nindasthuthi where one can rightfully get angry or disappointed with the Lord out of pure affection and esteem that He is not bestowing His blessings on the devoted suppliant!

I took an opportunity to visit Kalpakkam at the invitation of the Nuclear Physics group when I lectured on 'Concepts and Misconceptions in Probability' emphasising the meaning of Disparity and Activity. In the course of the lecture, I spoke on Inverse Probability when a problem suddenly confronted me. While in my work I had defined an inverse operator containing negative elements for getting the distribution backward in time, the Bayes theorem gave a positive definite matrix. How are we to reconcile these two? I did not raise this problem during the lecture since I thought it would only confuse the audience with my doubts when the other concepts I described were quite clear. The solution occurred to me after two days of active contemplation and I presented it at a seminar at Matscience entitled 'Duality in Stochastic Processes.'

Even a world used to miracles in science was astounded by the successful exploration of Saturn's rings by the Voyager, with its clean messages across a billion miles of void from the solid earth. Soon came the news of the Soviet cosmonauts Leonid Popov and Valery Ryumin returning to the earth after the incredible feat of making the longest space sojourn of sixteen months on the Salyut-Soyuz orbital complex in outer space!

On the 23rd November my elder brother Kuppuswami was sworn in as the Chief Justice of Andhra Pradesh, a fitting honour that the legal wards of my father's chambers sooner or later earned rewards of high judicial office.

It was a pleasant surprise when I was invited to Sriharikota (SHAR) again, this time to give a lecture on 'Einstein — Magic or Logic.' It was attended by the entire scientific community there and it was gratifying that the new look at Relativity was well appreciated.

On the 15th of October at the kind invitation of my good friend Kumar of the British Council, I delivered the C. P. Snow memorial lecture on the 'Integrated Life,' making a plea for the fusion of the two cultures, art and science. It was an extempore lecture which I recollected in tranquillity (see [1]). I had also the opportunity to talk on the same subject two months later at the International Theosophical Convention at the invitation of its President, Mrs. Radha Burnier, when I spoke on the role of freewill and its logical connection with the doctrine of Karma in Hindu thought.

On the 10th of November at the Tamil Nadu Academy of Sciences meeting I talked on 'New concepts in Probability Theory' when we had a distinguished visitor from Berkeley, Professor Joseph Cerny. We had a lecture by Professor Cerny at Matscience and later we took him on a guided tour round Ekamra Nivas just to bring home to him how through visits like his, we were introducing 'the heady atmosphere of Berkeley into the placid environment of my family home.'

One evening I was asked to speak at a music concert when I compared the nature of vocal with that of instrumental music. There has been a growing tendency which had to be restrained, of attempting to over-interpret the songs of Thyagaraja by conforming to the meaning of the song rather than to the metre and the rhythm. Poetry is something different from prose, so is music from the actual text and there is no necessity to make a conscious effort to convey the meaning which is obvious to any discerning listener. This is contrary to the original intention of the composer and disturbs the euphony, assonance and alliteration, characteristic of great music. Fortunately the instrumentalist does not assume this needless obligation since he is only concerned with the sound and rhythm while the sense dwells in the minds of both the listener and the artiste.

On the 2nd of December Madras was illumined by the golden gleam of the royal visit of Prince Charles. Even the clouds withdrew in the midst of the monsoon to let the Heavens smile in bluest bloom as the Prince enjoyed his visit to a local farm. His enchanting smile and sparkling wit captured the hearts and minds of the Indian population with greater ease than the stratagems and prowess of Clive and Hastings. A Commonwealth of hearts had replaced an Empire of conquest.

The Prince's visit was followed by that of Leonid Breznev, who symbolised the power and vitality of the Soviet Union, with its impregnable faith that it is the power of the state that safeguards the welfare of the individual, complementary to the Western view that it is individual initiative that ensures the prosperity of the larger society.

We left on the 9th December by car for Ooty for the Matscience Conference on 'Stochastic Processes and Applications.' The Conference with the HPF guest house and it was one of the most fruitful and pleasantest we had conducted. I lectured on Bhabha's contributions to stochastic theory, naming him the father of stochastic theory in India. I first described the history of product densities and later presented my new work in Probability Theory. We returned to Madras in time for the Music Academy festival the highlight of which was Semmangudi's concert as the audience rocked in raving ecstasy to the resonant sounds of Mukhari in the Thyagaraja kriti Ksheenamai which only he could achieve.

Chapter 46

A Year Of Divine Grace (1981)

Erdös lecturing at Matscience, June 1981. - Editor

The year started in a pleasant manner when I received a kind letter from Dharmasey Khatau, the eminent textile baron, inviting me to be a member of the local council of the Bharatiya Vidya Bhavan of which Mr. C. Subramaniam was the Chairman. My association with Mr. Munshi, its Founder, was very close ever since I watched him in animated conversation with my father during the great days of the Constituent Assembly. I had enjoyed his vivacious hospitality in Bombay in 1948 when he was speaking about his dreams of such an institution. So I accepted the invitation with pleasure.

At Matscience we had, as usual, distinguished visitors for the new year — Professors Marshall Stone from the United States, Beauregard from Paris, and Eliezer from Australia. We had a pleasant surprise in finding that Professor Gourdin of Orsay was in Madras on a short visit and was willing to give a lecture at Matscience.

Krishna and Mathura with Paul Erdös at Ekamra Nivas, June 1981. - Editor

January 12 was a significant day for Matscience when Professor Abdus Salam spent an evening at the Institute during his brief visit to India for a conference at Kalpakkam. It gave us the opportunity to congratulate him personally on the award of the 1979 Nobel Prize, for which he was almost a 'semi finalist' in previous years. Our patron Mr. C. Subramaniam was there and we naturally recalled with grateful pride the miraculous circumstances of the birth of Matscience and in fact its conception due to the joint interests of Abdus Salam and Niels Bohr on the one hand and C.S. and Nehru on the other. I remember Salam telling me that we had succeeded in getting an international institute within our own country while he was able to create such a centre only in Europe outside the domain of developing nations.

During Pongal we had a real surprise when an old tenant of ours, Herbert Herring (former Director of the Max Mueller Bhavan), visited the Alladi House with over fifty German tourists mainly to show them the lighted terrace, perhaps the only one of its kind in a private house in our city. That evening we enjoyed a music concert by the violinist T. N. Krishnan when I made a special request to him to render 'Na Jeevadhara' in Bilahari, a song which, as the legend goes, brought a dead man back to life.

On the 17th of January we celebrated the Nineteenth Anniversary of our Institute when the annual symposium was inaugurated by the Governor of

Madras. Mr. Christopher Sholes, Director of the U.S.I.S. and Mr. Ozaki, the Japanese Consul General unveiled the portraits of Schwinger and Segre, Yukawa and Tomonaga, the makers of modern Physics. Lalitha gave a concert that evening before the anniversary dinner. On the 21st January came the most astounding news of the release of American hostages in Teheran, ending a tension which gripped the world and rocked the American nation for four-hundred and forty four traumatic days. It turned out to be a most fortunate augury for Ronald Reagan who was being inducted as the President of the U.S.

At Matscience, we had the visit of Professor Gehlen, a member of the Bonn group closely associated with our Institute. His interest in Indian culture, music and sculpture was very flattering especially since he felt that Madras was a 'typically Indian' city.

Professor Dietz arrived from Germany in response to our invitation to the Mysore symposium. We left for Bangalore by car on 1st February and stayed overnight at the Indian Institute of Science guest house in Bangalore, enroute to Mysore. I took the opportunity to call on Rajagopala Iyengar in Bangalore and express to him my heartfelt thanks for his splendid biographical essay on my father (see [1]). Before leaving for Mysore, I gave a talk at the Indian Institute of Science in Bangalore to a combined group of physicists and applied mathematicians on my recent work on probability theory. We reached Mysore early afternoon and had enough time to ramble in the gardens of Krishnaraja Sagar.

The lectures at the summer school on 'Particle interactions and astrophysics' started the next day with Professors Dietz and Eliezer as the principal lecturers. That evening Lalitha gave a concert with the immortal song of Ramadas 'Nanu Brovu' in Kalyani as the main piece — the best way for a suppliant to reach a Royal King's heart is to entreat the Queen to plead for him in the ecstatic hour of an enchanted night!

At Matscience we had the visit of a Japanese number theorist, Professor Motohashi. We had a dinner in honour of Dietz and Eliezer who seem to have enjoyed their stay both 'on and off' the Mysore symposium.

I decided to leave for the US with Lalitha in response to invitations to visit various centres, particularly the National Accelerator Laboratory (NAL) in Batavia. We took the Indian Airlines Airbus for Bombay and after a pleasant stay at the Holiday Inn, watching the sunset at Juhu beach, we left by a Pan American Jumbo at the 'crack of dawn'. The flight through Frankfurt and London to New York was sheer delight but we found that we had arrived two hours late and had to rush to the TWA terminal to catch the plane to Detroit just before the doors closed! At the Detroit airport we were greeted by Krishna and Madhu.

On Monday, the 2nd March, Krishna phoned to us from the campus to give us the happy news that he was awarded the visiting membership at the Institute for Advanced Study at Princeton. Within half an hour came a phone call from India, when Vasu informed that I was the recipient of the national FICCI Award[1] in mathematical sciences for 1980 to be given by the President of India.

I left on the 15th of March for Chicago by an American Airlines 727 and reached the NAL at Batavia at 4:15 p.m. on Sunday, just at the time when it was being announced that the highest ever proton intensity was reached in the laboratory. I met Dr. Chris Quigg, the head of the theoretical physics department, who took me to the Director, Lederman. His ready smile and friendly informality, intended to put any visitor at ease, conceal (or reveal!) the impassioned dedication of the discoverer of the Upsilon particle, his irrepressible faith that present experiments will throw light on what happened during the first three minutes of the creation of the universe.

The 'High rise' building, the Robert Rathbun Wilson Hall, with its glass walled offices and the central atrium is a towering masterpiece of unconventional architecture designed by the first director of the Fermilab after whom it was named.

I hired a Chevette for the weekend and enjoyed the drive into Chicago and its environs in bright sunny weather. I visited the Science Museum where I walked through a new exhibit, a German U boat, wondering at the inhuman patience of the crew confined in that technically perfect dungeon lying in wait for the unwary victim, the surface ship.

I attended quite a few seminars at Batavia, the most interesting being the lecture of Eugene Parker on cosmic magnetic fields. Twenty five years ago I heard him talk at the University of Chicago on the same topic as a theoretical physicist of great promise which he fulfilled by becoming the Chairman of the department.

After a memorable stay at the Fermilab spending over fifty hours in the magnificent library, I left for O'Hare Airport in a Cadillac limousine, the driver of which refused a tip because he was the owner of the vehicle! I left for Denver by a United Airlines DC-10 and checked in at the very comfortable Holiday Inn. The next day I lectured on inverse probability at the University of Denver at the invitation of the Chairman, Stan Gudder. I dined that night with Professor Huzurbazar, a veteran Indian statistician with whom I had discussions about his early days in Cambridge.

I left Denver by a Republic Airlines DC-9 and was received at the Las Vegas airport by my good friend K. M. Rangaswamy. The same evening he took me to the Hoover Dam, a marvel of American engineering. It reminded me of the Bakra-Nangal and similar projects which Nehru planned immediately after the birth of Indian freedom. Las Vegas is stranger and more fantastic in reality than is described in tourist folders. It is incredible that the gambling instinct has such a hold over the human mind to the exclusion of all interests and reason. Dharmaraja gambled away his empire and his family yielding to that irresistible temptation!

I gave a talk on probability theory a subject relevant to gambling, at the University of Nevada, Las Vegas and walked around its magnificent buildings admiring the luscious lawns in the midst of a desert. From Las Vegas I flew by a Republic DC-9 to Burbank where I was greeted by Venki and that night we relaxed in a Mexican restaurant, talking on the insoluble problems of 'brain drain' from India.

The next day Richardson picked me up and took me out for lunch before my lecture at Rockwell where everyone was stricken with the excitement of the Space Shuttle which Rockwell was building for the American nation.

I moved to Irvine where I gave a talk at the invitation of Professor Gordon Shaw. My stay at Laguna beach was pleasant as usual and I returned to Northridge to spend a leisurely day with Venki before leaving for Dallas by an American Airlines 727 from Los Angeles airport.

Dallas had become almost a 'must' like Los Angeles during my visits to the U.S. After a pleasant lunch with my host Fenyves and the Nobel Prizeman Kusch, I gave my talk which was well attended by the combined group of mathematicians and theoretical physicists. I shifted to Arlington, watching with wonder during the cab ride, the Plaza of the Americas, an incredible tower of affluence and efficiency in affluent Dallas. The lecture at the University of Texas, Arlington was as usual well-organised and attended by the active group of mathematicians with Lakshmikantham. I returned to Ann Arbor the next day to spend a leisurely week with Krishna, Lalitha, and Madhu before leaving for Germany.

My week of leisure turned out to be the most momentous period in American history. The space shuttle, an amazing sequel to the Apollo flights, was to be launched on Friday the 10th April. It was an event of comparable significance to that of the release of Atomic Energy in 1945 and man's journey to the moon in 1969. For the first time America was building a space vehicle that was to circle the globe beyond the gravitational field do in like an ordinary plane, to be used again and again on similar projects, a miracle to make miracles common place! But nine minutes before count-down, there was a trouble in the matching of computers and the event was postpones by two days. So April 12th turned out to be one of the greatest days for American science and technology. The blast-off took place at 7 a.m. and we were fortunate enough, like millions of others, to watch it on television over toast and coffee, making this achievement a little different from the release of Atomic Energy which was done is sealed secrecy. The 14th of April was America's finest hour when the space shuttle landed perfectly on time. It is impossible to describe the thrill of witnessing the successful completion of so daring a venture, as John Young and Robert Crippin stepped out of the space shuttle. The smile on the lips of John Yardley, the director of the programme, expressed the triumph of the 'impossible' endeavour which involved the efforts of thousands of scientists and engineers.

I left for Europe by a Pan American 747 from Detroit and was seen off by Krishna, Madhu and Lalitha. After a most comfortable flight, I arrived at Frankfurt the next day and proceeded directly to Bonn by a Schnellzug.

At the University of Bonn, Professor Bleuler played a gracious host and I had very useful conversations with him and his colleagues. I met Don Zagier and discussed Krishna's work with him at lunch at an Indian restaurant.

I had the pleasure of meeting Dr. Pfeiffer, the chief of the Humboldt Foundation, and it was flattering when he spoke of primary role of my father in drafting the Indian Constitution. That night we had a delectable party at Bleuler's house in one of the most beautiful suburbs of Bonn, situated amidst a riot of flowers and foliage.

I left for Clausthal by Schnellzug and arrived at the lovely town of Goslar and checked in at the comfortable hotel near the station. My gracious host H. D. Doebner invited me for dinner to his house and took me next day to Clausthal where I lectured on stochastic processes. The winding drive up the Harz mountains in his sleek BMW convinced me why Doebner was so successful at attracting guests to his University.

On my return to Bonn I listened to a lecture by the veteran Amaldi on 'Gravitational waves'. Professor Gehlen invited me to his home where I heard the music of M.S. and taught my host the South Indian art of enjoying the 'Neraval' in the song 'Sarojadalanetri' in Sankarabharanam while admiring his collection of souvenirs brought from India.

I went by train to Frankfurt and checked in at the fabulous Frankfurt Airport Sheraton. I was to take the Pan American Flight 2 to India while Krishna, Lalitha and Madhu were to join me half an hour earlier by flight Pan Am flight 66 from New York. It was a touch and go connection and there was as much hair-raising excitement in expecting the arrival of Krishna's flight as in the landing of the space shuttle! We arrived in Bombay without our luggage and were taken to the Holiday Inn where we relaxed in idle comfort after the excitement of the 'Frankfurt connection'.

Krishna gave a series of lectures of Matscience on Probabilistic Number Theory in preparation for the visit of Erdös who flew straight from Canada to Madras for the Matscience Conference on Number Theory in Mysore immediately after the convocation at Waterloo where he was awarded an honorary doctorate. We had an evening party in his honour at Ekamra Nivas with Ramani's flute concert, followed by dinner.

We left for Mysore by car, staying over-night at the Indian Institute of Science guest house in Bangalore. Erdös, Ramachandra and Krishna were the principal lecturers at the conference. Introducing Erdös at the inauguration, I drew attention to the inexplicable situation that in Madras, the city of Ramanujan, there was yet no active school in analytic number theory. But soon Matscience would be initiating work in the subject and a beginning was made by organising symposia on number theory with the active cooperation from Ramachandra of the Tata Institute. Lalitha gave a concert in honour of our distinguished guests following which we had a dinner at Lalitha Mahal Palace Hotel. We visited the Chamundi temple up the hill with Erdös and left for the Mudumalai wildlife sanctuary for a tryst with wild life which Krishna had planned at Ann Arbor. We were lucky in seeing a wide variety, from elephants to red squirrels, peacock's galore, a pack of wild dogs and even a black bear amidst a herd of deer. When we had a flat tyre in the midst of the forest we had the excitement as in an African Safari augmented by watching an elephant taking a bath from a Machan. Our three days stay at the Kargudi Guest house looked too brief like a dream as we returned to Madras after a most pleasant car journey through Mysore and Bangalore. Upon return from Mysore, Krishna joined the Institute as a member of its permanent staff on the 11th of June 1981.

July 4, the American Independence Day, was a great day for American tennis when McEnroe beat Borg, a legend replacing another legend. The next day an article on the visit of Erdös to Madras appeared in The Hindu, a month after his visit.

On the 17th July my colleagues arranged a function at the Woodlands Hotel on the occasion of my receiving the FICCI award. Among these who participated were my well wishers C. Subramaniam, N. Mahalingam,

Sivalingam and Narayanaswamy. In my reply speech I suggested that some enterprising entrepreneur should create an organisation in India to conduct conferences like those at Oberwolfach in Germany to stimulate interaction between science and industry.

I presided over the function at the U.S.I.S. on 'American Education in the Eighties' and I lectured on 'New concepts in probability theory' at the Indian Physics Association at the invitation of Devanathan before leaving for the Matscience conference in Ooty.

We left for Ooty on the 20th August and after a delightful motor trip through Bangalore reached Mysore the same evening. The next morning as we entered the game sanctuary at Bandipur, we saw one of the largest herds of deer. After a picnic at Kargudi we drove to the guest house at the Manjushree plantations near Gudalur. The managing director offered us gracious hospitality and explained to us the challenges of the tea trade, the hazards on the plantations and the subtler arts of blending and tasting tea.

We reached Ooty in time for the inauguration of the Matscience Conference on Probability at the HPF Guest house. One evening we drove to Fernhill to see the house Ambal Vilas, where we had stayed forty five years ago, when the world and Ootacamund in particular was so different in the spacious days of the British Raj.

After the conference in Ooty, we left for Mudumalai on the 27th August by the Masanagudi route which is so steep that it is prudent only to descend via that route and not to ascend! We were lucky enough to see elephants and bisons across the valley of the Moyar water falls. The next morning as we drove through Theppakadu we were stopped and informed of a message by phone that a daughter was born to Madhu. What an incredible circumstance that we should receive the telegram when we were relaxing in the highest viewpoint in Kargudi!

We left Bandipur next morning and immediately on arrival at Madras we drove to the Willingdon Nursing Home to see the baby. We decided to name her Lalitha after my wife (her grandmother), but we would call her Vanitha, since we heard of her birth while we were in the forest (vanam).[2] We celebrated the event by attending a music concert of Semmangudi when he sang the Pallavi 'Chakkaga Nee Bajana' to satisfying my insatiable palate for Sankarabaranam.

I inaugurated a physics teachers workshop sponsored by the British Council on the 7th September, commenting that scientists in India, when invested with excessive administrative power, may become insensitive to creative talent.

We went to Tirupati on the 17th September to express our gratitude to Lord Venkateswara for His Divine grace for the birth of the child.

Krishna's dear friend G.P. arrived from Delhi to spend a few weeks with him. Navarathri was celebrated with additional zest and pleasure since Krishna Madhu and the baby Lalitha were with us in the Kolu, puja and Manthrapushpam, a tradition which started at Ekamra Nivas in 1933 in the golden era of my loving parents.

During Navarathri, Krishna left for Princeton on the 30th September by Pan American from Delhi with eagerness, expectation and enthusiasm to that haven of creative science. Madhu and baby Lalitha were to join him later. He phoned to us from Princeton that he was well settled in one of the garden apartments of the Institute campus and had started work in the most inspiring milieu for mathematical research in the world today.

At E.N. the Navarathri ran its usual course and on Vijayadasami, the final day, we had the privilege of receiving Professor George Andrews[3] of Penn State University who lectured at Matscience on Ramanujan's Lost Note Book. It was an inspiring lecture reviving the memory of the legendary mathematician. We then I took him to a Bharathanatyam performance, the best we could offer to a visitor from the affluent society.

Krishna had to go to Ann Arbor to bring his Pontiac loaded with baggage to Princeton. On the way back to Princeton from Ann Arbor, he gave a seminar at Penn State University where he was treated to generous hospitality by Professor George Andrews.

We had a visitor, Professor Nelson, the head of an active group of applied mathematicians of at Texas Tech University in Lubbock, who evinced keen interest in contacts with Matscience.

It was a pleasant surprise when I was invited to Kavali, a modest town near Nellore to deliver a lecture on 'Science in the service of mankind' on Nehru's birthday. It was particularly topical since the Space Shuttle Columbia was on its second journey round the earth with Astronauts John Engle and Richard Truly, when he were driving on the rough and tumble of the misnamed Grand Trunk Road to Nellore! We were treated to gracious hospitality in traditional style by our host, the Chairman of Visvodaya. I was able to keep my promise with my uncle Venkatanarayana who had been inviting me to Kavali for the past twenty years.

On my return to Madras I spoke at a function where Subbaroya Iyer's portrait was unveiled and paid my tribute to the dearest friend of my father. With Krishna well settled in Princeton and the baby with us, I indulged in contemplation at leisure on a problem on which I had been thinking without hope of solution for over forty years, ever since I read the chapter on relativity in 'Theoretical Physics' by Joos in my honours course. Is there a natural way of explaining the spatial term in the time transformation in as clear a manner as the temporal term in the space transformation which is obvious even in Galilean theory?

On the 22nd November, Madhu and baby entered Ekamra Nivas and four days later Madhu and the baby left for U.S.A. to join Krishna. They were accompanied by Madhu's father N. C. Krishnan and her uncle C. N. Srinivasan.

On the 29th November while relaxing in the passage at Ekamra Nivas there suddenly occurred to me a new procedure to solve the relativistic problem and I rushed to my study (father's office room) to work on the idea. It was found it worked when I used the concept of external relative velocity I had introduced earlier. I was able to obtain in a new way the Lorentz Transformation with only one assumption, the constancy of the velocity of light, and derive the invariance of the four dimensional length.

We then had distinguished visitors to Matscience, Professors Rene Thom,[4] the Field Medallist from Paris, and Lamberto Cesari from Ann Arbor. The music season had just started and we could offer them delightful fare. Around that time, I completed my paper on the 22nd December, Ramanujan's birthday, with the title 'Further Unnoticed Symmetries in special relativity' and the inscription 'bias is beautiful' to emphasise the hitherto unnoticed TA (Towards and Away) symmetry, inherent in the theory. External relative velocity was mentioned by Einstein and a Russian physicist but neither of them used it to derive the Lorentz transformation.

On the 26th I despatched my paper for publication and attended a concert of K. V. Narayanaswamy at the Mylapore Fine Arts Club. What an incredible coincidence that he should sing the song 'Intha soukhyamani ne cheppazala', which means "I cannot describe the magnitude of the ecstasy", in raga kapi just at 7:45 when I was thinking, with similar feelings about the Lorentz transformation. The musical delight continued with Semmangudi's concert the next day, the assonance, resonance and consonance of his unfailing music.

As the music season ran its festive course, I contemplated on the year of Divine Grace which bestowed its triple blessings, Princeton for Krishna, TA (Towards and Away) symmetry for me, and little Lalitha for all of us!

Notes

1) FICCI stands for the Federation of Indian Chamber of Commerce and Industry. Ramakrishnan received the 1981 FICCI Award in the Mathematical Sciences. The award was given in New Delhi by the President of India, Neelam Sanjeeva Reddy. Since Ramakrishnan was in America, he could not be in Delhi to receive the award. Professor N. R. Ranganathan of Matscience, Ramakrishnan's former PhD student, received the award for Ramakrishnan.

2) In the Hindu tradition, even though a grandchild is named after a grandparent, when the grandparent is alive, it is customary out of respect not to call the grandchild by the grandparent's name. Thus a different name is often given to address the child. In the case of Krishna's daughter Lalitha, she is called Vanitha at home.

3) Professor George Andrews of The Pennylvania State University is the discoverer of Ramanujan's Lost Notebook in 1976. We owe primarily to him our understanding of the contents of the Lost Notebook. He is the world's greatest authority on the work of Ramanujan and the theory of partitions combined. The visit to India in 1981 was his first, and he is deeply grateful to Ramakrishnan's hospitality in Madras and in particular that Ramakrishnan arranged a meeting for him with Mrs. Ramanujan, as he said in a letter.

4) Rene Thom, a world famous topologist, became especially well-known for study of singularities which led him to create *Catastrophy Theory*, for which he was awarded the Fields Medal in 1958. His talk at Matscience was on Catastrophy Theory.

Chapter 47

Trans-America (1982)

At the dawn of the year there was nothing to indicate that we were to realise a dream I had been nursing for years — travel by car across the U.S., watching the natural and man-made wonders of the 'Chosen Land'. The year started with the usual preparations for the Anniversary Symposium, the theme of which was 'Fundamental Science, the source and sustenance of technology'. In early January, I received a letter from Professor Dold of Springer Verlag agreeing to publish the proceedings of the Third Matscience Number Theory Conference 1981 in which Professor Erdös was a distinguished participant. On the 21st January at the Matscience Anniversary my distinguished (former) colleague, Professor Santhappa, unveiled the portraits of Erdös and Hardy,[1] one a living legend and the other a discoverer of an undying legend.

We had the privilege of receiving Professor Doebner from Clausthal, Germany, having a chance to repay his gracious hospitality. We arranged a dinner in his honour and invited Professor Schutte from Bonn and a few other visitors who had arrived for a Nuclear Physics conference organised at the University by my old student, Professor Devanathan.

I took our guests to the J-farm near Madras, the creation and pride of the great industrialist Anantharamakrishnan, now maintained by his distinguished son Sivasailam, as a research unit of the TAFE (Tractors and Farm Equipment Ltd.). It was so impressive that I could understand why it was chosen as the main attraction Madras could offer to Prince Charles when he visited our city a year before.

In February we arranged the Matscience conference at Mysore on Relativity and Cosmology. This time in Mysore, we had a delectable dinner at the the stately Hotel Metropole with its lingering (and vanishing) atmosphere of polite and efficient service by 'butlers' as in the British Raj.

We reached Madras after a day's halt in Bangalore where I gave a lecture on Relativity at the Indian Institute of Science.

We read with interest the description of Indira Gandhi's visit to England where she inaugurated the Festival of India at an impressive ceremony at the Royal Festival Hall. How appropriate it was for M.S. to sing, *Akshayalinga vibho Akhilanda koti prabho* as a homage to the King of Kings (Lord Siva) before Prince Charles and the Prime Ministers of England and India! I was also impressed by Indira Gandhi's speech when she said "It will be unfortunate if the dialogue between developed and developing nations was allowed to dry up on the desert sands of confrontation and lack of understanding." How true this was of the Indian scientific community, whose academic spirit was dried up by jealousy and mutual recrimination.

On the 15th April I spoke at the I. I. T. Jamuna Hostel Day on the source and stimulus for creative thought. The next day Krishna phoned to me that he was invited to the University of Texas, Austin, for a six week summer institute conference and so we included Austin, on our itinerary.

We left for the United States by taking the Indian Airlines Airbus to Delhi and G.P. metus at the Palam airport. We checked in at the fabulous Maurya Sheraton and left by the Pan Am flight on April 19. After a comfortable journey, with brief halts at Frankfurt and London, reached New York to be greeted by Krishna and Madhu and the smiling baby.

We drove to Princeton where we rested a day to get over the jet lag before leaving for Washington by car in response to an invitation from Professor Uberall to lecture at the Catholic University. It was my first opportunity to talk on 'Relativity for all' based on the new proof of the Lorentz transformation. We saw the White House and its lovely gardens and later the spectacular Air and Space Museum where the miracles of modern space age were presented to ambling visitors amidst the earthly comforts of a spacious modern building.

A month at Princeton during springtime was a tryst with an unforgettable past, as I recalled my stay with Lalitha twenty five years ago during the Oppenheimer era at the great centre of learning. Princeton in spring is heaven on earth, when all nature seems in love, a riot of flowers, foliage and lush green grass. It was like walking through paradise every day inhaling the vernal air with the scent of nascent flowers of fresh-quilted colours.

We responded to the call of Manhattan by spending a whole day there and visiting the World Trade Centre enjoying 'the view from the top.' This time we went by ferry to the Statue of Liberty and feasted our eyes on the Manhattan skyline at twilight.

It was total relaxation in Princeton for me. As we rambled on Nassau Street, I remembered the essay I wrote on America in 1958, sitting near Palmer Square, on my first reactions to the charm and dignity of the University town.

Krishna took us to Philadelphia by car where we saw the Independence Hall and Liberty bell symbolising the birth of the world's greatest democracy. The spirit of 1776 still dwells in the noble buildings where the environment is so well-preserved as to make the visitor feel the presence of Franklin and Hancock and hear the voices of Adams and Jefferson proclaiming the right to life, liberty and the pursuit of happiness.

I lectured at Rutgers University on 'Duality in stochastic processes' at the invitation of Professor Joel Lebowitz during a one day conference on Statistical Mechanics. I met Dr. Harry Woolf, the Director of the Institute for Advanced Study, and was flattered to embarrassment when he told me he had read with interest the chapter on 'The Princeton Experience' in the Alladi Diary which I had sent him sometime ago. He spoke with urbane charm and politeness which go with a strong mind and firm convictions, necessary for guiding the destinies of an Institute which attracts, fosters and propagates creative thought.

Krishna had a successful year at Princeton which brought him into contact with the leaders of mathematics like Enrico Bombieri, Atle Selberg and Harish Chandra.

We left Princeton for Austin in Krishna's Pontiac and on the first lap of the Trans-American journey, took the Penna turnpike to Harrisburg reaching Luray at the entrance to the Shenandoah National Forest before sunset. We saw the famous Luray Caverns, their stalactites and stalagmites, as impressive as those at Postagna in Yugoslavia. Next we took the scenic skyline drive through the Shenandoah enjoying the breathtaking beauty of the Blue Mountains. We saw a natural bridge, an example of erosion carving a natural sculpture. We then went to Chapel Hill where we enjoyed the warm and friendly hospitality of the Morrises in their lovely home nestling amidst stately pines and fragrant cedars.

Springtime in North Carolina is the best of seasons and it was a floral paradise in the gardens of the Duke University and the University of North Carolina. The cuisine at the Morrises had a South Indian flavour, brought in after their stay in Madras for many years.

We then took the four hundred mile drive on the Blue Ridge Parkway, the most beautiful road in the world along the verdurous spine of the Appalachian mountains, flattered by the hues of heaven impartially at roseate sunrise and crimson sunset. We reached Knoxville at sunset and stayed at our good friend Rajput's house.

The next day in Knoxville we visited the World's Fair which had as its theme the energy problems and triumphs of the day. Strangely, the most popular pavilion turned out to be the Chinese, where works of art claimed more attention than science or technology as visitors gazed at Chinese urns with Keatsian wonder. America projected the marvels of modern technology while Saudi Arabia displayed how such technology could harness God-given resources to desaline the oceans and make the deserts bloom!

We left Knoxville after breakfast by the arterial Highway 40 which took us through Nashville and Memphis. We reached the Holiday Inn in Forrest City just before a thunderstorm of great intensity broke out. We left Forrest City after a delectable breakfast and our journey took us through Little Rock, the capital of Arkansas, which had rocked the political world of the United States over a racial incident a few years ago. We had a delightful picnic before Texarkana and arrived in Dallas where we checked in at the Holiday Inn. Fenyves met us after breakfast the next morning and took us on a drive through downtown Dallas where we gazed at the dazzling affluence of the Plaza of Americas and visited the Kennedy Memorial before going to the University of Texas at Dallas for my lecture on special relativity.

We drove from Dallas to Austin through typical Texan country, flat expanse bounded only by the horizon, and entered our house in Austin a little before midnight. All through the journey, little Lalitha behaved perfectly as if to travel born, providing us full entertainment for a few spoonfuls of juice for her infant appetite.

The first week at Austin we spent in getting acquainted with the environment of the well-laid city. Krishna participated in the Number Theory conference and it was a fortunate circumstance that his guru Paul Erdös and his post-doctoral mentor Hugh Montgomery were there.

I left on my Californian trip by a Continental Airlines 727 for Los Angeles, while Lalitha stayed back with Krishna and his family. Venki (Professor Venkateswaran of UCLA) received me at the airport and we had as usual non-stop conversation of the Indian scientific scene over dinner at Ricardo's, a Mexican restaurant.

Next morning I was taken to UC Riverside, eighty miles away, where I gave a colloquium on 'New concepts in probability theory' at the invitation of Professor M. M. Rao. After a reception at his house and dinner at a Mexican restaurant, I returned to Venki's house at North Ridge. The next day we drove along the scenic coastal highway 101 to Santa Barbara recollecting my first visit there in 1956 by Greyhound. We strolled through the campus of the UC Santa Barbara, perhaps the most beautiful in the world, and returned via Santa Monica.

I went to lovely Laguna Beach in response to the invitation from Gordon Shaw. I lectured at Irvine on the Lorentz Transformation and returned to Venki's place by the comfortable Flydrive bus service watching with childish delight the fantastic reality of a city on a myriad wheels. I took an opportunity to visit UCLA and meet Krishna's professor Straus.

The next day Richardson took me to the Rockwell Science Centre where I gave a talk on 'Work at Matscience'. The centre is a perfect example of American initiative, enterprise, planning and forethought. Back at Venki's house, I relaxed for two days before returning to Austin.

The two months at Austin were one of the happiest periods Lalitha and I had spent in the United States during the past twenty five years. Taking advantage of the library in our rented house, I read the history of films from 1930, recalling with pleasant nostalgia those which I had seen in my early boyhood. We watched interesting shows on the TV but the most exciting were those relating to Wimbledon. For a full fortnight we feasted our eyes and tingled our nerves for over sixty hours watching the tennis matches and the titanic men's finals when Jimmy Connors emerged as the champion, with his newly burnished game of serve and volley against the holder, John McEnroe.

Davis Cup rivalled in thrill the Wimbledon when McEnroe won against young Wilander, a Swedish phenomenon of irrepressible talent invested with imperturbable coolness, after one of the toughest duels ever seen in tennis history. We played tennis on the hard courts near our house when I got the opportunity to show Krishna the merits of well-timed volleys I learnt at Manchester and watched at Wimbledon.

Following the summer conference, Krishna was offered a visiting assignment for the academic year 1982–83 year at Austin by the Chairman of the mathematics department on the suggestion of Professor Jeff Vaaler who was to spend his sabbatical at Princeton. Before Krishna was to begin his academic assignment in the Fall, we decided to make a car trip to the canyon lands as our principal destination on our trans-American journey.

We left for the Grand Canyon on the 9th August, my birthday, chosen deliberately by Krishna who wanted to satisfy my long nursed desire to travel by car across America. Our first stop was Van Horn where we stayed overnight at the Holiday Inn after a drive through typical Texan country, with its cattle studded ranches and prosperous looking barns and country houses. The next day we drove on Highway 10 through the deserts of New Mexico and the sun-drenched passes of the Dragoon mountains, once the stronghold of the proud and defiant Cochise, the Apache Chief during his vain struggle against the whelming tide of the White Man's conquest. We visited the Saguaro National Park with its cactus forests, examples of unbelievable botanical sculpture. We stayed at the Holiday Inn in Tucson and visited the Case Grande ruins, where we had an enjoyable picnic. From there we drove to the Montezuma Castle Monument, a spacious cave-dwelling in the mountains, carefully preserved as modern America's tribute to the Native American heritage. We reached Flagstaff at the entrance to the Grand Canyon and drove through the man made forest of stately pines. It was thrilling to see the Canyon emerge suddenly into sight as we reached the South Rim. After a few hours at the South Rim we drove through the Navajo desert, an awesome, terrifying waste of varied colours. The most spectacular was the view of the Vermilion cliffs which, true to their name, looked enchantingly crimson under resplendent twilight. We reached the North Rim through the Kabab National Forest and checked in at the Grand Canyon Lodge which has the best location in the world, with a breath-taking view of the gigantic Canyon with jutting cliffs and plunging gorges in its prodigious womb. Walking along the North Rim enraptured by a panorama of stupendous beauty that beggared description. I mused

'Where else on earth can Heaven be than here, for human eyes to see?'

We left for Zion through the Navajo painted desert having a spectacular view of another range of Vermilion cliffs. What a splendid sight of soaring mountains on either side of the winding road as we entered the Zion National Park! The stunning grandeur of the coloured mountain peaks has only to be seen to be believed. We checked in at a very comfortable motel, the Zion Rest, and shopped at a well stocked provision store nearby.

The next day we drove to Bryce Canyon, ninety miles north, an incredible expanse of natural sculpture, carved by the winds and shaped by the chastening hand of time. What fantastic shapes, as occur only in dreams, at once bizarre and regular, of myriad sizes and intricate structure, the enchantment of the crimson hues enhanced by the gleam of dawn, the blaze of noon and the glow of twilight. We returned to Zion and were in a great mood to relax over dinner at a Mexican restaurant.

We took the route through the Glen Canyon Dam, a man made marvel, a massive, mighty, interference with nature, an engineering feat of stupendous magnitude. On our way to Flagstaff we saw the Sunset Crater, a volcano which had burst into activity a few years ago. After a lovely picnic at the pine forest near the lava laden region, we visited the meteor crater, a huge dent on the earth due to the impact of a meteor, many times the size of a football field. After passing through the Petrified Forest, a lifeless expanse, a relic from the depths of time, we reached Holsbrook where we relaxed in a comfortable motel. It was a scenic drive through desert and mountain to Los Alamos where we stayed with the Viswanathans.

What a fortunate circumstance to celebrate little Lalitha's first birthday in my old student's house at the birth place of atomic energy! We heard the well-recorded *Ayushyahomam*, as recited by our family priests Nataraja and Pattabhirama Sastrigals at E.N. For a whole day we heard Madurai Mani's music enjoying his inimitable Neraval of 'Ragatalagathulanu' in the song 'Vasudevayani' in Kalyani.

We left Los Alamos after breakfast and after a drive through flat un-interesting country reached Lubbock, checked in at the very comfortable motel, La Quinta. At the invitation of Professor Nelson, I lectured at his mini-symposium on my new proof of the Lorentz transformation. What a pleasant feeling it was when a member of the audience asked the right question by answering which I was able to emphasise the real purpose of the lecture. Krishna and I were treated to a lavish lunch at a roof top restaurant by our gracious host.

One realises the magnitude of Nature's endowments and human achieve-ments that contributed to the all-pervasive power and prosperity of that happy land, only when travelling by car, feeling the throb of its mighty heart on the impregnable ribbon of concrete and steel that unites the states of America into one affluent society of different races with common ideals and aspirations. We had just a week to pack our bags for our return journey to India. Even when we were bound for Ekamra Nivas we felt we were leaving our home, Krishna, Madhu and little Lalitha who had enmeshed our hearts with her infant smile!

Notes

1) G. H. Hardy (1877–1947), was a British mathematician, who along with J. E. Lit-tlewood, led the resurgence of British mathematics in the first half of the twentieth century. Hardy is most famously known as the mentor for the Indian mathematical genius Srinivasa Ramanujan, and so Hardy has a special veneration in India and in par-ticular Madras. So it was very appropriate for a portrait of Hardy to be unveiled at Matscience.

Chapter 48

MATSCIENCE 21 (1982–83)

In November 1982 we had two Matscience conferences in Ooty (Nov 23–25) and in Mysore (Nov 29–Dec 1). So we left for Ooty on the 20th of November, halting in Mysore on the way. The next day we reached the HPF guest house in Ooty which has an enviable location near the golf links. This was also the venue of the conference whose theme was "Probability, Stochastic Processes and Applications". We descended from Ooty by the steep Masanagudi route and stayed three days and four nights at the Kargudi guest house right in the heart of the jungle. This time we had the curious experience of seeing snakes peeping out of the anthills at sunset! Back at the guest house, after dinner G.P. delighted us with Hindi songs of Saigal and Pankaj Mullick which recalled days before he was born. From Kargudi we reached Mysore for the conference on "Relativity, Gravitation and their ramifications".

The music season exploded in December on the city with a salvo of concerts and dances. As we emerged from the fantasy world of sonorous harmony, we found the New Year was already five days old and it was time for the 21st Anniversary of Matscience.

The symposium 'Matscience 21' was inaugurated by His Excellency Mr. S. L. Khurana who spoke of the brain drain from India and methods to arrest it. Welcoming the guests I said:

"Matscience has attained 21, the age of majority and what a glorious period it has been in the history of Indian science, nay, international science! These 7700 days have passed with close liaison with the world wide community of scientists from Harvard and Princeton, Caltech and Berkeley to Tokyo and Kyoto, Canberra and Melbourne. How else could it have been, for our Institute was blessed even before conception by Niels Bohr, Salam and Gell-Mann, at its conception by the great Nehru, our dear founder

C.S., Maurice Shapiro and Sir James Lighthill, during its growth from infancy to manhood by masters of mathematics and savants of science from over fifty countries in uninterrupted succession from Marshak and Schiff to Thom and Erdös with a generous sprinkling of Nobel Laureates and Fermi Prizemen. We acknowledge with gratitude the flood of greetings that has poured forth, just as it did at the dawn of the Institute in 1962, from those who had actively participated in our visiting scientists programme.

The four acres of land on which Matscience stands are in fact unofficial U.N. territory! Within weeks of its inception, Matscience witnessed Russian visitors shaking hands with American professors, discussing the origin of the universe and the apparent emptiness of inter-galactic space. How many times we saw German and English, Yugoslav and Italian, Australian and Japanese visitors to Matscience share the expense of a common taxi for travel to Mahabalipuram where they watched Arjuna's penance, discussing the possible existence of quarks and quasars, Black Holes and White Dwarfs?

It strikes me that the most fortunate feature of our Institute has been that we, the happy few at Matscience, learnt our first lessons in research from the great masters themselves in accordance with my own personal experience starting in 1947. I was introduced to stochastic processes by Professor Bhabha, the father of cascade theory and the architect of India's atomic energy programme. At Manchester, I learnt probability from the leaders of the Cambridge School, Bartlett and Kendall, Nuclear Structure from Rosenfeld and Cosmic Ray Theory from Janossy and Blackett. In 1954, I was inducted into Dirac algebra by Dirac at Madras and Radiation theory by Heitler at Zurich. Feynman taught me during my visit to Caltech, how electrons with negative energy travel back in time on Feynman graphs obeying Feynman rules. I understood the puzzle of nonconservation of parity from the lectures of Yang and Lee, and the role of the Gamma Five from Oppenheimer at Princeton, Fermi's view of the origin of Cosmic Rays from Chandra at Chicago, strangeness of new particles from Powell, and the Sakata model from Sakata in Kyoto and meson cascades from Heisenberg at Gottingen. Our group of students were ushered into the Gell-Mann era by Gell-Mann at Bangalore and we learnt the mystery of the Dalitz plot from Dalitz and the role of neutrino in weak interactions from Salam at Ekamra Nivas. Vasu was trained in invariant embedding by its creator Richard Bellmann in Los Angeles. Our theoretical physicists learnt about nuclear matter from Hans Bethe in this very hall and Krishna was initiated into number theory by Erdös in the Matscience guest room.

It has been a fusion of space and time through human goodwill and passion for knowledge, and what else could we ask for than the Grace of God to ensure the continuance of these salutary conditions at Matscience against the winds of change and ravages of time? It is not the occasion to speak of the inadequacy of grants from our sponsors but I cannot help warning against unkind voices that are raised to interfere with our academic freedom. Unhelpful criticism is the tribute that sterile mediocrity pays to creative talent. It is my earnest hope that His Excellency the Governor will transmit to the State and Central Governments, the sage counsel of our gracious Patron who, consistent with his love and affection for the members of Matscience, has set his standards as high as of those whose portraits he is unveiling to day — Wigner and Weyl."

Mr. C. Subramaniam unveiled the portraits of Hermann Weyl and Eugene Wigner, two Bhishmas of mathematical physics. In the evening we had the concert of Lalitha followed by a delightful dinner prepared by our institute staff.

I had a unique opportunity to pay a tribute to the great Russian nation, when I was invited to open an exhibition at the Russian cultural centre. I took the occasion to recall the famous reference to the Red Army by Winston Churchill "That once monstrous juggernaut engine of Germany military might and tyranny, was beaten and broken, outfought and outmaneuvred by Russian valour, generalship and science."

We left for Mysore on the 6th of March for the Matscience conference on 'Discrete and continuous in physics.' The passage from discrete to the continuous had intrigued me ever since I watched Bhabha work out with unconcealed excitement such an operation in detail in cascade theory. I gave lectures at the Mysore University on 'A new outlook in science' exhorting the youth to try new ideas to justify the creative tradition of Raman and Ramanujan which was being overwhelmed by mediocrity today.

My father's birth centenary was to be celebrated on the 14th May and to mark that memorable occasion we decided to create 'The Alladi Centenary Foundation' earlier on the 24th April at Ekamra Nivas. We were desirous of having Krishna with us and so he came all the way from Austin to participate in the inauguration. It was fortuitous that two days earlier at the Lakshmipuram Association, Governor Khurana unveiled the portrait of my father at a function so well-organised by K. Santhanam, the indefatigable secretary of the Lakshmipuram Young Men's Association (LYMA) and attended by admirers and octagenerian like Justice N. Rajagopala Iyengar. The 24th April was a great day at Ekamra Nivas when the Alladi Centenary

Foundation was inaugurated by C.S. on the lawns under the mango tree in the presence of about sixty invitees when I explained its aims and objects.

It was a pleasant surprise when I received a phone call from Delhi informing me that I could meet the Prime Minister on the 12th May! I flew to Delhi and took the opportunity to explain to her that the achievements of the Institute justified substantial increase in grants hitherto delayed for inexplicable reasons.

In honour of my father's birth centenary, The Hindu carried an excellent article by K. Chandrasekaran, a family friend of ours. On the 18th of May we registered the Trust deed of the Alladi Centenary Foundation.

I conducted a Tele-Communication conference for the All India Radio, the rehearsal for which was as interesting as the conference itself.

We made a to Tirumala to offer prayers to Lord Venkateswara after the completion of the Centenary functions and prior to my forthcoming visit to the United States in response to an invitation to participate in the American Mathematical Society conference at Boulder, Colorado.

I left for the United States on the 11th June, by an Indian Airlines Airbus to Bombay and then by Pan Am flight 73 to New York via Delhi. As before, at Delhi I stayed at the Maurya Palace Sheraton. It was a smooth flight all through via Frankfurt and I contacted Krishna and Fenyves by phone immediately on reaching Kennedy Airport. I arrived in Dallas the same day by American Airlines and checked in at the Holiday Inn. Next day after my lecture on special relativity at the University of Texas at Dallas, Fenyves took me to a leisurely lunch with his colleagues Ivor Robinson, Kusch and Johnson. I left for Austin to be received by Krishna and Madhu who were staying in the comfortable rented apartment of our distinguished friend C. V. Narasimhan.

On the 18th June we had the opportunity to see on the TV the twin triumphs of American technology and organisation — the blast off from Cape Canaveral of the Space Shuttle Challenger with a woman astronaut, and the launching of MX missile from California. This was followed by the Pope's historic visit to Poland and the kaleidoscopic extravaganza — the Wimbledon preview. For about a week we gorged ourselves with the feast of feats, Wimbledon tennis, in which the earlier rounds were more exciting than the finals. McEnroe's victory over Lendl was watched with tantalising excitement but the final was very tame since the New Zealander Chris Lewis, dazed at his earlier unbelievable triumph over Connors collapsed at the sight of the implacable conqueror, McEnroe!

On the 24th June the Space Shuttle landed with the smoothness of a Boeing 747 in Edwards Air Force Base in California and not at Canaveral as scheduled, a last minute change made with casual efficiency.

I left for Los Angeles by Pan Am from Houston and stayed at Northridge with Venki, my customary host in California. I gave a lecture at UC Riverside at the instance of Professor M. M. Rao on 'New concepts on probability.' After enjoying his gracious hospitality, I returned to Northridge for a day's rest at Venki's house before going to Rockwell Science Centre in response to the invitation from Richardson. The days of rest at Venki's were spent in perusing books, Einstein by Pais, Nuclear war by Clark and the autobiography of Casimir. I spent a few days with Gordon Shaw at Laguna Beach and lectured at UC Irvine on my new derivation of the Lorentz Transformation.

I phoned to Krishna who was leaving with Madhu and Lalitha for India for good, after his eight year sojourn in America which carried him through UCLA, Michigan, Princeton and Austin, a four-dimensional experience in American education. On his way back, he visited distinguished mathematicians in Germany and then went to Nairobi to fulfil his childhood dream to see the famous wildlife parks of Kenya.

I flew to Denver by United Airlines and reached the University of Colorado at Boulder by limousine and checked in at the Hallet hall. The campus is one of the most beautiful in the United States, looking more so in summer, with luscious lawns fringed by dense foliage against the background of light blue skies. The American Mathematical Society conference on Probability was one of the best organised I had attended, a typical example of American organisation and efficiency. It was a revelation to me that systematic studies were made in recent years on the role of Probability in Quantum Mechanics and a new breed of rigorous applied mathematicians had taken over the subject.

Immediately after the conference I flew to Dallas by American Airlines and took the Pan Am flight to New York and then to Frankfurt. I was able to spend two days with Krishna, Madhu and the child who were in Frankfurt before their departure to Nairobi. I was back in Madras on the 4th August, in time for my 60th birthday which was observed quietly in gratefulness to God by prayers at the Mylapore temple.

Chapter 49

Transition To Tranquillity (1983–84)

On arrival in Madras we phoned to Krishna who informed us of his 'wild life adventures' in the Safari parks of Amboseli and Masai Mara in Kenya. On the 17th of August we received Krishna, Madhu and little Lalitha at the Madras airport and it was indeed an excitement to watch the child settle down in Ekamra Nivas. Madhu joined Lalitha in the Varalakshmi Vratham on the 19th, an auspicious beginning for life at Madras.

We celebrated our wedding anniversary by a dinner at the Taj Coromandel. It was a pleasure to receive a phone call the same day from Fenyves that he was visiting India for the Cosmic Ray Conference in Bangalore. Fenyves gave the first lecture under the auspices of the Alladi Foundation on 'Computers in Environmental Studies.' Being one of our best academic friends in the U.S., he was the most suitable person to initiate the lecture programme of the Foundation. Meanwhile at Matscience we had Prof. P. L. Jain, my host in Buffalo, on a short visit to give a lecture.

We celebrated little Lalitha's first birthday with a sit-down dinner in out spacious Koodam for over seventy guests, proceeded by a Harikatha on Thyagaraja by Mannargudi Sambasiva Bhagavathar.

In connection with the Matscience conference on Probability, we left Madras before the crack of dawn so that we could reach Mudumalai the same evening. What a pleasant excitement to see a giant tusker at the entrance to the Bandipur forest! During our van drive we saw a small black panther dart across the Ombetta road, a true story to tell to unbelieving friends for years to come!

We left for Ooty and checked in at the HPF Guest house. The Matscience conference on 'Probability and applications' was inaugurated the next day when my article on 'How to stem Brain Drain' appeared in the Hindu. The problem had become so chronic that even constructive ideas to solve it are not taken seriously by a frustrated academic community.

I participated in the University post-centenary Silver Jubilee seminar on 'Science and Development' at the invitation of Professor P. M. Mathews, while Krishna spoke at the Ramanujan Institute on his recent work. Krishna received the happy news that the Hungarian Academy of Sciences was willing to support the travel of the Hungarian mathematicians to the Number Theory Conference he was organising in January 1984 for Matscience for the 70th birthday of Paul Erdös.

It was then I received a timely invitation from Venkateswara University to participate in a U.G.C. seminar on Stochastic Processes. It gave me a nice opportunity to also visit Tirumala and offer prayers to Lord Venkateswara?

At the Seva Chakra, I talked on Gandhi Jayanthi (October 2) emphasising that Gandhi, Hinduism and Ramayana are relevant to the modern age since they represent the unchanging values in transitory human life.

On the 14th October an Alladi Centenary Lecture was delivered by Parasaran, the Advocate General of India. It was a touching tribute when the highest legal official in India opened his lecture by stating that the very mention of the name Alladi was like the passing of an electric current through mind and body. Such was father's reputation for legal genius, matchless advocacy, profound learning, unconscious simplicity, natural generosity and trusting friendship.

On October 31st, I retired from the Directorship of Matscience completing a tenure of twenty one and a half years, the most challenging and eventful period of my academic life. There was of course the anxiety and concern whether the ideals which were formulated, fostered and safeguarded with ceaseless vigilance would be preserved or frittered away at the prospect of some apparent or illusory benefits. To me it was a transition to tranquillity after three decades of exacting struggle to establish an Institute which bears some semblance to the centres of excellence abroad.

On the second day of my retirement I spoke at the Engineering College on 'Chandrasekhar' and 'Brain Drain', obviously connected topics!

Krishna received an invitation to participate in the Number Theory Conference at Asilomar in memory of Professor Ernst Straus, his Guru and mentor. Just before his departure to the U.S. he received a letter from Professor Erdös accepting the invitation to be the Ramanujan Visiting Professor at Matscience.

On the 8th December Krishna left for Singapore enroute to the U.S., halting in Honolulu to give a lecture at the University of Hawaii at the invitation of Professor Bertram. At the Asilomar conference when he was given an opportunity to speak thrice at various sessions.

Soon after Krishna's return from the USA, Professors Richert and Halberstam arrived in December and settled down at the Woodlands Hotel. Pomerance from Georgia arrived on the 1st January, a welcome guest at the dawn of the New Year. We left for Ooty by car on January 2, making arrangements for receiving Erdős at Madras next day and 'transporting the precious cargo' to Ooty by car. After an overnight stay at Mysore we reached the HPF Guest house at Ooty. The next day Erdős and the Hungarian participants, Sarkozy and Vera Sos arrived and the inauguration took place at the Aranmore Palace. I presided over the first session when the inaugural lecture of Erdős set the style and standard for the conference.

We took our guests to Dodabetta, one of the most beautiful spots in the world, to the Government House with its vanished glory but well preserved gardens and to the Radio Astronomy Centre, where I lectured on relativity. We had a delightful excursion to the Glenmorgan tea estates when we found that Erdős was as much at work while walking through the wooded hills as at a lecture hall or a study room. On our way back to Madras we stayed for one night at Kargudi and took Erdős into the forest on a van ride. It was amusing to watch him enjoy the simplest and sweetest of human pleasures — spotting animals in their natural habitat.

We had the privilege of receiving the Governor at Ekamra Nivas on the 14th January when Erdős talked to a packed audience in our drawing room on 'Reminiscences of a mathematician' with deep emotion and high sentiment. How well he summarised a life charged with creative achievement for over six decades since he lisped in numbers at the age of four, writing a thousand papers during his visits to over a hundred centres of learning! It was a gracious gesture of our Governor to have invited the visitors for dinner at the Raj Bhavan when they had the opportunity to admire the architectural magnificence and horticultural profusion of the seats of power under the British Raj. We took our guests to Covelong Beach when they treated to a dinner hosted by Mathura's father N. C. Krishnan. The guests relaxed in five star comfort at the seaside resort hotel. Erdős spent many hours at E.N. and little Lalitha started imitating his walk, with head bent forward and hands folded behind, in deep thought.

I completed my second paper on 'Decomposition of intervals in special relativity, a conclusion of a five course sequence on special relativity.

We had twin lectures of the Alladi Foundation by Bruce Berndt on 'Ramanujan's Notebooks' and Heini Halberstam on 'How common is common sense — the role of mathematical in education.' Following this, the Halberstams graciously hosted a dinner at the Adyar Gate Hotel.

We had a distinguished visitor from Paris, Professor Choquet Bruhat who lectured at the A.C.F. on the mathematical traditions of France.

At a seminar on 'Science and Religion' organised by 'Samskriti' a cultural organisation in Mylapore, I expressed the view that the realisation by scientific argument that Space, Time and Matter have to be defined in an integrated way, reveals the true mystery of Creation. Inaugurating a U.G.C. seminar at the M.I.T., I stressed that creative work was the result of sudden inspiration on a background of prolonged contemplation demanding initiative and perseverance, restlessness and patience, in right and proper measure.

I responded to the kind invitation of Professor B. V. Sreekantan, the Director of TIFR, and lectured on special relativity in my Alma Mater. It was a 'tryst with the past', a tribute to the memory of my great teacher Bhabha. I met my old colleagues Daniel and Puthran and spent a few minutes in prayerful mood in Bhabha's office. After a dinner with Sreekantan and lunch with K. Ramachandra, I left for Madras musing on the transition of the Tata Institute from the sheltered elegance of 'Kenilworth' to the democratised affluence on the Colaba seafront.

On the 14th March (Einstein's birthday!) a felicitation address was presented to me at a function arranged by Professor Ramakarthikeyan under the auspices of the Seva Chakra, when Justice Sethuraman, Vaidyasubramania Iyer, Vasantlal Mehta. Hon'ble Hande, the Health Minister, Dr. B. Ramamurthi, C. Subramaniam, Professors Vittal, and K. V. Parthasarathy spoke with sincere affection and warmth.

I participated as a Chief Guest in a meeting of foreign students under the auspices of the Indian Council of Cultural Relations at the Ashoka Hotel. What a strange game time plays on human lives — 25 years ago at a similar function, before the same hosts, my frank and forthright speech on the prevalence of pride and prejudice both in science and society impressed C.S. so much that it resulted in the creation of Matscience. Today on retirement I was speaking pleasant platitudes to foster social and cultural relations between foreign and Indian students!

On the 31st of March I participated in a seminar 'On nuclear arms control', a subject as relevant to India after its entry into the Nuclear Club as to superpowers racing for supremacy both on earth and in space. In my talk, I emphasised that peace can be assured only on equality of strength and that readiness for war is the best way to prevent its incidence.

Chapter 50

Relativity For All (1984)

After the dawn of the New Year, I lectured at the University of Madras at the invitation of Professor P. M. Mathews. A simple idea of the decomposition of a space of time-like interval into an arbitrary number of space-like and time-like intervals occurred to me which gave a wholesome view of the past, present and future in the temporal domain. I completed this work on relativity and sent it for publication to the Journal of Mathematical Analysis and Applications. I went to Bangalore at the invitation of Professor P. S. Narayanan to present my results before a discerning audience and returned to Madras the same evening.

We attended the Swathi Thirunal festival in honour of the Royal Composer organised by the star singer Jesudas. I was reminded of the famous statement of General Wolfe, that he would have preferred to be the author of Gray's elegy rather than be the conqueror of Quebec. Swati Thirunal would be remembered for ever as the composer of that immortal song 'Pannagendra Sayana' than as the jewelled Maharajah of Travancore.

Krishna received the invitation for the Oklahoma conference on Number Theory to be held in July where he would have the opportunity to meet Selberg, Bombieri and Erdös. His trip was being supported by a travel grant from the Department of Science and Technology.

May 14 was the first anniversary of the Alladi Centenary Foundation. Krishna gave a lecture cum slide show on 'Centres of Excellence'. We had glimpses of the great campuses of the world, such as Harvard and Princeton, Cambridge and Oxford and the Californian 'multiplets' from Berkeley to San Diego with slides taken by me and Krishna on our academic trips.

On June 4 Professor Rothschild of UCLA was a guest of Krishna at MATSCIENCE where he gave a talk in the area of Combinatorics. That night and we had a dinner in his honour at Ekamra Nivas.

A few days later, Lalitha and Madhu gave a concert at their guru Mannargudi's son's wedding reception at Thyagaraja Vidwat Samajam.

Krishna left for the U.S. on the morning of the 20th June by Air India. What a pleasant surprise when he received a phone call just before leaving for the airport from Prof. Bertram of Honolulu offering him a visiting assignment at the University of Hawaii, a gesture so gracious that he accepted it without hesitation on the phone itself!

Krishna returned on the 23rd July after participating in the Oklahoma conference and left in August for his assignment in Hawaii by Japan Airlines via Tokyo. Madhu and little Lalitha followed him a month later because Madhu wanted to attend the marriage of her younger brother Chella Srinivasan in mid-September.

Just before Krishna's departure to Hawaii, on August 11, we had the privilege of having Professor John Thompson of Cambridge University, the great group theorist, visit MATSCIENCE and I attended a dinner in his honour at the Chola Sheraton.

My esteemed friend M. V. Arunachalam, Chairman of the Tube Investments (TI) group was starting a company called Cholamandalam Software, and he invited to be on its Board of Directors, top which I readily agreed.

Dasara was celebrated at Ekamra Nivas in the usual manner. We missed Krishna, Madhu and little Lalitha on such a festive occasion, but they were enjoying the gorgeous environs of Honolulu.

I was the Chief Guest at a Bank of Baroda function in Mylapore marking the opening of an evening counter. I stressed that the staff of the Bank should combine politeness with strictness since they were dealing with funds kept in their custody in high trust and confidence. It must be possible to return a defective cheque without hurting the feelings of the customer by a polite insistence on necessary formalities.

We had a pleasant surprise when we received a phone call from Professor Shoenberg of Brazil who was staying at the Taj Coromandel. Like me, he was a believer that the Feynman formalism, perfect as it was, could be retouched by splitting the propagator, a meaningful infraction of its relativistic purity. It was gratifying to find over our conversation at lunch that he and his wife were lovers of Indian art and admirers of ancient monuments.

We left for Bangalore on the 27th October, with G.P. accompanying us. At the Indian Institute of Science, I lectured on stochastic processes. The lunch at Windsor Manor was enjoyed in an atmosphere which created the illusion of the leisured elegance of the British Raj. Then we went to

Mysore where I lectured on relativity at the university. We had an open air dinner at Hotel Metropole, another reminder of the spacious comfort of the British days. We spent three days at Mudumalai, watching wildlife — forest birds, mongoose, spotted deer and elephants galore.

On the third day while relaxing under the mango tree at the Kargudi Guest House, we received the shocking news of the assassination of Indira Gandhi. We rushed back to Bangalore where we stayed at the Institute Guest House and watched on the TV Indira Gandhi's last journey from Teen Murthi Marg to Santivan attended by monarchs and presidents from all over the world.

The manner and magnitude with which nations paid tribute to the most powerful leader in the ranged democratic world revealed the measure of India's established status and the legendary aura of the Nehru family. The single voice that spoke on behalf of a billion people was now stilled for ever. Her natural heir was Rajiv Gandhi, who was unanimously elected leader and prime minister. His stoic dignity and composure, unusual at so young an age, was a natural acceptance of his triple inheritance from Indira, Jawaharlal and Motilal. The blood of the Nehrus throbs with the tides of democratic power.

I spoke on Western values at a symposium at the Bharatiya Vidya Bhavan drawing attention to the strange fact that America with its material wealth has as its motto. 'In God we trust' while India with its spiritual tradition was asserting with open emphasis the secular nature of its Republican constitution.

During these months I noticed 'nuances' in reflexion symmetry in special relativity which led to the realisation that the essence of relativity can be stated as 'the observer does not move, a contracted world slips by in contracted time.'

Relativity for all was the primary mission of my academic career — to comprehend the Lorentz Transformation as part of real life and experience. It was a date with destiny I made when I was seventeen in the honours course at the Presidency College, to understand the Lorentz transformation in as natural a manner as the Galilean.

I was in the right spirit to plan a Hawaiian holiday with Krishna and family at Honolulu at the dawn of the new year. The music festival was a pleasant sequel to the work on relativity and a festive prelude to the most exciting year in our life.

Chapter 51

Kahala Dawn And Waikiki Sunset (1985)

The New Year opened with an Alladi Centenary Foundation lecture by me on 'Profound Physics from Elementary Algebra' when Professor Santappa presided. It was a summary of my work on relativity.

It was a heart-warming sentimental function at Matscience on January 2 when my first student Professor P. M. Mathews, at the instance of Professor Sudarshan, unveiled my portrait with C. Subramaniam as Chairman.

We left for Singapore by Air India and checked in at the Oberoi Imperial. The next day we left for Honolulu by a Singapore Airlines 747 via Hong Kong and what a thrilling moment to land at Honolulu after a breathtaking view of the Hawaiian islands from the air and be received by Krishna, Madhu and little Lalitha who took us to their townhouse in Kahala. Krishna left the same evening for the mainland for the Annual Meeting of the American Mathematical Society at Anahein near Los Angeles.

The first evening we spent in the gorgeous environs of the Kahala Hilton, a paradise of affluent comfort within the island paradise. The next day we went to Waikiki, God's gift to the Affluent Society, by watching the sunset over the Pacific, a daily ritual for the tireless worshippers of sun and ocean.

Krishna's apartment was in Kahala, a charming suburb of Honolulu where every morning I enjoyed the scented air of verdurous hedges as I walked to the nearby Mall to browse through the books and magazines in the bookshop. We had our first picnic at the Kahala beach, the golden sands glistening against the blue sky and bluer waters, an idyllic setting as much for hedonic young honeymooners as for somnolent senior citizens.

Little Lalitha enjoyed her school just across the street and insisted on being escorted home with due acknowledgement of her visit to school as the significant achievement of the day. I spent many hours in the well-stocked library of the University on a wide range of books and journals on computer science, physics, mathematics and even medicine.

Following the conference in Los Angeles, Krishna visited the University of Arizona at Tucson and the University of Colorado at Boulder for colloquia on his recent work before returning to Honolulu on January 22.

One could never get satiated with the charms and attractions of Honolulu — Ala Moana shopping centre with its jostling crowds exuding the joy of life and the spirit of 'Aloha', the fragrance and loveliness of the flowers on the garlands of holiday makers redolent with Polynesian spirit of tranquillity and happiness, well-laid gardens of tropical trees and lush vegetation, towering hotels with inlaid lawns and lighted terraces, and above all the ravishing beauty of the golden beaches, the crystalline ocean and the verdant mountains basking under the radiance of heaven's blue smile.

We had plenty of tennis near the Waikiki on early mornings with the Diamond Head screening the warmth of the rising sun. It was pleasant to see the apartment towers glow in the golden light of the early sun as it rose over Diamond Head. We visited the Arizona Memorial in Pearl Harbor where well dressed Japanese tourists pay homage to American courage and valour, forgetful of the 'day of infamy' when Japan attacked Pearl Harbor.

I attended an evening party at the lovely home of Professor Jake Bear in honour of Professor W. K. Hayman, whose friendly interest twenty years ago was the starting point of mathematics at Matscience.

Krishna had an opportunity to visit the University Florida in Gainesville when he was invited to give a colloquium on his recent work. He enjoyed the hospitality of the entire department at the lovely campus. He was fortunate to meet Paul Erdös there, who expressed his willingness to visit Hawaii a week later.

We had the opportunity to receive Professor Erdös as our house guest for two weeks. It was a pleasure to take care of him for he was so simple, natural and easy to please. We watched how he is in intensive thought while conversing with us or even with the child. He enjoyed vegetarian food, slightly spiced and varied at every meal with considerable planning by Madhu and Lalitha. He admired the beauty of nature and showed interest in historical monuments, all the time thinking and working out some mathematical problem which was part of his pulse and heart beat. It was an unforgettably thrilling intellectual and emotional experience to play host to such a creative thinker with lovable human qualities.

We visited the Arboretum, one of the most beautiful spots in lovely Honolulu where an audible silence prevails amidst the profusion of nature under man's solicitous care. We made a day trip to the 'other side' of Oahu and spent a whole afternoon at the Sheraton Makaha Resort.

I left for Los Angeles by Singapore Airlines and was received at the redesigned airport by Sanka who took me in his new Mercedes to his spacious home in Encino overlooking the Valley. The next day I was taken to Thousand Oaks, where I lectured at the Rockwell Science Centre, at the invitation of Professor Richardson.

I took the Fly Away bus to Laguna Beach where Gordon Shaw received me at the John Wayne airport. He had rented an apartment right on the beach where I relaxed within sight and sound of the living ocean. I lectured at Irvine on 'New concepts in probability' and had a dinner with Gordon in a cosy Italian restaurant at Laguna Beach.

Gordon took me to San Diego where I spent a full day with my nephew Ramu in his comfortable studio apartment.

I left for Dallas by American Airlines DC-10 and Fenyves received me at the airport and took me to the comfortable Best Western Inn. I lectured at the Environmental Science Centre on 'Urban development in India' making a suggestion that one fast freeway from Madras to Mahabalipuram with suitable exits would not only accelerate urban development but relieve the stifling congestion and improve the quality of life in the metropolitan area. Next day I had a discussion on relativity with Professor Ivor Robinson before Fenyves took me to the airport.

I flew to Albuquerque, where I was received by V. K. Viswanathan, who took me to his lovely home in Los Alamos. It was a pleasant flight from Albuquerque to Los Angeles by American Airlines 727 and then by Singapore Airlines 747 to Honolulu where Krishna informed me that arrangements for our excursion to Kauai and Maui were complete.

We flew to Hawaiian Air DC-9 to Kauai where the luscious greenery and arboreal profusion were so enchanting that like lotus eaters we did not wish to leave the enchanted island! We hired a Lincoln Town Car at the airport itself and drove straight to the Wailua Falls before checking in at the luxurious condominium, the Hilton Kauai Beach Villas, a confluence of privacy, comfort and efficiency, a marvel of American business enterprise. The next day, after visiting the lovely Hanapepe valley we went round the rim of the Waimea Canyon which, with its spectacular views from well spaced look outs, approached the grandeur of the incomparable Grand Canyon. Equally impressive was the breathtaking view at the Kalalau Lookout of the ocean below through the plunging gorge between the Napali cliffs. After a Pizza lunch we visited the Sheraton Kauai and the spouting horn on the Poipu Beach, one of the most beautiful locations on earth. We had night tennis at the Condo and a morning walk on the lovely stretch of sand near the Villa, watching the spectacular sunrise.

The next morning after a delicious breakfast at Sheraton Coconut Beach, we drove through the entire circumference of the small island fringed by beaches each as beautiful as the other but with its own unique features. It was obvious why one of the beaches, the Lumahai, was chosen as the location of the film 'South Pacific.'

We left for Maui by Hawaiian Airlines and arrived just in time to see the spectacular sunset and the island of Molokai silhouetted against the crimson sky, as we drove in the sleek Mercury Cougar. We checked in at the Kaanapali Shores, a fabulous condominium, on a larger scale than that at Kauai. Breakfast at the Hyatt Regency Swan Court in a setting suitable for an Arabian oil prince is a typical example of the 'democratised affluence' of the American people. It is a paradise on earth, the Indraloka of our epics must be something like this, garden restaurants, lawns and patios, beside a swimming pool winding through well designed caves and waterfalls against the background of golden sands and the blue ocean glistening under blaze of noon or the mellow sheen of the waxing moon.

A few miles at the other end of the beach is the location of Sheraton Maui where, it is claimed, began the legend of Kaanapali beach, the most beautiful of all human settlements, heaven as we conceive it, with golf courses, super star hotels and plush condominiums amidst the lush lavishness of well laid gardens accentuating the natural beauty of the sundrenched and moon blanched beach and ocean.

We drove in an Oldsmobile Cutlass Ciera to the Haleakala crater, to gaze at the bizarre beauty of the awesome depression of outrageous barrenness, 10,500 feet above the sea. The drive was spectacularly beautiful with appearing and vanishing mists, revealing a 'panorama of the ocean' mountains and valleys with tantalising frequency. After a Mexican lunch at 'La Familia' overlooking the one mile Royal Kaanapali Golf Course, we flew by Hawaiian Airlines DC-9 back to Honolulu, just in time to prepare for our journey back to India.

I gave two lectures on Clifford Algebra at the mathematics department of the University of Hawaii at the suggestion of a visiting professor from Finland Dr. P. Lounesto. One of these was at a seminar and the other at a colloquium organised by Professor Jerry Yeh who hosted a dinner later at an Indian restaurant.

We had a last day ramble at Waikiki and a dip in the limpid turquoise blue waters of Hanauma Bay before leaving for India by Singapore Airlines on the 30th April.

Chapter 52

Will And Memory — Gifts Of God To Mortal Man (1985)

On arrival in Madras from Hawaii, I was informed by Professor Santappa that I was elected President of the Tamil Nadu Academy of Sciences, the creation of which is to be traced to my proposal read at the meeting in Adiseshiah's Institute in 1971.[1]

Suddenly, as if to interrupt the honeyed memories of the Hawaiian dream and unstate our happiness, fate struck an unkind blow at my health, reminding me that life is not all roses and sunshine. As I woke up early morning for my tennis on the 15th May, I felt an unusual numbness on my left cheek but ignoring it I played, only to find that the left half of my face was stricken with Bell's palsy — it became rigid and irresponsive to stimulus. I was instantly reminded of my father reading out to me in my boyhood the heroic struggle of Roosevelt against disabling paralysis. I resolved to summon confidence and maintain access to hope for the sake of the family, invoking the Grace of God who is at once a Bhayakritha and a Bhayanasana — the creator and dispeller of fear. We immediately contacted Dr. K. V. Thiruvengadam, the eminent physician who had always been kind to me. He was so generous that he agreed to meet me at the Fine Arts Club during a drama performance when he advised me to consult the nerve specialist Krishnamurthi Srinivas. I was assured that it was a palsy of the peripheral seventh nerve and would not impede my normal activities. By the grace of God I recovered just before Krishna and Madhu arrived from Honolulu.

As if to restore my spirits, the proofs of my papers on special relativity in the Journal of Mathematical Analysis and Applications arrived. It was the end of a forty year old quest to understand the distinction between space like and time like intervals.

Recovering under the affectionate care of Dr. Perumal, a physiotherapist friend, I was in a mood to accept an invitation from Professors Chisolm and

Common to participate in the International Conference on Clifford Algebra and its Applications at the University of Kent at Canterbury, to give a review lecture on our Madras work.

As I mused during idle hours on my favourite theme of relativity, I found that the Relativistic Trinity of quantities $v, c^2/v, (1 - v^2)/c^2$ are connected by an integral equation with an elegant interpretation which I presented at the monthly meeting of Tamil Nadu Academy of Sciences.

We offered prayers at the Tirumala temple to Lord Venkateswara to express our gratitude for my recovery from illness.

We left for Mysore in response to a handsome invitation to Krishna from Professor S. Bhargava of the University of Mysore. This time we made a trip to Nagerhole wildlife sanctuary, just as impressive as Mudumalai, with dense and luxuriant greenery.

On September 5, we had lecture at the ACF was by the noted French mathematician Professor Michel Waldschmidt who spoke on 'Diophantine equations' spanning the history of the subject at the rate of a century a minute! Following his lecture, we had him for dinner at Ekamra Nivas when our spiced Indian food satisfied his French palate! Waldschmidt was visiting MATSCIENCE as Krishna's guest, and gave three lectures there on transcendental number theory, his speciality.

The next day there was a delightful music concert by Semmangudi when he paid his tribute to his Guru Maharajapuram Viswanatha Iyer by singing with rapturous relevance 'Guruleka Etuvanti' in Gowri Manohari.[2]

My spirits were sufficiently restored for me to undertake the journey to Canterbury, England, for the Conference on Clifford Algebra. On September 14 I left for London by Air Canada L-1011 from Bombay and at Heathrow Airport I was received by Lalitha's cousin Jayaraman who took me to his comfortable house in West Drayton. I went by train to Canterbury and checked in at the very spacious and modern Darwin Hall. The conference was one of the best organised I have attended and it was a pleasant surprise to find that our Madras work had attracted the attention of Polish and German mathematicians! Canterbury is a tranquil haven of intellectual milieu and British tradition — England just as one imagines in 'Vicar of Wakefield' and 'Pride and Prejudice'. In the magnificent dining hall, the scientists met in relaxed comfort when excellent vegetarian food was provided for me with solicitous care. I presented a summary of the Madras work to a discerning and attentive audience.

During the excursion to Leeds Castle, pages of English history sprang to life at that residence of Henry VIII. No monarch had left on the history of England a mark more indelible than Henry VIII who promoted a revolution but while he lived, controlled it. Walking around the luscious green grounds of the Castle, one wonders whether Shakespearean diction, through the panegyric of John of Gaunt, is adequate to describe the serenity and loveliness of the English countryside.

I saw the famous Canterbury Cathedral, on enduring wonder of architectural splendour. At the reception at the Rutherford College, I met Professor Rickaysen, a visitor to Matscience in 1966 and a participant in our summer school. The return journey by Air Canada L-1011 from London was very comfortable and after a delightful breakfast at the Centaur Hotel in Bombay, I arrived in Madras by Indian Airlines.

It was then I had the pleasant surprise of receiving an invitation from the Principal, Women's College at Tirupati, to deliver a lecture on 'Indian Scientist — India or Science, which comes first?' I stressed my favourite theme imbibed from my father who read out a passage from Sir James Jeans on the Internationalism of Science. The best way for a scientist to serve India is to pursue science at an international level of excellence and not compromise standards for 'local approbation'.

Gordon Shaw and his wife Lorna visited Madras and we made every effort to meet the standards of their Golden State hospitality at Laguna Beach. Gordon gave a talk at the Alladi Foundation on 'Information processes in the Brain'.

We heard Sir Herman Bondi at the British Council when I had an opportunity to talk to him about my recent work on relativity in view of his own bias towards the K-factor which I called Bias in special relativity!

Notes

1) The idea to create the *Tamil Nadu Academy of Sciences* was proposed by Ramakrishnan in November 1971 in an address entitled "A Science Policy for Tamilnad" (see [1]) at the newly Madras Institute of Development Studies (MIDS) at the invitation of its Founder-Director Malcolm Adiseshiah, who had just retired as Director General of UNESCO. Adiseshiah invited leading members of the scientific scene in Madras to come up with suggestions to stimulate the scientific atmosphere in Tamil Nadu. Following Ramakrishnan's suggestion, the Tamil Nadu Academy of Sciences was formed in 1976 and Ramakrishnan served as its first Secretary. He was made President in 1985. It was due to Ramakrishnan's initiative that the Academy began the tradition of Monthly Meetings and lectures. Now this Academy is known as The Academy of Sciences, Chennai.

2) Maharajapuram Viswanatha Iyer was one of the greatest carnatic vocalists of the 20th century, and Semmangudi Srinivasa Iyer, another Carnatic music giant of the 20th century, was his pupil. Viswanatha Iyer's son, Maharajapuram Santhanam was an accomplished vocalist himself, and he started the Maharajapuram Trust on Viswanatha Iyer's 90th birthday. Semmangudi gave a magnificent concert on that occasion.

Chapter 53

Good Morning America (1986)

1986 was the year of the most incredible event in our domestic life. Krishna accepted a permanent position in the U.S. at the University of Florida to begin in January 1987. This had a direct impact on our personal life, even more than World War II!

The first event of the New Year was the lecture at the ACF on January 11 by my dear friend and reputed scientist Dr. A. Jayaraman of Bell-Labs. In the style of his Guru Sir C. V. Raman, he spoke on 'Bell-Labs, a centre of excellence'. Next week, we arranged his lecture at the Tamil Nadu Academy of Sciences on "High pressure physics".

At the Padma Seshadri School, at the kind invitation of its enterprising Principal, Mrs. Y. G. Parthasarathy, I lectured on 'special relativity' to an eager and attentive audience of teenage students.

On February 4, a young Russian visitor to Matscience, Dr. Marchuk, called on me at Ekamra Nivas to discuss some novel applications of Clifford Algebra. It is the greatest charm of creative science that it is international and impersonal, uniting workers across continents who are strangers to one another, by a single equation!

On February 5, Madras was blessed by a visit of His Holiness the Pope, who was received with universal warmth and enthusiasm as Prince Charles was, five years earlier. True religion transcends natural boundaries and racial prejudices, and a secular state respects all religions besides its own.

During the third week of February, we attended a Rotary Club lecture by Dr. McCormack, an American professor who readily shared and exchanged views with friendly frankness. We had him for tea at Ekamra Nivas when he evinced interest in my father's role in the making of the Indian Constitution, with India as a republic within the Commonwealth.

We had the U.S. Consul General John Stempel and my distinguished friend M. V. Arunachalam for dinner on the 26th February at Ekamra

Nivas. They had a guided tour of the old family home where the portraits and photographs revealed the history of Ekamra Nivas and the spirit of Mylapore in the thirties and forties.

We left for Guruvayoor by car on the 4th of March. We offered prayers at Guruvayur to Lord Krishna and had Thulabharam performed for me, little Lalitha and Madhu. After visiting our niece Geetha and her husband at Palghat on the way, we returned to Madras with the satisfaction of having performed Thulabharam as we had resolved earlier during my illness.

We saw a very educative video film at the USIS on the training of an astronaut when I recalled the cryptic statement of Alan Bean who visited Matscience in 1976, that it was a process of natural 'selection' or 'elimination' by vigorous tests!

At the Rotary Club in Connemara Hotel on the 18th March, at the invitation of Dr. A. L. Mudaliar, I talked on 'Instant vs Traditional Excellence'. Tradition is a source of inspiration to set standards of achievement but it can stifle progress if it thwarts enterprise and initiative. Instant excellence is possible in new institutions if the experience of traditional centres is used and transplanted with care and caution.

We heard a magnificent concert of Semmangudi under the auspices of the Maharajapuram Trust with the Pallavi 'Chakkaga Ni Bhajana Chesevariki Thakkuvagalada' in Sankarabaranam that he sang at our request. The message of this immortal verse is — 'There is no desire that cannot be fulfilled for one who chants the mellifluous name of Sri Rama.'

On the 6th May Madhu gave a delightful dance performance at a local sabha, and the next day we had the Vasantotsavam at the Kapaleeswar Temple, a tradition begun by my dear mother.

During his stay in Hawaii in 1985, Krishna had received an offer from the University of Florida of a permanent position in the Department of Mathematics. During the long period of waiting till September 1986 in India for the Immigrant Visa, he wished to spend the months of July and August in an academically useful manner. He decided to participate in the International Congress of Mathematicians at Berkeley August 3–5 and lecture at various centres in U.S. and Europe and at the Oberwolfach conference in Germany. So he took a round the world ticket through his friend K. L. M. Subbu, who made perfect arrangements for his travel.

We enjoyed the excitement over the Cricket Test between Australia and India when a miracle happened — the match ended in a tie — the second example in 109 years of Test Cricket, the earlier one at Brisbane in 1961! Who could imagine when Australia declared in the second innings at 170 for 5, feeling secure with its 574 for 7 in the first, that India with its initial 397 would score just 374 in the next innings with the last man out three balls before the close of play?

As expected we received a message in September that all the formalities for the issue of the visas were completed and we informed Krishna by phone of these developments. He was in New Hamphsire staying with my cousin Sivaprasad and his wife Indira. He told us he had a fruitful time at the International Congress of Mathematicians in Berkeley followed by visits to UCLA, the University of Colorado at Boulder, The University of Georgia at Athens, and the University of Florida at Gainesville. He later visited Europe and lectured at Bordeaux, Nancy, Paris, Stuttgart and Oberwolfach, before returning to India, just in time to receive the US Immigrant Visa for him and his family.

Before Krishna and his family were to depart for the U.S., we decided to offer prayers to Lord Venkateswara at Thirumala. After a good Darshan of the Lord we left by the pivturesque Chittoor route for Bangalore where G.P. joined us on our trip to Mysore. He made arrangements for our stay at the new luxury hotel, Southern Star, which true to its name is a blend of elegance and taste, planning and efficiency, a proud addition to the facilities of the tourist paradise of Mysore owned by our gracious host Basavaraj.

We had an ACF lecture by John Stempel, the US Consul General on 'Indo-US Cooperation in Trade and Technology', a clear and frank exposition, spiced with his characteristic humour. M. V. Arunachalam, the distinguished industrialist presided.

Krishna, Madhu and little Lalitha left for the US on the 27th November by Malaysian Airlines 747 to Singapore enroute to Hong Kong, Honolulu and Los Angeles by Singapore Airlines.

On December 15 we had two distinguished visitors at Ekamra Nivas — Fields Medallist Professor John Thompson, a Cambridge mathematician who had just accepted a half-time assignment at the University of Florida in Gainesville, and Professor Bryce DeWitt, a relativist from the University of Texas, Austin. After giving them a tour of Ekamra Nivas, we took them to dinner at the Hotel Savera. At the instance of Professor Mathews, I presided over DeWitt's lecture at the Madras University when I noticed that his interest in the nuances of the Lorentz transformation were similar to mine. Years seem like seconds when we consider a unit distance as a light year! I spoke about this in my lecture to the Association of Mathematics Teachers of India a week later.

Chapter 54

Sunshine In Florida And
A New Light In Our Lives (1987)

The year started with a surprise gift on New Year's Day — of the January issue of the Delhi-based magazine 'Link', containing excerpts from the 'Alladi Diary' on my association with Professor Bhabha. It was a reproduction of the entire chapter on my life in Bombay, the 'Kenilworth' period of TIFR when Bhabha's Institute was situated in his aunt's house. I was pleased that these reminiscences were brought to light by a news magazine.

On the 19th January, at the request of my old student Devanathan, I inaugurated an International Symposium on Nuclear Physics organised by him in which many American scientists participated. The next day I gave a talk at the Sarvepalli Radhakrishnan Symposium at the kind invitation of my classmate S. Gopal, paying my tribute to the Philosopher-President whose very presence inspired me from my boyhood in my academic career. There has never been, and never will be, an orator like Radhakrishnan — a torrent of diction carrying the tides of philosophical thought. My talk had a pleasant sequel, a TV interview along with Professor Richard Gregory, the famous neuro-physiologist from Bristol, on 'Science and Humanity'.

There was an ACF lecture at Ekamra Nivas by the French mathematician Jean-Marc Deshouillers on the January 23rd evening after which I left for Calcutta in response to the gracious invitation from Professors B. K. Dutta and M. Dutta for the Dirac Symposium. It was a superb example of free enterprise and co-operation by the mathematicians and physicists in Calcutta, zealous of the high academic traditions of the Calcutta Mathematical Society. I stayed in the comfortable guest house of Guest Keen and Williams arranged through my distinguished friend A. L. Mudaliar. I summarised my work on *L*-Matrix Theory as a tribute to the greatest theoretical physicist of the century who had graced Ekamra Nivas with his presence in 1954 and encouraged our initial efforts in relativistic quantum mechanics in Madras. It was then in Calcutta that my attention was drawn to a paper by Beauregard who had made a detailed reference to my work on special relativity (see Notes of Chapter 39). I returned to Madras just in time for the Raj Bhavan reception on Republic Day to which I took my good friend Nagi from the Catholic University at Washington.

I wrote up my ideas on 'The Reflection Principle' in mathematical reasoning as a paper, applying it to Special Relativity and obtaining the Lorentz Transformation as its natural and inevitable consequence. Later in February I had an opportunity to speak on the reflection principle at the Mathematics Festival at the Loyola College.

Since it was the Ramanujan Centenary year, I gave a lecture on the 12th February on 'Lessons from Ramanujan's Life' emphasising the need to publish, since this is the only way in which the immortal work of a genius, whose life like those of other humans is finite, can be made available to posterity. As Erdös likes to emphasise, theorems are the enduring legacy of mathematicians to future generations.

There was a function at the Meenakshi College to celebrate the Science Day in honour of the late Sir C. V. Raman. I also talked at the Tamil Nadu Science Academy when my distinguished friend Dr. A. L. Mudaliar presided, on Bhabha and Raman, the twin heroes of my youth, and the twin leaders of Indian science, who influenced my career.

I decided to go to the U.S. with Lalitha to spend a few months with Krishna and his family in Florida and fulfil lecture engagements at various centres. At the suggestion of my friend Venkatesan, I consulted D. Shanmugam and his assistant Dr. Jaysankar at the K. K. Nagar Hospital for a chronic foot ailment which was interfering with my regular tennis. Their advice that I should use arch support for my heels proved very useful during my travel abroad.

All was set for our trans-Pacific travel with arrangements made efficiently by KLM Subbu and Singapore Airlines Sampath. We left for Singapore by Singapore Airlines Boeing 747 on March 31. It was a delightful stay at the luxurious Hotel Meridien on Orchard Road with its garden atrium typifying French elegance, Eastern opulence, and American efficiency.

The flight to Hong Kong was sheer delight with service on the 'Big Top' lounge, superb on any standard. At the Kaitak airport we were received by a hospitable young Indian executive, Nilakantan, who took us to the Holiday Inn, Golden Mile, in the heart of the teeming city where the square inch is the suitable unit for urban real estate just as carats are for diamond trade! Yet the customer is king in that haven of commerce where nothing matters except the willingness to part with money and the desire to acquire it.

We flew in supreme comfort to Honolulu to be received by our gracious hosts, the 'Hawaii' Ramanathans, with whom we stayed at their home in Hawaii Kai. The holiday atmosphere of the island paradise is only to be experienced to be understood, for everyone is in love with life, learning from the native Hawaiians the pleasures of watching the colours of the sunset or listening to the sounds of the living sea and the tireless waves. Then to Los Angeles, the city of 'Perpetual Motion' on a million wheels where distances are measured in transit times!

While in Los Angeles, I went to Rockwell to lecture at the Science Centre at the invitation of my distinguished friend John Richardson, after a delightful lunch hosted by him at the Westlake Inn. We flew by a Delta 727 to Oklahoma where my good friend Professor Swamy had made efficient arrangements for stay and lecture at the Oklahoma State University. We moved from Oklahoma city to Dallas to be welcomed by my gracious host, Erwin Fenyves. After my lecture the next day at the University of Texas at Dallas, we left by an American Airlines flight for Austin.

At Austin, at the invitation of Professor Bryce DeWitt, I gave a seminar at the Institute for Gravitation on my recent work on 'The Reflection Principle in Special Relativity' before a group of general relativists. The lunch at the Faculty Club and the dinner at the Indian restaurant were arranged by my solicitous host to satisfy my vegetarian palate.

Our next stop was Houston where we stayed with Dr. Parameswaran, a flourishing cardiologist.

We left by a Delta 727 for Atlanta and arrived at Gainesville by an Eastern DC-9 to be received by Krishna, Madhu and my grand-daughter Lalitha who took us to their comfortable apartment in their new Oldsmobile Cutlass Ciera. It was sunshine in Florida, all the five months, the happiest sojourn in all our travels, with the pleasant prospect of an addition to the family in October.

Gainesville is a pretty University town with the lushest greenery, well planned neighbourhoods, where every house is set amidst a profusion of stately semi-tropical trees. The city offers well laid tennis courts in vedurous setting where Krishna and I played regularly with his colleagues Louis Block and James Keesling, as much for good company as for exercise.

I spent long hours in the new air-conditioned library building, browsing through current literature or through reference books on topics from mathematics to medicine, from super symmetries to heart surgery. My grand-daughter Lalitha loved to watch wildlife shows as they reminded her of our visits to the Mudumalai and Bandipur sanctuaries.

It was a pleasure to receive as our house guest Professor Vittal, Krishna's teacher at Vivekananda College in Mylapore, who was on a visit to the U.S. to participate in a conference on stochastic processes.

Shortly before we left Florida, Professor and Mrs. Thompson[1] took us to the University Centre Hotel for dinner at the roof top restaurant with a breath-taking view of the Florida sunset.

During the last week of our stay, it was thrill and suspense in James Bond style as Krishna made road-dashes in his Oldsmobile Cutlass amidst sunshine and flash floods to receive Madhu's mother at Orlando without knowing the arrival flight, bring her to Gainesville, fulfil a lunch engagement, take us to Dunedin and arrive there one minute before Ramu's marriage ceremony! We attended the wedding dinner with our sister Meenakshi at the country club and returned to Gainesville the same night to enable me and Lalitha to catch the Eastern flight the next day to New York connecting the KLM 747 to Amsterdam!

The KLM provided its share of excitement for, as we were sipping orange juice waiting for the plane to take off, there was an announcement that the flight was cancelled due to failure of the electrical system. At that late hour at night, the four hundred disappointed passengers scrambled at the airport KLM office to make new reservations when fortunately we were confirmed on the corresponding flight next day and provided accommodation for the night in the comfortable Viscount hotel near the airport.

On arrival at Amsterdam, we were taken to the Airport Hilton where we relaxed for two days before leaving for Delhi. It was a pleasant surprise to be received by G.P. at the Delhi Airport, and taken to the Hyatt Regency. We arrived in Madras in time for Navarathri which was celebrated in the customary style in Ekamra Nivas as a social and a religious festival.

On October 4, we received by phone the best news of the year, the birth of Amritha[2] at Gainesville, a new light in our lives, two weeks ahead of time, and one day before Krishna's birthday October 5.

Notes

1) John Griggs Thompson, the greatest living group theorist, has been recognised with numerous prestigious awards including the Fields Medal in 1970 and the Abel Prize in 2008. When he visited Matscience in 1986, he appreciated the hospitality of Ramakrishnan who invited him to Ekamra Nivas. To reciprocate that hospitality, John and Diane Thompson took Ramakrishnan and Lalitha to dinner in Gainesville in 1987.

2) Krishna and Mathura named their second daughter as Amritha, inspired by *Amrit*, the holy nectar that the Gods drink for immortality.

Chapter 55

The Ramanujan Centenary And Pacific Serenades (1987–88)

Evan Pugh Professor George Andrews (Penn State Univ.) lecturing at the Alladi Foundation for Ramanujan's Centenary, 23 December, 1987. - Author

The 1987 Nobel Prize in Physics was awarded jointly to George Bednorz and Alex Muller of IBM in Zurich for their discovery of super-conductivity at high temperature, an almost instant recognition, as in the case of Lee and Yang thirty years earlier. The possibility of industrial revolution through applications set Governments competing with one another for release of funds, with excitement similar to that over harnessing of atomic energy after the war.

At a reception at the U.S. Consul-General's house, we met a lady Astronaut, Dr. Mary Cleave, whose modesty and poise were as impressive as her achievements. We were reminded of the aphoristic statement of Alan

International visitors at Ekamra Nivas for the Ramanujan Centenary. L to R: David Bressoud, Bruce Berndt, Mrs. Askey, Alladi Ramakrishnan, George Andrews, Krishna Alladi, Richard Askey, Basil Gordon, M. V. Subbarao, Dec 1987. - Editor

Bean, an earlier astronaut visitor to Madras, that the selection of the best is just the elimination of the rest by a series of rigorous tests!

At Florida Krishna was busy making plans for the Symposium on Number Theory to be conducted at Anna University in Madras on Dec 21, the eve of Ramanujan's birth centenary. He phoned to us from the Mandarin Hotel on his arrival in Singapore and it was a pleasure to receive him, Madhu, Lalitha and the new addition Amritha, at the Madras airport! Professor Basil Gordon was with them and we took him directly to the Woodlands hotel where accommodation was provided for the foreign participants to the Symposium. The next day professors George Andrews and David Bressoud arrived and we promised them a feast of music and dance besides the Ramanujan fare! It was a generous gesture on the part of the U.S. Consul General, John Stempel, to have arranged a reception to the delegates at his splendid house.

Krishna was invited to address the Rotary Club at Connemara on 'Ramanujan' on the 22nd December, the mathematician's hundredth birthday!

The Ramanujan Centenary was celebrated almost as a festival with total participation by the enlightened citizens of Madras. Eminent mathematicians from around the world gathered in Madras to pay homage to

Ramanujan. Krishna's symposium turned out to be a success thanks to the participation by Richard Askey, George Andrews, Bruce Berndt, M. V. Subbarao, David Bressoud, Basil Gordon and Imre Katai who were keen workers on, and ardent admirers of, Ramanujan's mathematics.

We thought of Professor Erdös, Krishna's guru and mentor, who should have joined us but could not because of an eye operation he had to undergo at that time. It was gracious of him to have sent a twenty page manuscript in his own hand of his lecture entitled 'Ramanujan and I' to be included in the proceedings of the number theory symposium organised by Krishna.

What a privilege it was to receive Askey and Andrews at Ekamra Nivas where they gave lectures at the Alladi Centenary Foundation on 'Thoughts on Ramanujan' (Askey on Dec 21) and 'The Lost Notebook of Ramanujan' (Andrews on Dec 23).[1] Krishna feted his visitors with dinners and lunches at the some of the finest restaurants of Madras and with delightful music at the Academy. Krishna and his family left for the U.S. on the 8th January and his TV interview on 'Ramanujan' was shown on 18th January as part of the science programme.

Time Magazine presented Michael Gorbachev as the man of the Year (1987) who had the courage as the new leader of Soviet Russia, to talk of *glasnost* (openness) and *perestroika* (restructuring) in the land of Stalin and Kruschev. Reagan had answered him with a similar gesture as he signed the Intermediate Missile Ban Treaty, in reversal of his well-known criticism of Soviet policies, just as Nixon reversed the course of history by initiating friendly relations with communist China.

Political events is Madras took a sad and tragic turn after the passing of MGR.[1] The Raj Bhavan reception had a strange atmosphere in such a chaotic situation of a leaderless government. There was no alternative except the dismissal of the Assembly and the introduction of The President's rule — at once the tragedy and triumph of democracy.

At home I had the pleasant surprise of a whole day visit by Sundaravaradan, one of my most intimate friends from boyhood whom we called 'Sundara', after long years of absence. He was my constant companion when I planned the MATSCIENCE Institute in the anxious years of watchful waiting 1958–62 after my return from Princeton. We recalled with sentimental pleasure, mixed with a tinge of sadness at the inexorable passage of time, how we spent many hours on the Madras beach when I expressed my dreams and hopes which looked meaningless till we found ourselves on the 'track of destiny' with the fortuitous visits of Niels Bohr and Abdus Salam in that 'miracle month' of January 1960.

It was a busy, interesting and eventful week before our departure to the U.S. I organised a symposium at the Alladi Foundation on 'The Ramanujan Centenary, the Aftermath' in which leading administrators and businessman participated.

We left for Singapore on the 16th February by a Singapore Airlines 747. It was a memorable stay at Westin Stamford, the tallest hotel in the world, with its shopping plaza in Raffles City. I was wondering how the independent city-state honoured the founder, an Englishman, Sir Stamford Raffles, by perpetuating both the Christian and family names, while in India we were erasing historical memories by a false sense of patriotism, as each new government and party came into power. Just three hours away from Madras is the cleanest city in the world rivalling in attractions and facilities, the most modern cities of Europe and America. The work ethics has taken possession of the island city, proud of its multiracial immigrant population, a superb example of how harmony, discipline and hard work, generate prosperity for all. Government control and private enterprise go hand in hand to make it the Jewel of the East.

It was a wonderful flight to Honolulu via Taipei and we were received at the airport by Ramanathan who took us to his comfortable home in lovely Hawaii Kai. The flight from Honolulu to Los Angeles was a beauty and we were received by Lalitha's grand-niece Radhika and her husband Arun who took us to their comfortable house in Granada Hills. The next day I gave a talk at the Rockwell Science Centre in Thousand Oaks at the instance of my distinguished friend, Richardson. The affluence and efficiency of the great industrial enterprises of the U.S. like Rockwell, were apparent from the glasswalled elegance of the offices overlooking the valley and the total commitment to achievement of the scientific staff working therein.

I had the opportunity to lecture at UC Santa Barbara on my recent work on 'Approach to stationarity in stochastic processes' at the invitation of my old friend Joe Gani,[2] now chairman of the Statistics Program on Applied Probability. What a lovely location for a University — right on the Pacific coast with sun-drenched beaches and golden sunsets over the shimmering ocean! As I drove by the sleek 'airbus' along the Pacific Highway with the deep blue ocean on one side and tawny cliffs on the other, I mused how we are wasting our coastline in Madras by lack of access and facilities for residences or pleasure resorts. A well-planned highway from Besant Nagar to Mahabalipuram would transform Madras from a congested city to a sprawling suburbia.

It was a smooth flight by American Airlines DC-10 to Dallas where we were received by our gracious host, Professor Fenyves. I lectured on stochastic processes at the University of Texas at Dallas and spent the evening discussing relativity with professor Ivor Robinson.

We flew to Atlanta where we took the Eastern flight to Gainesville to be received by Krishna and his family. Weather in Gainesville was glorious, so cool and sunny, that I played tennis with Krishna and his friends, a proper start to the Gainesville routine.

The next day we left for Orlando where Krishna delivered a lecture on Ramanujan before the Mathematical Association of America at its regional meeting which I attended at Rollins College, the venue of the Conference. We stayed at the comfortable Langford Resort Hotel, and after the conference visited the Epcot Center in the Walt Disney World, the showplace of the nation. It was gorgeous weather and we romped through the major exhibits in true holiday mood jostling with the smiling crowds. The Disney World is the best example of American enterprise, organisation, lust for life and leisured affluence. Can there be a greater missionary of peace than Walt Disney, who created this fantasy in a real world where millions gather, each for his own and his family's enjoyment of leisure at his pleasure?

Our next excursion was during the Easter break to South Florida, a thousand mile trip by car in glorious weather, a springtime holiday as described in folders. We drove to Everglades National Park, which true to its name, offered the lushest greenery and bird life in insatiable profusion. We saw plenty of alligators of all sizes and ages in the well preserved ponds and swamps and had a delightful picnic in Flamingo, the southern most part of Continental United States. Watching the ocean, we had a feeling of exaltation similar to that at Cape Comorin, enjoying the sunset as attractive and colourful as at the lands end of India.

The next day we saw the Viscaya, the gracious old Spanish style residence of an American millionaire, with its sprawling gardens right on the Atlantic coast, reminding us of the lovely castle Miramare in Trieste. From there we drove to Key Biscayne, the enchanting island where the Lipton International tennis tournaments are held and then to Miami Beach lined by luxury hotels like the Fontainbleu Hilton overlooking the ocean and harbour.

We returned to Gainesville after driving through Palm Beach with its country clubs and million dollar homes where Rolls Royces are dime a dozen. The American Republic has its own breed of royalty, not based on birth but on wealth, which in principle can be acquired by enterprise by anyone — the American dream of a ball-boy becoming a Wimbledon Champion or a waiter aspiring to own a hotel chain.

During one of our walks in Gainesville, I met a distinguished and venerable American, Professor Richard Eberhart,[3] an eminent poet, whose poems were recently published by the Oxford University Press. I had the privilege of perusing the book and I was thrilled to find many of my thoughts so well expressed in his delightful poems — that despite the finiteness of human life, man can exceed himself and that is why God has blessed him with judgement, discernment and creative faculties. In his words, the richness of life is beyond saying and the loss of time is beyond redeeming, a most valuable message my great father left behind. I was flattered when he perused my Alladi Diary with considerable interest. An individual's role in human destiny is consistent with the faith in the immanence of God so well-expressed through Pascal's argument.

He saw a watch in the mud
He deduced it was made by some one,
He looked at the world
and thought the same.

We watched a science feature film on TV 'Grand Unification Theory' of the fundamental forces of nature, how Einstein's dream was being realised in a different way by the architects of new physics — Salam, Weinberg, Hawking and Glashow. How purposeful such shows will be in India where there is not enough awareness of the pains and pleasures, the strains and triumphs, of creative effort! The desire for achievement is too much with us without the willingness to work, persevere and prevail over the unavoidable hazards, disappointments and challenges in scientific research.

In response to the invitations from my old student, Professor Mathews and my cousin Professor Sivaprasad, I flew to Boston by Eastern Airlines. Mathews hosted a dinner when I had the opportunity to meet Professor I. Shapiro, the eminent director of the Harvard-Smithsonian astrophysical laboratory, an institute as zealous of established tradition as of innovative research. The competition between the instant and traditional centres of excellence is the characteristic feature of American science — Stonybrook and Harvard, Rockwell and Caltech, vie with one another in establishing the American tradition of achievement through competitive endeavour.

We enjoyed our visit to the lovely country home of a flourishing young doctor couple, our relatives Ravi and Rajani, in neighbouring Lake City, a good example of how immigrant professional talent prospers in the land of unlimited opportunities and free enterprise.

Lalitha and I left for India on 8th June by Eastern to Los Angeles and by Singapore Airlines across the Pacific. Krishna and his family joined us at the Westin Stamford Hotel in Singapore for a delightful stay. We arrived in Madras on the 13th just in time for the religious and social festivities relating to the sixtieth birthday of Madhu's father.

Wimbledon was the major event after our arrival and we watched on the TV Becker's victory over Lendl in the semi-final and Staffi Graf's uninterrupted march of triumph to the world's most coveted trophy. Edberg won the crown in a serve and volley duel with Becker on July 4. That evening we attended the American Independence Day Reception at the U.S. Consul-General's house where there was a traditional picnic and fire works on its lovely grounds.

We left for Mysore on the 12th of July. G.P. joined us at Bangalore and we checked in at the Hotel Southern Star. The next day we shifted to the Kargudi Guest House in the Mudumalai Wildlife Sanctuary, watching this time porcupines and elephants, sambur and gaur (Indian bison) and a pack of wild dogs on their savage prowl. During our stay at Kargudi, after jungle rides in the morning, Krishna relaxed under the mango tree of the guest house and read the recent PhD thesis of Frank Garvan, who had established some exciting results relating to Ramanujan's congruences for partitions. This led Krishna to suggest to his department to recruit Garvan and the two are now colleagues in Florida.

We left for Mudumalai for Ooty and reached the Southern Star Hotel. Ooty was in the grip of monsoon weather and so we spent most of the time in the cosy comfort of the well furnished rooms. On return from Ooty, Krishna gave a few lectures at Matscience while I inaugurated the 'Conference on Mathematical modelling' at I.I.T.

We made a trip to the Venkateswara temple in Tirupati on the eve of Krishna's departure to U.S. with his family. This time he travelled via Bangkok after visiting Singapore and phoned to us from the Bangkok Hilton that he was enjoying the exotic city, visiting the famous Buddhist shrines with his family. After returning to Florida, Krishna made a short trip to Germany for the Oberwolfach Conference in Analytic Number Theory.

On September 16, there was a pleasant function at Matscience when C.S. unveiled the portrait of Marshak and I recalled the visit in 1963 of the distinguished architect of the 'Rochester conferences on High Energy Physics', an organiser par excellence, who enjoyed the music festival in

Madras and visits to Ekamra Nivas with enthusiasm. Marshak was also the PhD advisor of Sudarshan, my successor as Matscience Director.

I inaugurated a cultural fiesta at the P. S. High School emphasising the need for a pleasant environment for school children who should imbibe qualities of cleanliness, civic sense and a taste for elegance and beauty which are necessary ingredients of true culture.

It was then that Gorbachev visited India and was received with the same acclamation and enthusiasm as Kruschev was in 1956. The visit concluded with a spectacular cultural show in Delhi which we watched on the TV wondering how the bonds between countries transcend the personality of the leaders. Every reigning King is treated as a Prince Charming and every ruling Prime Minister a messenger of goodwill and friendship. Such is the magic breath of power and affluence.

I spoke at a function at the Sanskrit College paying a tribute to the memory of K. Chandrasekharan,[4] a true Mylaporean in the purest sense of the term, to whom cultural refinement was as ardent a pursuit as professional success, loyalty to friends, pleasures of family life and enlightened use of leisure meant more than an ungentle emphasis on rank and status in a demanding competitive world.

At the High Energy Symposium in I.I.T., I gave an evening lecture on 'Fifty years of Cascade Theory', recalling my close association with Bhabha and the best gift he gave me — the unsolved but solvable fluctuation problem of cosmic radiation which took me to Manchester and many centers around the world on lectures on stochastic theory.

Krishna phoned to us from Florida in early December that he entered his new house with his family on the day and hour suggested as auspicious for Grihapravesam.

The Consul-General Timberman and his wife came for dinner to Ekamra Nivas when they evinced interest in the picture gallery in our drawing room, of the contemporaries of our father during the British Raj and his associates in the drafting of the Constitution.

The music season had a triple significance this time — Semmangudi had completed his eightieth birthday and was singing like a maestro in his prime, the Academy was entering its sixty-first year and Professor Marshall Stone, despite his feeble health was there to attend the season with his customary zest.[5] It was a splendid season resounding with the rhythm of Semmangudi's 'Pankaja lochana' in Kalyani at its most ecstatic hour.

Notes

1) M. G. Ramachandran, a Tamil film hero, was the Chief Minister of Madras and the head of the ruling ADMK party. He was in failing health and he died on the night of December 23, one day after the Ramanujan Centenary and the night after Andrews' lecture at Ekamra Nivas. By December 23, fortunately all events Krishna and Ramakrishnan organised for the Ramanujan Centenary were over, but the events scheduled for the next few days suffered because not just Madras, but the whole state of Tamil Nadu came to a standstill after MGR's death. Such was the following he had among the millions of his fans.

2) Joseph Gani (1924–2016) was a famous applied mathematician, probabilist and statistician. During his long career, he held faculty and administrative positions at various universities in Europe, Australia and America. From 1985–94 he was the Head of the Applied Mathematics and Statistics Program at UC Santa Barbara. In 1980, Gani edited a book entitled "The Making of Statisticians" in which several leading statisticians describe their careers. In that book, M. S. Bartlett talks about his academic life, and in speaking of this PhD students, makes special mention that it was Ramakrishnan who introduced Product Densities: "J. E. Moyal had joined me on the statistics staff at Manchester, and among our research students Alladi Ramakrishnan, who had worked with H. J. Bhabha on the theory of cosmic ray showers, came over from India to work in the theory of point processes. These processes had been introduced in the context of time series by Herman Wold, and in effect by David Kendall in connection with the characteristic functional for the age variable in his paper on population processes for the 1949 RSS Symposium, but it was Ramakrishnan who emphasised the special rules to be employed like $dN(t) - N(t + dt) - N(t)$, as distinct from a continuous process $X(t)$, introducing the concept of *product densities*." (M. S. Bartlett, "Chance and Change" in *The Making of Statisticians, J., Gani Ed.*, Springer Verlag, New York (1982), p. 50.)

3) Richard Eberhart (1904–2005) was a leading American poet who won the Pulitzer Prize in 1966 and the National Book Award for Poetry in 1977. He was Poet Laureate Consultant to the US Library of Congress during 1959–61.

4) The late K. Chandrasekharan, the second son of the great V. Krishnaswami Iyer, was a lawyer by profession. He played a significant role in the cultural scene of Madras. He played a prominent role in the Madras Sanskrit College, and served with distinction as a Secretary of the Madras Music Academy. He was known for his fine English, and he wrote a beautiful article on Sir Alladi's remarkable life (see [1]).

5) Professor Marshall Stone actually died in Madras on January 9, 1989, a week after the 1988 Madras Music Academy Season ended. He attended the Music Academy concerts with Ramakrishnan and Lalitha, and after the music season, went to Thiruvannamalai to visit the Ashram of Ramana Maharishi, a few days before he passed away. He loved Madras — its culture and its people, and he breathed his last there at the Woodlands Hotel where he stayed regularly.

Chapter 56

Round The World By SIA And TWA, To And Across The U.S.A. (1989)[1]

At the beginning of the New Year, I had an opportunity to talk to the famous novelist R. K. Narayanan.[2] I told him that the purpose of the Alladi Diary was to illustrate that fact was stranger than fiction, that sometimes there is no need to transmit reality through the medium of fiction.

We attended the Sathabishekam[3] of our dear C.S. It is hard to believe he could grow old! It was one of the best organised functions I have ever attended — an emotional experience for him, his family, friends and admirers, an expression of thankfulness to God for the gift of a life, full of years and honours, earning the affection and esteem of everyone who came into contact with him.

On 6th February we held a symposium under the auspices of the Alladi Foundation on 'American Mathematics Centenary — lessons for India' with the US Consul-General T. Timberman as the Chief Guest and the venerable Chitra Narayanaswamy as Chairman. I drew attention to how even uncompromising pure mathematicians were accepting the role of computers in the development of mathematics. In America, a single organisation held its sessions at several centres for the benefit to the wider community while in India there were uncorrelated and even conflicting efforts by competing organisations to mark the centenary of the legendary Ramanujan, yielding no beneficial results to the community at large.

I spoke on the 'Sir C. V. Raman tradition' at the Rotary meeting at the Connemara on January 31 emphasising its triple feature — faith in one's research, recognition of collaborators' work and merits, desire to publish and disseminate knowledge.

I gave a TV talk on 'Special Relativity for all' as part of my mission to propagate the idea Einstein is a 'natural completion' of Newton and relativity is in fact a 'logical extension' of 'Classical Mechanics'.

We were ready to leave for the U.S. via Singapore on a SQ-TWA round the world ticket, on the night of 19th February. As the Singapore Airlines aircraft was about to take off at Meenambakkam (Madras airport), we were informed that the flight was cancelled due to engine trouble. We were taken to the Adyar Park Sheraton for stay till the next flight was arranged. It was a strange feeling to be still in Madras when all our friends and relatives were under the impression we were in Singapore!

We arrived in Singapore next night and checked in at the Imperial Hotel where the reservation was kept open despite the delay. We offered prayers at the Murugan Temple in Singapore before our departure to Honolulu.

The flight by Singapore Airlines 747 on the upper deck from Singapore to Honolulu via Taipei was sheer luxury where we were pampered with attention and delicious food. The stay in Honolulu rivalled in enjoyment that at Singapore with its recurrent delights — feeling the enchantment of the Hanauma Bayand watching sunsets at Waikiki. Our hosts, the Ramanathans, provided us delightful company as we drove round the island gazing at the blue skies and bluer waters, golden beaches and swaying palms amidst the lushest greenery on the surrounding mountains.

At Los Angeles we were received by Krishna's friend and UCLA collegemate Nadadur Kumar (alias Sampath) who took us to his cosy home in Santa Monica. I lectured at Rockwell at the invitation of my distinguished host John Richardson, hospitable and friendly as ever. I also gave a colloquium at the University of California, Riverside at the instance of Professors Gokhale and M. M. Rao, on 'Approach to Stationarity'. It was a gracious gesture by Professor Basil Gordon, Krishna's teacher during his Ph.D. study at UCLA, to drive us to Riverside. He also hosted a lunch at the UCLA Faculty Club inviting Krishna's colleagues and took us to the Huntingdon Library and botanical gardens at Pasadena. It was a thrilling experience to see original Shakespearean texts so well preserved and displayed in that stately mansion of Huntingdon. Even America, whose affluence is built on the new world tradition of change, invention and innovation, was zealous of remembering its past and the roots of its language and culture. It strengthened my conviction that the compensation for finite human life is the Divine gift of memory, best preserved through man-made paintings, sculpture, monuments and enduring literature, so precious that Kings would wish to die to leave such a legacy. The gardens look lovelier under gorgeous California sunshine.

Early in the morning Sampath drove us to the Los Angeles airport where we took the TWA flight to Orlando via St. Louis. We had to abandon our trip to Dallas as originally planned since it was hit by a snow storm, unusual in the sunbelt! At the Orlando airport, we were greeted by Krishna, Madhu, and our grand daughters Lalitha and Amritha and were taken to their lovely new home in Gainesville in their new Pontiac Bonneville. The two hour drive was spent listening to a full concert of Semmangudi, a wonderful prelude to our entry into Krishna's new house.

It was early spring weather in Gainesville, ideal for tennis, picnics and weekend excursions. I enjoyed tennis with Krishna and his colleagues, Block and Keesling, fine sportsmen in the best sense of the term.

We made a weekend trip to St. Augustine, the first European settlement in America and the oldest town on the Atlantic coast. The Spanish fortress was preserved as a national monument to satisfy the curiosity of holiday makers and rambling residents of the beach front condominiums. The town had a flavour of a resort with its beautiful beaches under golden sunshine though it could not rival in strength and magnitude the 'paradise' atmosphere of the Hawaiian islands.

It was a most enjoyable, exhilarating and memorable visit to the Kennedy Space Centre, similar to the experience at Los Alamos, the birth place of Atomic Energy. The thrill was even greater since it was the scene of current endeavour where the space-shuttles are being launched in man's adventure into outer space.

A two hour drive, listening to Madura Mani's lilting music, took us to the Space-Centre where we were just in time for a thrilling film show of a Space Shuttle orbiting round the earth — a fantasy which is reckoned next best to being in space! There was a strange feeling as I mused how at the dawn of this century the Wright brother's flight in Kitty Hawk of 120 feet in 12 seconds was hailed by an astounded world as an incredible feat. The century was closing with the possibility of space stations and even space colonies to ease the congestion on the planet earth.

The three hour bus tour took us to the sequence of launch pads, from the first Gemini to the present space-shuttle, Discovery. We could not believe our eyes to see the shuttle on the launch pad getting ready for the launch in April. How mortal man whose breath is so frail as to succumb to a thousand ills, is able to achieve so much through co-operative enterprise!

We celebrated Rama Navami and Tamil New Year's Day with a mini-concert by Lalitha and Madhu at a pleasant get-together of our Indian friends. The power and sanctity of Rama Mantra are transmitted to mellifluous verse by Ramadas and Purandaradas, in style and spirit transcending space and time. It was ringing in Florida fifteen thousand miles from the land of its birth, exciting the same emotions perhaps with greater intensity.

American science achieved a miracle among miracles — the space craft Magellan was projected into space on a 180 day Venus probe mission from the Space Shuttle Atlantis which was launched on the 4th May. The manner in which the launch was first suspended on the 28th April, 31 seconds before zero hour, and resumed after the defect was rectified is as much a part of the incredible achievement as the projection of Magellan spacecraft whose orbit was directed and corrected from the space shuttle itself.

The best show on TV was the celebration of Bob Hope's eighty sixth birthday in Paris coinciding with the two hundredth anniversary of the French revolution! The parody of Louis XIV by the versatile comedian is a clear instance of how history can be presented without perpetuating conflicts and prejudices. Age had not touched the exuberant humour of the indestructibles Hope, typifying the American zest for life, desire for friendship and penchant for generosity. Watching him pour an inundation of humorous skits and anecdotes, I was reminded of the vitality of the living South Indian legend Semmangudi whose immaculate 'swaras' are an octogenarian's tribute to Divine music.

For the long memorial day weekend we went to Orlando on a visit to the new Disney-MGM studio, a real fantasy of engineering ingenuity, creative art, economic enterprise and incredible efficiency of organisation. We spent the evening at the fabulous Hyatt Grand Cypress Resort with its 500 acre site of golf courses, tennis courts, swimming pools and tree studded gardens, nature's profusion amidst manmade wonders. We stayed with our friends, the Sivamoggis, in their new home in Orlando. The two hour drive back to Gainesville was made hearing Madura Mani's concert, the incomparable swaras contributing to the safety of the drive by keeping Krishna wakeful!

In early June we enjoyed the French open on the TV day after day. We could not believe our eyes as we watched one of the most incredible events in tennis history, an early round victory of 17 year old Michael Chang over Ivan Lendl, the world's number one and thrice French open champion, in a five set duel. The manner in which it happened is even more incredible than the result itself! Lendl leading two sets to love, lost the next two and the fifth set was played as Chang was almost exhausted with cramps. He alternately tossed the balls high and hit them hard at sharp angles upsetting the Champions aplomb and rhythm. Once he served under hand like a school girl but it was the final point that looked like Chinese magic. The teenager stood at the service line — David intimidating Goliath and indeed the miracle worked — Lendl double faulted yielding the match as Chang raised his arms and rolled on the ground in ecstatic joy! The Chinese magic continued till Chang's final victory over Stefan Edberg, making him the youngest ever to win the French open.

Lalitha and I had to take the TWA flight at Orlando on the 15th June on our eastward journey. Krishna decided to leave for India the same day from Orlando via the Pacific.

Our flight 814 to Amsterdam from New York was delayed due to cyclonic weather at JFK and there was considerable suspense and anxiety whether we would be able to connect the Singapore Airlines flight to Bombay at Amsterdam. At my request the captain of the TWA flight sent a message to Amsterdam and what a pleasant feeling to be told on arrival at Schipol Airport that the Singapore Airlines flight was waiting for the group of passengers from New York!

It was a superb flight on the upper deck of the 'Big Top' 747 to Bombay, where at midnight we were greeted by Mr. Iyer (G.P.'s friend) who took us to his comfortable home in Mulund. It was 'airport to airport' hospitality offered by Mr. Iyer and we arrived in Madras by the Indian Airlines flight. Within a few days Krishna and his family joined us after a most enjoyable stay in Singapore and it was a pleasure to watch my grand daughters Lalitha and Amritha romp about Ekamra Nivas.

Notes

1) Ramakrishnan was motivated to choose this title for Chapter 55 because there was a catchy slogan by TWA in the sixties "Fly by TWA to and across the USA". The SIA in the title of Chapter 55 stands for Singapore (International) Airlines.

2) R. K. Narayanan (1906–2001) was a leading Indian writer whose most famous works involved the fictitious South Indian town of Malgudi as the setting. His mentor was Graham Greene. He was recognised with Padma Vibhushan, the second highest National Award in India.

3) *Sathabishekam* is a Hindu religious ceremony to mark the completion of eighty years. The ceremony is performed after the 80th birthday and before the 81st to signify that the person has lived long enough to witness one thousand crescent moons! It is an occasion when friends and relatives who are younger pay their respects to the octogenarian.

Chapter 57

Time, The Supreme Alchemist
(1989–90)

In the summer of 1989, when Krishna and his family visited Madras, we did not make any trips within India. It was then that I read detailed accounts of the celebration in Paris of the 200th anniversary of the storming of the Bastille, when leaders of thirty five nations were received with official ritual and ceremony to remind the nation of the blood, toil and tears of the masses who struggled for redemption against the unfeeling tyranny of the French monarchy. Time is not only a great healer but a supreme alchemist, transforming emotions as memory takes strange shapes when events and actors just become legends in recorded history. Even the torture chambers of the past become tourist attractions for curious customers staying at Paris Hilton and the Frankfurt Sheraton!

After the celebration of my birthday on August 9, Krishna and his family left for U.S. via the Pacific. Krishna attended a conference of the Canadian Number Theory Association in Vancouver on their way back to Gainesville. The next few months for me were a period of tranquil leisure at Ekamra Nivas but there were momentous events causing incredible changes in the national and international scene.

Rajiv Gandhi, the Prime Minister, called for elections which, to his surprise, resulted in the collapse of the Congress and the rise of the opposition parties, united as a National Front. V. P. Singh became the Prime Minister, not through an absolute majority but with the support of parties in opposition to the Congress.

Even stranger things happened at the State level, where the Southern States, with the exception of Tamil Nadu, voted for the Congress and the Northern for the National Front. It was a test for democracy and India withstood the trial, yielding to the people's will, as the transfer of power to the National Front took place at the Centre consistent with the Congress Governments in Andhra, Kerala and Mysore.

On the international scene, the world witnessed the most incredible event since the Second World War — the opening of the Berlin Wall. Time, the supreme arbiter sways human destiny according to its decrees — inscrutable, irresistible, inevitable.

At the invitation of my enlightened friend Madhavan, I delivered the Balasubramania Iyer Memorial Lecture at the Sanskrit College on 'Ramayana — its relevance to real life'. We must consider ourselves fortunate to have been born to a tradition with its roots in this immortal epic enshrining ideals to guide us through our mortal lives.

It was indeed a happy circumstance that one of my 'Manchester Gurus', Sir James Lighthill, visited Madras when Lalitha and I had the opportunity of meeting him at leisure at the Hotel Savera. We recalled with grateful pride his role in the creation of MATSCIENCE when at a dinner in Ekamra Nivas in December 1961, he endorsed the idea of an Institute of Advanced Study in our city in the presence of the other distinguished guest, C. Subramaniam. The hundred lectures of his on methods of Mathematical Physics at Manchester were reproduced by me to the Madras students at my personal initiative at various institutions in Madras, which provided me a forum for propagating of the 'Lighthill spirit' to enliven the languid and listless atmosphere of mathematical research in Madras.

Krishna and his family returned to Madras in December for a three week stay. This time they made the transpacific journey by Cathay Pacific from San Francisco to Hong Kong and then to Bombay. We enjoyed the entire music series, particularly Semmangudi's concert — hard to believe the rhythm and vitality of the music of this remarkable octogenarian. Madhu had an opportunity to give a Bharatha Natyam performance at the Nungambakkam Cultural Academy, thus keeping alive her interest in this traditional art. At a delightful function at the Anna University, the Volume on the Ramanujan Centenary Symposium on Number Theory edited by Krishna and published by Springer Verlag was released by C. Subramaniam. At the dawn of the New Year, we had an ACF lecture by Krishna on 'Ramanujan — the Second Century' with Dr. A. Jayaraman of Bell-Labs chairing the session. Krishna expressed his firm belief that the Second Century of Ramanujan will be as significant as the first.

After Krishna and his family left for the U.S., there was a heartwarming function at the Department of Nuclear Physics to mark the completion of twenty years of fruitful activity under the leadership of Professor Devanathan, one of my earliest students. It was a grateful gesture by Devanathan to have arranged the unveiling of the portraits of Sir A. L. Mudaliar, Dr. Sundaravadivelu and me, as his old teacher.

Lalitha and I had a memorable trip to Bangalore to attend the marriage of Jairaj, the son of our good friend Basavaraj, a prominent industrialist of initiative and enterprise. We enjoyed Basavaraj's princely hospitality staying at the Taj Residency as his guests, with G.P. as our companion. We were deeply touched by the frequent references to my father by Basavaraj as he introduced us to his distinguished guests at the reception.

I participated in a Nuclear Physics Symposium in honour of Professor Devanathan, my old student who was retiring at the age of sixty. It was a scientific meeting, well-organised by his associates, and as a tribute to him, I presented my work on special relativity and also gave a lecture on 'The birth, growth and future of theoretical physics in Madras', tracing the march of my students from short-term stipends to tenured professorships.

I had the opportunity to talk at the Lion's Club on 'Ramanujan — the Second Century' and later at the Rotary Annual Convention on 'Rotary — the timeless concept' at the instance of my friend Gangadharan before an audience of a thousand delegates. I drew attention that the 'Integrated Life' of a truly educated person implied Rotarian ideals — individual achievement consistent with service to those less fortunately placed, devotion to one's profession with an appreciation of other pursuits, singleness of purpose with an understanding of other's objectives.

We left for the U.S. via Singapore from Madras on the 24th April and at Singapore we stayed at the Pan-Pacific, a real fantasy of luxury and elegance with its super-size atrium and spraying fountains in its spacious foyer. After enjoying its Oriental opulence and Western efficiency, we left for Honolulu by Singapore Airlines via Taipei. At Honolulu we were received by our good friend (Hawaii) Ramanathan who took us to his comfortable apartment in Waikiki — the throbbing heart of the Island Paradise. The breakfast at the Ocean Terrace of the Sheraton Waikiki was particularly memorable for we could enjoy the beauty of natural environment amidst the man-made wonderland of elegance and luxury.

We flew to Los Angeles to be received by our gracious host Sampath, an affluent lawyer and taken to his beautiful home not far from the airport. The next day Professor Basil Gordon took me to Santa Barbara where I lectured on 'Duality in stochastic processes' after a delightful lunch at the Faculty Club with Professor J. S. Rao as our kind and solicitous host. I admired the location of the University, right on the Pacific Ocean from where we could watch resplendent sunsets musing how we are 'wasting' our coastline in Madras and what is worse, quite unaware of such waste!

We spent a whole day with Professor Gordon who took us round Los Angeles through its famed attractions on a hundred mile drive before visiting the famous County Museum with its superb collection of paintings and sculptures. We lunched at the Calender's recalling similar visits with Dick Bellman, who like Gordon, loved Los Angeles as (Samuel) Johnson loved London. The next day I lectured at the Rockwell International Science Centre in Malibu with Richardson and Vijay Sankar as my gracious hosts.

Our next stop was Dallas with our tireless host Fenyves at the University of Texas, where I spoke on 'Feynman, the legend of American Science' with emotion and enthusiasm tracing the origins of MATSCIENCE to my meeting him in 1956 when I received the 'Upadesam'[1] of 'Space-Time approach' from the sole creator himself.

The four months at Gainesville were a period of tranquil leisure and domestic pleasure but incredible changes were taking place in world affairs — the winds of freedom blowing across Eastern Europe and Russia — Nelson Mandela being greeted with open arms in New York and California, Gorbachev declaring an end to cold war, above all the unification of East and West Germany, a miracle of miracles, Russia seeking technical aid from a united Germany, the victor, a suppliant to the vanquished.

I spent many hours in the well-equipped air-conditioned library of the University of Florida browsing through current literature and books. I enjoyed reading the lives and achievements of the 'makers of Twentieth Century Physics' from Lorentz to Hawking. Conflicts of opinion and intellectual confrontation among scientists result in the discovery of truths and the laws of Nature. Progress is the 'path-integral' over all possible errors and mistakes which are eschewed and rejected by the 'principle of natural selection'. The magic of Maxwell is the logic of Lorentz. The result of Michelson leads to the triumph of Einstein. The disbelief of Einstein is the principle of Heisenberg. The Black hole of Oppenheimer reveals the Big-Bang to Hawking. Such is the endless frontier of Science despite hopes of a 'Theory of Everything' (TOE). No wonder Einstein remarked 'The most incomprehensible thing about Nature is that it is comprehensible'. That comprehension has no bounds!

We had an occasion to enjoy live Carnatic music at a concert at Jacksonville by Maharajapuram Santhanam. There was an air of unreality as he created a Music Academy atmosphere by singing 'Endaro Mahanubhavulu' and 'Seethamma Mayamma' ten thousand miles from Mylapore.[2]

I had an interesting experience in academic ethics which may be a helpful guide to young aspirants to a research career. While perusing journals in Santa Barbara on my way to Florida, I noted a paper by Robert Burton of Oregon State University in the 'Annals of Probability' where he introduced functions identical with 'product-densities' I had formulated forty years ago! What is more, he called them 'product-densities' as I did in my paper in the Proceedings of the Cambridge Philosophical Society. I drew the attention of the author and the editor Peter Ney to my earlier work and on a reference by the editor, the author prepared an addendum stating that the functions were defined by me in 1950. However the Editor without any explanation was unwilling to publish the 'Addendum' though I wrote to him that no reputed journal should allow mistakes when discovered to go uncorrected!

Eternal vigilance is the price an author has to pay for the proliferation of journals and permissiveness in publication without regard to priority and authenticity. Science is the pursuit of truth but scientists are subject to human frailties of vanity and lapses from professional ethics.

It was a curious coincidence that at that time I was reading the biography of Columbus to my grand daughter Lalitha explaining how America was named after Amerigo Vespucci who reached the West Indies only in 1499, seven years after Columbus, the first discoverer of the New World. Amerigo was fortunate in a biographer who mentioned him and omitted Columbus! Still, truth prevailed and the world knows that Columbus discovered America though it was not called after him. The nation made amends by naming its capital district, some cities, a University and a space shuttle to honour his name.

Lalitha and I left for India by United Airlines from Orlando to Los Angeles and by Singapore Airlines from Los Angeles. We spent a delightful day in Singapore at the Pan Pacific and arrived in Madras on the 10th of September just in time for the Navarathri festival.

Notes

1) *Upadesam* in Sanskrit means spiritual instruction from a Guru. To Ramakrishnan, the private lecture that Feynman gave to him at Caltech in 1956 was like an Upadesam.

2) Maharajapuram Santhanam (1928–2002) was a leading Carnatic vocalist, who in the eighties rose to the top by captivating audiences young and old with a style of singing that had both power and *bhava* (emotion). Before the start of his concert in Jacksonville, as Ramakrishnan and Lalitha entered the hall and sat in the front row, he and his accompanying artistes were overjoyed to see them and exclaimed "With your presence, the Madras Music Academy has been transported here!"

Chapter 58

A Year Of Miracles (1990–91)

The events we heard and saw on TV in Florida were precursors to the sequence of miracles the world had to witness and experience in the months to follow — the most incredible being the fall of communism in Soviet Russia, the Union of East and West Germany, and the relaxation of Apartheid in South Africa, which seemed impossible for decades till now.

It was hard to believe that our dear Chitra Narayanaswamy was no more! He was the quintessence of the Mylaporean way of life, treating social and moral obligations, friendship and human relations, on a par with, nay even more important than, the demands of big business and Board meetings. He conformed to the ideal of a true gentleman with a winning smile and affable speech, consistent with free and frank expression of opinions, a lovable simplicity which enhanced his natural ability for wise leadership. He never missed an Alladi Foundation function and in reply to my piquant exhortation that he should treat it as significant as a Board meeting, he would remark with disarming candour that Ekamra Nivas was the hallowed home of his revered Guru 'The Great Alladi'.

October is the month for the announcement of the Nobel Prizes. As expected, Gorbachev, the architect of the greatest miracle of the twentieth century, the freedom for Eastern Europe and the union of West and East Germany, was awarded the Nobel Prize for Peace.

On October 2nd (Gandhiji's birthday), United Germany elected Helmut Kohl as its Chancellor and the united parliament met for the first time since World War II on October 4th.

In mid-October, without the slightest warning, Madhu's father became seriously ill and was admitted to hospital for heart trouble. I made arrangements for Madhu and children to travel to Madras and they arrived in time to see him recover to normalcy by the Grace of God and the skill of the doctors. Therefore she could participate in the pleasant routine of Mylapore life. In November, Lalitha and Madhu had an opportunity to sing the compositions of Papanasam Sivan in a 'Harikatha' by their guru Mannargudi Sambasiva Bhagavatar.[1]

Krishna was invited to deliver the Ramanujan Endowment lecture at the Anna University on the 20th December. The Ramanujan Endowment lecture was well-organised by the Vice-Chancellor, Dr. Ananda Krishnan in the Vivekananda Auditorium. Krishna spoke on 'Ramanujan, continued fractions and partitions' to a large audience ranging from experts to enlightened citizens fascinated by the mystery of Ramanujan's genius.

The New Year 1991 dawned and Krishna and his family left for U.S. via Singapore and Honolulu on the 2nd January.

I inaugurated a Mathematics exhibit at the P.S. Senior Secondary School. A biographical essay on my academic career appeared in a popular children's journal 'Bal Vihar'. I was particularly pleased that the writer brought out the unique circumstances that directed my scientific career.

It was a pleasant surprise that Dr. Uppuluri, a great friend of mine, a veteran statistician from the U.S., whom I had not seen for fifteen years, called on me at Ekamra Nivas. I handed him the Alladi diary which he read with such interest that he had an index prepared of the scientists I had met during my travels in pursuance of my research career!

In early March, in response to an invitation to lecture at the Indian Institute of Science, I left for Bangalore by car with Lalitha. I also lectured at the Rotary Club of Bangalore East at the Ashoka Hotel making a plea for the internationalisation of prizes and awards to generate a true competitive spirit in Indian science and ensure the quality of creative work.

Meanwhile India's a political scene reached a crisis with the President dissolving the Assembly on the 13th March and ordering national elections. Two months of uneasy calm followed and suddenly the whole country was shocked by the assassination of Rajiv Gandhi on the 21st May at Sriperumbudur near Madras on the eve of national elections. Actually we received the news by phone from Krishna who heard it on American TV! The world leaders paid their homage and tribute to the handsome heir of the Nehru tradition, by attending the funeral in person representing their countries. When elections were held, the Congress emerged as the largest single party and with the cooperation of the other parties, formed the Government.

Krishna informed us that he would be able to spend six weeks with his family in Madras during the summer vacation. He and his family travelled by Malaysian Airlines across the Pacific, choosing Penang and Kuala Lumpur for their stop-over holidays, and arrived in Madras on June 25.

We went to Tirupati to perform two Kalyana Utsavams, one by Lalitha and myself and the other by Krishna and Madhu. It was a purifying experience, a thrill expressed only through the the joy as we watched the marriage of Lord Venkateswara, with the chanting of Vedic hymns. To satisfy Krishna's fascination for wildlife we left for Mysore enroute to the Mudumalai sanctuary where we had reserved accommodation at the Kargudi Guest House, with its splendid secluded location overlooking the lush green forests in the undulatmg valleys and hills. I lectured at the University of Mysore on 'Feynman, the legend of American physics'.

It was a fantastic stay at Mudumalai wildlife sanctuary, this time with my grand daughters Lalitha and Amritha spotting with excitement the deer, bison, clephants and langur in the dense foliage, where heavens breath smells wooingly over the verdant wilderness.

A day after our arrival in Madras, Krishna left for Bombay to give a course of lectures on his recent work at the Tata Institute of Fundamental Research, my alma mater, at the kind invitation of Professor K. Ramachandra and Dean Sridharan. During his stay he had breakfast at the Raj Bhavan with C. Subramaniam, who then was the Governor of Maharashtra, and who had evinced paternal interest in his career. After his return from Bombay, he left for the US with his family in August.

Notes

1) *Harikatha* is a discourse of a devotional nature interspersed with classical music. To give a good Harikatha, one should not only know the meaning and philosophy of the sanskrit scriptures, but should be able to sing as well. Lalitha's music teacher Mannargudi Sambasiva Bhagavatar was a leader in Harikatha. Sometimes in a Harikatha performance, the vocal music is rendered by accompanying artistes. In this instance, Lalitha and Mathura provided the music for Sambasiva Iyer's Harikatha.

2) One important event Ramakrishnan forgot to mention in his narrative here is the surprise visit of Granville Austin and his wife to Ekamra Nivas in April 1991. Austin is the historian of the Indian Constitution and has written an authoritative and comprehensive book covering all the deliberations of the Constituent Assembly. Sir Alladi's role on the Drafting Committee and in the Constituent Assembly is well reported in Austin's book. While in Ekamra Nivas, he took photographs of the *Easy Chair* (= reclining chair) in the Office Room that Sir Alladi used to take rest. On that easy chair is a mark made by the hair oil that Sir Alladi used. Granville Austin told Ramakrishnan that the mark should not be painted over, but left as it is for historical value! Austin wrote the following fine inscription in the Alladi Foundation Visitors Book: "To visit Ekamra Nivas is to reawaken the admiration and respect I have had for Sir Alladi ever since I began to study him."

Chapter 59

The Ramanujan Connection (1991–92)

During the period 1991–92, 'Ramanujan' dominated the mind and heart of Krishna, directing the course of his career and this is turn influenced our travels to and from the United States.

Krishna's article on 'Ramanujan, the second century' appeared in The Hindu on the 22nd December, the birthday of the legendary mathematician, emphasising that the unfolding of Ramanujan's genius is a process continuing into 'the second century'.

On the 10th January, I inaugurated the 30th anniversary symposium at Matscience with a lecture on my favourite theme 'Feynman — the legend of America', this time emphasising our successful attempt to retouch the master by splitting the Feynman propagator into positive and negative energy parts. It was a pleasure to be back at Matscience and speak at an Anniversary Symposium.

We left for Singapore on the 2nd of March and after a delightful stay at the Imperial Hotel, we took the Singapore Airlines 747 flight to Honolulu. The paradise of the Pacific offers its perennial charms which leave every visitor with an insatiable desire to visit it again at every possible opportunity. Our hosts, the Ramanathans, were hospitable as ever and took us round the island with its famous attractions — Hanauma Bay, Waimanalo Beach, Kahana Bay, China Man's Hat, and of course Waikiki Beach with its gorgeous sunsets. After a delightful flight from Honolulu, we arrived in Los Angeles when Krishna's friend Sampath took us to his new Bel-Air home and offered us Semmangudi's music to satisfy our South Indian palate. The next day, at the invitation of Dr. Vijay Sankar, I lectured at Rockwell on 'The passage from the discrete to the continuous' (in stochastic processes) explaining the nuances of the technique, I learned from Bhabha, as Arjuna from Drona, in my early research career.

The flight from Los Angeles to Orlando by Delta Lockheed 1011 was a beauty and we took the connection to Gainesville where we were greeted by Krishna, Madhu and the kids.

In March Krishna received a handsome invitation from Professor George Andrews to spend his Sabbatical at Penn State University, The 'Ramanujan Connection' which started with the visit of Andrews to Madras would continue with Krishna's association with him at Penn State.

It was then we heard the announcement of the great astrophysical discovery supporting the Big Bang, by George Smoot of Berkeley, at the Washington meeting of the Physical Society. Around the same time, we watched a thrilling live TV programme of the miracle of space walk to repair a defective satellite, an incredible ultra human achievements by a human being. Such is the power of creative technology, a product of the human brain.

One weekend we drove to the beaches of Jacksonville and saw the parade of luxury homes of exquisite architectural elegance, in good taste for gracious living. That weekend, in the spacious house of our hosts Ganesh Kumars we had a musical evening when Lalitha and Madhu gave a musical concert before a private circle of friends.

Krishna had the opportunity to review Robert Kanigel's book "The Man Who Knew Infinity" on Ramanujan for the 'American Scientist', drawing attention to the common Hindu faith in the Divine origin of the transcendent creative powers of Ramanujan whose work stretches beyond the bounds of Hardy's estimate.

Lalitha and I left Florida in June for Los Angeles where we took the Singapore Airlines flight via Honolulu and Taipei to Singapore and checked in at the Hotel Imperial. We were back in Madras at midnight the next day. Krishna and his family arrived a week later by Air India from Singapore. Krishna had lecture engagements at the Ramanujan Institute, Matscience, Anna University and the Tamil Nadu Academy of Sciences.

It was gratifying that my old student Professor Mathews was offered a visiting assignment at Harvard immediately on his retirement from the Madras University. I inaugurated the symposium in his honour in the Raman Auditorium, an exhilarating experience for a teacher to watch his student at the peak of his career. I arranged a lecture by P. M. Mathews at the Alladi Centenary Foundation when he explained how the interior of the earth can be studied by astrophysical data!

Krishna and his family left for U.S. on the 9th of August via Singapore where they had a fruitful stay as guests of the University. A week later, they informed us that they were well settled in State College, Pennsylvania.

Krishna started his academic work at Penn State University in right earnest interacting with Professors George Andrews and David Bressoud.

Here in Madras, it was exciting to hear that the Nobel Prize for Physics was awarded to Charpak of CERN, who was a visiting professor at Matscience when Prime Minister Indira Gandhi visited the Institute in 1967! The Hindu published the photograph taken at that time of the prime minister with the visiting scientists, drawing attention to the necessity of inviting scientists at the peak of their performance in the prime of their career.

In November I entered 'a second boyhood' by giving an elegant proof of the 'converse bisector theorem' using the distinction between two types of congruence of plane figures — 'ordinary' and 'reflexive' (see Appendix 14). It was a problem which puzzled me for fifty five years, ever since my teacher Narasimhachari challenged me to solve it in 1937! It was gratifying to receive an appreciative letter about my proof from a veteran geometer Professor Coxeter[1] of Toronto, Canada.

We received letters regularly from Krishna and Madhu from Penn State where Krishna was having a rewarding sabbatical collaborating with George Andrews and Basil Gordon in the area of Rogers-Ramanujan type identities.

On December 11, our dear N. C. Krishnan, Madhu's father, passed away after a heart attack following a strenuous trip to Bombay to attend three Board meetings within a week. He could never accept that he was too ill to continue his feverish activity which had brought him to the top of his profession with well-deserved recognition for public service. What a wide circle of clients, friends and relatives he had who admired his ceaseless zest for work and achievement! There were more than a thousand persons who came to pay their last respects to him. He was blessed with a devoted wife and loving children who felt that he should have taken greater care of his health, at least for the sake of his family. Madhu rushed to Madras by air arriving on December 16 to be at the side of her mother and brothers and participate in the Hindu rites for her departed father.

Notes

1) H. S. M. Coxeter, FRS and FRSC (1907–2003), a British born Canadian mathematician, is considered to be one of the greatest geometers of the 20th century. Although he did fundamental work in algebra, Coxeter was a strong advocate for the classical approach to geometry during an era when it was the norm and the fashion to approach geometry through algebra. He thus appreciated Ramakrishnan's elegant and novel proof of the Converse Bisector Theorem. Whenever Ramakrishnan came up with a new idea, he always communicated with the the masters of the field to get their critical assessment.

Chapter 60

Springtime In Our Hearts (1993)

After Madhu's return to Penn-State, Krishna and his family wanted us to visit them there. So we decided to leave for the U.S. via the Pacific. Winter at Penn State is long and severe but it will be spring time in our hearts in the company of Krishna, Madhu and our grand daughters Lalitha and Amritha.

We left for Singapore on the 12th February and checked in on arrival at Hotel Imperial, which as usual offered efficient and courteous service. We relaxed in comfort in the lively atmosphere of the tourist paradise of cleanliness and affluence. We took the United Airlines flight via Tokyo to Los Angeles where we were received by our friendly hosts the Sampaths who took us to their comfortable Bel Air home. The next day we flew to Dallas via Denver. Professor Fenyves arranged a special seminar when I spoke on the 'Legacy of Bhabha — passage from the discrete to continuous in stochastic processes'. It was flattering to see Dr. F. Johnson, former president of the University, among the audience.

We arrived at State College by United Airlines flight via Washington and were received by Krishna, Madhu and the children, all wrapped up in winter clothing with 'parkas' ready for us to wear, as we emerged from the small propeller plane, shivering even in our overcoats! Madhu has prepared a delectable dinner which we enjoyed in their warm comfortable apartment on the 20th floor of Parkway Plaza. What a gracious gesture by Professor Andrews, Krishna's host, to send a lovely flower bouquet to welcome us to Penn State!

Though winter lingered chilling the lap of March, the two months stay with Krishna and his family was one of the happiest sojourns in our annual visits to the U.S. since 1956. Krishna and Madhu had a wide circle of friends and we exchanged dinners and social visits with them, starting with Prof. C. R. Rao, my distinguished friend, now at Penn State, Professor Vedam, an emeritus physicist in material science. Members of the Indian community met once a week for a classical musical evening called "Raaga" in which Lalitha, Madhu and Krishna too were active participants; it was organised by an undergraduate student by name Mahalanobis, a very talented vocalist who could sing both Hindustani and Carnatic music. There was also a story hour once a fortnight when tales from the Mahabharatha and Ramayana were narrated to attentive young Indian children by one Dr. Gandhi. The parents also attended this since there was a get together over lunch following Dr. Gandhi's discourse.

Despite the unrelenting winter and persistent snow fall, Krishna took us on motor trips during the weekends. We offered prayers at Lord Sri Venkateswara temple in Pittsburgh, the first of the many Venkateswara temples in the United States. We saw the Carnegie Science Centre, a typical American enterprise to stimulate in the citizen an awareness of the role of science and technology in the evolution of society and culture.

We visited the Corning Glass Works and saw the impressive exhibition of the story of glass and a variety of artistic objects sculpted, carved, or etched in glass. I was reminded of my visit there in 1963 with McCrea Hazzlett. We visited the historic town of Gettysburg where the battlefield, preserved as a national park, was the principal attraction; Lincoln's address is part of the live heritage of the world's most powerful democracy.

We attended the Thyagaraja festival in Cleveland, a tribute as much to the nation of immigrants, respecting freedom of religion and culture of each component of its multiracial society, as to the Indian community, zealous of preserving its ancient heritage. Thyagaraja is relevant to the modern age since the human mind, stricken by the fitful fever of the competitive world, needs the tranquil peace of devotional music.

Within the halls of Cleveland, it was like watching Mylapore inside the New World for the crowd of a thousand was mostly Indian with fewer Americans than in Mylapore during the Academy season! No effort seems to have been made by the Indian professional and intellectual class — to transmit the spirit of Hindu culture to a Western audience who could certainly appreciate Carnatic music if they were properly introduced to its spirit and substance. My own experience with enlightened American visitors during the Golden Age of Matscience made me believe that the 'Intellectual West' is responsive to the aesthetic and emotional content of Carnatic music. Who will not be overwhelmed with triple tides of Hindu culture in the song 'Enduku peddala' in the raga Sankarabharanam — the profound meaning of the Vedas, the universality of Vedanta, and the rapturous nuances of classical music?

Veda Sastra Thathvarthamulu Telisi, Bheda Rahita Vedatamunu Telisi, Nada Vidya Marmamulu Telisi

V. K. Viswanathan, my former student, was there in Cleveland. He had initiated the induction of Carnatic music to the U.S.

Lalitha and Madhu had the unique opportunity to give a concert in Carnatic Music with explanationary remarks to an eager audience of 150 American students of the music class in the Penn State University, arranged by young Mahalanobis. Krishna's sabbatical turned out to be his most productive year in research in close association with Professor George Andrews at Penn State and Basil Gordon of UCLA. I used to go frequently to the University library to browse through mathematical literature. At home I enjoyed reading Kanigel's Ramanujan, over and over again, comparing it with a splendid biography of Thomas Alva Edison, the wizard of inventors.

Through George Andrews, Penn-State had become a 'Ramanujan Centre' and the spirit of Ramanujan reigned there as that of Einstein at the Institute for Advanced Study in Princeton. It was a pleasant evening with George Andrews in his lovely home in the countryside when his wife prepared a delicious vegetarian dinner not only to satisfy our palate, but in deference to Ramanujan's vegetarianism! Professor Andrews recalled with obvious pleasure his visits to Ekamra Nivas and Matscience, his lectures at the Alladi Centenary Foundation and the symposium organised by Krishna at the Anna University.

After our Cleveland trip we had just enough time to pack our suitcases and leave for Los Angeles. It was gracious of Professor Andrews to see us off at the airport along with Krishna, Madhu and our grand daughters. At Los Angeles we enjoyed being taken round the city by Prof. Basil Gordon, after a lunch at the UCLA Faculty Club hosted by Venki. Gordon took us to the Mathematics 'Hall of Fame' in the Science Museum and it was thrilling to find Ramanujan and Erdös among the Makers of Mathematics!

At Rockwell after the retirement of Dr. Richardson, Vijay Sankar, the Director of Computer Centre, became my new host. Addressing a discerning audience of high level professionals on the "Legacy of Bhabha", I emphasised that it is an occupational hazard in creative work to differ form one's hero and this does not affect the admiration of a devoted student for his reversed teacher. Bhabha appreciated my independence by accepting the membership of the Board of Governors of Matscience.

We left for Honolulu by United Airlines flight 59, to be received by Dr. Jayaraman, who had settled down in the Hawaiian paradise after his retirement from Bell-Labs. Jayaraman and Kamala are friendly hosts and we stayed in their modern apartment located on the 26th floor overlooking the enchanting city, gleaming under golden sunrise and crimson sunset.

After two days we shifted to the comfortable home in Hawaii Kai of our customary hosts, the Ramanathans, gracious and hospitable as ever. This time we enjoyed the I-Max theatre show on the 'Ring of Fire' a technical marvel projecting the awesome beauty of the volcanic ring round the Pacific. After a delightful stay in Honolulu, we took the United Airlines flight to Singapore via Tokyo, and spent a night at Hotel Imperial, Singapore, before boarding the Singapore Airlines flight to Madras.

This time Krishna and Madhu were visiting Europe in June, first to the International Centre for Theoretical Physics, Trieste in response to Salam's invitation, and then to France on a series of lecture engagements in Paris, Nancy, Lyon and Nice. Following that, he visited Hawaii with his family in August where at the instance of our friend Ramanathan, Madhu was invited to perform Bharathanatyam at the East-West Centre. It was a great experience for my grand daughter Lalitha who joined Madhu in a few dances. In view of the travel to Europe and Hawaii, they did not come to India in the summer.

I called on C. Subramaniam at his home, when he presented me with his newly published autobiography in which he himself had described the birth of Matscience in a chapter entitled 'Mathematics — the mother of sciences.' While I captioned a chapter in the 'Alladi Diary' describing the incredible events leading to the creation of MATSCIENCE as the 'Track of Destiny', he titled his entire book 'The Hand of Destiny'. Since facts are stranger than fiction, even in a rational world we have to realise that there is a Divinity, an unseen hand that shapes our ends.

On June 23, the mathematical world received the stunning news that the three hundred year old Fermat's last theorem was proved by Andrew Wiles of Princeton University. His announcement at a seminar in Cambridge before a small but enlightened group was flashed across continents by newspapers exciting adoration and admiration, with a trace of disbelief and scepticism.

Here in Madras, the main excitement for us in the tranquil life of Mylapore was the Alladi Centenary Foundation lecture of Michel Waldschmidt, on 'The History of π — Archimedes to Ramanujan'. We invited Professors Ramachandra of the Tata Institute and Akio Fujii of Japan to join us at breakfast in honour of our guest before the lecture.

Since I was completing seventy on August 9, I decided to mark my entry into the seventies by giving a series of seven lectures on 'Smiles and Tears of Creative Science' summarising my experiences in research for nearly half a century. The first lecture on special relativity was delivered at MATSCIENCE on August 6 with C. Subramaniam, its founding father presiding over the function. On 9th August, my colleagues and former students arranged a dinner at the Mathura restaurant which reminded me of the glorious days of MATSCIENCE when visiting scientists were entertained frequently at the Woodlands.

Early next morning we received the sad news that my brother-in-law, our dear Raghavarama Bava passed away. Sixty years of close association passed through my mind as we paid our last tribute to him. He was the supreme example of a person who placed domestic happiness above fame and honour. I had spent more than a thousand hours in conversation with him since my boyhood.

The second lecture on 'Smiles and Tears of Creative Science' was at the Tamil Nadu Science Academy when I spoke on 'Wave Particle Dualism'. I concluded the lecture by saying that a mathematician had to sometimes contemplate a thousand hours for solving even an apparently simple problem and therein lay the 'smiles and tears' of creative science.

The third lecture entitled 'Pauli and Dirac Matrices' was delivered at the I.I.T. Madras at the invitation of Professors S. M. Johri and Maji.

The fourth lecture was at the Padma Seshadri Higher Secondary School at the invitation of its enlightened principal Mrs. Y. G. Parthasarathy, on my most favourite topic 'Feynman Graphs'. I emphasised that the best tribute to creative genius is an attempt at improvements rather than meek worship and admiration. What a pleasant surprise the Hindu published a report with the caption 'The Charm of the Feynman formalism' a tribute as much to the enlightened newspaper as to its discerning readership.

It was a surprise to me that I was able to solve a deceptively simple looking problem in plane geometry proposed by Paul Erdös in the American Mathematical Monthly! The solution occurred to me in a flash without the ordeal of long hours of trial and error. What was more surprising the geometric problem led to a theorem of great generality! This motivated me to deliver the fifth lecture on 'Are elementary problems really elementary' at MATSCIENCE.

The sixth lecture on 'The legacy of Bhabha' was at the Department of Theoretical Physics at the University of Madras and the seventh on 'New Concepts in Probability' at the Anna University.

I was asked to repeat some of these lectures at the Rajaji Vidyashram at the suggestion of our doctor Jagannath and at the Ramanujan Institute at the instance of its Director Professor P. S. Rema.

We made a trip to Tirupati to invoke the blessings of Lord Venkateswara for my completing seventy years. G.P. accompanied us and we stayed at the comfortable guest cottage of Basavaraj.

Madhu and the children arrived on the 25th November two weeks ahead of Krishna. Madhu gave a dance performance at the Bharat Kalachar at the invitation of Mrs. Y. G. Parthasarathy. Krishna had his customary schedule of lectures — at the Tamil Nadu Science Academy, at Padma Seshadri High School, the Rajaji Vidyashram, the Ramanujan Institute and MATSCIENCE.

The music season ran its festive course with the concerts of K. V. Narayanaswamy and Mandolin Srinivas as the highlights. During the afternoon sessions I gave a lecture[1] 'On the Spiritual Diction of Thyagaraja', which was attended by music vidwans like Semmangudi and K. V. Narayanaswamy.

On 22nd December, Ramanujan's birthday, I inaugurated a Mathematics festival at the Ramanujan Museum, organised by the veteran votary of Ramanujan, Mr. P. K. Srinivasan.

Madhu had the opportunity to give a Bharatanatyam recital at the Nungambakkam Cultural Academy while Krishna gave a talk at the Alladi Foundation on on 'Fermat and Ramanujan — a comparison' with C. Subramaniam presiding over the function.

Notes

1) The Music Academy season is actually called the "Annual Conference" because each day there are a number of lectures on various aspects of Indian classical music and dance. Ramakrishnan was passionately interested in Carnatic music, and with his sound knowledge of Telugu, he made a deep analysis of the spiritual and philosophical content in the compositions of Saint Thyagaraja (the greatest of all Carnatic music composers), whose songs were in Telugu. Thus Ramakrishnan was invited to give a talk on "The Spiritual Diction of Thyagaraja" at the Annual Conference of the Music Academy in December 1993.

Chapter 61

A Great Year For Human Destiny, And Springtime in Florida (1994)

Twentieth Century is the most momentous period in human civilisation — technological advances in aviation, electronics, communication, two World Wars the rise and fall of dictatorships, the birth of the atomic age and above all the birth, growth and demise of global communism.

1994 witnessed an event more incredible than any of these, the disappearance of Apartheid and the election of a native African, Nelson Mandela, as the President of South Africa, as spectacular as the installation of Nehru as Prime Minister after three hundred years of British Rule. It was like the rising of the sun on the Western horizon and it happened on the 11th May, Tuesday, the day of the total eclipse of the sun!

On New Year's Day, Krishna was interviewed on television on the theme "Fermat and Ramanujan" in view of the excitement of the announcement of the proof of Fermat's Last Theorem a few months before. The next day, Madhu's dance program was broadcast on the television. Krishna and family left for Florida on January 5 after an eventful stay in Madras.

On the 28th of January, I inaugurated a mathematics festival at the I.I.T., when I spoke about my solution of an Erdös problem in plane geometry. In early February, at the invitation of Mrs. Y. G. Parthasarathy, I spoke about Thyagaraja's music at Bharat Kalachar, her sabha.

We left for Singapore by SQ409 on the night of the 23rd February and arrived next morning to check in at the Hotel Imperial. We took the United Airlines flight 890 and after a delightful journey via Tokyo arrived in Honolulu. We enjoyed our Hawaiian stay in glorious weather — rambling at the international market place, relaxing over snacks and coffee at the sea side promenades and watching the golden sunset at Waikiki.

We left for San Francisco by a United DC-10 and took the connecting flight to Dallas, Texas. My lecture at Dallas on 'Are Elementary problems really elementary?' was well-arranged by my distinguished host, Professor Fenyves and was well attended by physicists and mathematicians. We left Dallas on March 10 for Gainesville for a three month stay.

Weather in Gainesville was heavenly in early spring — cool afternoons and cooler nights, bright sunshine and mild breezes. I went to the University Library with Krishna every morning. This time I was reading more biographical literature on the makers of 20th century physics like Dirac and Niels Bohr on the one hand and Cosmology and Astrophysics on the other.

Krishna arranged a splendid musical get together at home when we had over sixty guests from Orlando, Jacksonville, Tallahassee and Miami, with music by Lalitha, Madhu and at least a dozen of our guests.

On Monday, March 21, at the invitation of Krishna, Bruce Berndt gave a Colloquium at the University of Florida on the theme "The Unorganized Portions of Ramanujan's Notebooks", and followed this with two seminars the next day. I attended all his lectures wondering how Ramanujan had become the rage of America with the additional realisation that the work he had done in India before leaving for England was yet to be fully investigated! He seems to have anticipated the dimension 24 of hyperspace, which dominates the minds of speculative theoretical physicists today. My grand daughter Lalitha had the opportunity to give a short dance recital at the international festival at Santa Fe College when we took Bruce Berndt with us for the event.

On Lalitha's birthday, April 4, we went to Amelia island and spent a whole afternoon at the Ritz-Carlton which has one of the finest locations in the world — standing in lone splendour overlooking the Atlantic ocean.

We enjoyed the concerts of Mandolin Srinivas at Jacksonville on April 9 and of T. N. Krishnan at Gainesville on May 9. Krishnan was our house guest and we invited over thirty friends for dinner at home to have the pleasure of leisurely conversation with the vidwan.

I received a thoughtful gift of a book of Roger Penrose from my student P. S. Chandrasekaran, who was now flourishing in Detroit working for the General Motors research division. It arrived at the correct time when I was trying to understand the impact of elementary particle physics on cosmology and astrophysics. I was intrigued by the remarks of Roger Penrose on Inverse Probability in quantum mechanics. Many paradoxes can be resolved if the distinction between analytic and Bayes inverses is understood as was shown in one of my earlier papers in 1955. I had lectured on this topic at the Belgaum meeting of the Indian Academy of Sciences when I was elected a Fellow of the Academy at the instance of Professor C. V. Raman.

Chapter 62

An Eventful Quarter,
June–October 1994

The third quarter of 1994 was eventful with travel in South India interspersed with academic programmes.

We left Gainesville for Orlando on the 15th June and took the United Airlines flight to San Francisco, where Lalitha and I took United Airlines while Krishna and his family travelled by Northwest Airlines. Interestingly, we met again in Narita Airport, Tokyo, where after a few hours transit, we reached Singapore where we celebrated Father's Day! We went on an excursion to Sentosa Island by cable car to see a fine museum devoted to the history of Singapore, and emphasising how the island country was influenced by the turning tides of World War II.

From Singapore, Krishna left with his family for Indonesia for a holiday in Bali — a trip that he and Madhu had been planning for years. They stayed at the newly opened Intercontinental Resort on Jimbaran Bay, and enjoyed visiting many of the temples in that enchanting island where Hinduism in the dominant religion. Lalitha and I arrived in Madras straight from Singapore. Upon return from America, I gave a report to The Hindu on 'Ramanujan, the rage of America'. Krishna arrived with his family on the 27th June and he immediately started his schedule of lectures.

We left for Palghat on the July 5 by the Alleppey Express, and stayed in the comfortable home of Lalitha's niece Geetha and her husband Chellappa, who own and run the Venkatesa hospital. Though it was the rainy season in Kerala, luckily there was a break in the monsoon, and so we enjoyed the drive to Guruvayur watching the verdant beauty of the Kerala countryside. We had a good darshan of Lord Krishna and performed Thulabaram for me and the children, in accordance with our 'Prarthana'.

We reached Kalamasserry after a comfortable journey where we stayed for two days with Lalitha's sister Janu and her husband Jayaram. We made a trip to Ernakulam and visited the Malabar Hotel to recall my stay there with my father in 1938. The city had grown into a seething metropolis since that visit but the hotel had preserved the colonial style of the older rooms in proximity to the newer addition. The visit to the Kerala Museum was particularly enjoyable for it was a rare instance where memories of previous rulers were preserved by exhibits and 'a light and shadow' show. We returned to Palghat to catch the Alleppey Express and were back in Ekamra Nivas on the 9th July.

On the 10th July, Krishna gave his lecture under the auspices of The Alladi Centenary Foundation on 'Roses from the Ramanujan Garden' with C. Subramaniam presiding over the function. It was a pleasant surprise that Timothy Hauser, the Consul General of the U.S. attended the lecture!

Next we left for Bangalore and Mysore where the arrangements made by our generous hosts Basavaraj and his son Jayaraj were excellent. G.P. was with us as their representative and our family friend to take care of our comforts. After the monsoons, it was delightful to visit the Krishnara-jasagar Dam and its lovely gardens, and the Ranganathittu Bird Sanctuary that attracts birds from as far as Russia. I gave a lecture at Manasagan-gotri (Mysore University) at the invitation of my old student Professor G. Ramachandran.

We left for Mudumalai and stayed at the Kargudi Guest House with its ecstatic location, overlooking an expanse of valleys and hills. We took a van ride into the forest, watching elephants, deer, sambur, and peacocks to the delight of my grand daughters. The Moyar Waterfall was impressive with its foaming waters after the season's rainfall. After a brief but very enjoyable visit to Ooty, we returned to Madras.

On return from our trip, I gave my first lecture on the 'Spiritual diction of Thyagaraja' at the Alladi Centenary Foundation with the eminent violin-ist T. N. Krishnan presiding. It was attended by the venerable Semmangudi and other vidwans, and was well-reported in The Hindu.

After his sabbatical at Penn State University in 1992–93 which was so productive, Krishna wanted to spend some time there in Fall 1994 to continue his research on partitions with George Andrews. So he made a special arrangement with the University of Florida to be free of teaching in Fall 1994. This also gave him time to be in Europe and in India in Fall 94.

Krishna left for Europe and the United States on the 1st of August. He attended the International Congress of Mathematicians in Zurich. At Penn State he had a two month academic assignment with Prof. George Andrews, the greatest living authority of Ramanujan and then participated in the Mathfest of the Mathematical Association of America in Minnesota. After his return to Madras in early October, he gave his second talk on the theme "Roses from the Ramanujan Garden" for the Alladi Foundation. He was also able to with us for the Navarathri festival which was celebrated in traditional style in Ekamra Nivas, with 'Sahasranama puja' followed by Mantra Pushpam twice a day. It was a thrilling experience for my grand daughters to hear the Mantra Pushpam as we joined in the recital by Pattabhirama and Nataraja Sastrigals.[1] Krishna and his family left for Florida after the conclusion of the Navarathri celebrations.

After Navarathri, I was invited by Justice Natarajan to inaugurate the M.Sc. course at Ethiraj College. I suggested to a large gathering of teachers and students of regular monthly programme of seminars to stimulate creative thought by frank and free discussion. On October 9, I lectured on "Sanskrit and Science" at a symposium at the Sanskrit College. At the invitation of my good friend B. Madhavan, the President of the Sanskrit Academy, I delivered the Justice T. L. Venkatarama Iyer Endowment Lecture[2] on December 3 as part of my series on Thyagaraja. Vidwan K. V. Narayanaswamy presided over my lecture.

In mid-December, Professors Johri and Mahji of the Mathematics Department of the I.I.T. convened an International Conference on the 'Early Universe' and I participated in it by giving a talk on 'Exterior Relative Velocity' a new concept in special relativity. This velocity of separation of two particles has a range from 0 to $2c$ (c is the velocity of light) consistent with the condition that the velocity of a single particle with respect to the observer cannot be greater than c.

Notes

1) Nataraja Ghanapatigal and Pattabirama Sastrigal were our family priests. A *Ghanapati* is one who has mastered the vedas including the recitation of Ghanam, which is very difficult. During each Navarathri, the two priests recite the entire Yajur Veda over a nine day period. Twice each day, in the morning and evening, they would recite various sections of the Yajur Veda sequentially for about an hour each time. Ramakrishnan would often sit in front of the priests to hear their recitation. Following this, there was Manthrapushpam for which all family members joined. Ramakrishnan and Lalitha used to invite some friends each day to join the Manthrapushpam in the morning followed by an elaborate lunch served on a plantain leaf.

2) Justice T. L. Venkatarama Iyer (1892–1971) was not only a judge of the Supreme Court of India, but a great Sanskrit scholar. He was deeply interested in Carnatic music, and owing to his knowledge of Sanskrit, he made a special study of the compositions of Muthuswami Dikshitar (universally regarded as second only to Saint Thyagaraja as a Carnatic composer), whose compositions were primarily in Sanskrit. For his knowledge and contributions to Carnatic Music, he was recognised with the title of Sangeetha Kalanidhi (means the repository of music), the highest award in Carnatic Music and given by the Madras Music Academy, and indeed was the first recipient of this award.

Chapter 63

Three Significant Developments (1995)

In 1995, it was a blessing that Ekamra Nivas, the home built by father, was given to us in a family partition. Secondly, Krishna assumed the Chief Editorship of the Ramanujan Journal. Thirdly, I was able to interpret the Lorentz Transformation in a manner as to extend it to many variables.

Following the settlement of Ekamra Nivas in late January, we made the trip to Tirumala to invoke the blessings of Lord Venkateswara and returned to Madras after visiting our village Pudur.

It happened early morning, as it did in 1967 — a new idea, the cubic generalisation of the Lorentz Transformation occurred to me suddenly after years of contemplation. It was an application of the 'principle of same behaviour' to three variables instead of two as in special relativity. I immediately obtained the transformation for many variables. I communicated the result to Professor T. W. Kibble (FRS),[1] who acknowledged its novelty in a letter (see Appendix 12).

It was pleasant surprise to hear from Krishna that he would be arriving in Honolulu with his family to greet us on the Island Paradise. He made perfect arrangements for our trip to the Big Island to see the Volcanic National Park where the volcanic activity was of just the right intensity to excite the visitors without endangering their safety.

We left for Singapore on the night of 27th February, and after a night's stay at the Hotel Imperial, we departed by United Airlines for Honolulu via Tokyo. Krishna and his family joined us in Honolulu and we made the usual jaunts to our favourite haunts in the island resort.

At the University of Hawaii, Krishna gave a seminar after which we had dinner in the lovely home of the Bertrams, driving via the Pali look out. The next day we left for Hilo where we hired a luxury Lincoln Town Car for our excursion round the Island. Our first stop was the Lava Tube, a natural hydrodynamic marvel of a giant tunnel formed by the differential flow of lava through a solidifying crust. We then drove to the volcanic site where we saw the breath taking spectacle of red hot lava flowing into the foaming ocean — a miracle of God's creation. In our excitement we summoned courage to walk half a mile across solidified lava to see red hot molten lava oozing out of the earth at several places like writhing pythons!

At sunset we watched with wonder the crimson glow of lava amidst swirling clouds of steam as it poured into the ocean at several places on the jagged coast skirted by black sand beaches, a fantastic sight not only worth seeing but worth going to see by crossing the oceans. We drove back to Hilo and checked in at the magnificent Naniloa Surf Hotel in one of the finest locations in the world with coconut palm-fringed lawns and private beaches right on the sheltered bay of the Pacific Ocean. This is where Krishna and Madhu had their honeymoon in 1978 and Krishna wanted his parents and his children to see this idyllic place.

We flew back to Honolulu where Lalitha and I took the United Airlines flight to Florida via Chicago while Krishna and his family travelled by Northwest, arriving almost at the same time at the Orlando airport, to take us to Gainesville.

Among the visitors to the mathematics department were Professors Erdös and Abhyankar who reminisced on their visits to Matscience and Ekamra Nivas with nostalgic pleasure. We had them for parties at home which were attended by several mathematics faculty.

In late March, Krishna took us on a trip to Orlando where he had to meet John Martindale of Kluwer to discuss the creation of The Ramanujan Journal. Later in June, Krishna received the official letter from Kluwer appointing him Chief Editor of the new Ramanujan Journal. It was a magnanimous gesture on the part of the 'Big Three' — Professors George Andrews, Bruce Berndt and Richard Askey, and many senior mathematicians to accept membership of the Editorial Board.[2]

On 6th April I had the opportunity to lecture at the Mathematics Department on my latest work on special relativity introducing the 'Principle of same behaviour' of two variables — transit and basic times which can be extended to 3 and n variables. Professors John Thompson and John Klauder[3] attended the seminar.

In May Krishna made a trip to Los Angeles to work with Professor Basil Gordon and stayed at the comfortable Bel Air home of his friend Sampath. On return from Los Angeles, he made a trip to Urbana to speak at a conference in honour of Heini Halberstam.

Amritha was the unique opportunity to deliver a 'Nobel Lecture' acting as Einstein at her school! I knew of her precocious interest in science since she joined Lalitha in presenting me with a book of 'Special Relativity' by Einstein on Father's Day!

30th May was an important day for my grand daughter Lalitha when we attended her 'graduation' from middle to high school. I presented her the book 'Tale of Two Cities' by Charles Dickens with the inscription that her life is a pleasant tale of two cities — Madras and Gainesville!

Krishna left for Toronto on the 14th June for an International Conference on q-series at the Fields Institute where he gave a plenary lecture at the invitation of Professor George Andrews. After his return from Toronto, we left for Orlando the next day for our departure to India. Krishna and his family left for Europe while we returned directly to India via the Pacific.

After spending two weeks in Europe visiting France, Switzerland and the Netherlands, Krishna arrived with his family in Madras on July 8. He gave lectures at the Paris Number Theory Seminar, and at the Universities of Lyon and Nancy. In Amsterdam at the suggestion of his older daughter Lalitha, they visited the home of Anne Frank, a young girl who maintained a diary of what happened when she and family hid in the attic for months together to avoid capture by the Nazis. Krishna's second daughter Amritha was so touched by the story of Anne Frank, that she wrote an article on this during her stay in Madras. I was so impressed with Amritha's article, that I sent it to The Hindu which published it in their Sunday Magazine section. So already at the age of seven, Amritha showed her talent in writing.

Krishna had a busy schedule of lectures in Madras at the Tamil Nadu Science Academy, Anna University, and the Ramanujan Institute, as well as at various schools and colleges in the city.

On the 23rd July I arranged the first of the series of 'Cambridge style' discussion seminars under the auspices of the ACF on 'Quantum Mechanics and Relativity'. In these seminars, a major scientific topic is introduced by a few experts, and this is followed by an interactive session with the audience. For this first seminar, Devanathan of Madras University, Parthasarathy of Matscience, and I led the discussion in which an audience of more than fifty participated for about an hour.

On my birthday 9th August, Krishna delivered the ACF lecture on 'Sylvester and Ramanujan' presided over by Prof. M. S. Swaminathan and this was followed by a second on 'Schur and Ramanujan' with Prof. M. Ananda Krishnan in the chair.

Krishna and family left on the 19th of August via Delhi. They were shown the great sights of the capital such as The Red Fort, the Qutb Minar, and the Rashtrapathi Bhavan by GP's friend Appaiah and his family before leaving USA via Amsterdam by KLM the next day.

What an intense emotional and intellectual experience it was to read a splendid article on my father by Justice V. R. Krishna Iyer in the Sunday Hindu of October 8, 1995! His matchless diction was so powerful, sincere and vivid that my father stood right before me as I read the article, comparing it with the earlier sequence of superb essays of Professor K. R. Srinivasa Iyengar and Justices M. Seshachalapathi, N. Rajagopala Iyengar and Sethuraman and Sri K. Chandrasekharan (see [1] for the collection).

We had a series of 'Cambridge-Style' seminars at the Alladi Centenary Foundation on 'Scales of Space, Time and Matter', 'Hyper-space, why we need higher dimensions', 'Fractional charges and new quantum numbers', 'Force in classical and quantum mechanics', 'New Mathematics from old problems' and 'Ramanujan congruences and Galois theory'.

Krishna decided to spend three weeks in Madras with his family during Christmas and New Year, since he was requested to organise an International Symposium on Number Theory by Dr. Ananda Krishnan, the Vice-Chancellor of the Anna University for the 'Ramanujan Season' which coincides with the Music Festival in Madras. He then had an a busy schedule of lectures as a prelude to the International Symposium in early January. At the invitation of Professor M. S. Swaminathan, Krishna delivered a lecture on 'Ramanujan and the Theory of Partitions — Past, Present and Future' on December 22, Ramanujan's birthday.

Notes

1) Sir T. W. B. Kibble (1932–2016), FRS, was a noted British theoretical physicist who made fundamental contributions to particle physics and cosmology. He was a senior researcher at the Blackett Laboratory at Imperial College, London. He was born in Madras in 1932, and his father Walter F. Kibble, a statistician, was a Professor at the Presidency College where Ramakrishnan studied. He received several awards — the Hughes Medal (1981), the Rutherford Medal (1984), the Dirac Medal (2013), the Albert Einstein Medal (2014), and the Isaac Newton Medal (2016) just before he died.

2) From 2005, the Ramanujan Journal, in expanded form, is published by Springer.

3) John Klauder is an eminent mathematical physicist who joined the University of Florida in 1988 after several years at Bell Labs in New Jersey where he was the Head of Theoretical Physics and Solid State Spectroscopy departments. Klauder held a joint appointment as Professor both in the mathematics and physics departments at the University of Florida. While in Florida, he was honoured with the Lars Onsager Medal in 2006. Ramakrishnan first met Klauder at Bell Labs in the sixties. It was a pleasure for them to renew contact in Florida. After Ramakrishnan passed away, Klauder coedited the book "The Legacy of Alladi Ramakrishnan in the Mathematical Sciences" with Krishna and C. R. Rao.

Chapter 64

Golden Memories (1996)

Lalitha and I with our family during our 50th Wedding Anniversary in Madras (1996).
- Author

1996 was the year of Golden memories with Divine Grace — for, half a century ago I chose both my partner in life, Lalitha, and my career for life, Theoretical Physics.

What better way of celebrating the dawn of the golden year than by a feast of mathematics at Ekamra Nivas following the International Conference on Number Theory organised by Krishna at the Anna University on the 1st and 2nd of January featuring several eminent speakers including Professors Wolfgang Schmidt of Colorado, David Brownawell of Penn State, Michel Waldschmidt of Paris — leaders in the area of irrational and

With the Doyen of Carnatic Music — Semmangudi Srinivasier, the architect of my marriage in 1946 (in Ekamra Nivas in 1996). - Author

Alladi Ramakrishnan lecturing on the "Spritual Diction of Thyagaraja" at Ekamra Nivas in 1996. Semmangudi in the audience. - Editor

transcendental numbers.[1] Professor Ramachandra was there along with his former TIFR colleague Professor Raghavan, who was now settled in Madras after retirement from the Tata Institute. Immediately after the Anna University conference, the foreign delegates and Krishna left for Trichy to attend a number theory conference held in conjunction with the 10th Anniversary of the Ramanujan Mathematical Society. Upon return from Trichy, Krishna and his family left for the USA on January 6 in time for classes to begin at the University of Florida after the Christmas-New Year break.

Professors Brownawell, Waldschmidt and Wolfgang Schmidt spent a few days in Madras after the Trichy conference and so I invited each of them to give lectures at the Alladi Foundation — Brownawell on January 10 on "The noted Indian mathematician Sarvadaman Chowla", Waldschmidt of January 11 on "Ramanujan's τ function and transcendence", and Wolfgang Schmidt on January 12 on "A survey of Diophantine Approximations". Although these lectures were held on three consecutive days, there was a large audience for each, with faculty of the Ramanujan Institute and Professor Raghavan[2] attending all three talks.

At a sentimental function to mark the Birth Centenary of Justice Ch. Raghava Rao, a chela of my father, I talked about the relevance of our scriptures and Thyagaraja's message in song to human lives to relieve and redeem us from the fitful fever of competition in our professional pursuits for material pleasures.

It was a pleasant evening at the Taj Fisherman's Cove Hotel under a moonlit sky when we met Nobel Laureate Norman Borlaug[3] at a reception arranged by Professor M. S. Swaminathan.[4]

I lectured on special relativity at a symposium at the Tamil Nadu Science Academy on the February 10 before leaving for the United States.

As usual, our transpacific travel was via Singapore and Tokyo to Honolulu. After a day's stay with the Jayaramans, we left for Los Angeles. Professor Basil Gordon received us and took us to the spacious new house of Sampath in fashionable Brentwood. I lectured at Rockwell at the invitation of Dr. Vijay Sankar and attended a Kalyana Utsavam of Lord Venkateswara in Sampath's Brentwood house when Lalitha was asked to sing before the Lord. From Los Angeles we flew to Florida to spend a few months with Krishna and his family.

Professor Erdős was in Gainesville and I attended his lecture on 'Open problems in Prime Number Theory'. We invited him for dinner and next day we had a party in his honour arranged by Krishna and Madhu at our home, attended by colleagues in the Mathematics Department.

We made a delightful trip to North Carolina to visit our friends the Morrises in their lovely retirement retreat — the Carol Woods in Chapel Hill. On our way we offered prayers at the Venkateswara Temple in Atlanta on Lalitha's Hindu birthday. In Atlanta we enjoyed the friendly hospitality of Professor and Mrs. Navathe in their spacious home. We left for Charlotte to spend the afternoon with our friends, the Bhojas, and attended a well conducted Bhajan, in the Hindu Temple before driving to Chapel Hill.

It was a sentimental stay with the Morrises in the heart of America because our association started at Madras when Morris was the Director of the USIS and I was the director of MATSCIENCE. Sentiment and memories added delicious flavour to the delightful breakfast 'Mrs. Morris style', where everything was in its place and dishes were served in proper sequence.

Lalitha and I made a separate trip to Dallas in response to the kind invitation of our distinguished friend Professor Fenyves. We travelled by United Airlines via Chicago making a 'touch and go' connection at the busiest airport in the world, where the concourses are separated by large distances. My lecture on 'Special Relativity' was attended by Professors Rindler and Robinson who earlier joined us at lunch in an Indian restaurant. We prayed at the Mahalakshmi Temple at Dallas, an opportunity for us in the United States to offer worship as at home. After return from Dallas, at the invitation of Professor Kunisi, I lectured at the University of North Florida attended by the mathematics, physics, and chemistry faculty.

During my stay in America, I arrived at a fundamental theorem in special relativity, unnoticed since its birth in 1905 but embedded in the Lorentz Transformation. Of course I was contemplating for fifty five years on the meaning of non-simultaneity of events in one frame but simultaneity in another! The idea of taking a moving rod of same length in motion as a stationary rod suddenly flashed at midnight of the 4th May and I spent a pleasant sleepless night verifying the theorem, so easy to understand, once the suitable choice of rods is made.

On May 3rd we attended an impressive Baccalaureate function of the University of Florida in which Dean Harrison in his speech made a handsome reference to Krishna's contributions to number theory and his founding of The Ramanujan Journal.

At Gainesville we had the Vedams of Penn State as our house guests when Krishna and Madhu arranged a musical get-together in their honour. Music was rendered by several friends from around Florida. Krishna rendered a few songs in Semmangudi style while Lalitha and Madhu gave a mini-concert for an hour.

In June, Lalitha and I left for India by the United Airlines via San Francisco and Tokyo to Singapore. We stayed at Westin Stanford and enjoyed the spectacular views of the bustling city from our room on the 64th floor. We arrived in Madras on the night of the 17th June.

Krishna and his family arrived in Madras two weeks later for a six week stay. He first lecture was at the Tamil Nadu Science Academy at the invitation of Professor Devanathan, followed by a talk at the P. S. High School in response to the Principal Mrs. Alamelu, his former high school mathematics teacher.

Professor Basil Gordon arrived from U.S. on July 17 night for a two week visit to make up for his absence at the January conference at Anna University where he was supposed to speak. He enjoyed attending the concerts and Bharata Natyam performances, the best that Madras can offer to a visitor from California. On July 22, he gave a magnificent lecture at the Tamil Nadu Academy of Sciences on "Dilogarithms, hypergeometric series and modular forms" followed by Krishna's talk on partitions.

24th July was a day of dedication for me when at Ekemra Nivas, I recalled my association with scientists and savants during fifty years of my research career. Prof. M. S. Swaminathan presided over the function and Mr. C. Subramaniam, the founding father of MATSCIENCE graced the occasion with his presence. Professors Mathews and Devanathan offered their greetings while Professor Basil Gordon participated by giving a special lecture 'The Final Problem — Ramanujan Mock Theta Functions'.[4] The very next day, Gordon left for Mysore and Ooty accompanied by GP as his companion and guide, and returned on July 29.

On the 30th July, Krishna and Madhu arranged an elegant dinner at the Hotel Savera as a forward celebration of our 50th Wedding Anniversary, attended by Professor Basil Gordon and distinguished friends like N. Ravi of The Hindu and Dr. Anandakrishnan, Vice-Chancellor of Anna University, Dr. Jagannath, and Dr. Devanathan.

To express my gratitude to God on our Golden Wedding Anniversary, I lectured on August 2 at Ekamra Nivas on "The spiritual diction of Thyagaraja" to a discerning audience. Semmangudi Srinivasier, the Bhishma of Carnatic music, presided.

Krishna and family along with Mathura's mother left for the US on the morning of the 11th of August, staying at Westin Stamford at Singapore and with the Ganesh Kumars in Tokyo. Ganesh and Prema Kumar played gracious hosts and arranged Madhu's Bharatha Natyam performance in the Japanese capital in a Ginza Auditorium as part of the celebrations on Indian

Independence Day. My grand daughter Lalitha accompanied Mathura on a few dance items. The day of our Golden Wedding Anniversary (August 21) was celebrated by a quiet dinner at the Woodlands.

On the 22nd September we heard the sad news from Krishna that our dear Erdös was no more! What an inconsolable loss to Krishna to whom he was mentor, guide, inspiration and hero, like Lord Krishna to Arjuna. We attempted to assuage our grief by the thought that such a mathematician is immortal through his theorems which are eternal truths. The Hindu promptly published Krishna's tribute to Erdös whose visit in 1974 was the origin of number theory at MATSCIENCE.

Krishna informed us that he was coming to Madras with his family in December for the music season. Professor M. S. Swaminathan arranged a function on the 30th December at his foundation when the first issue of Ramanujan Journal was released by Sri. C. Subramaniam and Krishna spoke on 'Erdös — a legend in Mathematics', an excellent report of which was published in The Hindu the next morning.

Notes

1) Professor Wolfgang Schmidt of the University of Colorado is the one of the world's greatest experts in the area of Diophantine Approximations in Number Theory. His most famous result is the extension of deep Thue-Siegel Roth Theorem to higher dimensions.

Professors Michel Waldschmidt (Paris) and Dale Brownawell (Penn State) are leading authorities in the area of transcendental number theory. Waldschmidt visits India regularly and is involved in organising workshops with Indo-French collaboration.

2) Professor S. Raghavan (1934–2014) was an eminent mathematician at the Tata Institute, who did fundamental work in the theory of modular forms and on quadratic forms. After retiring from TIFR in 1994, he settled in Madras. He regularly attended lectures arranged by the Alladi Centenary Foundation and actively participated in the discussions following each lecture that he attended.

3) Dr. M. S. Swaminathan is one of the world's most eminent agricultural scientists. Norman Borlaug, who won the Nobel Prize for peace in 1970, is credited for having contributed to the extensive increases in agricultural production, termed as the *Green Revolution*. Swaminathan who worked closely with Borlaug, brought the Green Revolution to India during the period when C. Subramaniam was the Food Minister. After serving as Director of the International Rice Research Institute in Manila during 1982–88, Swaminathan returned to India and founded the M. S. Swaminathan Research Foundation in Madras in the Taramani Campus where Matscience is located. He and Ramakrishnan were close friends. Dr. Swaminathan is the winner of numerous awards including India's Padma Vibhushan (1989), The Albert Einstein World Award of Science (1986), and the Indira Gandhi Peace Prize (1999).

4) Professor Basil Gordon of UCLA, was known for fundamental contributions to both number theory and combinatorics. One of his most famous results is his generalisation in the sixties of the celebrated Rogers-Ramanujan identities to all odd moduli. In the nineties, he started working on Ramanujan's mock theta functions. Gordon was one of Krishna's teachers at UCLA in the seventies, and he and Krishna began a fruitful collaboration in the nineties on partitions.

Chapter 65

Golden Anniversary Of India's Independence (1997)

Krishna left with his elder daughter Lalitha for the U.S. on the 3rd January after his lectures at the Tamil Nadu Science Academy and Anna University. Madhu and Amritha followed them two days later.

In late February, I discovered to my surprise a triangular matrix with elegant properties when raised to higher powers. I presented this in a lecture at the Theoretical Physics Department of the University of Madras demonstrating how a simple triangular matrix when raised to higher powers generates binomial coefficients.

We left for Singapore on 1st March and checked in at the Westin Stamford. Singapore has truly become a gastronomic capital and we enjoyed delicious international cuisine during our two day stay.

We left for Los Angeles by the United Airlines via Tokyo. Professor Basil Gordon met us at the airport and with gracious generosity hosted a lavish lunch at the luxurious airport restaurant atop the theme building of the Los Angeles Airport (LAX). After a few hours transit at LAX, we then took the flight to Orlando to be received by Krishna and his family.

Amritha took part in a French Congres at the fabulous Hotel Omni in Orlando. She received an award for outstanding performance, a note about which was published later in the Gainesville Sun, to our pleasant surprise.

In April Krishna received an official invitation to speak at the International Congress on Algebra in Hong Kong in August. It offered an opportunity to visit the fascinating city after its transition from British to Chinese rule on 1st July, an event the world would be watching with bated breath. The question on everyone's mind was: Will communist China become a free country or will Hong Kong lose its free enterprise?

We played tennis frequently at the 300 Club with its lovely courts amidst the lush environment of tall stately trees, the pride of Gainesville.

We went to Orlando for the Thyagaraja Aradhana at the Hindu temple when Lalitha and Madhu gave a mini concert with 'Nidhi chala sukhama' my favourite in Kalyani as the main piece. Later at home we had a musical get-together with seventy guests for dinner following a marathon six hour concert session that featured singers from around North Central Florida.

Lalitha and I decided to go to Dallas as a side trip from Florida. At Dallas I lectured on my latest work on Special Relativity at the kind invitation of Professor Fenyves who hosted us a delightful lunch at an Indian restaurant. My lecture on 'Non-simultaneity' dealt with the unification of three concepts, Time Dilatation, Lorentz Contraction and Non-Simultaneity, connecting space-like and time-like events in a single example of the crossing of rods. The mystery of space-time unity is resolved if we treat every point as the end of an imaginary rod and define an event as the crossing of points.

A proper understanding of space time unity is necessary not only for the study of sub-nuclear matter but also comprehending the age and size of the universe and its components. The amazing connecting between the microscopic and macroscopic worlds has been the basis of the modern Big-Bang theory of the origin of the universe with mind boggling contrasts in scales of space and time.

We visited Tallahassee, where Madhu took an active part in a music session of the Sargam group by leading a chorus, which received a standing ovation. We stayed overnight with Professor Sethuraman who was my host in 1974.

There was a pleasant function at the Bahai Annual Meeting when Krishna was honoured for his role as an efficient teacher. It was a spontaneous gesture by the organisation when at the suggestion of a young Indian mathematics teacher Kamath, I was extended a similar recognition. And what is more, a detailed report appeared in the local newspaper 'The Gainesville Sun'!

We made a trip to Jacksonville when Krishna delivered a lecture on Thyagaraja at the Hindu Temple drawing attention to the devotional aspect of the Kirthanas of the saint composer, considered as the incarnation of Valmiki, to redeem the human race by propagating the spirit and content of Ramayana through the 'swara raga laya sudha rasam' of the Tharaka nama of Sri Rama.[1]

Prior to our return to India from Florida, we visited Lake City in response to the kind invitation for tennis and dinner at the delightful country home of our doctor friends, Ravi and Rajani.

We left for Orlando from where Lalitha and I, and Krishna's family travelled by United and Northwest Airlines respectively at about the same time next morning. Krishna and his family wanted to have a Hawaiian holiday before arrival in Madras. Lalitha and I flew to Singapore via Los Angeles and Tokyo and stayed two days at the Westin Stamford. We had reached a stage when leisure gave us more pleasure than visits to tourist haunts and attractions.

Krishna arrived with his family on 27th June after a fabulous stay at the Sheraton Princeville Resort Hotel in Kauai, and visits to Maui and Honolulu. We celebrated Krishna and Madhu's wedding anniversary on the 29th June by a lunch at Taj Hotel in Covelong right on the beach twenty miles south of Madras — our version of Hawaii!

We decided to make a road trip to the Mysore area visiting the man-made marvels of Belur and Halebid, and the natural attractions of Sivasamudram's roaring waterfalls and the Mudumalai forests with their abundant wild life. We hired an air-conditioned Tempo Traveller, a spacious van, and left for Bangalore where we stayed at the Raman Institute guest house before leaving for Mysore. We made a day trip to Somnathpur and Sivasamudram, and were fortunate in witnessing the twin falls at Sivasamudram in glorious weather — the foaming cataracts looking dazzling against the resplendent evening sun — a photographer's dream.

The trip to Belur and Halebid via Hassan was also made in splendid weather and we feasted our eyes on the intricate sculptures of anonymous artists who had carved the epics Ramayana and Mahabharata in stone. What passion, what patience, what faith must have sustained the artisans working under the masterplan of a master mind who could visualise the whole with its myriad intricate components. We returned to Mysore via Sravanabelagola where the Gomateshwar statue inspires awe and wonder at its stupendous size and splendid location, symbolising spiritual tranquillity and universal compassion.

It was a three day tryst with wild life amidst profuse natural beauty of Bandipur and Mudumalai forests when we stayed at the comfortable guest houses at Abhayaranyam and Kargudi on either side of the hills affording spectacular views.

In Mysore we visited 'Rangana thittu', the bird sanctuary, where we took a boat trip to watch the crocodiles at close but safe distance and the fantastic variety of birds amidst verdurous surroundings. We offered prayers at the Chamundiswari Temple and what a fortunate circumstances that we could watch the Kalyana Utsavam of Siva and Parvati hearing the ecstatic strains of devotional Kalyana keerthanams on Nadaswaram. We received Prasadams from the venerable priest Seetharamiah, who had ensured us satisfactory Darshan of the Deities on all our visits.

At Bangalore Krishna lectured at the Raman Research Institute at the kind invitation of Professor N. Kumar, its director. At the end of the well-attended lecture, I was asked to say a few words. I recalled my father's association with Sri C. V. Raman at the time of the concept and birth of the Raman Institute. Prof. K. Ramachandra who attended Krishna's lecture invited us to the new 'National Institute for Advanced Study' before we left for Madras.

Krishna left for the U.S. on the 27th July to participate in two international conferences on Number Theory, one at Penn State in honour of his former teacher Professor Basil Gordon and the other at Atlanta organised by the A.M.S. He returned to Madras on the 8th August just in time for my birthday on August 9th. The next day I delivered the Balasubramania Iyer Endowment lecture on Thyagaraja at the Srinivasa Sastri Hall at the kind invitation of Raghunandan.

I took Krishna to the Raj Bhavan reception on August 15, the Golden Anniversary of our Independence. After his lectures at Anna University and the I.I.T., Krishna left for Hong Kong to attend the International Conference on Algebra while Madhu and children left a few days later to join him at Singapore. On reaching Florida, Krishna informed me that my papers on non-simultaneity in Relativity were published in the September issue of the Journal of Mathematical Analysis and Applications, a gratifying conclusion to my long effort to understand a ninety year old puzzle.

We enjoyed Navarathri conducted in a traditional way at Ekamra Nivas. Sadly on Vijayadsami, the last day of Navarathri, my sister Kalpakam passed away after a prolonged illness. But for the last year of suffering, she had a long and tranquil life of supreme domestic happiness and contentment in the house that my father had built for her opposite Ekamra Nivas, and rightly called "Lakshmi Nivas" after our dear mother. In view of my sister's passing, we did not celebrate Deepavali at Ekamra Nivas.

Notes

1) Since it was a lecture on Thyagaraja, Krishna benefited from the notes Ramakrishnan had prepared for his lecture series on Thyagaraja.

Chapter 66

Annus Mirabilis (1998)

1998 turned out to be the Annus Mirabilis both for me and for Krishna in our academic life. I realised a fifty-seven year old dream — to understand the distinction between length and spatial separation of events in Special Relativity, while Krishna became the Chairman of the Mathematics Department at the University of Florida which gave him an opportunity to introduce several new programmes in the next decade that would bring worldwide visibility and recognition for his department.

In early January, I gave the keynote address "On the birth, growth and future of stochastic processes in Madras" at the International Conference in Anna University at the invitation of the Vice-Chancellor R. M. Vasagam. The next day the Proceedings of the 1996 Anna University Conference on Number Theory organised by Krishna was released as the December 1997 issue of the Ramanujan Journal.

On the 19th February, I lectured at MATSCIENCE on 'Non-Simultaneity' when one of my oldest students, Dr. V. K. Viswanathan from Los Alamos was present. I received a handsome letter from Professor Kibble, FRS, at the Imperial College recognising that my theorem on 'non-simultaneity' threw new light on an old and well established theory. The Hindu published a reliable report on my new work incorporating the comments of Professor Kibble.

We departed Madras on February 21 for Singapore and stayed this time at the luxurious Hilton on Orchard Road. We left Singapore for the island paradise of Hawaii by United Airlines on the 23rd via Tokyo and were received by our generous host Ramanathan. The charms of the Hawaiian isles are insatiable and we enjoyed relaxing under the shade of the banyan tree over snacks and coffee at the Moana hotel in the Banyan Court overlooking the Pacific ocean. Why not time stand-still as we watched the lapping waves, with moving ribbons of foam glistering under glorious sunshine brightening the expense of cloudless skies?

From Honolulu we had a comfortable flight to Dallas via Los Angeles and Denver by United Airlines DC 10. Lalitha's nephew Hari received us at the airport and drove us to the elegant hotel where Krishna had made reservations for our stay. Professor Fenyves entertained us at lunch at the Indian restaurant before my seminar at the Kusch auditorium on 'My affair with Special Relativity'. I introduced the concept of the 'rod of light' with transformation property of length different from that of an 'ordinary rod' since light has no rest system.

From Dallas we flew to Orlando via Chicago to be received by Krishna and his family. This time my grand daughter Lalitha drove us from Orlando to Gainesville to demonstrate that she was a calm, composed driver even as a teenager!

Krishna became the Chairman of the Cultural Council of the ICEC (India Cultural and Educational Centre) and so we had the opportunity to meet many members of the Indian Community in different professions during social gatherings. The million dollar auditorium, well situated near the University, was formally dedicated, with Dr. Kumar Patel, the eminent scientist from UCLA, as the Chief Guest. We attended the banquet at the auditorium when Krishna welcomed the Chief Guest and Madhu gave a short Bharatanatyam performance. Madhu's students including my grand daughters participated in an earlier youth programme in a lively Tharana (Tillana) and Vande Matharam ballet. Dr. Ramanathan, the president of the ICEC, invited us to his elegant home in nearby Crystal River.

And suddenly the incredible happened — on April 9, at night, it suddenly occurred to me that the geometrical representation of 'non-simultaneity' was just a particular case for $t = 0$ of a rod of more general length $x - vt$ moving with a velocity v across a stationary rod of length x. It was a fifty-seven year old dream come true! Why did this elude me for six decades? Why did not full blooded relativists think about it? There is no mention of crossing rods even by the creators Lorentz and Einstein! So I quickly wrote up the result for publication (see Appendix 13).

The concept of 'Lorentz Contraction' of rods must precede Lorentz Transformation which deals with spatial and temporal intervals between 'events'. An 'event' has to be defined as the coincidence of the end points of crossing rods and so must follow Lorentz contraction. But that was what Lorentz attempted by postulating contraction through 'ether'. In a sense we are getting the 'ether' back to life in a mathematical sense!

We had a musical get together for six hours at home on the 2nd May with seventy guests from Gainesville, Jacksonville, Orlando and Tampa who stayed for dinner. The festive mood was a prelude to Krishna's appointment as Chairman of the Mathematics department of the University of Florida giving him an opportunity to serve an institution which had recognised his creative talents by providing him a tenured professorship early in his research career.

On May 11th when Krishna received the formal letter of appointment we heard the stunning news that India exploded the nuclear device in five tests in the deserts of Rajasthan entering the Nuclear Club, till now the exclusive privilege of only five nations. It provided, as expected, severe international reaction. It was not possible to predict the repercussions, economic, political and diplomatic in and outside India. Pakistan replied by performing nuclear tests two weeks later.

We attended a graduation party arranged by our friend Dr. Pericherla in Ocala when I was asked to address the young school graduates as they entered the Universities. In doing so I referred to the three generations of Indo-U.S. relationship — the first by highly qualified academics with permanent positions in India visiting the U.S. as temporary visitors under Exchange Visitor programmes — the J visa period. The next generation became migrants taking advantage of the hospitable atmosphere of America in institutions responsive to talent and initiative. The new generation of their children born in the U.S. were American citizens who by their achievement at school could aspire for education in leading universities in their 'mother country' cherishing of course the cultural values of their 'grandmother' country — the mother country of their mothers, India. I concluded by saying "It is fortunate to be talented and successful — but it is even more so to have talented and successful children." May God bless the young graduates.

Professor Venkat Kunisi, Chairman of the Natural Sciences Department of the University of North Florida, invited me to give a seminar presenting my new work on the geometrical representation of the Lorentz Transformation on the 29th June and in response to his gracious invitation we went to Jacksonville. The lecture was well-attended and received by a group of senior members of the faculties of Mathematics, Physics and Chemistry. I emphasised that this method was the best way to introduce special relativity to undergraduates.

It was then that I read the autobiography of Bill Gates — the living legend of American enterprise, describing how he formed a small two-man private company in 1975 on a shoe string budget after leaving Harvard, built Microsoft into a multi-billion dollar enterprise making him richer than the Queen of England in less than two decades. All this due to the right man at the right moment with the right idea — the conjunction is a greater miracle than the achievement which is its natural consequence. The theme of the book of Bill Gates was well-summarised by the author himself — it is important to be able to compete and cooperate at the same time but that calls for a lot of maturity.

On the 5th of June we attended the 'graduation' function at the Hidden Oak Elementary School when Amritha received an award for excellence as she 'graduated' from the fifth grade to the middle school.

Just before our departure to India, Krishna attended a conference of new Chairs in an elegant garden resort at Howie in the Hills, presumably to train them to run their departments efficiently! He then took us to Orlando on the 13th June where we stayed overnight with our friends the Shivamoggis before departing for India.

Krishna and his family arrived on the 30th June after a delightful holiday in Thailand visiting Bangkok and the exotic resort of Phukhet enjoying the comforts of Sheraton Grand Laguna amidst the idyllic environment of the seagirt island. Their stay in Madras was short but eventful. It started with a splendid dinner at Ekamra Nivas attended by over seventy guests. After Krishna's lectures at the Anna University and the Tamil Nadu Academy of Sciences, we went on a trip to the state of Karnataka.

We left for Bangalore and checked in at the guest house of the Raman Research Institute where Krishna gave a lecture at the invitation of its Director, Professor N. Kumar. At the instance of Professor K. Ramachandra he also gave a seminar at the TIFR Bangalore Centre.

The next day we left for Mangalore via Hassan and travelled through the Western Ghats in glorious weather, for to our pleasant surprise, the monsoon abated just for a week as though in answer to our fervent desire to offer prayers at the Mookambika Temple at Kollur near Mangalore to fulfill our prarthanas. The winding road through verdant valleys surrounded by soaring hills reminded us of the Blue Ridge Parkway on the Appalachian mountains. From Mangalore we drove to Kollur where we had a thrilling darshan of Goddess Mookambika at the temple consecrated by Adi Sankara. On our return journey we had a delectable picnic beside a limpid stream and lovely cascade before reaching Udipi where we offered prayers to Lord Krishna at the famous temple. We returned to Madras on the 18th July after an overnight stay in Bangalore enjoying the generous hospitality of our distinguished friend, the industrialist Basavaraj.

We arranged two lectures under the auspices of the Alladi Centenary Foundation, one on 'Lorentz and Einstein' by me and the other on 'Erdös and Ramanujan' by Krishna, well-attended by a discerning audience.

Krishna and his elder daughter Lalitha left for the U.S. on the 20th July followed by Madhu and Amritha two weeks later after the performance of Varalakshmi Viratham at Ekamra Nivas.

August 19 turned out to be significant since Mr. K. S. Padmanabhan, the Managing Director of East-West Books, expressed his desire to publish the Alladi Diary 1930–65 describing the true story of two miracles — the birth of MATSCIENCE from the womb of Ekamra Nivas and the 'Revelation in Relativity', a new geometric interpretation of the Lorentz Transformation.

On my wedding anniversary August 21, Krishna conveyed by phone from Florida that a letter had arrived stating that my paper on the extension of the Lorentz Transformation was accepted for publication in the Journal of Mathematical Analysis and Applications (Academic Press).

We were happy to hear that Krishna participated in the Conference on Number Theory in honour of George Andrews (for his 60th birthday) in Maratea, Italy, amidst spectacularly beautiful surroundings. It was then that Krishna mailed to us a copy of the book 'My brain is open' on the life of Paul Erdös by Bruce Schechter in which it was explicitly mentioned in the first chapter that Erdös rerouted his trip from Calcutta to Sydney to fly via Madras to meet Krishna. Meanwhile I received another letter from Prof. Kibble, FRS, of Imperial College, favourably commenting my work on Special Relativity.

We had the pleasure and privilege of receiving in Madras Prof. Keesling, Krishna's colleague in Florida, as the invited lecturer at the Anna University under the auspices of the Ramanujan Endowment.

Chapter 67

A New Lease Of Life
With Divine Grace (1999)

Lalitha and Amritha dancing for their Arangetram at the Madras Music Academy, 24 July 1999. - Author

Party for Ronald L. Graham at Krishna's home in Gainesville. L to R: Mathura Alladi, Ron Graham, Vera Sos, Jean-Louis Nicolas, Krishna Alladi, Alladi Ramakrishnan, and Lalitha Ramakrishnan (1999). - Author

On the campus of the University of Florida when Krishna was recognised at the Baccalaureate for the Editorship of The Ramanujan Journal (1996). - Author

The year turned out to be significant because of the most precious gift of God's Grace — a fresh lease of life for me after an unanticipated major heart surgery in the United States!

The year started with an Alladi Centenary Foundation function at Ekamra Nivas when I initiated a project 'Special Relativity for Y2K' and Prof. Keesling gave a lecture on his recent research. On January 3 we took Prof. Keesling for a dinner hosted by the Indian Science Congress where I introduced him to many of my old friends from other institutions in India.

Later in January, at the Sanskrit College, at the kind invitation of my friend Madhavan, I lectured on 'Glory of Sanskrit through Thyagaraja's diction', which conveyed the tides of Hindu philosophy and religion.

I decided to publish the revelation in relativity that rod of length $x - vt$ moving with velocity v unlocks space time in the Science Supplement of The Hindu to avoid delay in the dissemination of the startlingly simple new idea. I had communicated the paper earlier to the JMAA but for some inexplicable reason I did not receive any response for over nine months though the Journal had published all my earlier papers in a timely manner. Eventually the paper did appear in the JMAA, but my preliminary announcement of the result was published on Thursday January 21 in the Science supplement of The Hindu.

We left for Singapore on February 11 enroute to the United States and after a delightful stay there, took the United Airlines flight to Los Angeles and then to Florida for a four month stay. My academic routine started by correspondence with eminent theorists on my recent work in Relativity.

At the India Cultural and Educational Center (ICEC) in Gainesville, there was a Carnatic Music day featuring singers from North Central Florida, when Lalitha and Madhu gave a mini concert.

March was an eventful month since the American Mathematical Society Regional Meeting was held at the University of Florida with over four hundred delegates from all over the country. Krishna as Chairman had the pleasant and onerous responsibility of playing the chief host. I attended the sessions which started with John Thompson's one hour address and the splendid reception to the delegates followed by dinner. I attended the Erdős Colloquium (newly launched by Krishna) delivered by Ron Graham[1] on 'Erdős and his problems'. Attending the colloquium were several distinguished visitors — Professors Vera Sos (Budapest), Jean-Louis Nicolas (Lyon), and John Selfridge of the Number Theory Foundation. Krishna and Mathura hosted a party at home for his department colleagues to meet Ron Graham and the visitors.

I had a lunch meeting with Pierre Ramond,[2] a frontline physicist, leading the Institute of Fundamental Theory at the University. He evinced interest in my new work on Special Relativity.

Suddenly on the 7th April, as it often happens in frail human life, there was an unexpected setback in my health — a stomach upset followed by extreme weakness. I was admitted to the nearby North Florida Regional Hospital in the emergency ward. Though I apparently recovered under medical attention within twenty four hours, the detailed tests both revealed blocks in my heart which required immediate bypass surgery. The operation was performed the next week by Dr. Wesley, a very competent and considerate surgeon on the advice of my close friend Dr. Ravindra, an eminent cardiologist, and Dr. Shaheeda Qaiyumi, our physician. By the Grace of God and the skill of the surgeon, I recovered rapidly and returned home in two weeks to the relief of Lalitha, Krishna and his family.

During my convalescence, Lalitha acted as a ministering angel waiting on me twenty-four hours a day while Krishna took care of me as a dutiful son. Madhu and my grand daughters were of great help and encouragement to me. We offered prayers at the Lord Venkateswara Temples in Atlanta and Orlando in gratitude for my recovery from heart surgery before leaving for India on the 8th of June after getting the approval of the doctors. A day before our departure Krishna arranged a reception at the Holiday Inn for his friends and colleagues to celebrate the graduation of Lalitha from school and her admission to the University of Florida.

We left for Singapore by the United Airlines from Orlando via Los Angeles and Tokyo and as customary stayed at the Westin Stamford. Madhu and children joined us the next day and all of us arrived in Madras by Singapore Airlines on the 11th of June. Krishna arrived a week later in time for the marriage of Madhu's brother and we attended all the functions relating to the celebration.

Krishna left Madras for Hungary on the 3rd of July by the KLM from Bombay to Budapest via Amsterdam to attend the Erdös Memorial conference. There were over 300 delegates for the conference which started with the lecture of Bollobas of Cambridge University, who paid tribute to the legend of Number Theory by stating that 'after many years, the world will wonder whether a genius like Erdös had ever walked on this earth' — and indeed he roamed from centre to centre without a home, with mathematics as his passion and his mission, as problems and their solutions took shape in his incisive brain, shared by the world wide community in an incredible sage of collaborative effort.

On Krishna's return to Madras from Hungary, The Hindu published his report of the Erdös conference in its Thursday Science Supplement.

In gratefulness to God for my recovery from heart surgery, we performed the triple Homam (havan) in Ekamra Nivas — Ganapathi, Mrithyunjaya and Navagraha homams — expressing our gratitude to God for His blessings particularly during the past few months.

The *Arangetram*[3] of Lalitha and Amritha at the Kasturi Srinivasan Hall of the Music Academy took place on the 24th July when the girls danced splendidly in perfect unison in the presence of their two Gurus — Kowsalya and Madhu, our family friends, Bharatha Rathna, C. Subramaniam and Prof. M. S. Swaminathan. T. T. Vasu, the President of the Music Academy, presided over the function.

Krishna left for the United States followed by his family two weeks later.

I continued my new project 'Einstein for schools in Y2K'. The new geometrical interpretation could form the basis of lectures at schools introducing talented teenagers to the Special Theory of Relativity. The response from the schools in Mylapore was very enthusiastic and purposeful. The Hindu under the distinguished editorship of N. Ravi supported this project by publishing my articles on the Lorentz Transformation in its Thursday Science supplement with the titles. 'Lorentz Transformation and Cleopatra's beauty' and 'Lorentz Transformation for the Millennium' on September 2nd and 16th.

Meanwhile Prof. Devanathan and young Hariprasad delivered lectures on the New Rod Approach to Relativity at local schools and colleges. It was then a new idea suggested to me to derive the velocity transformation formula directly from Lorentz contraction. This led to the article on 'Light for the New Millennium', which I offered for the eve of the millennium issue of The Hindu on December 23rd 1999.

India went through the excitement of elections resulting in the formation of a BJP Government with the cooperation of the allies in the National Democratic Alliance.

The Alladi Diary Vol. I was nearing completion and K. S. Padmanabhan informed me that we could have the book release function on December 22 at Ekamra Nivas. We were delighted when Krishna phoned to us that he would be visiting Madras for about a week in December and participating in the function.

In November Krishna made a visit to China at the invitation of the Academy of Sciences to deliver lectures on his recent work on partitions in Beijing and Shanghai. Madhu and Amritha accompanied him while his elder daughter Lalitha stayed behind at Gainesville because of her busy college schedule. Both in Beijing and in Shanghai he and his family enjoyed the lavish hospitality of the hosts, the cleanliness of the environment and the keen interest on Ramanujan among Chinese mathematicians.

In December, I delivered an 'Endowment Lecture' at the Tamil Nadu Academy of Sciences at the invitation of Prof. Devanathan in memory of his brother. The topic was 'Spectacular symmetries of the Lorentz Transformation'.

On December 9, I despatched the full paper on the 'New Rod Approach to Special Relativity' to Prof. George Lietmann, Editor of JMAA, for the Bellman Millennium issue August 2000 (see Appendix 13).

Krishna arrived from Singapore by the Air India on the 16th December. Krishna's annual article for The Hindu on Ramanujan appeared on the 19th with the title 'Ramanujan for the New Millennium'. He lectured at various colleges before delivering the Millennium Lecture at the M. S. Swaminathan Foundation. On the 24th December, the Alladi Diary Vol. I was released at Ekamra Nivas when N. Ravi of The Hindu received the first copy. K. S. Padmanabhan, the Managing Director of East-West Books, agreed to have authentic documents included as appendices in the Diary.

Notes

1) Ronald Graham, one of the world's most eminent combinatorialists, was the one who managed all of Erdös' finances. Graham was head of the Mathematics Division at Bell Labs for many years and Erdös would visit Bell Labs frequently on his global travels. Graham was a "discovery" of Paul Erdös. Graham served as President of the American Mathematical Society. He has received numerous honours and awards including the Steele Prize of the AMS in (2003), the very first Euler Medal, the Lester Ford award of the Mathematical Association of America, and the Polya Prize.

2) Pierre Ramond, a Distinguished Professor of Physics at the University of Florida, is world renowned for his work in super string theory. He received his PhD from Syracuse University in 1969 under the direction of A. P. Balachandran, who was Ramakrishnan's PhD student in the early sixties. Thus Ramond is Ramakrishnan's grand student.

3) After several years of rigorous training, students of Bharathanatyam perform an *Arangetram*, which is a full two hour graduation performance expressing gratitude to the Guru for having taught this art form. In India, Bharathanatyam teachers will permit their students to perform in public only after the Arangetram is completed. Arangetram is a Tamil word meaning "ascending the stage".

Chapter 68

Dawn Of The New Millennium (2000)

N. Ravi, editor of The Hindu, receiving the first copy of *The Alladi Diary* published by East West Press, 24 December 1999. - Author

Alladi Ramakrishnan with his son Krishna at the Mathematics Department of the University of Florida when Krishna was Chair (2000). - Editor

Krishna left for U.S. right after Christmas to reach Gainesville in time to leave for Hawaii with his family. He and his family were keen to watch the dawn of the new Millennium at Kaanapali Beach in the paradise island of Maui. On our part we enjoyed the music concerts at the Academy on the eve of the Millennium.

We attended the Raj Bhavan reception on Republic Day — twice significant for it was the 50th Anniversary of the Indian Republic and the first year of the new Millennium.

On the 19th February we left for Singapore and we checked in at the Westin Plaza for a delightful one day stay. Prasad and his wife entertained us at their elegant vegetarian restaurant — The Bombay Woodlands — a pleasant gesture of gracious hospitality. We had a smooth trans-pacific journey to Florida for a four month stay in Gainesville.

We made the usual weekend trips to Jacksonville for Madhu's dance classes. Of special note was 'Mahasivarathri' when we we offered prayers at the Jacksonville Temple.

In March 2000, Krishna in collaboration with George Andrews and Alexander Berkovich completed an important extension of Göllnitz Theorem, a breakthrough in the Theory of Partitions, by solving a problem posed by George Andrews thirty years before. They intended to announce their result during the Millennial Number Theory Conference in Urbana in April.

To our pleasant surprise Krishna's research work with Andrews and Berkovich was announced as front page news with a prominent photograph in the 'The Gainesville Sun', the city newspaper! The children had their share of excitement when a fine photograph of Lalitha appeared in the college newspaper The Alligator after her dance performance, organised by the Hare Krishna group on Rama Navami Day!

While reading a book on Fermat's theorem, it suddenly suggested to me that there was an alternative generalisation of the Pythagoras Triplets which has a positive conclusion. If we treat the difference of squares as a determinant of a two dimensional circulant, we find that the determinants of integer valued circulants can be powers of an integer!

I wrote an article for The Hindu on 'Pythagoras to Lorentz via Fermat' (see Note 1 below) with the conclusion in poetic refrain in Churchillian spirit:

"It is by the Fermat window only that the integers are out of sight.

Look through Lorentz and Einstein, and the integers come into sight."

Krishna invited fifty guests to his house on the 7th May to mark the publication of the Alladi Diary, Vol. I by East West Books. I spoke to an attentive audience summarising the momentous events in India during the decades 1930–65 projected in the Alladi Diary — the transition from a life with father to life as a father.

We attended a spiritually elevating 'Vigraha Stapana' function at the Jacksonville Temple when my grand daughters Lalitha and Amritha had the good fortune to perform Bharata Natyam.

Krishna made a trip to Urbana for the Millennium Conference on Number Theory where he and his collaborations Berkovich and Professor George Andrews had the opportunity to present their new results. A picture of them taken at the Millennium Number Theory Conference accompanied an article on their work in the University of Florida Research Magazine "Explore".

It was then we heard of the announcement that the Clay Foundation was instituting million dollar prizes for the solution of seven outstanding problems in mathematics — the Riemann hypothesis, the Birch and Swinnerton-Dyer conjecture, the Poincare conjecture, the Navier-Stokes equation, P-NP problem, the Hodge conjecture and the Yang-Mills theory. It is an open question whether prizes for specific problems are better incentives to higher research than awards for any new discoveries.

I started reading Vishnu Sahasranamam regularly with Sankara's commentary on the thousand names explaining their significance. The culture of our ancient land of saints and sacred rivers and mountains is embedded in the Sanskritic diction of the sacred verse. We went to Atlanta to offer prayers at the Lord Venkateswara temple and stayed overnight at Professor Navathe's comfortable home.

We made a trip to Jacksonville where I lectured on 'Pythagoras to Lorentz via Fermat' to the science departments at the University of North Florida at the invitation of Professor Venkat Kunisi when I presented my new results on the positive extension of the Pythagoras theorem. I also lectured on the same topic at the University of Florida.

We left for Orlando on the 14th June. After a delightful two day stay at the Westin Stamford in Singapore, we arrived in Madras on the 18th June evening.

Krishna and his family left for Europe from Orlando in late June. He had his wedding anniversary in Vienna, where he also lectured at the University. He made a trip to Hungary where he participated in an International Conference on Number Theory in Debrecen while his family enjoyed the sights of beautiful Budapest.

After Krishna and his family arrived in Madras, we left on a trip in late July for Bangalore and Mysore combining academic engagements with a

'Tryst with Wild Life' at the Mudumalai Sanctuary. In Bangalore Krishna lectured at the Raman Institute and the TIFR Centre. In Mysore we stayed as guests of our industrialist friend Jairaj at the splendid Southern Star Hotel with its incomparable location and verdant environment.

It was a memorable stay at the Kargudi guest house in Mudumalai overlooking the tree studded hills and valleys, a fragment of heaven at sunrise and sunset. During the jeep journeys we feasted our eyes and our curiosity by watching wild life. We returned to Madras just in time for a function to mark the 50th Anniversary of my formulation of product densities. I recalled my association with Professors Bhabha, Bartlett and Kendall which resulted in the publication of my paper in the August 1950 issue of the Cambridge Philosophical Society simultaneous with Professor Bhabha's paper in the Proceeding of the Royal Society (see Appendices 4B and 4C).

Krishna and his family left the United States in early August.

In September, I was pleased to receive a set of reprints of my article on the New Rod Approach to Special Relativity in the September 2000 issue of the Journal of Mathematical Analysis and Applications (Academic Press). After 58 years of contemplation I had realised the solution of the distinction between the concepts of 'length' and the 'distance' between simultaneous events, which haunted me for six decades.

It was a pleasure to receive a phone call from Krishna that he was arriving in Madras on the 22nd September albeit just by himself for a short visit — to participate in a conference at Matscience and to lecture at an International Conference on Number Theory at the Panjab University in Chandigarh. During his stay in Madras, we had Professor Krattenthaler, his host at Vienna, for dinner at Ekamra Nivas. Krattenthaler was one of the main speakers at the Matscience conference.

We had an Alladi Centenary Foundation lecture by Professor Bruce Berndt on 'Ramanujan' at Ekamra Nivas before he left with Krishna for the conference at Chandigarh. After his participation at the Chandigarh conference, Krishna left for the U.S. We had the pleasure of receiving Prof. Waldschmidt of Paris for dinner at Ekamra Nivas when he came to Madras after the conference in Chandigarh.[1]

The 7th of November was a sad day when our dear C.S. passed away. We called at his residence to pay our last respects to a person who had played the role of a father, patron and guide for over fifty years in my academic life.

In mid-November, Krishna conveyed the exciting news that his distinguished colleague Professor John Thompson was to be awarded the National Medal of Science by the President Clinton of the United States. As Chairman of the Mathematics Department, Krishna was invited to attend the Award Ceremony in early December and a dinner in Washington D.C. arranged to honour Professor Thompson and other winners of the National Medal of Science.

Notes

1) It was during this visit to Ekamra Nivas that Ramakrishnan told Waldschmidt about his conjecture concerning integer power values of circulants related to Fermat's Last Theorem. Waldschmidt proved this conjecture and communicated it in a letter to Ramakrishnan dated June 8, 2000. The precise form of Ramakrishnan's conjecture and Waldschmidt's solution, were published in 2010 in the volume in memory of Ramakrishnan (see [2], pp. 329–334). A detailed article on "Pythagoras to Lorentz via Fermat" appeared in his book: "Special Relativity", East-West Books, Madras (2003), 90–97.

Professor Phillip Griffiths, Director, Institute for Advanced Study, Princeton, at Ekamra Nivas (2001). - Author

Chapter 69

Paradise Revisited (2001)

At the dawn of the new year I had the privilege of talking on the 'Efficacy of Vishnu Sahasranamam' in the sacred precincts of the Kesava Perumal Temple in Mylapore at the invitation of Mallikarjuna Rao, an active member of a religious society.

We had the opportunity to receive Professor Phillip Griffiths,[1] Director of the Institute for Advanced Study, Princeton, in Ekamra Nivas. He was amused when I told him that our hall at Ekamra Nivas was of just the same size as the seminar hall at Princeton and was being used for a similar purpose since my visit to the Institute in 1957.

On 31st January we had Professor Kanemitsu[2] of Japan for breakfast at E.N. when he informed us that he was inviting Krishna to Japan for an International Conference in early March.

We left on March 17 for Singapore, our favourite city, an affluent blend of Western comfort and Eastern bustle. After a comfortable flight by Northwest Airlines through Tokyo, we arrived in Honolulu to be greeted by Krishna and Madhu who arrived at the same time from Osaka, Japan, after attending the International Conference organised by Kanemitsu.

It was 'Paradise revisited' at Honolulu as we romped on the promenades of the sun drenched beaches, and driving round the island. We stayed with the Ramanathans, our customary hosts, hospitable and friendly as ever. We flew by Northwest to Jacksonville and reached Gainesville the next day after an overnight stay with the Kunisis. It was early spring in Gainesville, cool and pleasant, with the University town looking delightfully green as ever with its profusion of trees and lush foliage.

We made a pleasant trip to Fort Myers when Amritha was chosen to compete at State Level in a Science Fair. From Ft. Myers, Krishna left for Washington for a conference of chairmen organised at the US National Academy of Sciences, while we returned to Gainesville.

Our stay in Gainesville was characterised by tranquil leisure enjoying the company of friends and making frequent trips to Jacksonville and Orlando, offering prayers at the temples there, and participating in musical events. when Lalitha, Madhu gave mini music concerts, such as on Composers Day in Orlando. In Gainesville there was a musical get together at the home of our friends Bhavani Sankar and Mira where Lalitha and Mathura and Krishna sang before an attentive audience.

There was the Annual Talent Show at the ICEC, a pleasant social function in which Amritha was an active participant. On my part, I had the opportunity to discuss with Professors Klauder and Ramond the 'New Approach to Special Relativity'.

Our Gainesville stay came to a close on the 28th June, when Krishna took us to Orlando where we stayed with the Shivamoggis before leaving for San Francisco the next day.

We left for Tokyo by NW27 on the 1st July and it was a pleasure to meet Krishna and his family at Narita airport where they had just arrived by a direct flight from Detroit. We travelled together to Singapore, spent a day there and returned to Madras.

Two days later Krishna and his family on a holiday to visit the Taj Mahal in Agra, the palaces of Jaipur, and the Sariska wild life sanctuary. They enjoyed their stay at the Moghul Sheraton in Agra, the Rambhagh Palace in Jaipur and a palace resort in the game sanctuary.

Krishna left for France from Delhi to attend the Joint Meeting of the American and French Mathematical Societies in Lyon while his family returned to Chennai from Delhi. He returned to Chennai on the 21st July and spent a fortnight before leaving for the United States on the 6th August with his family. On their way they had a short holiday in Malaysia — enjoying the luxurious comfort of the Hotel Pan Pacific in Kuala Lumpur and at the Sheraton resort on the island of Langkawi. Madhu's mother Mrs. Gomathi Krishnan accompanied them to the United States on this transpacific journey.

Krishna informed me by phone that he saw a front page article on the 'Twin Paradox in Special Relativity' in the 'American Mathematical Monthly'. I wrote to the Journal that the 'New Rod Approach to Special Relativity' dispels such needless paradoxes which vitiate the proper understanding of a 'clean and clear' theory of space time unity.

Suddenly, on September 11th, the most incredible tragedy in the eventful history of the United States occurred when the Twin Towers of the World Trade Centre crumbled to dust after an explosion caused by the crashing of two hijacked U.S. planes piloted by suicide terrorists. We could not believe that this was actually happening as we watched the tragedy on TV at home in Madras. The world shared the agony of America for the loss of innocent lives and the destruction of the massive buildings, marvels of architecture and engineering, expressing its sympathy through the media, Television, Radio and the Press reaching the hearts of millions.

Madhu arrived at the end of November with her mother and attended the music festival in full, while Krishna arrived during the third week of December for a one week stay. He delivered lectures at the Tamil Nadu Science Academy the I.I.T., the Anna University and the M. S. Swaminathan Foundation, before leaving for the US with Madhu on December 31.

Notes

1) Phillip Griffiths is an algebraic geometer of world repute. After getting his PhD in 1962 from Princeton, he held faculty positions in Berkeley, Princeton, Harvard and Duke, before his term at Director of the Institute for Advanced Study from 1991 to 2003. He has received several prestigious awards for his fundamental contributions such as the Wolf Prize (2008), the Chern Medal (2014), and the AMS Steele Prize for Lifetime Achievement (2014). In 2010, when Krishna met Griffiths at the 80th Anniversary of the Institute for Advanced Study (two years after Ramakrishnan had passed away), he remembered his visit to Ekamra Nivas in 2001.

2) Shigeru Kanemitsu is a Japanese number theorist at the University of Fukuoka. Besides being active in research in analytic number theory, he played a leading role in organising about a dozen conferences both in Japan and in China. Kanemitsu visited Matscience regularly from the nineties, and always made it a point to call on Ramakrishnan at Ekamra Nivas during each of his visits.

Chapter 70

Another Paradise Visited (2002)

The author with his wife, son and daughter-in-law in Princeville, Kauai (2002). - Author

The author with his wife and grand daughters in Honolulu, Hawaii (2002). - Author

The year 2002 offered us an incomparable opportunity for an ecstatic holiday in the paradise resort of Kauai in the Hawaiian islands.

The year started with Prof. Kanemitsu's lecture at Krishna's school, Vidya Mandir. Later I lectured at Vidya Mandir on February 28th, the anniversary of the Raman Effect, explaining how creative work is due to sudden inspiration based on years of deep contemplation!

We called on the new Governor of Madras, Ram Mohan Rao[1] who was aware of my father's stature as a lawyer and jurist of legendary reputation. I presented him with a copy of the Alladi Diary and I was pleasantly surprised when at the Republic Day reception at the Raj Bhavan he walked towards me and Lalitha exclaiming that he read the Diary with keen interest and learnt more about the story of Indian Freedom from it than from other sources!

We made reservations for our departure to the United States on the 16th of March so that at the end of our four month stay, we can accompany Krishna and his family to Hawaii in early July on our way back to India. We left for Singapore on the 16th March, and this time checked in at Meridien Changi near the airport, since we were to leave for the USA the same night.

I attended lectures and colloquia at the University of Florida and gave a talk to graduate students on 'Plane Geometry as a source of creative thought' attended even by seniors professors like John Thompson, the Field Medallist. Among the distinguished visitors to the department were David Bressoud from Minnesota, Daniel Quillen from Oxford and Sergei Novikov from Russia, both Field Medallists.[2] We also had the pleasure of receiving in Gainesville Prof. M. M. Rao, my host at Riverside, California, and I attended both his lecture and the dinner in honour.

We attended the Graduation ceremony at the O'Connell Centre of the University of Florida when Lalitha received her Bachelors degree with honours, qualifying her for a law course at Stetson University in St. Petersburg, the very first law school in Florida.

On the cultural side, Madhu conducted the Arangetram of two of her students in Jacksonville attended by three hundred guests at the impressive Moroccan Auditorium, Krishna acting as the master of ceremonies. This was followed by a concert by Lalitha and Madhu at Sethuraman's house attended by a large group of friends and rasikas in and around Gainesville. Father's Day was celebrated by a delightful lunch at the exquisitely beautiful home of our dear friends, the Ramayyas.

On the 22nd June, Krishna arranged a Graduation party for Lalitha at the Holiday Inn when she gave a delightful Bharatha Natyam performance before dinner in the presence of over a hundred guests. Live music was provided by Madhu, her mother and guru, and by Amritha, while Hariprasad played the Mridangam.

We left for Orlando on 26th June for Honolulu. While Lalitha and I stayed with the Ramanathans, Krishna and family stayed at the Sheraton Princess Kauilani. During the three day Honolulu serenade, we went around the island enjoying breathtaking views of its coastline, especially Hanamuma Bay which is a natural wonder. We celebrated Krishna and Madhu's wedding anniversary on the 29th June by a lunch at Ramanathans and dinner at a Waikiki restaurant. Celebrating their wedding anniversary in Hawaii was appropriate because it was in Hawaii that Krishna and Madhu had their honeymoon in 1978.

We left for Kauai on the 30th June and checked in at the luxury condominium 'Pahio at Shearwater' in the fabulous Princeville area with a patio overlooking the beach and ocean. The next day we visited the various beaches, the most spectacular being the Lumahai, so attractive as to be chosen as the location for the film South Pacific, and Poipu Beach where the Sheraton Kauai resort is located.

We look the motor launch ride on the ocean enjoying the spectacular view of soaring mountain cliffs and plunging canyons of the Na Pali coastline against the deep blue ocean and above all the crimson sunset of inexpressible beauty when the shimmering ocean meets the luminous sky apparelled in celestial light.

The most thrilling experience was on the last day when we visited and offered prayers at the temple of Lord Siva set amidst the profusion of the verdant beauty of nature. As we stood in obeisance before the sacred precincts of the Deity of Shiva, it was like being in 'Bhooloka Kailas' — the abode of Shiva on earth! It was incredible that a temple was built with such faith and devotion in the midst of a Hawaiian island, on American soil. I recited slokas from Manu Charithra, the story of a devout priest wandering in the Himalayan heights of breathtaking natural beauty.

We had breakfast at the Sheraton Princeville Resort Hotel with its incomparable location on Hanalei Bay. Later we dined at the Hyatt Regency in Poipu Beach, which rivalled the Sheraton Princeville in luxury and elegance, after a whole day to the Kalalau Lookout of fantastic beauty, the mountains meeting the ocean with steep plunging gorges, and the inspiring expanse of the Waimea Canyon — the Grand Canyon of the Pacific — with its varied hues in light and shade under sunny skies. How could one be satiated with Nature's boundless bounty of transcendent loveliness ranging from bizarre barrenness to vernal profusion!

We left Kauai feeling as though we had an ecstatic dream and reached Honolulu to take the Northwest flight to Tokyo and then to Singapore, from where we flew Madras on July 9. Krishna and family stayed for a few days in Honolulu before visiting China and Japan, both for sightseeing and for lectures, and returning to Gainesville. So they did not visit India in the summer of 2002. But they came to Madras in December to enjoy the Platinum Jubilee festival of the Music Academy.

During the months of idle contemplation, it suddenly occurred to me one early morning that the ellipse is as relevant to special relativity as a hyberbola if we assume t (time) and x (space) as coordinates different from the conventional Cartesian as was done for a century since the birth of relativity! Why should this occur to me at my age, sixty years after I got fascinated by space time unity? It is just the Grace of God, a result of my life long faith in Gayathri Manthram, Vishnu Sahasranamam, Mantra Pushpam and the Ramayana in which the saint preceptor Agasthya conveys his benediction to Lord Sri Rama through Aditya Hridayam:

Rama Rama Maha Baho! Srinuguhyam Sanathanam
Enasarvanarinvatsa Samare Vijayishyathi
Aditya Hridayam Punyam Sava Satru Vinasanam
Jayavaham Jape Nityam Akshayyam Paramam Sivam
Sarva Mangala Mangalyam Sarva Papa Pranasanam

'It will grant you victory, peace and happiness.'

Lalitha and I wish to transmit this message to Krishna and his family at the dawn of 2003.

Notes

1) His Excellency Ram Mohan Rao served as the Governor of Tamil Nadu during 2002–04. He was a student of Alladi Ramakrishnan in the Masters class at the Presidency College Madras in the fifties. Ram Mohan Rao mentioned that to Ramakrishnan when the two met at the Raj Bhavan.

2) It was Fields Medallists Week in the Mathematics Department of the University of Florida with the presence of Daniel Quillen (Oxford), Sergei Novikov (Moscow) and John Thompson, the resident Fields Medallist. Novikov was awarded the Fields Medal in 1970 along with John Thompson at the International Congress of Mathematicians in Nice. But since Novikov was not allowed to travel outside the USSR, the medal was given to him in 1971 at a meeting of the International Mathematical Union in Moscow.

Lalitha and Amritha organised a surprise party in Gainesville for the 25th Wedding Anniversary of Krishna and Mathura (2003). - Author

It was also a surprise party for my 80th birthday! - Author

Epilogue By The Editor

My father concluded his narrative with the year 2002. I now briefly summarise some highlights starting from his 80th birthday year 2003.

In May 2003, my daughters Lalitha and Amritha arranged a surprise party for Mathura and my 25th Wedding Anniversary AND for the 80th birthday year of my father. Two pictures of this event chosen by my father for the Alladi Diary, Vol. II, are included here in Part II.

In Madras, in August 2003, an "International Conference on Point Processes and Product Densities" in honour of my father for his 80th birthday was organised at Anna University by Professors S. K. Srinivasan (his second PhD student), and A. Vijayakumar (his last PhD student), and the refereed proceedings was published by Narosa (New Delhi).

My father had the pleasure of witnessing the wedding of my first daughter Lalitha in February 2007 in Florida. While speaking at the wedding reception to a large gathering, he emphasised how important it is to spend time with family and friends, and that such events are among the happiest moments of his life. Several friends specially told me how much they appreciated his touching speech.

My father died at our home in Gainesville at night on June 7, 2008. A few hours prior to that he enjoyed attending an Annual Recital of my wife Mathura's Jathiswara School of Dance and Music. The final item of that program was a dance by Mathura and my daughters Lalitha and Amritha on Lord Panduranga, which he was looking forward to all evening. After we came home from the event, he said that he had some difficulty breathing, and died peacefully a few minutes later surrounded by his family members. With his faith in God that he so often mentions in his narrative, the final event he witnessed was the prayer to Lord Panduranga in the form of dance and music. The Lord beckoned him. May his Soul Rest in Peace.

In July 2008, the eminent scientist Professor M. S. Swaminathan, a close friend of my father, arranged a memorial function at his Foundation in Madras. Shortly after that, Matscience named a new lecture hall they had built as the "Alladi Ramakrishnan Hall" since the very mention of Ramakrishnan's name brings to mind the "seminar spirit" that he emphasised throughout his life. I had the privilege of cutting the ribbon to inaugurate that lecture hall at Matscience and give the very first seminar there.

In the summer of 2009, I arranged a function at the Presidency College of Madras University, in the *Old English Lecture Hall* where Matscience was inaugurated in 1962, and presented a portrait of my father to be hung alongside the portraits of two other distinguished alumni — Nobel Laureates Sir C. V. Raman and Subramanyam Chandrasekhar. Justice Mohan who presided over the function said that the speech "The Miracle has happened" that my father gave at the inauguration of Matscience in the same hall 47 years ago, is of such fine diction, that it must be made as compulsory reading in English classes for all high school students in India! Mr. N. Ravi, Editor of The Hindu, attended the function and had a complete report of it published in The Hindu.

On January 2, 2013, at a conference to mark the completion of the Golden Jubilee Year of Matscience, I gave the Opening Lecture entitled "Alladi Ramakrishnan's Theoretical Physics Seminar and the creation of Matscience". For my article based on that talk, see [1].

In Fall 2008, I was invited by Professor Peter Goddard, Director of the Institute for Advanced Study in Princeton to write an Obituary on my father to be published in the Institute Letter emphasising how my father's visit to Princeton inspired him to launch the Theoretical Physics Seminar in Ekamra Nivas which led to the creation of Matscience. The obituary article appeared in the May 2009 Institute Letter.

In 2010, the book "The Legacy of Alladi Ramakrishnan in the Mathematical Sciences" was published by Springer; it was edited by me, and by Professors C. R. Rao, a great statistician, and John Klauder, an eminent mathematical physicist. "The Alladi Diary" by World Scientific is another major international publication to foster my father's memory, but this provides a complete account of his life and reveals both his academic and human sides. I am most appreciative of World Scientific for publishing the edited version of my father's autobiography.

Krishnaswami Alladi

Appendices

Appendix 1

Letter From Prime Minister Nehru (1963)

PRIME MINISTER'S HOUSE
NEW DELHI

No. 2436-PMH/63

October 9, 1963

My dear Alladi Ramakrishnan,

  Thank you for your letter of the 6th October and for the quarterly report of your Institute. I am very glad to learn that this Institute is doing well and making good progress.

       Yours sincerely,

       Jawaharlal Nehru

Professor Alladi Ramakrishnan,
Director,
Institute of Mathematical Sciences,
Madras 4.

Appendix 2

Letter From Prime Minister Shastri (1965)

प्रधान मंत्री भवन

NO. 3514- PMO/65   PRIME MINISTER'S HOUSE
NEW DELHI

August 23, 1965.

Dear Dr. Alladi Ramakrishnan,

I am sorry for the delay in acknowledging receipt of your letter of the 11th August, 1965. Please excuse me.

I am glad to know that the Institute of Mathematical Sciences is holding its second annual summer school at Bangalore. You have all my good wishes for the success of the Summer School.

As I mentioned to you, I will be glad to comply with your wishes about my being a Patron of the Institute. You may, if you so wish, announce this at a time of your choosing.

Yours sincerely,

(Lal Bahadur)

Dr. Alladi Ramakrishnan,
Director,
Institute of Mathematical Sciences,
Central Polytechnic Buildings,
Madras-20.

Appendix 3
Letter From Prof. Aage Bohr

UNIVERSITETETS INSTITUT   BLEGDAMSVEJ 17, KOBENHAN
    FOR   TELEFON: TRIA 1616
TEORETISK FYSIK   TELEGRAMADR:PHYSICUM, KOBENHAVN
                    DEN

28TH NOVEMBER, 1962

The Institute of Mathematical Sciences Madras

I wish to express my sincere gratitude for the kind messages of sympathy at the occasion of the death of my father, and for the tribute paid to his memory. He often told about the deep impression he received at the visit to your institute, whose development he followed with the greatest interest and expectations.

Sincerely,
(sd) Aage Bohr

Appendix 4

Announcement By K. Srinivasan, I.A.S., Education Secretary, Government of Madras

It has been decided by the Board of Governors to institute two annual visiting professorships entitled
(i) The Niels Bohr Visiting Professorship
and
(ii) The Ramanujan Visiting Professorship.

The first is a tribute to the memory of the creator of modern physics and the founder of quantum theory, whose life has been a glorious example of the universality of science and the eternal quest for the laws of nature. His benign interest in the advancement of Indian science and in particular the work of the young group of theoretical physicists at Madras was the immediate stimulus for the creation of the Institute.

We are extremely grateful to the gracious consent of Mrs. Niels Bohr and Professor Aage Bohr, his son, now the Director of the Bohr Institute of Theoretical Physics at Copenhagen, to this proposal. In a personal letter to the Director, he writes:

"I wish to assure you that both my mother and I would consider such a gesture as a beautiful tribute to my father's memory and one which he would himself, we feel sure, have deeply appreciated in view of his general interest in your institute, his warm sympathy for its endeavours, and in view of the great importance he attached to international co-operation in science both as a means of furthering science and international understanding."

On behalf of the Board of Governors of the Institute, I request Professor Marshak to accept the Niels Bohr Visiting Professorship for this year.

The name of the person who will be accepting the Ramanujan Professorship for this year will be announced later.

Professor Robert Marshak of the University of Rochester was the First Niels Bohr Visiting Professor in January 1963. Marshall Stone, Distinguished Service Professor at the University of Chicago, was the First Ramanujan Visiting Professor in 1964. - Editor

Appendix 5

Letter From Prof. Abdus Salam (1964)

INTERNATIONAL CENTRE FOR THEORETICAL PHYSICS

Director: Piazza Oberdan 6,
ABDUS SALAM   Trieste, ITALY
                                                   5th December 1964

Dear Ramakrishnan,

     I am glad to receive your letter and I am delighted to hear of the excellent work which your Institute is doing in building up theoretical physics in India.

     We shall be happy to have a visit from you of two months duration, for which we shall pay you a salary of $900 per month and your own air travel. Kindly let me know as soon as possible of the times which are convenient for you.

              With best regards

                         Yours sincerely,

                         (sd.) Abdus Salam

Dr. Alladi Ramakrishnan.
Director,
MATSCIENCE
Institute of Mathematical Sciences,
Madras-4,
India.

The International Center for Theoretical Physics (ICTP) was founded by Abdus Salam in 1964. Soon after its founding, Salam invited Ramakrishnan to the ICTP. Ramakrishnan accepted Salam's invitation and visited the ICTP for two months in the summer of 1965, his first of several visits to the ICTP.

Appendix 6

Letter From Governor Mark Oliphant (1973)

GOVERNMENT HOUSE

ADELAIDE

SOUTH AUSTRALIA

May 4th, 1973.

Dear Alladi,

It is good to know that you may be here in
Adelaide at some future time which you do not specify.
If you do come to South Australia, we would consider
it a privilege if you would stay with us here at
Government House for at least part of your time.   I
am quite sure that Professor Hurst will wish to have
you speak to a seminar.   For my part, we shall be
happy just to have you with us for a day or two.

I saw Professor Mahler a few days ago, in
Canberra.   He was about to leave for U. S. A.
temporarily, but doubtless he will be back by the time
your son arrives.

With warm regards,

Yours sincerely,

Professor Alladi Ramakrishnan,
The Institute of Mathematical Sciences,
Madras - 20,
INDIA.

Appendix 7

Paper On Feynman Graphs

JOURNAL OF MATHEMATICAL ANALYSIS AND APPLICATIONS 17, 68-91 (1967)

Some New Topological Features of
Feynman Graphs

ALLADI RAMAKRISHNAN

MATSCIENCE, Institute of Mathematical Sciences, Madras, India

In this contribution we bring to focus some unnoticed topological features of Feynman graphs hitherto obscured by the extraordinary emphasis on manifest covariance at every stage of the description of a quantum mechanical process. It is well known that the equality of the perturbation expansions in the Feynman formalism to those in covariant field theory was achieved only after painstaking efforts through laborious and longwinded arguments. The aim in such attempts had been to establish the equality of a single term corresponding to a Feynman diagram with n vertices to the sum of $n!$ terms in field theory. Such equality is not so "manifest" as is the correspondence between the Feynman propagators and the field theoretical commutators.

Our object now is to show that if only we decompose the propagators into positive and negative energy "arms," a single Feynman diagram splits into 2^{n-1} diagrams which we call *patterns* and the correspondence between the 2^{n-1} *patterns* and $n!$ terms of field theory can be made "manifest" in a manner as to enhance the "topological beauty" of a Feynman diagram. *The fact that the terms corresponding to patterns are not covariant should not worry us any more than the noncovariance of $n!$ terms since the covariance is preserved for the sums in both the cases.*

This idea of decomposition of the propagator was suggested and used by this author and his collaborators earlier in a series of papers [1, 2, 3, 4], but we were deterred in the pursuit of our attempts when confronted by some puzzling features. We shall now show that considerable insight can be gained if we recognize that the four-dimensional transforms of singular functions occurring in perturbation theory can be obtained in *two stages*, a three-dimensional transformation over space, followed by a transformation over time. *This quite naturally leads to the decomposition of the propagator.* The use of a simple lemma in complex variable theory combined with the conservation law of energy resolves the "puzzle" and strikingly brings to light new facets of the topological structure of Feynman diagrams. The energy conservation is expressed as the vanishing of the sum of energy *imbalances* associated with vertices rather than with propagators, an idea introduced by this author 3 years ago [5].

68

Appendix 8

First Paper On L-Matrix Theory

JOURNAL OF MATHEMATICAL ANALYSIS AND APPLICATIONS **20**, 9-16 (1967)

The Dirac Hamiltonian as a Member of a Hierarchy of Matrices*

ALLADI RAMAKRISHNAN

MATSCIENCE, Madras, India

"Of strange combinations out of common things" — Shelley

We shall give a method of generating a hierarchy of square matrices L_m involving m independent continuous parameters λ_1, λ_2,..., λ_m such that

$$L_m{}^2 = (\lambda_1{}^2 + \lambda_2{}^2 + \cdots + \lambda_m{}^2) I, \tag{1}$$

as m takes values 2, 3,... . We shall show that the L matrices can be expressed as a linear combination of m 'generator' matrices independent of the parameters. The L matrices fall into one of two classes, *saturated* or *unsaturated* according as m is odd or even.

One of the most interesting features of this hierarchy is that the Pauli matrices are recognized to be the generator matrices which saturate L_2, while the Dirac Hamiltonian is an unsaturated L_4.

We start by writing

$$L_2 = \begin{bmatrix} a & b \\ c & d \end{bmatrix} \tag{2}$$

and requiring that

$$L_2{}^2 = \begin{bmatrix} a^2 + bc & (a+d)\,b \\ (a+d)\,c & d^2 + bc \end{bmatrix} = \begin{bmatrix} \lambda_1{}^2 + \lambda_2{}^2 & 0 \\ 0 & \lambda_1{}^2 + \lambda_2{}^2 \end{bmatrix}. \tag{3}$$

L_2 then falls into *canonical forms* of two distinct types.

Type I

$$L_2 = \begin{bmatrix} 0 & \lambda_1 - i\lambda_2 \\ \lambda_1 + i\lambda_2 & 0 \end{bmatrix} \tag{4}$$

or

Type II

$$L_2 = \begin{bmatrix} \lambda_2 & \lambda_1 \\ \lambda_1 & -\lambda_2 \end{bmatrix}. \tag{5}$$

---

* Read at the Sixth Anniversary Symposium January 2-12, 1967 at the Institute of Mathematical Sciences, Madras.

9

Editor's Note for Appendices 9–16

Ramakrishnan's dream since his student days was to understand Einstein's Theory of Special Relativity. This dream was realised decades later. Starting from 1973 when he was fifty years old, until the year 2000 when he was seventy seven, he published ten papers in the Journal of Mathematical Analysis and Applications, and some more as announcements in the Science Supplement of The Hindu, about his elegant and simplified approach to Special Relativity by explaining symmetries associated with the Lorentz transformation in novel ways. He was justly proud of his new approach to Special Relativity which kept him intellectually alert into his eighties. So I have included in these appendices the scans of the opening pages of Ramakrishnan's first and last papers on Special Relativity. Also included are scans of MIT mathematician Norman Levinson's paper on "Ramakrishnan's approach to Relativity" and physicist Costa de Beauregard's paper on the Lorentz transformation where he emphasises the elegance of Ramakrishnan's approach. Following this, the 1995 letter of T. Kibble (FRS) complimenting Ramakrishnan on his higher dimensional generalisation of the Lorentz transformation is included.

Ramakrishnan believed in propagating science at the highest level to school students and he lectured on advanced scientific topics to high schools in Madras regularly. In these lectures he often showed how symmetries in geometry can be used to explain the Lorentz transformation. He published a paper on this theme in the journal Physics Education, and a scan of this is included. In the course of all these investigations, he came up with a new proof of Converse Bisector Theorem in classical Euclidean Geometry. This proof is also included in the Appendices since it will be of interest to anyone who has had a course on Euclidean geometry in school.

Finally, Ramakrishnan published a book (East-West Books, Madras, 2005) for the Centenary of the birth of Special Relativity collecting all his refereed research publications in relativity as well as his articles in The Hindu on this topic. Appendix 16 contains the scan of the title page of this book.

Appendix 9

Ramakrishnan's First Paper On Special Relativity

JOURNAL OF MATHEMATICAL ANALYSIS AND APPLICATIONS 42, 377–380 (1973)

Einstein–A Natural Completion of Newton

ALLADI RAMAKRISHNAN

MATSCIENCE, The Institute of Mathematical Sciences, Madras-20, India

It is generally accepted that the theory of relativity is the greatest contribution to scientific thought that has emanated from a single mind since Newton's formulation of the laws of motion. At the same time Einstein's contributions have been interpreted as a departure from Newtonian ideas and this belief is essentially due to the fact that new concepts like the equivalence of mass and energy and the symmetry of spacetime were not envisioned in the Newtonian universe. We shall now present the relativistic theory as a natural continuance and completion of Newtonian ideas. The mansion of relativity has many entrances and the most suitable one for entering it from the Newtonian structure is the *velocity transformation formula*.

We start with the following simple argument which takes us right into the heart of the theory of special relativity. Considering one dimensional motion if v is real and is a possible velocity of a point particle so is $-v$ since it merely implies a reversal in direction. Accepting the Newtonian definition of relative velocity and the axiomatic principle of no preference for any particular frame we find $2v$ is a realizable velocity. Therefore $2^n v$ is also realizable, where n can be chosen as large as we please. If we make the postulate that such a world admitting velocities as large as we please would be "chaotic," then an upper limit l has to be prescribed for the relative velocity. Thus if v_a and v_b are velocities of two point particles a and b, then the relative velocity is assumed to be

$$v_r = (v_a - v_b)/f(v_a, v_b). \tag{1}$$

We now require that

$$v_r < l \quad \text{if} \quad v_a < l \quad \text{and} \quad v_b < l \tag{2}$$

and that

$$\begin{aligned} v_r &\to v_a - v_b \quad &\text{as} \quad v_a \text{ and } \quad v_b &\to 0 \\ v_r &\to l \quad &\text{as} \quad v_a \text{ and (or)} \quad v_b &\to l. \end{aligned} \tag{3}$$

The only choice of $f(v_a, v_b)$ turns out to be

$$1 - v_a v_b / l^2. \tag{4}$$

377

Appendix 10

Norman Levinson's Paper

JOURNAL OF MATHEMATICAL ANALYSIS AND APPLICATIONS 47, 222–225 (1974)

On Ramakrishnan's Approach to Relativity

NORMAN LEVINSON*

Massachusetts Institute of Technology, Cambridge, Massachusetts 02139

In an interesting approach, A. Ramakrishnan [1] presents "the relativistic theory as a natural continuance and completion of Newtonian ideas." By adding the single requirement of an upper bound on the magnitude of the velocity, he shows how Newtonian theory can be modified in a natural way that leads to relativity which he states is the unique consequence. Actually, his simple and natural formulation does not have a unique outcome. However, the relativistic result is by far the simplest one that fits his postulates and therefore, is the most natural consequence of Ramakrishnan's approach.

Here is Ramakrishnan's formulation modified for convenience by rescaling. The key assumption is that the magnitude of the velocity has an upper limit L. By dividing the velocity by L, the theory becomes dimensionless and the limiting magnitude is now 1. Consider one dimensional motion and let two point particles be moving with velocities u and v, respectively. Then $|u| < 1$ and $|v| < 1$. Moreover, the relative velocity w is now also required to satisfy $|w| < 1$. This latter condition cannot be met without modifying the Newtonian formula $w = u - v$, and therefore Ramakrishnan postulates that

$$w = \frac{u - v}{f(u, v)}, \qquad (1)$$

where f is such that

$$|u| < 1 \quad \text{and} \quad |v| < 1 \quad \text{implies} \quad |w| < 1, \qquad (2)$$

$$w \to 1 \quad \text{if} \quad u \to 1 \quad \text{or} \quad v \to -1, \qquad (3)$$

$$f(0, 0) = 1. \qquad (4)$$

The requirement (4) states that the Newtonian relation holds as the velocities approach zero. In (1) it is assumed that $f \geqslant 0$. Because of (4), it is natural to set $f(u, v) = 1 - g(u, v)$ where

$$g \leqslant 1; \quad g(0, 0) = 0. \qquad (4')$$

* Supported in part by the National Science Foundation NSF P 22928.

222

Appendix 11

Costa de Beauregard's Paper

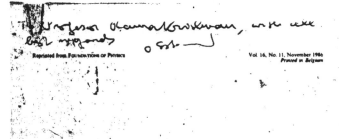

Reprinted from Foundations of Physics

Vol. 16, No. 11, November 1986
Printed in Belgium

On Carmeli's Exotic Use of the Lorentz Transformation and on the Velocity Composition Approach to Special Relativity

O. Costa de Beauregard[1]

Received August 30, 1985; revised March 20, 1986

As shown by Ramakrishnan, the faithful mapping, in the sense of Lie groups, of the real line onto the finite segment $-1 < u < +1$ is $u = \tanh A$, from which follows the "relativistic velocity composition law" $w = (u + v)/(1 + uv)$ and the Lorentz–Poincaré transformation formulas. Composition of translations is merely one application of this. Carmeli has shown that composition of rotations is another one. There may be still others.

1. INTRODUCTION

M. Carmeli [1][2] has recently used the Lorentz–Poincaré group for composing rotations instead of translations: see his formulas (10) and (46). At first sight this looks quite surprising, but much less at second sight. A theorem by Ramakrishnan,[3] which he used for deriving the relativistic velocity composition law, and then the Lorentz–Poincaré formulas, states that the faithful mapping, in the sense of Lie groups, of the real line $-\infty < A < +\infty$ onto a finite segment $-1 < u < +1$ is given by $u = \tanh A$. Section 2 below presents a compact derivation of Ramakrishnan's result.

As in the history of Alpine climbing, Ramakrishnan's 1973 straight way up to the Lorentz formulas via the velocity composition law has come after a long succession of preliminary explorations. It all started in 1818, when Fresnel formalized Arago's null result in an ether-wind experiment by his "ether-drag formula." Section 3 outlines very briefly this exciting story.

---

[1] Institut Henri Poincaré, 11, Rue P. et M. Curie, 75231 Paris Cedex 05, France.

1153

0015-9018/86/1100-1153$05.00/0 © 1986 Plenum Publishing Corporation

Appendix 12

Professor Kibble's Letter

IMPERIAL COLLEGE OF SCIENCE, TECHNOLOGY AND MEDICINE

T.W.B. Kibble FRS
Professor of Theoretical Physics

· Blackett Laboratory, Prince Consort Road, London SW7 2BZ, UK

☎ +44-171-594 7845 Fax. +44 171-594 7777 Telex: 929 48 IMPCOL G
Electronic mail: kibble @ ic.ac

Professor Alladi Ramakrishnan
c/o Professor Krishnaswamy Alladi
Department of Mathematics
201 Walker Hall
University of Florida
Gainesville FL32611–2082
U.S.A.

28 April 1995

Dear Professor Ramakrishnan

I am sorry that it has taken me so long to respond to your letter of 10 March.

You have certainly discovered a very intriguing generalization of the Lorentz transformation. It is a most interesting piece of mathematics, though I have not been able to think of any particular physical interpretation. I kept thinking that if I delayed a little more I would be able to understand it more thoroughly, but I am afraid that I have not succeeded. I do agree that something so elegant *ought* to have an application of some kind!

You are of course right that I am the son of Professor W.F. Kibble of Madras Christian College. Your name is very familiar to me.

With best wishes,

Yours sincerely,

Tom Kibble

T.W.B. Kibble

Appendix 13

Paper On "Rod Approach to Relativity"

Journal of Mathematical Analysis and Applications **249**, 243–251 (2000)
doi:10.1006/jmaa.2000.6929, available online at http://www.idealibrary.com on IDEAL

A New "Rod" Approach to the Special Theory of Relativity

Alladi Ramakrishnan

Alladi Centenary Foundation, 62 Luz Church Road, Madras 600 004, India

Received January 1, 2000

DEDICATED TO PROFESSOR DICK BELLMAN, MY DISTINGUISHED FRIEND
WHO SHARED MY FAITH, "MATHEMATICS IS AN UNERRING
GUIDE TO PHYSICAL THOUGHT"

INTRODUCTION

Einstein, the author of special relativity did not define an event explicitly as the crossing of the end points of rods, as we do in our new approach. He states in his book "The Meaning of Relativity," "The experiences of an individual are arranged in a series of events; in this series the single events which we remember appear to be ordered according to the criterion *earlier* and *later* which cannot be analysed further."

Lorentz transformation deals with the separation in space and time of two "events." From this Einstein derives the contraction of length of moving bodies.

On the contrary, here we start with Lorentz contraction as fundamental and derive the Lorentz transformation by defining events precisely as the crossing of the end points of rods. This eliminates misconceptions and paradoxes like virtual or real contraction, faster than light particles, and differential aging of twins. Actually this is what Lorentz attempted to do but he needlessly invoked the medium of ether which is dragged along with particles. Though ether is nonexistent, the distance between "equivalent" observers is meaningful and so is its contraction, which is postulated in much the same way as curvature in space is postulated in general relativity.

Special relativity deals with changes in intervals in space and time only in the direction of motion. Therefore we need consider only two coordi-

243

Appendix 14

Proof Of The Converse Bisector Theorem

ALLADI RAMAKRISHNAN

Theorem: *If the bisectors of two angles of a triangle are equal, then the triangle is isosceles.*

Proof. Let ABC be a triangle and let the bisectors BD and CE of angles B and C be equal. (For simplification, we sometimes say angle B for angle ABC.)

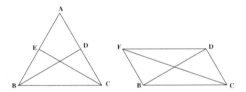

Choose a point F such that $FB = BE$ and $FD = BC$. Since $EC = BD$, the two triangles EBC and FBD are congruent in the 'reflexive' sense; that is if we 'flip' the surface of EBC, it can be superposed on BDC.

Now compare triangles FDC and FBC. Note that the side FC is common and that $BC = FD$. Next,

$$\text{(for Angles)} \quad FDC = FDB + BDC = \frac{C}{2} + \left\{ A + \frac{B}{2} \right\}$$

and

$$\text{(for Angles)} \quad FBC = DBC + FBD = \frac{B}{2} + \left\{ A + \frac{C}{2} \right\}.$$

The non-included angles FBC and FDC are equal, but they are obtuse. Thus triangles FBC and FBC are congruent. Hence $FB = DC = BE$. Hence triangles EBC and DBC are congruent which implies that the base angles $B = C$. Hence the triangle ABC is isosceles. $\square$

---

* Both in high school and in college, Ramakrishnan showed his mastery over Euclidean geometry and won the Adivaharan prize in Loyola College (see Ch. 3). In the course of investigating various reflection symmetries in his study of the Lorentz transformation, he found this beautiful proof of the converse bisector theorem and was quite proud of it. Thus he included it as an Appendix in Volume II of the Alladi Diary in 2003. The great geometer H. S. M. Coxeter expressed his appreciation of the novelty of this proof in a letter to Ramakrishnan - Editor

Appendix 15

Paper in Physics Education

Physics Education, Vol 30 (1995), 204-205

LETTERS TO THE EDITOR

The Reflection Principle

The following has been called the simplest, fastest and most direct derivation of the Lorentz Transformation. It can be understood easily by high school students or undergraduates.

We shall define a 'reflection' principle by its use in the derivation of the Pythagoras Theorem and demonstrate its profound consequence in physics and mathematics.

Let BC in the figure be the hypotenuse, AB and AC the sides and AD the altitude. Without loss of generality we write

$$BC^2 = k(AB^2 + AC^2).$$

Since ABD, ADC and ABC are similar

$$AB^2 = k(AD^2 + BD^2)$$
$$AC^2 = k(AD^2 + DC^2)$$
$$BD. DC = AD^2.$$

Therefore

$$AB^2 + AC^2 = kBC^2.$$

If this reflected relation is to be consistent with the first, k should be equal to unity.

What we have done is to apply the initial assumption about k twice to get back to the original relation and this we call the 'Reflection' principle. This simple principle lies behind the secret of spacetime unity, the basis of modern physics.

If a particle A moves with a velocity v with respect to B, we make the 'obvious' assumption that B moves with velocity $-v$ with respect to A. This is consistent since, by applying the assumption twice, the relative velocity of A with respect to B is $-(-v) = v$.

This leads to the non-trivial, in fact super-profound, Lorentz Transformation, for which the world had to wait 300 years after Newton! Newton would write the spacetime transformation as

$$x' = x - vt, \quad t' = t.$$

He would equally write, using the reflection principle,

$$x = x' + vt', \quad t = t'.$$

which is consistent with the first.

However, if we make the assumption that x/c and t (c is the velocity of light) have the same transformation properties, we should write

$$\frac{x'}{c} = \frac{x}{c} - \frac{v}{c}t, \quad t' = t - \frac{v}{c}.\frac{x}{c}.$$

Applying the reflection principle we can also write

$$\frac{x}{c} = \frac{x'}{c} + \frac{v}{c}t', \quad t = t' + \frac{v}{c}.\frac{x'}{c}$$

The two sets are *not consistent* unless the right-hand side is multiplied by a constant $k = \sqrt{1 - v^2/c^2}$. This is just the Lorentz Transformation. The author has also applied this principle in another method of derivation of the Lorentz Transformation. The Reflection principle can be remembered through the couplet

> John looks at Smith just as
> Smith looks at John
> So each looks at himself ever
> and anon.

Appendix

When light signals emanate from events separated in space and time by x and t, x/c is the transit time and t the basic time difference between events. We can equally interpret the signals as emanating from events separated by x_0 and t_0 with $x_0/c = t$ and $t_0 = x/c$ where x_0/c is the transit time and t_0 the basic time difference. Thus transit and basic time differences are interchangeable in relativity and therefore their transformation properties are the same.

Alladi Ramakrishnan
Alladi Centenary Foundation,
Mylapore, Madras, India

Further reading

[1] 1995 Tantalising asymmetries in special relativity *J. Math. Anal. Appl.* 110 222

[2] 1983 A fresh look at special relativity *Alladi Centenary Foundation, Creative Thought Series 1*

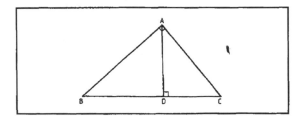

Appendix 16

Centenary Celebration of
Einstein's Special Relativity

SPECIAL RELATIVITY

PROFESSOR ALLADI RAMAKRISHNAN
(Founder-Director (Retd.) Matscience)

'*It is sheer ecstasy to notice unnoticed
nuances in Space-Time Unity*'.

EastWest Books (Madras) Pvt. Ltd
• Chennai • Bangalore • Hyderabad • New Delhi

Appendix 17

Institutions Visited By Ramakrishnan

List of Institutions Abroad
where Professor Alladi Ramakrishnan
lectured on his research work

UNITED STATES OF AMERICA**

Arizona State University, Tempe (1970)

Bell Telephone Laboratories, Murray Hill, New Jersey (1963, 67)

Boeing Research Laboratories, Seattle, Washington (1968, 69)

Boston University, Boston, Massachusetts (1967)

University of California, Berkeley (1962, 66, 71)

University of California, Irvine (1966, 71, 78, 80 to 83, 85)

University of California, Los Angeles (1962, 69)

University of California, Riverside (1971, 72, 74, 82)

University of California, Santa Barbara (1977)

*University of California, San Diego (1985)

*California Institute of Technology, Pasadena (1956, 67, 68)

Case Institute of Technology, Cleveland, Ohio (1958, 80)

University of Chicago, Chicago (1956)

University of Chicago, Yerkes Observatory (1956)

University of Colorado, Boulder (1967, 83)

Colorado State University, Fort Collins (1976, 77)

Catholic University of America, Washington D.C.
    (1973, 76, 77, 78, 80, 82)

Cornell University, Ithaca, New York (1967, 68, 69)

Courant Institute of Mathematical Sciences, New York (1968)

University of Denver, Denver, Colorado (1980, 81)

Duke University, Durham, North Carolina (1980)

Douglas Aircraft Company, Long Beach, California (1966, 67, 68, 69)

Florida State University, Tallahassee (1974)

University of Florida, Gainesville (1980)

University of North Florida, Jacksonville (1997, 98)

Fusion Research Laboratory, Princeton, New Jersey (1976)

General Motors Research Laboratories, Detroit, Michigan (1969)

George Washington University, Washington D.C. (1973)

University of Hawaii, Honolulu (1966, 67, 68, 69, 85)

Hughes Research Laboratories, Malibu, California (1962)

Howard University, Washington D.C. (1971)

*Harvard University, Cambridge, Massachusetts (1956, 57, 58, 74)

Illinois Institute of Technology, Chicago (1958)

University of Illinois, Urbana (1967, 70)

Institute for Advanced Study, Princeton, New Jersey (1957, 58, 82)

Iowa State University, Ames (1971)

Kent State University, Kent, Ohio (1975)

Lawrence Radiation Laboratory, Berkeley, California (1962, 67)

Lockheed Aircraft Corporation, Burbank, California (1969)

St. Louis University, St. Louis, Missouri (1968, 70)

Los Alamos Laboratory, Los Alamos, New Mexico (1978)

University of Nevada, Las Vegas (1981)

Massachusetts Institute of Technology (MIT), Cambridge (Physics, 1968)

Massachusetts Institute of Technology (MIT), Cambridge
    (Engineering, 1966)

Massachusetts Institute of Technology (MIT), Cambridge
    (Mathematics, 1977)

University of Michigan, Ann Arbor (1973, 78, 79)

University of Maryland, College Park (1958)

State University of New York (SUNY), Buffalo (1967, 68, 69, 70, 72)

State University of New York (SUNY), Albany (1976, 78)

City University of New York (CUNY), New York (1977)

US Naval Research Laboratory, Washington D.C. (1958, 59, 62, 66–74)

National Accelerator Laboratory, Batavia, Illinois (1981)

North Texas University, Denton (1970, 71)

University of New Hampshire, Dover (1975)

University of North Carolina, Chapel Hill (1971, 80)

University of New Mexico, Albuquerque (1977)

National Bureau of Standards, Washington D.C. (1962, 75, 78)

Oklahoma State University, Stillwater (1975, 78)

Ohio State University, Columbus (1974)

Oakridge National Laboratory, Oakridge, Tennessee (1970, 74)

The Pennsylvania State University, University Park (1972, 73)

*Princeton University, Princeton, New Jersey (1957, 58, 72)

Purdue University, Lafayette, Indiana (1968, 70)

Rutgers University, New Brunswick, New Jersey (1971, 72)
University of Rochester, Rochester, New York (1956, 63, 66, 67)
University of Rhode Island, Kingston, Rhode Island (1971, 72)
RAND Corporation, Santa Monica, California (1956, 62, 71)
Rockwell Science Center, Thousand Oaks, California
    (1977, 78, 80 to 83, 85–96)
Rensselaer Polytechnic Institute, Troy, New York (1978)
Stanford University, Palo Alto, California (Physics, 1962, 66, 67, 71)
Stanford University, Palo Alto, California (Elec. Eng. 1971)
Stanford University, Palo Alto, California (Statistics, 1972)
Stanford Linear Accelerator Center (SLAC), California (1977, 78)
Long Beach State College, Long Beach, California (1966 to 70)
San Jose State University, San Jose, California (1974, 75)
Syracuse University, Syracuse, New York (1966, 69, 72)
University of Southern California, Los Angeles (1966 to 68, 71, 72)
Systems Development Corporation, California (1969)
Southern Methodist University, Dallas, Texas (1975, 77, 81)
Southern Illinois University, Carbondale (1971, 73, 77, 79)
University of Texas at Dallas, Richardson
    (1970, 71, 73 to 75, 77 to 83, 85–98)
University of Texas, Arlington (1975, 77, 81)
University of Texas, Austin (1970, 82)
University of Tennessee, Knoxville (1980)
IBM Thomas J. Watson Center, Yorktown Heights, New York (1971, 73)
Utah State University, Logan (1971, 72)
University of Washington, Seattle (1967 to 69)
University of Wisconsin, Madison (1966 to 70, 80)
University of Wisconsin, Milwaukee (1966, 67, 71)
Wright Patterson Air Force Base, Dayton, Ohio (1968, 69, 70)
Wright State University, Dayton, Ohio (1971, 73)
University of Wyoming, Laramie (1972)
Yeshiva University, New York (1967–73, 75)

ENGLAND

University of Edinburgh, Edinburgh (1949)
Imperial College of Science and Technology, London (1960, 63, 67, 69)
University of Manchester, Manchester (1949, 50, 56)
*Cambridge University, Cambridge (1949–50, 75)
*Oxford University, Oxford (1950, 60)

Physical Society of Great Britain, Birmingham (1950)
University of York, York (1974, 75)
University of Kent, Canterbury (Mathematics, 1985)
University of Kent, Canterbury (Physics, 1985)

AUSTRALIA

Australian National University, Canberra (1954, 71, 73, 78)
La Trobe University, Melbourne (1971, 73, 78)
University of Melbourne, Melbourne (1954, 1971, 73, 78)
University of Sydney, Sydney (1954, 71, 73, 78)
Monash University, Melbourne (1973)
University of Adelaide, Adelaide (1973)
University of Western Australia, Perth (1973)
*CSIR Laboratories, Sydney (1978)
*CSIR Laboratories, Geelong (1978)
*University of New South Wales, Sydney (1978)

CANADA

University of Alberta, Edmonton (1969, 78, 80)
Air Canada Research Center, Montreal (1967)
Carleton University, Ottawa (1969)
University of Ottawa, Ottawa (1958)
National Research Council of Canada, Ottawa (1958)
Sir George Williams University, Montreal (1971, 73, 75)
McGill University, Montreal (1971, 72)
University of Montreal, Montreal (1968, 71, 72, 75)
Simon Fraser University, Vancouver (1967)
University of Toronto, Toronto (1968)
University of Manitoba, Winnipeg (1973 to 78, 80)
University of Calgary, Calgary (1976)

WEST GERMANY

University of Bonn, Bonn (1971, 77, 80, 81)
University of Gottingen, Gottingen (1956)
University of Heidelberg, Heidelberg (1956)
University of Marburg, Marburg (1956, 57, 60)
University of Wurzburg, Wurzburg (1975, 77)
University of Clausthal, Clausthal (1977, 81)

University of Stuttgart, Stuttgart (1956)
University of Frankfurt, Frankfurt (1977)
Mathematisches Institut, Oberwolfach (1977, 78)

U.S.S.R

Institute of Nuclear Research, Dubna (1964)
Academy of Sciences, Moscow (1968)
Moscow State University, Moscow (1968)
Physical Technical Institute, Academy of Sciences, Leningrad (1968)

JAPAN

Yukawa Hall, Kyoto (1956)
University of Kyoto, Kyoto (1956)
University of Osaka, Osaka (1956)
Tokyo University of Education, Tokyo (1962, 68, 69, 70)

BELGIUM

University of Liege, Liege (1971, 75)

IRELAND

Institute for Advanced Study, Dublin (1949)
University of Dublin, Dublin (1949, 74)

SWEDEN

Cramer's Institute, Stockholm (1950)
University of Uppsala, Uppsala (1950)

NORWAY

University of Oslo, Oslo (1950)

SWITZERLAND

University of Berne, Berne (1960, 69)
CERN, Geneva (1960, 62, 66, 74)
Swiss Federal Institute of Technology (E. T. H.), Zurich (1950)
University of Geneva, Geneva (1967)
Swiss Physical Society, Winterthur (1960)
University of Zurich, Zurich (1950, 56)

FRANCE

University of Paris, Orsay (1966)
Institute Henri Poincare, Paris (1960, 66, 77)
C. E. N. Saclay (1964, 65, 67, 68, 69)

DENMARK

Bohr Institute, Copenhagen (1950, 60)

ITALY

International Centre for Theoretical Physics (ICTP), Trieste
    (1963, 65, 67, 68, 70, 74, 75)
University of Naples, Naples (1965, 67)
University of Padua, Padua (1965, 69)
University of Rome, Rome (1962, 65, 66, 69)

SINGAPORE

University of Singapore, Singapore (1967)

HONG KONG

University of Hong Kong, Hong Kong (1975)

NEW ZEALAND

University of Canterbury, Christchurch (1971)

IRAN

Arya-Mehr University, Teheran (1968)

Notes

*Visits and discussion only.

**If a US Academic Institution bears the name of a State, then the State is not mentioned
for its location. - Editor

Appendix 18

List Of PhD Students Of Ramakrishnan

The following is the list of students who obtained their PhD under the supervision of Professor Alladi Ramakrishnan at the University of Madras and at MATSCIENCE, The Institute of Mathematical Sciences. The year PhD was granted is given at the beginning. All PhD degrees were granted by the University of Madras. - Editor

1) (1956) **P. M. Mathews**
2) (1957) **S. K. Srinivasan**
3) (1960) **R. Vasudevan**
4) (1961) **N. R. Ranganathan**
5) (1962) **T. K. Radha**
6) (1962) **R. Thunga**
7) (1962) **A. P. Balachandran**
8) (1963) **V. Devanathan**
9) (1963) **K. Venkatesan**
10) (1963) **G. Bhamathi**
11) (1963) **S. Indumathi**
12) (1963) **G. Ramachandran**
13) (1964) **K. Raman**
14) (1964) **R. K. Umerjee**
15) (1965) **K. Ananthanarayanan**
16) (1969) **T. S. Shankara**
17) (1970) **T. S. Santhanam**
18) (1970) **K. Srinivasa Rao**
19) (1971) **P. S. Chandrasekharan**
20) (1971) **A. Sundaram**
21) (1972) **Nalini B. Menon**
22) (1975) **A. R. Tekumalla**
23) (1976) **R. Jagannathan**
24) (1979) **A. Vijayakumar**

Appendix 19

Publications of Alladi Ramakrishnan

Research Publications

1) (with H. J. Bhabha) "The mean-square deviation of the number of electrons and quanta in cascade theory", *Proc. Indian Acad. Sci.*, **32** (1950), 141–153.

2) "Stochastic processes relating to particles distributed in a continuous infinity of states", *Proc. Cambridge Phil. Soc.*, **46** (1950), 595–602.

3) "A note on the size frequency distribution of penetrating showers", *Proc. Phys. Soc. London*, **63A** (1950), 861–863.

4) "Stochastic processes and their applications to physical problems", *PhD Thesis, Univ. Manchester* (1951).

5) "Some simple stochastic processes", *J. Royal Stat. Soc.*, **13** (1951), 131–140.

6) "A note on Janossy's mathematical model of a nucleon cascade", *Proc. Cambridge Phil. Soc.*, **48** (1952), 451–456.

7) "On an integral equation of Chandrasekhar and Munsch", *Astrophysical J.*, **115** (1952), 141–144.

8) "Stochastic processes associated with random divisions of a line", *Proc. Cambridge Phil. Soc.*, **49** (1953), 473–485.

9) (with P. M. Mathews) "A stochastic problem relating to counters", *Phil. Mag.*, **44** (1953), 1122–1127.

10) (with P. M. Mathews) "Numerical work on the fluctuation problem of electron cascades", *Prog. Theor. Phys.*, **9** (1953), 679–681.

11) (with P. M. Mathews) "On a class of stochastic integro-differential equations", *Proc. Indian Acad. Sci. Ser A* **38** (1953), 450–466.

12) (with P. M. Mathews) "On the solution of an integral equation of Chandrasekhar and Munsch", *Astrophysical J.*, **119** (1954), 81–90.

13) "A stochastic model of a fluctuating density field", *Astrophysical J.*, **119** (1954), 443–455.

14) "A stochastic model of a fluctuating density field - II", *Astrophysical J.*, **119** (1954), 682–685.

15) "On the molecular distribution functions of a one dimensional fluid - I", *Phil. Mag.* Ser 7 **45** (1954), 401–410.

16) "On counters with random dead time", *Phil. Mag.*, **45** (1954), 1050–1052.

17) (with P. M. Mathews) "On the molecular distribution functions of a one dimensional fluid - II", *Phil. Mag.*, **45** (1954), 1053–1058.

18) (with P. M. Mathews) "Studies in the stochastic problem of electron-photon cascades", *Prog. in Theor. Phys.*, **11** (1954), 95–117.

19) (with S. K. Srinivasan) "Two simple stochastic models of cascade multiplication", *Prog. in Theor. Phys.*, **11** (1954), 595–603.

20) "On stellar statistics", *Astrophysical J.*, **122** (1955), 24–31.

21) "Inverse probability and evolutionary Markov stochastic processes", *Proc. Indian Acad. Sci.*, **XLI** (1955), 145–153. (Read at the Annual Meeting of the Academy in Belgaum, Dec 1954.)

22) (with S. K. Srinivasan) "Fluctuations in the number of photons in an electron-photon cascade", *Prog. in Theor. Phys.*, **13** (1955), 93–99.

23) (with P. M. Mathews) "Straggling of the range of fast particles as a stochastic process", *Proc. Indian Acad. Sci.*, **XLI** (1955), 202–209. (Read at the Annual Meeting of the Academy in Belgaum, Dec 1954.)

24) "Phenomenological interpretation of the integrals of a class of random functions", *Proc. Koninkl. Netherlands Akad.*, **58** (=*Indag. Math.*, **17**) (1955), 470–482.

25) "Phenomenological interpretation of the integrals of a class of random functions - II", *Proc. Koninkl. Netherlands Akad.*, **58** (=*Indag. Math.*, **17**) (1955), 634–645.

26) "Inverse probability and evolutionary Markoff stochastic processes", *Proc. Indian Acad. Sci.*, Ser A **41** (1955), 145–153. (Read at the Annual Meeting of the Academy in Belgaum, Dec. 1954.)

27) (with S. K. Srinivasan) "Correlation problems in the study of brightness of the Milky Way", *Astrophysical J.*, **123** (1956), 479–485.

28) "Processes represented as integrals of a class of random functions", *Proc. Koninkl. Netherlands Akad.*, **59** (=*Indag. Math.*, **18**) (1956), 121–127.

29) (with P. M. Mathews) "Stochastic processes associated with a symmetric oscillatory Poisson process", *Proc. Indian Acad. Sci.*, **43A** (1956), 84–98.

30) (with S. K. Srinivasan) "A new approach to cascade theory", *Proc. Indian Acad. Sci.*, Ser. A. **XLIV 44** (1956), 263–273.

31) "A physical approach to stochastic processes", *Proc. Indian Acad. Sci.* Ser A, **44** (1956), 428–450.

32) (with S. K. Srinivasan) "Stochastic integrals associated with point processes" (in French), *Publ. Inst. Stat. Univ. Paris*, **5** (1956), 95.

33) (with R. Vasudevan) "On the distribution of visible stars", *Astrophysical J.*, **126** (1957), 573–578.

34) "Ergodic properties of some simple stochastic processes", *Z. Angew. Math. Mech.*, **37** (1957), 336–344. (Read at the GAMM Conference in May 1956 in Stuttgart.)

35) (with S. K. Srinivasan) "A note on cascade theory with ionisation loss", *Proc. Indian Acad. Sci.*, Ser. A **45** (1957), 133–138.

36) (with N. R. Ranganathan, S. K. Srinivasan and R. Vasudevan) "Multiple processes in electron-photon cascades", *Proc. Indian Acad. Sci.*, Ser. A, **45** (1957), 311–326.

37) (with S. K. Srinivasan) "On age distribution in population growth", *Bull. Math. Biophys.*, **20** (1958), 289–303.

38) "Ambigenous stochastic processes", *Z. Angew Math. Mech*, **39** (1959), 389–390.

39) (with N. R. Ranganathan and S. K. Srinivasan) "Meson production in nucleon-nucleon collisions", *Nucl. Phys.*, **10** (1959), 160 165.

40) (with N. R. Ranganathan, S. K. Srinivasan and K. Venkatesan) "Photo-mesons from polarized nucleons", *Proc. Indian Acad. Sci. Ser. A* **49** (1959), 302–306.

41) (with N. R. Ranganathan and S. K. Srinivasan) "A note on the interaction between nucleon and anti-nucleon", *Proc. Indian Acad. Sci.*, **50** (1959), 91–94.

42) "Probability and stochastic processes" in *Handbuch der Physik* (S. Flugge, Ed.) **III/2** (1959), Springer-Verlag, Berlin (1959), 524–651.

43) (with N. R. Ranganathan, S. K. Srinivasan, and R. Vasudevan) "A note on dispersion relations", *Nuclear Phys.*, **15** (1960), 516–518.

44) "Perturbation expansions and kernel functions associated with single particle wave functions" in *Studies in Theor. Phys., Proc. 1959 Mussoorie Summer School* **1** (1960), 1–14.

45)"Quantum mechanics of the photon" in *Studies in Theor. Phys., Proc. 1959 Mussoorie Summer School* **1** (1960), 15–18.

46) "Applications of the theory of stochastic processes to physical problems", in *Studies in Theor. Phys., Proc. 1959 Mussoorie Summer School* **2** (1960), 239–253.

47) (with R. Vasudevan) "A physical approach to some limiting stochastic operation", *J. Indian Math. Soc.* **XXIV** (Golden Jubilee Volume, 1960), 458–477. (work done at Institute for Advanced Study in 1957–58; paper read by AR at Int'l Congress of Math. Edinburgh 1958)

48) (with A. P. Balachandran and N. R. Ranganathan) "Some remarks on the structure of elementary particle interactions", *Proc. Indian Acad. Sci.*, **52** (1960), 1–11.

49) (with T. K. Radha and R. Thunga) "On the decomposition of the Feynman propagator", *Proc. Indian Acad. Sci.*, Ser. A, **52** (1960), 228–239.

50) (with P. Rajagopal and R. Vasudevan) "Ambigenous stochastic processes", *J. Math. Analysis and Appl.*, **1** (1960), 145–162. (Read by AR at the GAMM Conf., Hanover in May 1959)

51) (with T. K. Radha) "Correlation problems in evolutionary stochastic processes", *Proc. Cambridge Phil. Soc.*, **57** (1961), 843–847.

52) (with A. P. Balachandran, N. G. Deshpande, and N. R. Ranganathan) "On an isobaric spin scheme for leptons and leptonic decays of strange particles", *Nucl. Phys.*, **26** (1961), 52–56.

53) (with R. Vasudevan) "A physical approach to limiting stochastic operations", *J. Indian Math. Soc. (N.S)* **24** (1961), 457–4777.

54) (with G. Bhamathi and S. Indumathi) "A limiting process in quantum electrodynamics", *Proc. Indian Acad. Sci. Ser A*, **LIII 53** (1961), 206–213.

55) (with V. Devanathan and G. Ramachandran) "A time dependent approach to rearrangement collisions", *Il Nuovo Cimento*, **21** (1961), 145.

56) (with S. K. Srinivasan) "A note on electron photon showers", *Nucl. Phys.*, **25** 1961), 152–154.

57) (with G. Bhamathi, S. Indumathi, T. K. Radha, and R. Thunga) "Some consequences of spin $\frac{3}{2}$ for $\equiv$", *Il Nuovo Cimento*, **22** (1961), 604–609.

58) (with V. Devanathan and G. Ramachandran) "Elastic photo production of neutral pions from dueterium", *Nucl. Phys.*, **24** (1961), 163–168.

59) (with N. R. Ranganathan) "Stochastic models in quantum mechanics", *J. Math. Analysis and Appl.*, **3** (1961), 261–294. (Presented at the Conference on Elementary Particles, Trieste 1960)

60) (with K. Venkatesan) "Some new stochastic aspects in cascade theory", *in Proc. 7th Annual Cosmic Ray Symposium, Chandigarh* (1961), 59–61.

61) (with T. K. Radha) "Essay on symmetries", *Lectures at the Kodaikanal Summer School*, **2** (1961), 1–77.

62) (with N. R. Ranganathan) "Stochastic methods in quantum mechanics", *J. Math. Analysis and Appl.*, **3** (1961), 261–294.

63) (with T. K. Radha and R. Thunga) "The physical basis of quantum field theory", *J. Math. Analysis and Appl.*, **4** (1962), 494–526.

64) (with T. K. Radha and R. Thunga) "On the concept of virtual states", *J. Math. Analysis and Appl.*, **5** (1962), 225–236.

65) (with G. Ramachandran) "Magnetic bremsstrahlung in nucleon-electron collisions", *Rand Corporation Preprint*, Los Angeles (1962).

66) "New perspectives on the Dirac Hamiltonian and the Feynman propagator", *in High Energy Phys. and Fundamental Particles*, Gordon and Breach, NY (1962), 665–672.

67) "A new form of the Feynman propagator", *J. Math. Phys. Sci.*, **1** (1967), 57–64.

68) (with A. P. Balachandran and K. Raman) "Low energy K^+-nucleon scattering", *Il Nuovo Cimento*, **24** (1962), 369–378.

69) (with V. Devanathan and K. Venkatesan) "On the scattering of pions by dueterons", *Nucl. Phys.*, **29** (1962), 680–686.

70) (with T. K. Radha and R. Thunga) "Possible resonances in $\equiv_p$ reactions", *Nucl. Phys.*, **29** (1962), 517–523.

71) (with A. P. Balachandran) "Partial wave dispersion relations for Λ-nucleon scattering", *Il Nuovo Cimento*, **24** (1962), 980–999.

72) (with A. P. Balachandran, T. K. Radha, and R. Thunga) "On the Y^* resonances", *Il Nuovo Cimento*, **24** (1962), 1006–1012.

73) (with A. P. Balachandran, T. K. Radha, and R. Thunga) "On the spin and parity of Y^* resonances", *Il Nuovo Cimento*, **25** (1962), 723–729.

74) (with A. P. Balachandran, T. K. Radha, and R. Thunga) "Photo production of pions and Λ-hyperons", *Il Nuovo Cimento*, **25** (1962), 939–942.

75) (with G. Bhamathi, S. Indumathi, T. K. Radha, and R. Thunga) "Dispersion analysis of $\equiv$ production in KN collisions", *Nucl. Phys.*, **37** (1962), 585–593.

76) (with T. K. Radha, K. Raman, and R. Thunga) "quantum numbers and decay models of resonances", *Rand Corp. Preprint*, Los Angeles (1962).

77) "An unconventional view of perturbation expansions", *in Proc. Seminar on Unified Theories of Elem. Particles, Univ. Rochester, D. Lurie and N. Mukunda, Eds.* (1963), 411–421.

78) (with V. Devanathan and K. Venkatesan) "A note on the use of Wick's theorem", *J. Math. Analysis and Appl.*, **8** (1964), 345–349.

79) (with K. Raman and R. K. Umergee) "Isobar production in nucleon-nucleon scattering", *Nuclear Phys.*, **60** (1964), 401–426.

80) (with K. Raman and R. K. Umergee) "Isobar production in nucleon-nucleon scattering - II, Polarization effects", *Nuclear Phys.*, **66** (1965), 609–631.

81) (with S. K. Srinivasan and R. Vasudevan) "Some new mathematical features in cascade theory", *J. Math. Analysis and Appl.*, **11** (1965), 278–289. (Presented at the Int'l Conf. on Cosmic Rays, **5** (1964), TIFR, Bombay, 458–501.)

82) (with T. S. Shankara and K. Venkatesan) "Sensitivity of the vector coupling constant to μ-nuetrino mass and T-invariance", *Il Nuovo Cimento*, **37** (1965), 1046–1048.

83) (with R. Vasudevan and S. K. Srinivasan) "Scattering phase shifts in stochastic fields", *Z. fur Phys.*, **196** (1966), 112–122.

84) "Fundamental multiplets" in *Symp. in Theor. Phys. and Maths., Alladi Ramakrishnan Ed.* **5** Plenum Press, NY (1967), 85–92.

85) "New perspectives on the Dirac Hamiltonian and the Feynman propagator", in *High energy physics and fundamental particles* (1967), Gordon and Breach, New York, 665–672. (Presented at the Theoretical Physics Institute, University of Colorado, Boulder, 1967.)

86) (with S. K. Srinivasan and R. Vasudevan) "Angular correlations in the brightness of the Milky Way", *J. Math. Phys. Sci.*, **1** (1967), 75–84.

87) "Some new topological features in Feynman graphs", *J. Math. Analysis and Appl.*, **17** (1967), 68–91.

88) "A new form of the Feynman propagator", *J. Math. Phys. Sci.*, **1** (1967), 57–64.

89) "L-matrix hierarchy and the higher dimensional Dirac Hamiltonian", *J. Math. Phys. Sci.*, **1** (1967), 190–193.

90) (with S. K. Srinivasan and R. Vasudevan) "Multiple product densities", *J. Math. Phys. Sci.*, **1** (1967), 275–279.

91) "Graphical representation of CPT", *J. Math. Analysis and Appl.*, **17** (1967), 147–150.

92) (with I. V. V. Raghavacharyulu) "A new combinatorial feature of Feynman graphs", *J. Math. Analysis and Appl.*, **18** (1967), 175–181.

93) "The Dirac Hamiltonian as a member of a hierarchy of matrices", *J. Math. Analysis and Appl.*, **20** (1967), 9–16.

94) "Helicity and energy as members of a hierarchy of eigenvalues", *J. Math. Analysis and Appl.*, **20** (1967), 397–401.

95) "Fundamental multiplets" in *Symposia on Theor. Phys.* (Alladi Ramakrishnan Ed.) **5** (1967), Plenum Press, NY, 85–92.

96) "Symmetry operations on a hierarchy of matrices", *J. Math. Analysis and Appl.*, **21** (1968), 39–42.

97) "On the relationship between L-matrix hierarchy and Cartan spinors", *J. Math. Analysis and Appl.*, **22** (1968), 570–576.

98) (with P. S. Chandrasekharan, N. R. Ranganathan, and R. Vasudevan) "A generalization of the L-Matrix hierarchy", *J. Math. Analysis and Appl.*, **23** (1968), 10–14.

99) "L-Matrices, quaternions and propagators", *J. Math. Analysis and Appl.*, **23** (1968), 250–253.

100) (with I. V. V. Raghavacharyulu) "A note on the representation of Dirac groups", *in Symposia on Theor. Phys. and Math.* (Alladi Ramakrishnan, Ed.), **8** (1968), Plenum Press, NY, 25–32.

101) "Should we revise our notions about spin and parity in relativistic quantum theory?" *J. Math. Phys. Sci.*, **3** (1969), 213–219.

102) (with P. S. Chandrasekharan and T. S. Santhanam) "On representations of generalised Clifford algebras", *J. Math. Phys. Sci.*, **3** (1969), 301–313.

103) "Symmetries associated with roots of the unit matrix", *J. Math. Phys. Sci.*, **3** (1969), 317–318.

104) "Generalized helicity matrices", *J. Math. Analysis and Appl.*, **26** (1969), 88–91.

105) (with P. S. Chandrasekharan, T. S. Santhanam, and A. Sundaram) "Helicity matrices for generalized Clifford algebra", *J. Math. Analysis and Appl.*, **26** (1969), 275–278.

106) (with P. S. Chandrasekharan, N. R. Ranganathan, T. S. Santhanam, and R. Vasudevan) "The generalized Clifford algebra and the unitary groups", *J. Math. Analysis and Appl.*, **27** (1969), 164–170.

107) (with P. S. Chandrasekharan, N. R. Ranganathan, T. S. Santhanam, and R. Vasudevan) "Idempotent matrices from a generalized Clifford algebra", *J. Math. Analysis and Appl.*, **27** (1969), 563–564.

108) (with P. S. Chandrasekharan, N. R. Ranganathan, and R. Vasudevan) "Kemmer algebra from generalized Clifford elements", *J. Math. Analysis and Appl.*, **28** (1969), 108–110.

109) "On the algebra of L-matrices", in *Symposia on Theor. Phys. and Math.* (Alladi Ramakrishnan, Ed.), **9** (1969), Plenum Press, NY, 73–78.

110) "*L*-matrices and propagators with imaginary parameters", *in Symposia on Theor. Phys. and Math.* (Alladi Ramakrishnan, Ed.), **9** (1969), Plenum Press, NY, 79–84.

111) (with R. Vasudevan) "A hierarchy of idempotent matrices", *in Symposia on Theor. Phys. and Math.* (Alladi Ramakrishnan, Ed.), **9** (1969), Plenum Press, NY, 85–88.

112) (with P. S. Chandrasekharan and R. Vasudevan) "Representation of para-Fermi rings and generalized Clifford algebra", *J. Math. Analysis and Appl.*, **31** (1970), 1–5.

113) "On the composition of generalized helicity matrices", *J. Math. Analysis and Appl.*, **31** (1970), 254–258.

114) (with R. Vasudevan) "On generalized idempotent matrices", *J. Math. Analysis and Appl.*, **32** (1970), 414–423.

115) "New generalizations of Pauli matrices", *in Proc. Int'l Conf. on Symmetries and Quark Models, Wayne State Univ. 1969*, Gordon and Breach, NY (1970), 133–138.

116) "Unitary generalization of Pauli matrices", *in Symposia on Theor. Phys. and Math.* (Alladi Ramakrishnan, Ed.), **10** (1970), Plenum Press, NY, 51–57.

117) (with I. V. V. Raghavacharyulu) "Generalized Clifford basis and infinitesimal generators of the unitary group", *in Symposia on Theor. Phys. and Math.* (Alladi Ramakrishnan, Ed.), **10** (1970), Plenum Press, NY, 59–62.

118) (with P. S. Chandrasekharan and T. S. Santhanam) *"L-matrices and the fundamental theorem of spinor theory"*, *in Symposia on Theor. Phys. and Math.* (Alladi Ramakrishnan, Ed.), **10** (1970), Plenum Press, NY, 63–68.

119) "Stochastic theory of evolutionary processes (1937–71)" in *Stochastic Point Processes, Proc. IBM Conf. on Point Processes*, (J. Lewis, Ed.), Interscience, John Wiley (1971), 533–548.

120) "A new approach to quantum numbers in elementary particle physics", *in Proc. Rutherford Centennial Conf., Christchurch* (1971), 150–156.

121) (with P. S. Chandrasekharan and R. Vasudevan) "Algebras derived from polynomial conditions", *J. Math. Analysis and Appl.*, **35** (1971), 131–134.

122) (with P. S. Chandrasekharan and R. Vasudevan) "Para-Fermi operators and special unitary algebras", *J. Math. Analysis and Appl.*, **35** (1971), 249–254.

123) "The weak interaction Hamiltonian in L-Matrix theory", *J. Math. Analysis and Appl.*, **37** (1972), 432–434.

124) "On the shell structure of an L-Matrix", *J. Math. Analysis and Appl.*, **38** (1972), 106–108.

125) "A matrix decomposition theorem", *J. Math. Analysis and Appl.*, **40** (1972), 36–38.

126) "Einstein - a natural completion of Newton", *J. Math. Analysis and Appl.*, **42** (1973), 377–380.

127) "The generalized Gell-Mann-Nishijima relation", in *Proc. Conf. on Nucl. Phys., MATSCIENCE Report* **78**, (1973) 1–4.

128) (with R. Jaganathan) "A new approach to matrix theory or many facets of the matrix decomposition theorem", in *Topics in Numerical Analysis, Proc. 1974 Int'l Conf. on Numerical Analysis, Dublin* (J. H. Miller Ed.) **133** (1976), 133–139.

129) "New concepts in matrix theory", *J. Math. Analysis and Appl.*, **60** (1977), 255–258.

130) "Unnoticed symmetries in Einstein's special relativity", *J. Math. Analysis and Appl.*, **63** (1978), 335–338.

131) "Lorentz to Gell-Mann - can the masters be retouched?" *Proc. Tamil Nadu Acad. Sci.*, **1** (1978), 29–33.

132) "A new concept in special relativity - exterior relative velocity", *Proc. Tamil Nadu Acad. Sci.*, **1** (1978), 67.

133) "A new look at matrix operations", *in Proc. Oberwolfach Conf. on Math. Phys. Dec. 1977* (Methoden und Verfahren der Mathematischen Physik - B. Brosowski and E. Martenson, Eds.), Peter Lang Publ. (1978), 49–53.

134) "New facets and new concepts in the special theory of relativity", *in Proc. Oberwolfach Conf. on Math. Phys. Dec. 1977* (Methoden und Verfahren der Mathematischen Physik - B. Brosowski and E. Martenson, eds.), Peter Lang Publ. (1978), 55–61.

135) "Approach to stationarity in stochastic processes", *Proc. Tamil Nadu Acad. Sci.*, **2** (1979), 189–190.

136) "On the generalization of the Gell-Mann-Nishijima relation", in *Symmetries in Science* (B. Gruber and R. S. Millman, Eds.), Plenum Publ. Corp. (1980), 323–325.

137) "Mathematical features of evolutionary stochastic processes", *Methods of Operations Research*, **36** (1980), 239–240.

138) "A new concept in probability theory", *J. Math. Analysis and Appl.*, **83** (1981), 408–410.

139) "Duality in stochastic processes", *J. Math. Analysis and Appl.*, **84** (1981), 483–485.

140) "Further unnoticed symmetries in special relativity", *J. Math. Analysis and Appl.*, **94** (1983), 237–241.

141) "Tantalizing asymmetries in special relativity", *J. Math. Analysis and Appl.*, **110** (1985), 222–224.

142) "Decomposition of intervals in special relativity", *J. Math. Analysis and Appl.*, **110** (1985), 225–226.

143) "A reflection principle", *Phys. Education*, **30** (1995), 204–205.

144) "A remarkable unnoticed theorem in special relativity", *J. Math. Analysis and Appl.*, **213** (1997), 155–159.

145) "Theorem on non-simultaneity - extension to an external observer", *J. Math. Analysis and Appl.*, **213** (1997), 354–356.

146) "Cubic and general extensions of the Lorentz transformation", *J. Math. Analysis and Appl.*, **229** (1999), 88–92.

147) "A new Rod approach to the special theory of relativity", *J. Math. Analysis and Appl.*, **249** (2000), 243–251. (Special Millennium issue dedicated to Richard Bellman)

Books/Monographs

1) "Elementary particles and cosmic rays", Pergamon Press, Oxford (1962), 580 p.

2) "*L*-Matrix theory or the grammar of Dirac matrices", Tata-McGraw Hill, Bombay-New Delhi (1972).

3) "The Alladi Diary", Vol 1, East-West Books, Madras (2000).

4) "The Alladi Diary", Vol 2, East-West Books, Madras (2003).

5) "Special relativity", East West Books, Madras, India (2005).

Book Series Edited by Alladi Ramakrishnan

1a) *Matscience Symposia on Theoretical Physics* 5 Volumes, Plenum Press, New York - Vol 1 (1966), Vol 2 (1966), Vol 3 (1967), Vol 4 (1967), Vol 5 (1967).

1b) *Symposia on Theoretical Physics and Mathematics*, 5 Volumes, Plenum Press, New York - Vol 6 (1968), Vol 7 (1968), Vol 8 (1968), Vol 9 (1969), Vol 10 (1970).

Other Articles and Texts of Speeches on Science, Carnatic Music, and Hindu Religion/Philosophy

(These articles and speeches could be found in [1] or could be obtained from the Editor)

1) "Theoretical Physics in the USA", *Current Sci.* **27** (1958), 469–471.

2) "A science policy for Tamil Nadu" (Presented at the Institute of Development Studies, Madras, 1971).

3) "The impact of science on society" (Text of Talk delivered at the All India Radio, Madras on Sept. 4, 1973.)

4) "Einstein's contribution to human thought" (Based on talk delivered at the Rotary Club of Madras, Oct 5, 1976.)

5) "A fresh look at special relativity", (Nov. 1983).

6) "Special relativity and wave particle dualism", (Oct. 1984).

7) "Einstein and Lorentz - together in perfection", (1995).

8) "Cubic and general extensions of the Lorentz transformation".

9) "Seven lamps of mathematical architecture" Text of Commemoration Address, Founder's Day, P. S. High School, Madras, on 28 March 1971).

10) "C.R. and Alladi - unison in contrast" (Appeared in the C. Rajagopalachari 93rd Birthday Volume, Kalki Press, Madras, 1971).

11) "Ramayana - its direct relevance to real life" (Text of K. Balasubramania Iyer Endowment Lecture, Sanskrit College, Madras, 10 Nov 1989).

12) "Swathi Thirunal - the musician prince" (Text of Talk given at the Swathi Thirunal Day, Thyagaraja Vidwat Samajam, Madras, (1974)).

13) "The song of the saints" (Appeared in the First Anniversary Souvenir of *Nadopasana* Sabha, 1969).

14) "The bridge of sound and light" (Talk given on his own 25th Wedding Anniversary on 21 Aug, 1971 when several leading Carnatic musicians performed at Ekamra Nivas).

15) "Divine swells from common diction" (Appeared in the Indian Fine Arts Society Souvenir for the 30th South Indian Music Conference and Festival, Dec 1971–Jan 1972, in Madras).

16) "C. S. - can he excel himself?" (Appeared in the C. Subramaniam 60th Birthday Volume, (1970), Madras).

17) "P. C. Mahalanobis - a tribute" (Text of Talk delivered at the All India Radio Madras, 28 June 1977; appeared in The Indian Review, July 1977, pp. 59–62).

18) "The Integrated Life" (Text of the C. P. Snow Memorial Lecture delivered at the British Council, Madras, on 15 Oct, 1980. Appeared in the Bulletin of the Theosophy Group of India, Vol 19, (1981), pp. 25–27).

19) "Indian contribution to mathematical sciences during the past twenty five years" (Text of Talk given at the All India Radio, Madras, 8 Jan, 1975).

20) "Research in mathematical sciences in India" (Talk given at the All India Radio, Madras, 9 Aug, 1976).

21) "Competitive spirit in Science and Industry" (Based on talk at the Rotary Club of Madras in 1969; Appeared in the Bulletin of the Rotary Club of Madras, August, 1969).

22) "Matscience - the haven of freedom" (Text of Talk given at the All India Radio, Madras, on 23 Aug, 1974).

References

1) K. Alladi, *The Alladi Diary — Letters, Documents and Photographs* — see https://www.worldscientific.com/worldscibooks/10.1142/11346. World Scientific Press, Singapore (2017).

2) K. Alladi, J. H. Klauder, and C. R. Rao, *The Legacy of Alladi Ramakrishnan in the Mathematical Sciences*, Springer, New York (2010).

3) Granville Austin, *The Indian Constitution: Cornerstone of a Nation*, Oxford Univ. Press (1966).

4) Suresh Balakrishnan, *Great Judges and Lawyers of Madras*, Law Book Sellers, Publishers and Distributors, Chennai (2012).

5) Alladi Krishnaswami Aiyar, *The Constitution and Fundamental Rights*, Lectures at the Srinivasa Sastri Institute of Politics, MLJ Press, Madras (1955).

6) Alladi Ramakrishnan, *The Alladi Diary, Vols. I and II*, East-West Press, Madras (2000) and (2003).

7) T. R. Venkatarama Sastri, *Sir Alladi Krishnaswami Iyer Sashtyabdhapoorthi Volume*, MLJ Press, Madras (1943).

8) C. Subramaniam, *The Hand of Destiny, Vols. I and II*, Bharatiya Vidya Bhavan, Bombay (1993).

Index